W0230483

COSMOCHEMICAL EVOLUTION AND THE ORIGINS OF LIFE

VOLUME I

INVITED PAPERS

COSMOCHEMICAL EVOLUTION AND THE ORIGINS OF LIFE

PROCEEDINGS OF THE FOURTH INTERNATIONAL CONFERENCE ON THE ORIGIN OF LIFE AND THE FIRST MEETING OF THE INTERNATIONAL SOCIETY FOR THE STUDY OF THE ORIGIN OF LIFE, BARCELONA, JUNE 25–28, 1973

VOLUME I: INVITED PAPERS

Edited by

J. ORÓ

Departments of Biophysical Sciences and Chemistry, University of Houston, Houston, Texas, and Universidad Autónoma de Barcelona, Barcelona

S. L. MILLER

Department of Chemistry, University of California, San Diego, California

C. PONNAMPERUMA

Laboratory of Chemical Evolution, Department of Chemistry, University of Maryland, College Park, Maryland

and

R. S. YOUNG

Head of Planetary Biology, NASA, Washington, D.C.

D. REIDEL PUBLISHING COMPANY

DORDRECHT-HOLLAND / BOSTON-U.S.A.

Library of Congress Catalog Card Number 74–77967

ISBN-13: 978-94-010-2241-5 e-ISBN-13: 978-94-010-2239-2
DOI: 10.1007/978-94-010-2239-2

Published by D. Reidel Publishing Company,
P.O. Box 17, Dordrecht, Holland

Sold and distributed in the U.S.A., Canada and Mexico
by D. Reidel Publishing Company, Inc.
306 Dartmouth Street, Boston,
Mass. 02116, U.S.A.

Softcover reprint of the hardcover 1st edition 1974

In Honour of Professors
A. I. Oparin, Melvin Calvin and H. C. Urey
in recognition of their pioneering studies
on the problem of
the origin of life on Earth

Dedicated to the Memory of
J. B. S. Haldane – 1964
J. D. Bernal – 1971
A. Katzir-Katchalsky – 1972
W. Vishniac – 1973

TABLE OF CONTENTS

PREFACE 1

PART I / COSMOCHEMISTRY

GEORGE WALD / Fitness in the Universe: Choices and Necessities 7
DAVID BUHL / Galactic Clouds of Organic Molecules 29
TOBIAS OWEN / The Outer Solar System: Perspectives for Exobiology 41
EDWARD ANDERS, RYOICHI HAYATSU, and MARTIN H. STUDIER / Catalytic Reactions in the Solar Nebula: Implications for Interstellar Molecules and Organic Compounds in Meteorites 57

PART II / PALEOBIOLOGY

KEITH A. KVENVOLDEN / Natural Evidence for Chemical and Early Biological Evolution 71
HEINRICH D. HOLLAND / Aspects of the Geologic History of Seawater 87
JAMES E. LOVELOCK and LYNN MARGULIS / Homeostatic Tendencies of the Earth's Atmosphere 93
M. D. MUIR / Microfossils from the Middle Precambrian McArthur Group, Northern Territory, Australia 105
J. WILLIAM SCHOPF / The Development and Diversification of Precambrian Life 119

PART III / PRIMORDIAL ORGANIC CHEMISTRY

STANLEY L. MILLER / The Atmosphere of the Primitive Earth and the Prebiotic Synthesis of Amino Acids 139
J. P. FERRIS, J. D. WOS, T. J. RYAN, A. P. LOBO, and D. B. DONNER / Biomolecules from HCN 153
J. ORÓ and E. STEPHEN-SHERWOOD / The Prebiotic Synthesis of Oligonucleotides 159
MELLA PAECHT-HOROWITZ / The Possible Role of Clays in Prebiotic Peptide Synthesis 173
CARL SAXINGER and CYRIL PONNAMPERUMA / Interactions Between Amino Acids and Nucleotides in the Prebiotic Milieu 189
T. N. EVREINOVA, T. W. MAMONTOVA, V. N. KARNAUHOV, S. B. STEPHANOV, and U. R. HRUST / Coacervate Systems and Origin of Life 201
ALEXANDER RICH / Transfer RNA and the Translation Apparatus in the Origin of Life 207

PART IV / PRECELLULAR ORGANIZATION

A. I. OPARIN / A Hypothetic Scheme for Evolution of Probionts 223
SIDNEY W. FOX, JOHN R. JUNGCK, and TADAYOSHI NAKASHIMA / From Proteinoid Microsphere to Contemporary Cell: Formation of Internucleotide and Peptide Bonds by Proteinoid Particles 227
KLAUS DOSE / Chemical and Catalytical Properties of Thermal Polymers of Amino Acids (Proteinoids) 239
R. BUVET, L. LE PORT, and F. STOETZEL / Pre-Enzymic Origin of Metabolic Redox Processes and of the Energy Storage Processes 253
W. THIEMANN and W. DARGE / Experimental Attempts for the Study of the Origin of Optical Activity on Earth 263
JOHN KEOSIAN / Life's Beginnings – Origin or Evolution? 285

PART V / EARLY BIOCHEMICAL EVOLUTION

L. BIERMANN and G. DIERCKSEN / On the Chemical Constitution of Cometary Nuclei 297
T. E. PAVLOVSKAYA and T. A. TELEGINA / Photochemical Conversions of Lower Aldehydes in Aqueous Solutions and in Fog 303
M. O. DAYHOFF, W. C. BARKER, and P. J. MCLAUGHLIN / Inferences from Protein and Nucleic Acid Sequences: Early Molecular Evolution, Divergence of Kingdoms and Rates of Change 311
THOMAS H. JUKES / On the Possible Origin and Evolution of the Genetic Code 331
G. W. R. WALKER / Genetics and the Origin of the Genetic Code 351
JOSEPH NAGYVARY and JANOS H. FENDLER / Origin of the Genetic Code: A Physical-Chemical Model of Primitive Codon Assignments 357
D. O. HALL, R. CAMMACK, and K. R. RAO / The Iron-Sulphur Proteins: Evolution of a Ubiquitous Protein from Model Systems to Higher Organisms 363
HERRICK BALTSCHEFFSKY / A New Hypothesis for the Evolution of Biological Electron Transport 387
A. A. KRASNOVSKY / Pathways of Chemical Evolution of Photosynthesis 397
FUJIO EGAMI / Inorganic Types of Fermentation and Anaerobic Respirations in the Evolution of Energy-Yielding Metabolism 405

PART VI / EXOBIOLOGY

K. BIEMANN / Test Results on the Viking Gas Chromatograph-Mass Spectrometer Experiment 417
HAROLD P. KLEIN / Automated Life-Detection Experiments for the Viking Mission to Mars 431
DONALD A. FLORY, JOHN ORÓ, and PAUL V. FENNESSEY / Organic Contamination Problems in the Viking Molecular Analysis Experiment 443

M. A. MITZ / Model Systems for Life Processes on Mars 457
G. EGLINTON and S. TONKIN / An Automatically-Returned Martian Sample by 1985? 463
W. F. LIBBY / Life on Jupiter? 483
V. A. OTROSHCHENKO and YU. A. SURKOV / The Possibility of Organic Molecule Formation in the Venus Atmosphere 487
SHIV S. KUMAR / Planetary Systems and Extraterrestrial Life 491
CARL SAGAN / The Origin of Life in a Cosmic Context 497

LIST OF PARTICIPANTS 507

INDEX OF SUBJECTS 517

PREFACE

This publication, in two volumes, includes most of the scientific papers presented at the first meeting of the International Society for the Study of the Origin of Life (ISSOL), held on June 25–28, 1973 in Barcelona, Spain. The first volume contains the invited articles and the second volume the contributed papers, which also appear in the 1974 and 1975 issues, respectively, of the new journal *Origins of Life*, published by D. Reidel.

A relatively large number of meetings on the subject of the origin of life have been held in different places since 1957. In terms of its organization, scope, and number and nationality of participants, the Conference celebrated last year in Barcelona closely followed the three international conferences held earlier in Moscow, U.S.S.R., 1957, Wakulla Springs, U.S.A., 1963, and Pont-à-Mousson, France, 1970. For this reason the first ISSOL meeting was also named the 4th International Conference on the Origin of Life.

As an introduction to the papers published herein it seems appropriate to make some remarks concerning the inception and development of this Conference. It was during the previous conference at Pont-à-Mousson that the organizational meeting for the founding of ISSOL took place. There a request was made by one of us (J.O.), on behalf of the Autonomous University of Barcelona, to celebrate the first meeting of the new Society in Barcelona in honor of Professors Alexander I. Oparin, Melvin Calvin and Harold C. Urey, in recognition of their pioneering studies on the problem of the origin of life on Earth. Upon approval of the executive committee of the ISSOL the Conference was organized under the honorary chairmanship of Professor Severo Ochoa, by two of us (J.O. and R.S.Y.) with the help of the Spanish Organizing Committee, with Dr Vicente Villar Palasi as chairman and Dr Jaime Palau as secretary general.

The Conference was held at the Congress Hall of the city of Barcelona for four consecutive days. On the first day, after a brief introductory session where Professors Oparin, Calvin, Urey and Miller were conferred the medal and title of Honorary Councilors of the Higher Council for Scientific Research of Spain, the scientific presentations of the Conference got underway. Basically the Conference was planned according to a scheme that approximately followed the different stages of the evolutionary process. It was arranged into six colloquia of invited papers and corresponding sessions of contributed papers. The titles and brief contents of the six colloquia were: (1) *Organic Cosmochemistry*: organic composition of interstellar clouds, comets, planets and meteorites; (2) *Primitive Earth and Paleobiology*: primitive Earth's atmosphere, hydrosphere and lithosphere, and Precambrian microfossils; (3) *Abiotic Organic Synthesis and Interactions*: abiotic syntheses of amino acids, peptides, nucleotides, etc., and molecular interaction, replication, and information transfer processes of biochemical molecules; (4) *Structural and Thermodynamic Considerations on the Origin of Life*: theories and models for the self-organization and evolution of probionts, and related thermodynamic, structural, catalytical, and optical activity aspects; (5) *Early Biochemical Evolution*: fitness in the universe, and evolution of proteins and of the genetic code, electron transport and other energy-yielding processes; (6) *Exobiology and Planetary Exploration*: exploration of the moon and planets of the solar system in the search for extraterrestrial life, and the possibility of life's existence beyond the solar system.

As part of Colloquium (6) there were a group of papers specifically concerned with the possibilities of life on Mars, and with the current status of the U.S.A. NASA Viking project for the organic and biological analysis of this planet, as well as with the planning of possible future Martian missions. The interesting discussion which ensued was centered on how much information is needed from prior missions in order to eliminate or reduce to a minimum the potential risk of back contamination before samples can be brought back from Mars.

The sequence of themes of the Conference has been kept in these volumes although some minor modifications have been made in the names of the different sections and a few papers have been rearranged. For instance, because of the wide scope in the treatment of evolutionary processes Professor G. Wald's paper has been moved to the very front.

Looking in retrospect at the Conference as a whole it was encouraging to observe the steady growth of interest in this field of science manifested there. It was almost 50 years ago that Oparin's book *The Origin of Life* was published in the U.S.S.R., and more than two decades ago that Miller demonstrated the synthesis of amino acids under possible primitive earth conditions, observations which gave the initial theoretical and experimental impulses to the evolutionary chemical theory of the origin of life. However, it was not until 1957 that the first serious meeting on the subject was organized in Moscow, where concern was expressed for the lack of interest in this field of science during the first half of this century. Thus it was gratifying to see some one hundred and fifty actively working research scientists coming from all parts of the world to meet in Barcelona and present their most recent observations and experimental results, pushing forward different lines of research and providing further support for the theory. Of particular interest was the participation of the Nobel laureates, M. Calvin, S. Ochoa, Sir R. Robinson, H. C. Urey and G. Wald, who together with their colleagues discussed their new ideas and potential future developments in this challenging field of science. It became apparent that although the experimental research in this area has had a very slow start, the advances made during the last two decades are beginning to provide a real understanding of the problem of the origin and early development of life and suggests that its solution may be reached before the end of this century.

We would like to express our gratitude first to the more than three hundred members registered for this Conference, who by their active participation made the success of this scientific meeting possible. The Conference was held under the auspices of the Ministry of Education and Science, the Higher Council for Scientific Research of Spain, and the City Hall of Barcelona, with the collaboration of the Spanish Society of Biochemistry and other governmental and private institutions of Spain. We acknowledge these departments and institutions and their respective heads, namely, Julio Rodrīguez Martinez, Jose Luis Villar Palasi, Enrique Masó Vazquez and Federico Mayor Zaragoza for their invaluable assistance.

Special acknowledgements are due to the Institute of Fundamental Biology of the Autonomous University of Barcelona, and the Department of Biophysical Sciences of the University of Houston for organizing the Conference. We also acknowledge the National Aeronautics and Space Administration for making possible the participation of many scientists to this Conference and for making available display material of the results obtained from the Mariner 9 space mission to Mars as well as the plans for the forthcoming Viking mission to the same planet.

Our special thanks go to the eminent Spanish painter, Salvador Dali, for his unselfish contribution of a symbolic painting for the poster of the Conference and to P. Puig Muset and his technical staff for its faithful reproduction. The initial developmental stages and organization of the Conference in Spain would not have been possible without the help of Jose Ma de Porcioles (ex-mayor of the city of Barcelona), Luis Miravitlles and A. Romaña, S. J., as well as Allen H. Bartel, Daniel Navarro, Marianna O'Rourke and Maria Vallés de Palau.

We would not like to close these introductory remarks without expressing our sorrow for the untimely deaths of Professors Aharon Katzir-Katchalsky and Wolf Vishniac. Dr A. Katzir-Katchalsky was a victim of the terrorist attack at the Lodd Airport of Tel-Aviv in May of 1972. Dr Katzir-Katchalsky had been expected to participate in the Barcelona conference, the date of which nearly coincided with the first anniversary of his death. Dr Wolf Vishniac, who had actively participated in the Conference, died in the Antarctica in December, 1973, while exploring the very cold and dry habitats of terrestrial microorganisms in order to obtain a better understanding of the possibilities of life under extreme environments, as may be found on Mars. Two other scientists, well known for their early studies on the problem of the origin of life, also passed away during this past decade: J. B. S. Haldane in 1964, and J. D. Bernal in 1971. As an appreciation of their significant contributions to this field of science we are dedicating to these four outstanding scientists the published papers from the 1st International Meeting of the Society for the Study of the Origin of Life.

J. ORÓ

S. L. MILLER

C. PONNAMPERUMA

R. S. YOUNG

PART I

COSMOCHEMISTRY

FITNESS IN THE UNIVERSE: CHOICES AND NECESSITIES

GEORGE WALD*

Biological Laboratories of Harvard University, Cambridge, Mass., U.S.A.

Freedom is the recognition of necessity.
Hegel

Abstract. We live in a universe of chance, but not of accident. Repeatedly in the course of its development choices have been made for which one can ask the reasons. One such choice is fundamental: if the proton had not so much greater mass than the electron, all matter would be fluid; and if the proton did not have exactly the same numerical charge as the electron – or some simple multiple of that charge – virtually all matter would be charged. If a universe were started with charged hydrogen, it could expand, but probably nothing more. Hydrogen, carbon, nitrogen and oxygen play as fundamental – and irreplaceable – roles in the metabolism of stars as of living organisms. Both metabolisms are coupled, through radiation from the stars providing the energy on which life must come ultimately to run on the planets. In the course of their evolution on the Earth, living organisms have found their way repeatedly and exclusively to certain types of organic molecule to perform specific functions; so, for example, the chlorophylls for photosynthesis, and carotenoids for plant phototropism and for vision. It is argued that some measure of necessity has governed these choices; and that an extended principle of natural selection has operated at all levels of material organization to produce such elements of order and compatibility in the universe.

One can hardly ask about the origin of life on the Earth without raising broader questions. We live in an inhabited universe, one in which, given enough time, life arises wherever it can, as a kind of culmination of the spontaneous organization of the lighter elements. Eddington has given us the formula: 10^{11} stars = 1 galaxy; 10^{11} galaxies = 1 universe. Our own galaxy, the Milky Way, contains some 10^{11} stars; and it is estimated that 1–5% of them – hence at least one billion (10^9) – have planets that might support life. About 10^9 such galaxies are within reach of our most powerful telescopes; so the already observed universe contains at least 10^{18} planets fit for life. Life may not have arisen in all those places, yet surely it must have done so in many of them.** Inevitably what we learn of how life began here must tell us something of how it arises elsewhere.

* I have discussed some of the matters of this paper earlier and in greater detail: the bioelements in the light of their positions in the Periodic System (1962, 1964); light and life: photosynthesis, phototropism and vision (1943, 1945, 1946, 1959, 1965, 1968a); and relations between stars and living organisms (1968b).

** Obviously any such statement requires a time frame, for as one looks deeply into space, one is also seeing far backward into time. The light that reaches us now from the furthest boundaries of the visible universe has been some 6 billion years on the way. Stars even within our own galaxy may be 10^5 light years away, and within the local group that includes our sister galaxy Andromeda 10^6 light years away. What do we mean asking whether life has arisen or what its state may be in such distant places? I think that the only reasonable meaning to attach to such statements is in terms of time, not here, but there – say the time a star has been on its Main Sequence, which I assume has some resemblance to our Main Sequence, wherever it is in the universe. By the time we observe them, many such stars may have long since evolved further or vanished, and life on their planets long since perished.

It is a popular fallacy that everything is relative. A few important things are universal – meaning that they hold throughout this universe. Thus it is made everywhere of four kinds of elementary particle – protons, neutrons, electrons and photons; or perhaps in some places the antiparticles of the first three (photons are their own antiparticles). The Periodic System of the elements – a simple number series – also is universal. The laws of chemical reaction, kinetic and thermodynamic, seem also to be universal. They govern universally what molecules exist, depending on the conditions.

And so it is with life, for life also involves universal aspects. It is a precarious development wherever it occurs. This universe is fit for it; we can imagine others that would not be. Indeed this universe is only *just* fit for it. Of the 92 natural elements, only four – hydrogen, oxygen, nitrogen and carbon – possess the properties upon which life principally depends. They have this status in part because of their position in the Periodic System. They are the lightest four elements that achieve stable electronic configurations by gaining one to four electrons. Gaining electrons by sharing them with other atoms is of course the mechanism of forming chemical bonds, and hence molecules. These lightest elements make the tightest bonds, hence the most stable molecules; and they – specifically C, N and O – are the only elements in existence that regularly form *multiple bonds*. The importance of that property becomes clear when we compare carbon with silicon. Silicon lies just below carbon in the Periodic System and shares many fundamental properties with it, but cannot form multiple bonds.

Thus in carbon dioxide the carbon is bound to the oxygens by double bonds, $O{=}C{=}O$, so satisfying the combining tendencies of all three atoms. Hence carbon dioxide is an independent molecule, that goes off into the air as a gas and dissolves in the waters of the Earth, the states in which organisms circulate carbon. In silicon dioxide, however, the oxygens can join with silicon only by single bonds, $\begin{array}{ccccc} & & | & & | \\ O & — & Si & — & O \\ & & | & & | \end{array}$, leaving four unpaired electrons ready to pair with those of neighboring silicon dioxide molecules, and those with others, over and over. What results is a huge polymer such as quartz, literally a supermolecule, so hard because it can be broken only by breaking chemical bonds. The accessible portions of the Earth contain about 135 times as many silicon as carbon atoms, yet for this and other reasons silicon cannot replace carbon in living organisms; nor can sulfur replace oxygen, nor phosphorus nitrogen; nor can any other elements in nature replace any of these four.

These unique elements make unique molecules. I doubt that life is possible anywhere without liquid water, the strangest molecule in all chemistry. Among its strangest properties is that ice floats. If ice did not float – and hydrogen sulfide and ammonia ices do not – it is highly unlikely that life could arise, or long survive; for it is only that circumstance that permits the bulk of water to remain liquid under a surface skin of ice, rather than freezing solidly during any protracted cold spell, and then remaining frozen indefinitely.*

This universe breeds life inevitably only because H, C, N and O have unique properties not shared with other elements; because carbon combines readily with itself; because only C, N and O readily form multiple bonds; because hydrogen and oxygen form so odd a molecule as water; because ice floats. Prior to life, a universe existed that was fit for it. Back of the origin of life lies the origin of the molecules that make life possible; back of them, the origin of the elements that alone can constitute such molecules; and still further back, the origin of the elementary particles as we find them, for even there we tend to take too much for granted.

We live in a world of chance, but not of accident. "God gambles, but He does not cheat." In a world of chance, anything can happen, but only some things persist. When they do, we can ask why.

Putting such questions to Nature, and wringing answers from her is often a difficult business, as every scientist knows. Yet sometimes it is as though Nature were trying to tell us something, almost to shake us into listening.

I wish to discuss a few such instances: two on the level of highly specialized organic molecules that serve basic biological functions; one on the level of elements, coupling the lives of stars with those of organisms; and lastly one involving elementary particles, for strangely enough they too raise a problem of fitness.

How to introduce fitness among so many possibilities for random variation and change? I think always in the same way, through the operation of an extended principle of natural selection; hence always retroactively, throug hcompetition among the variations that have arisen, on the basis of stability and compatibility. We are used to thinking in such terms about living organisms, because under the conditions familiar to us they vary, whereas atoms and elementary particles are stable. At the much higher temperatures in the stars, however, the elements also vary and evolve; and in still other circumstances elementary particles become unstable, vary and interact. We can imagine starting such a universe as this with neutrons; for within minutes free neutrons would have decayed in part to protons, electrons and photons, the other components of our universe. Conversely, in dying stars gravitational collapse can yield densities at which protons and electrons fuse to neutrons; the pulsars appear to be neutron stars.

* I first became aware of the importance of ice floating through Henderson's "Fitness of the Environment" (1913, 1958), pp. 106–110. Looking back recently at that 'golden book', I see that the thought is much older, having assumed some importance in the so-called Bridgewater Treatises of the early nineteenth century, in which very considerable scientists cited it as an evidence of the existence of a Creator. Thus Whewell in his Bridgewater Treatise discussed this, citing an experiment of Rumford in which a vessel of water containing ice confined at the bottom could be heated and even boiled at the top without melting the ice. William Prout writes in his Bridgewater Treatise: "The above anomalous properties of the expansion of water and its consequences have always struck us as presenting the most remarkable instances of design in the whole order of nature – an instance of something done expressly, and almost (could we indeed conceive such a thing of the Deity), at second thought, to accomplish a particular object." Many scientists since, including Henderson, have been so anxious to repudiate the deism that they disparage the design. Perhaps the principal point of the present paper is to emphasize the design, but derive it, not *a priori*, through the specifications of a Designer, but *a posteriori*, through an extended principle of natural selection.

At various times and places in the universe conditions arise that promote natural selection at every level of material organization. I think that this process accounts ultimately for the order that we observe in Nature, and for the peculiar compatibilities that it displays.

1. Chlorophylls and Photosynthesis

Photosynthesis represents the assembly of what were probably two earlier metabolic systems: the hexose-pentose-phosphate cycle, an anaerobic process that supplies hydrogen attached to TPN for reductions; and photophosphorylation, a process that uses the energy of sunlight to make ATP. $TPN-H_2$ and ATP together constitute what Arnon calls *assimilative power*, the factors needed to reduce CO_2 to carbohydrate (cf. Clayton, 1970).

With the exception of a few exotic chemosyntheses, all life on the Earth has come to run ultimately on sunlight, on photosynthesis performed by bacteria and plants. We find a wide variety of photosyntheses, and at least 8 different chlorophylls to mediate their light reactions; yet no photosynthesis without chlorophyll.

Photosynthesis is by far the dominant organic process on the Earth, not only because of its importance for life, but in sheer bulk. Each year it fixes about 2×10^{11} tons of carbon in organic linkage. Hence no other organic molecule bears as stringent a burden as chlorophyll, and no other biochemical relationship is as critical as that between chlorophyll and light. We have every reason to expect that natural selection has done its utmost to optimize this relationship, to select pigments that accomplish it most effectively.

That is what makes it so significant that the absorption spectra of the chlorophylls are so grossly wrong for the job. The chlorophylls absorb light most poorly where the spectrum of sunlight is most intense, and have their peak absorbances to both sides, where the sunlight spectrum has fallen to low values (Figure 1).

This situation is so paradoxical that I have no doubt that it is telling us something very important: that for photosynthesis, and presumably photophosphorylation before it, organisms could find no better choice; that the chlorophylls possess other properties so compelling as to override their inappropriate absorption spectra. What are those other properties?

I would note first the peculiar position of the magnesium ion. This is the only instance I know in which magnesium serves as the nucleus of an organic co-ordination complex.* Ordinarily we find a transition element in this position, such as iron in the closely related hemins (Figure 2). Transition elements lend themselves to such uses in two ways: since they tend to add many electrons to approximate the nearest inert gas configuration, they readily nucleate coordination complexes; and having variable valence, they readily trade electrons. The choice of magnesium ion in the chlorophylls

* Magnesium activates a number of enzyme systems involving phosphates, among them ATP; thus for example carboxylase is a thiamine pyrophosphate – Mg^{++} – apoprotein. The role of Mg in such complexes is not well understood, but very likely involves some degree of co-ordinate linkage.

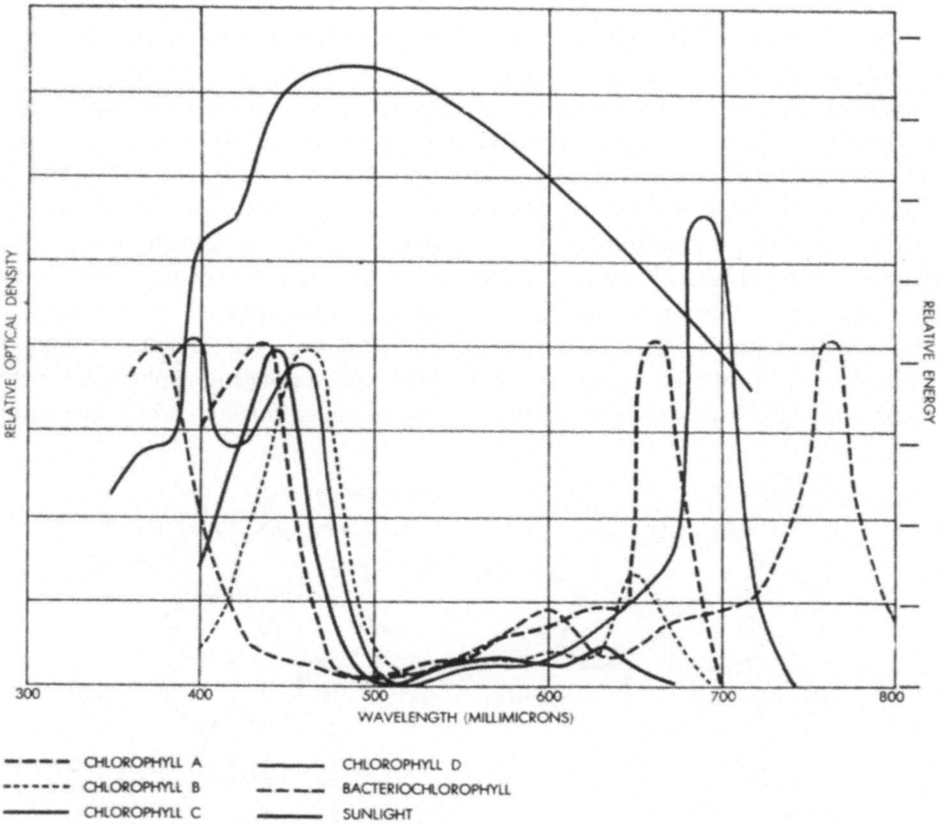

RELATIVE OPTICAL DENSITY

RELATIVE ENERGY

300 400 500 600 700 800

WAVELENGTH (MILLIMICRONS)

- - - -	CHLOROPHYLL A	——	CHLOROPHYLL D
·······	CHLOROPHYLL B	— — —	BACTERIOCHLOROPHYLL
——	CHLOROPHYLL C	——	SUNLIGHT

Fig. 1. Absorption spectra of various types of chlorophyll compared with the energy distribution of sunlight at the Earth's surface (from Wald, 1959, 1965).

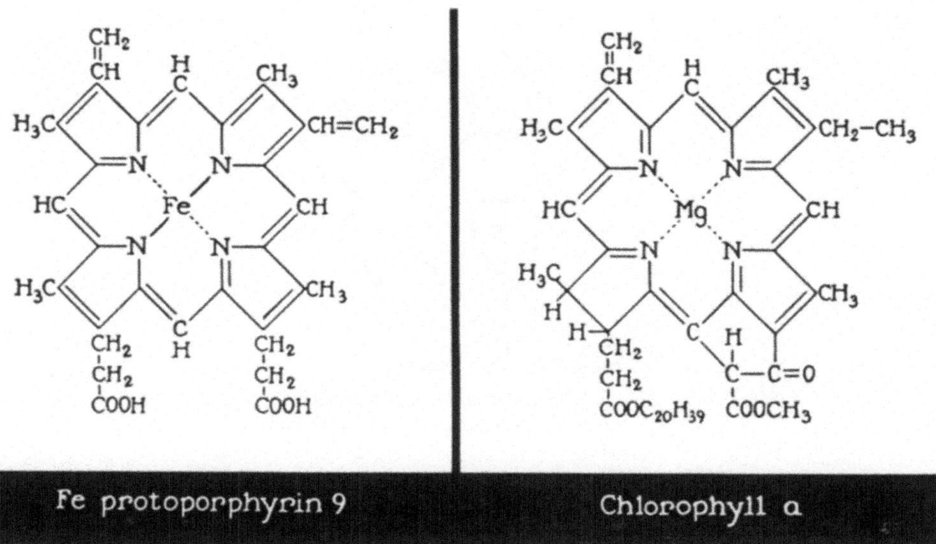

Fig. 2. Structures of Fe protoporphyrin 9, (heme) the prosthetic group of b-cytochromes, catalase, peroxidase and the hemoglobins; and of chlorophyll a.

is as though to tell us that what is wanted here is specifically *not* a transition element, that here its job is *not* to pass electrons.

Would calcium or zinc have done as well? Possibly; but for reasons that are touched upon below, they are much rarer elements than magnesium. In the cosmic distribution of elements, Mg:Ca:Zn = 3400:260:1; and, more relevantly, in sea water Mg:Ca: Zn = 378000:66500:1. (These are the proportionate numbers of atoms, not their weights.) Also magnesium ion has a special association in biochemistry with enzyme catalyses involving phosphates, particularly ATP. Is it that association that brought Mg^{++} into photophosphorylation, and eventually photosynthesis?

All organic pigments owe their color to the presence of a so-called conjugated system of regularly alternating single and double bonds. In the chlorophylls this comes around upon itself to form a rigid, planar ring of rings, through which the mobile

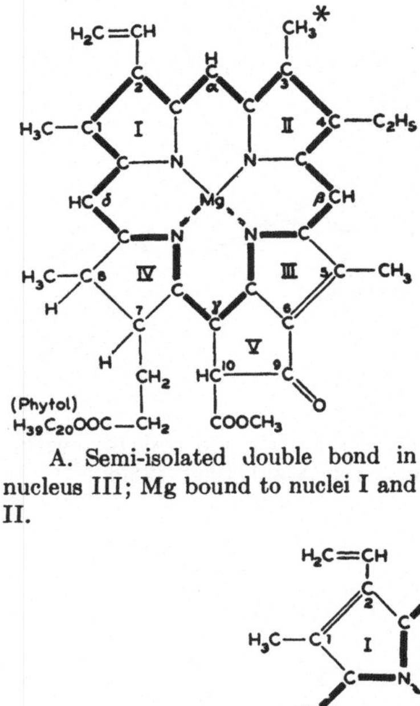

A. Semi-isolated double bond in nucleus III; Mg bound to nuclei I and II.

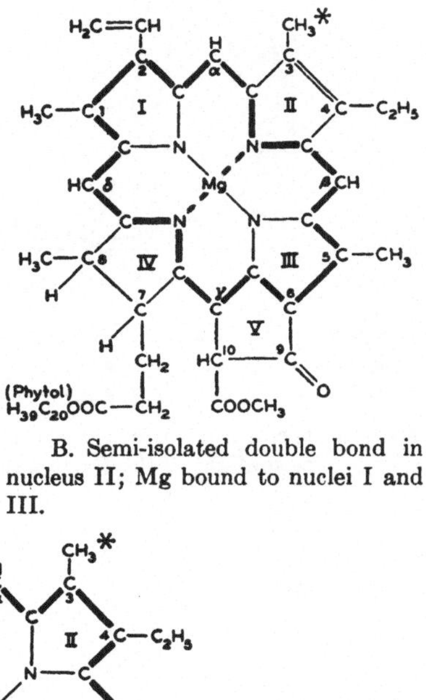

B. Semi-isolated double bond in nucleus II; Mg bound to nuclei I and III.

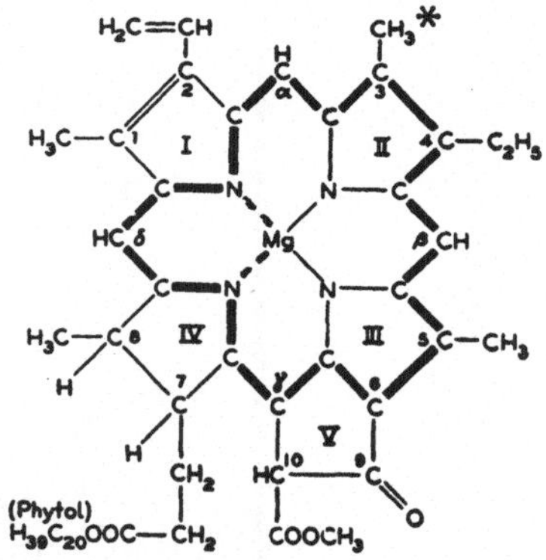

C. Semi-isolated double bond in nucleus I; Mg bound to nuclei II and III.

Fig. 3. Chlorophyll a structure, showing three limiting resonance (mesomeric) forms, each a different routing of conjugation around the ring system (heavy lines), and the corresponding changes in position of the covalent linkages to Mg^{++} and of the 'semi-isolated' double bonds. Figure 2 presents still another resonance form of this molecule.

pi-electrons can reverberate and circulate. In the higher plant chlorophylls there are also alternate pathways of conjugation. All of this permits multiple rearrangements of the conjugated system, each of which represents a limiting resonance form. The chlorophylls are all resonance hybrids of such possible rearrangements (Figure 3). One consequence of this structure is their great stability. The porphyrins are among the most inert organic molecules; porphyrins, apparently derived from chlorophylls, have been found in petroleums, oil shales and soft coals some 400 million years old (Treibs, 1936; Dunning, 1963).

This rigid, highly resonating structure probably makes another decisive contribution to photosynthesis; for this process demands of the chlorophylls not only that they absorb photons, but keep quanta of the absorbed energy intact – their strong red fluorescense is evidence of this – and pass it on without degradation to neighboring chlorophyll molecules by resonance transfer, until eventually it reaches a reaction center ready to use it in photosynthesis. In this way some 50 to 500 molecules of chlorophyll harvest light for each reaction center. In the alga Chlorella the energy of an absorbed photon may be transferred an average of 200 times before being used.

So only minor fractions of chlorophyll, whether in bacteria, algae or higher plants, are photochemically active. The great bulk of chlorophyll molecules act only as antennae, absorbing light and passing the energy quanta about until they reach reaction centers. In the latter, the chlorophyll is in a special state that absorbs at considerably longer wavelengths than the light-harvesting forms, so serving as an energy sink. Quanta are fed into such reaction centers, not only by other chlorophylls, but by other light-harvesting pigments: carotenoids, and in certain algae also phycobilins – phycoerythrin and phycocyanin. These latter pigments help to fill in the chlorophyll spectrum (cf. Clayton, 1970), thus compensating in part for its deficiency as an absorber of sunlight, yet still leaving important lacunae. For the rest, chlorophyll makes up for its poor spectrum by increasing its density, forming on occasion layers of concentrated pigment so deep as to absorb virtually all the incident light. One has only to look upward under a tree to see that this is so.

Coupled with all the geometric rigidity and general inertness of chlorophyll, there must go also the potentiality to react: somewhere chlorophyll contains a reactive site. James Franck suggested that that is the 5-membered ring – ring V in Figure 3 – with its conjugated keto group, the structural feature that most distinguishes the chlorophylls from the other natural porphyrins.

These are as yet no more than hints toward answering our question. One can hardly ask more as yet, for we still do not understand how chlorophylls work in photosynthesis. When we do, we shall also understand better how these molecules came to play an exclusive role in photosynthesis on this planet despite their inappropriate spectra. It looks as though organisms on the Earth had no preferable choice.

There must come a stage in the development of life wherever in the universe it persists when it must go over from feeding upon the prebiotic accumulation of organic matter to synthesizing its own with light from its star. When that time comes it seems to me likely that the same factors that governed the exclusive choice of the chlorophylls for photosynthesis on the Earth might prove equally compelling elsewhere.

2. Carotenoids: Phototropism and Vision

With all life contingent on photosynthesis, it is of the highest importance that plants find the light, that herbivores find plants, and that carnivores find herbivores. These are the uses of phototropism in plants and vision in animals.

As with photosynthesis, this entire class of reactions seems to have settled upon a single, closely knit family of organic pigments to mediate the light reactions, in this case the carotenoids. That plant phototropisms, like vision, depend mainly or entirely upon carotenoids has lately been challenged – of which more below; but I think that the weight of the evidence still favors this view.

These are very different uses of light from photosynthesis. The business of photosynthesis is to use the energy absorbed as light to do chemical work, to turn it into the bond energies of organic molecules. The more efficiently that is done, the better the process is working; and so much hangs upon it that natural selection has probably exercised sharper discrimination there than in any other aspect of metabolism. That is why the aberrant spectra of the chlorophylls make such a problem. As already noted, the carotenoids help with that problem by absorbing light in the blue and blue-green where the chlorophylls are transparent – and the sunlight spectrum is more intense – and transferring the energy absorbed to chlorophylls for use in photosynthesis.

The business of light in phototropism and vision however is not to do work, but to excite: by exciting the absorbing pigment, to trigger a response, ultimately the bending of a plant, by unbalancing the rates of growth on the darker and lighter sides; or the nervous excitation that stimulates vision in an animal. To say this in another way: the role of light in these processes is kinetic rather than thermodynamic, to *activate* a chemical reaction rather than to fuel it, by surmounting a potential energy barrier.

In such molds as *Phycomyces* and such higher plants as the oat shoot *Avena*, phototropism seems to be mediated through conventional plant carotenoids, C_{40} molecules, principally β-carotene ($C_{40}H_{56}$) in Phycomyces, and leaf xanthophyll (lutein, $C_{40}H_{54}(OH)_2$) in Avena (Figure 4) (cf. Wald, 1943; Thimann and Curry, 1960; Delbrück and Shropshire, 1960; Hager, 1970). But by the time animals developed well organized eyes, two things had changed: those animals, unlike plants, cannot synthesize carotenoids *de novo*, and so must obtain them in their diets, as vitamins; and visual photoreception has come to depend upon C_{20} molecules, haplo-carotenoids, the aldehydes of the vitamins A (Wald, 1945, 1946).

And again the message: In the course of evolution on this planet, three animal phyla have developed highly organized, image-resolving eyes: Arthropods (insects, crabs), Molluscs (squid, octopus) and Vertebrates.* These seem to represent wholly independent developments, not linked to one another in any way, anatomical,

* In a fourth phylum, the Annelids, certain polychaete worms have developed very well-formed cyes; but nothing is yet known of their chemistry.

embryological or evolutionary. Yet all of them have the same chemistry. In all of them a vitamin A (retinol) in the form of its aldehyde (retinal) is joined as chromophore to an opsin, a specific type of membrane protein found in visual receptors, to form a visual pigment. Every visual pigment that we yet know, including represen-

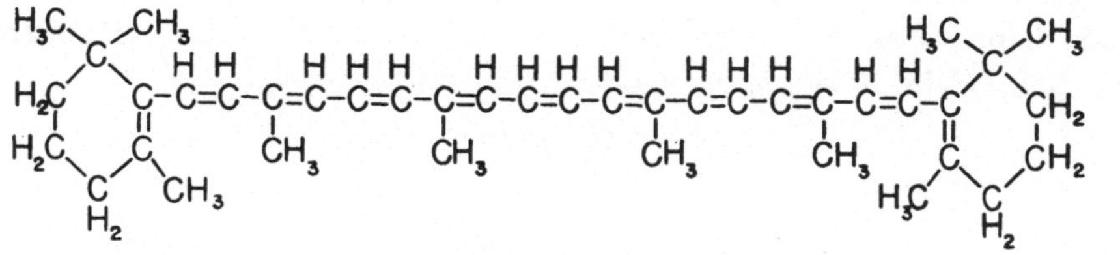

β-carotene, $C_{40}H_{56}$

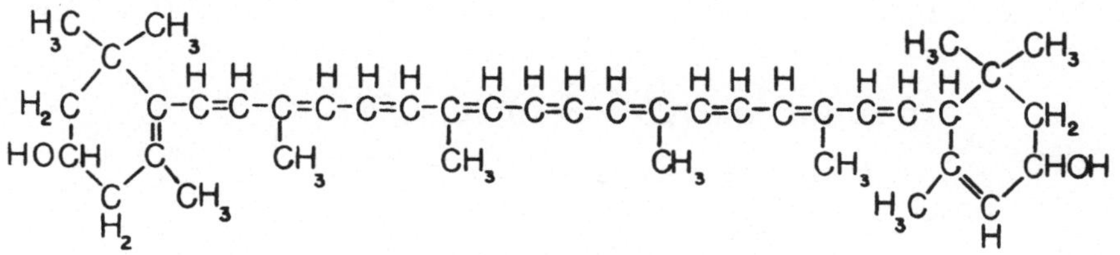

lutein, $C_{40}H_{54}(OH)_2$

Fig. 4. Structures of the plant carotenoids, *β*-carotene and lutein (leaf xanthophyll).

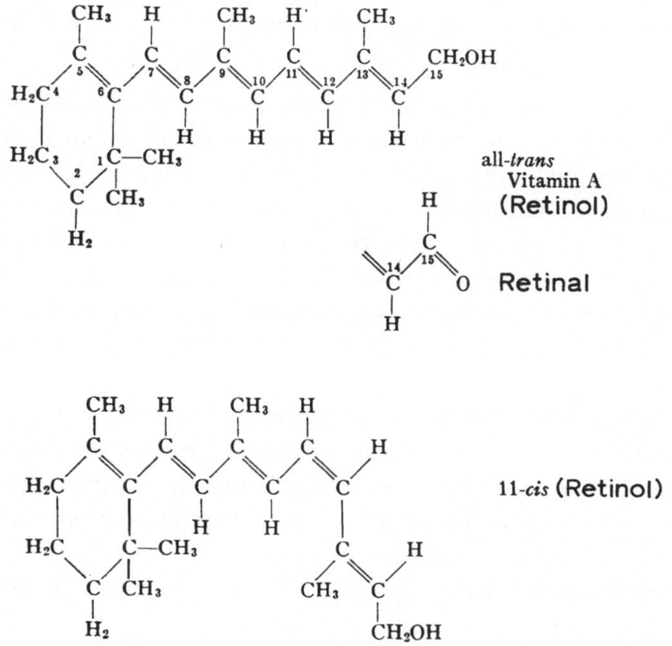

all-*trans* Vitamin A (Retinol)

Retinal

11-*cis* (Retinol)

Fig. 5. Structures of all-*trans* and 11-*cis* vitamin A (retinol) and of the corresponding aldehyde, retinal.

tatives from all three phyla, has as chromophore the same, sterically hindered *cis* configuration of retinal, the 11-*cis* isomer (Figure 5). In every visual system yet known, the only effect of light in initiating visual excitation is to isomerize the chromophore of a visual pigment from 11-*cis* to *trans*; and on occasion it may isomerize the *trans* chromophore back to 11-*cis*, as part of recovery (Figures 5 and 6). Everything else in vision happens as dark consequences of this one light reaction (Hubbard and Kropf, 1958; Wald, 1968).

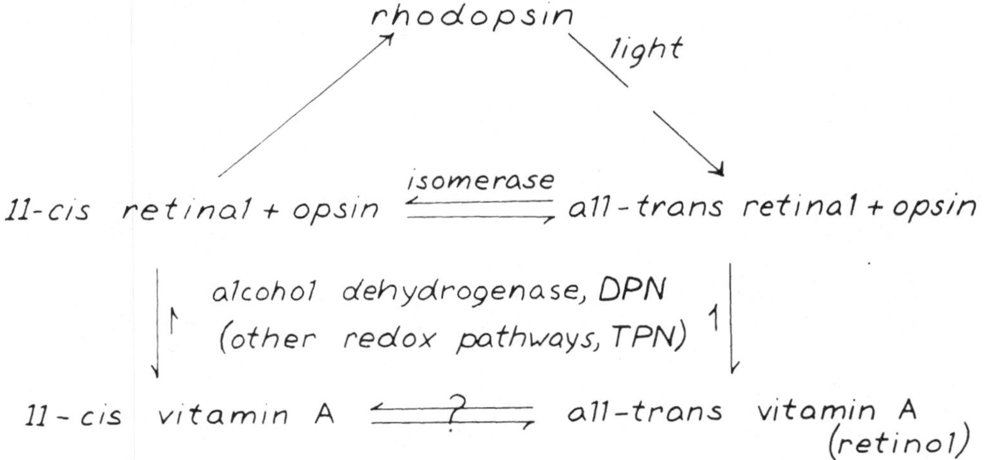

Fig. 6. The rhodopsin cycle, as it occurs in vertebrate rods. All known visual pigments, vertebrate and invertebrate, have as chromophore the 11-*cis* isomer of retinal or retinal₂, and the only action of light upon them is to isomerize this chromophore to all-*trans*. (From Wald, 1968).

That this same arrangement should turned have arisen three times independently in the course of animal evolution on the Earth seems to me to imply that 11-*cis* retinal is so peculiarly suited for mediating visual excitation as to offer a signal advantage over other types of organic molecule that might otherwise have performed this function.

What gives it this advantage? The essential problem in vision is to find a photochemical change of high quantum efficiency in an organic pigment, that is readily and economically reversed, so that vision can continue at little cost to the organism. I think that this demand resolved into the call for *a pigment that would change its shape in the light*.

An organic molecule becomes a pigment through possessing a conjugated system of alternate single and double bonds. In almost all organic pigments, however, natural or artificial, the conjugated systems are rigidly bound up in rings, as in the chlorophylls. For such a system to change its shape, it must be *straight-chain*. Then light can readily change its shape by *cis-trans* isomerization.

A simple, straight-chain polyene, $CH_3\,(CH{=}CH)_n\,X$, might do. In the carotenoids and vitamins A, however, composed of isoprene units $(-CH{=}CH-\overset{\overset{\displaystyle CH_3}{|}}{C}{=}CH-)$, methyl groups project at regular intervals along the chain. In 11-*cis* retinal it is the projecting methyl group on C_{13}, overlapping the hydrogen atom on C_{10}, that causes steric hindrance (molecular overcrowding), forcing a twist in the molecule at this point, so interfering with resonance and causing an intrinsic instability. That projecting methyl group puts this molecule, so to speak, on hair trigger: 11-*cis* retinal is isomerized by light with the highest quantum efficiency of all the geometrical isomers of retinal.

So if animals on this planet came to need pigments that would change their shape efficiently in the light, there was much to lead them to 11-*cis* retinal. The combination of this molecule with opsins yields highly photosensitive pigments whose absorption maxima for the most part lie close to where the spectrum of sunlight is most intense. This seems to have offered an optimal solution to the problem of visual excitation, whenever encountered.

We have just had another striking instance of this narrowing of choices. Vitamin A, and hence retinal, has heretofore been found only in the higher animal phyla, never in lower invertebrates or in a plant. Now retinal has appeared as the chromophore of a rhodopsin-like pigment in the purple membrane of the saline bacterium, *Halobacterium halobium* (Oesterhelt and Stoeckenius, 1971). What it does, and whether again 11-*cis*, are uncertain; yet this discovery promises to extend our view of these relationships in new directions.

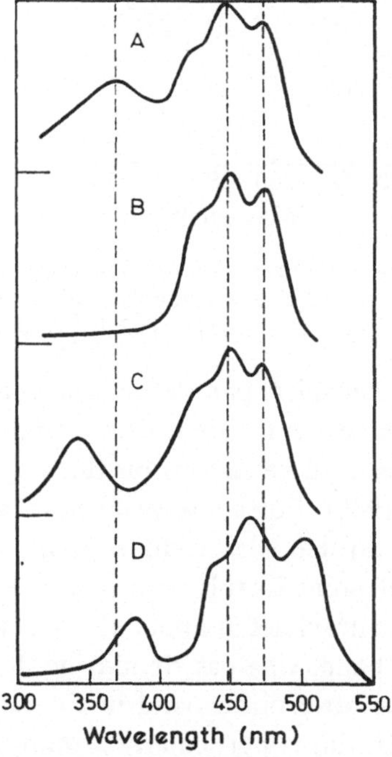

Fig. 7. Comparison of the action spectrum for the first positive phototropic bending of *Avena* coleoptiles (A: Thimann and Curry, 1960); with the absorption spectrum in hexane of all-*trans* β-carotene (B); and 11,11′-di*cis*-β-carotene (C); and with the absorption spectrum of riboflavine in castor oil (Bara and Galston, 1968). From Hager (1970).

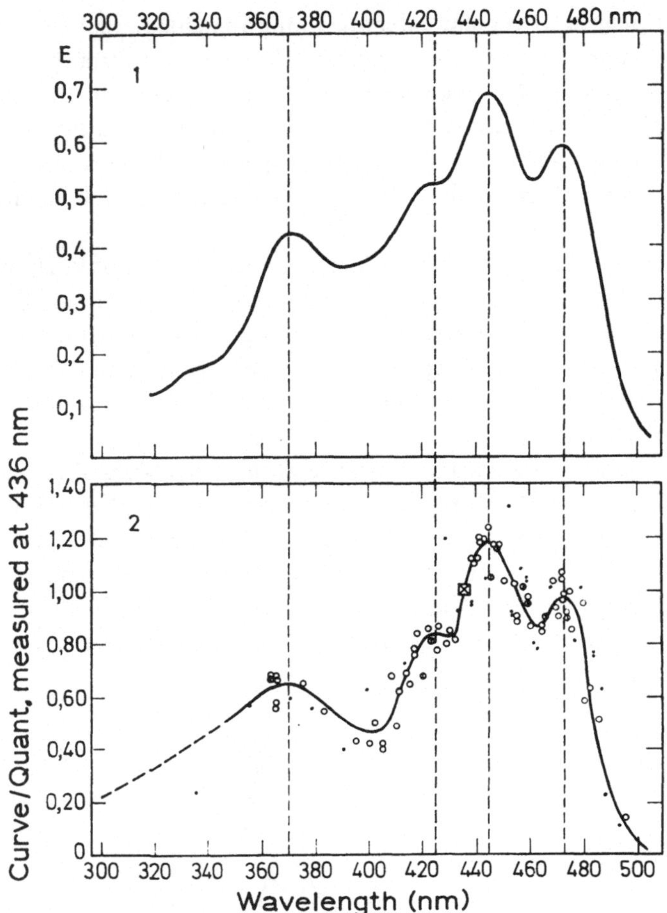

Fig. 8. Comparison of the action spectrum for the first positive phototropic bending of *Avena* coleoptiles (below: Thimann and Curry, 1960); with the absorption spectrum of lutein (leaf xanthophyll, 4.52 μgm ml^{-1}) in an ethanol-water mixture, 54.7:45.3 by volume (above). From Hager (1970).

To what degree plant phototropism parallels vision is still in doubt. I have supposed that phototropisms, phototaxes and vision, from molds to man, are involved in a profound chemical homology, all variations on the single theme of carotenoid pigments (Wald, 1943, 1945, 1946). For some years past, however, Galston has urged that flavines rather than carotenoids mediate plant phototropism (cf. Bara and Galston, 1968). A principal point in this controversy is a peak in the phototropic action spectrum in the near ultraviolet at about 370 nm, where carotenoids ordinarily do not have an absorption band, whereas riboflavine does. *Cis* forms of the carotenoids, however, do have an absorption band in the near ultraviolet, the so-called *cis*-peak (Zechmeister, 1962); this is particularly prominent in central-*cis* carotenoids (e.g., 15,15′-β-carotene) and hindered *cis* carotenoids (e.g., 11,11′-β-carotene). To be sure, in the free carotenoids in hexane solution such *cis* peaks come at considerably shorter wavelengths, at 330–340 nm. It has seemed possible, however, that the ultraviolet band in the phototropic action spectrum might represent such a *cis*-peak, shifted about 30 nm toward the red through binding of the carotenoid to protein, and its particular molecular environment in the plant. In that case plant phototropism, like vision, might depend on the photoisomerization of a *cis* carotenoid to the *trans* configuration. Though there may be other ways of achieving a near ultraviolet absorption band in carotenoid spectra (cf. Hager, 1970), this seems to me still to represent a highly significant possibility.

I should like to summarize the viewpoint of the preceding pages. It reduces to asking how much of biochemistry we can believe to be so general – to represent either necessary or such highly favored choices – as to approach universality.

For the reasons briefly mentioned here and discussed at greater length elsewhere, I believe that life, wherever in the universe it arises, must be composed primarily of H, C, N and O; and that liquid water is as necessary to it, as component and environment, everywhere else as here. I do not feel equally sure of the necessity for amino acids, still less nucleotides. Yet Urey (1952) has made a convincing argument that planets in general begin with reducing atmospheres. He has calculated that as long as they continue to contain more than 1.5×10^{-3} atm pressure of hydrogen gas, their atmospheres will remain reduced and will contain mixtures of water vapor, ammonia, methane, and hydrogen sulfide and cyanide. Miller, Abelson, Oro, Ponnamperuma and others have shown how readily such mixtures of gases generate amino acids, short-chain organic acids and nucleotide bases. So we may take the presence of such unit molecules on planets to be highly probable. It seems less probable to me that the nitrogenous bases need form nucleotides; there the ribose seems to me the most problematical component. In any case, once the problem is solved of polymerizing amino acids and nucleotides or their analogues in the presence of water, there is some likelihood that macromolecule formation will take the direction of polypeptides, proteins and nucleic acids or closely similar structures.

This then may be the prevalent if not universal background for the origin of life; the background therefore for solving such much later problems as those discussed above. To achieve a chlorophyll for photosynthesis – probably initially for photophosphorylation – would mean starting, not with some random assortment of organic molecules out of Beilstein's Handbook, but with glycine and acetic acid, components by that time plentiful in living organisms, and from which they synthesize chlorophylls at present. Similarly carotenoids would have been developed beginning with acetate, the component from which organisms synthesize them now.

I think that is the way the problem shapes up: given that organisms could achieve an enormous advantage by developing photophosphorylation, and already possessed stores of amino and fatty acids, what was the probable outcome? Very likely a chlorophyll, even to magnesium, of which I shall have more to say below. Similarly carotenoids for photoexcitation – though they may well have already been available, having made a prior appearance as auxiliary pigments in photophosphorylation and photosynthesis. My argument is that had they not already existed, organisms may have had to invent them for phototropism and vision, even to 11-*cis* retinal. That is the way it has come out on the Earth, for reasons that we are just beginning to understand. If those reasons hold here, why not elsewhere?

It is an invitation to formulate a universal biochemistry. That seems an over-pretentious term; perhaps *theoretical biochemistry* would be both more modest and more realistic. We are not likely soon if ever to learn how organisms are composed in other solar systems. What this is really asking is why they are made as they are here. It would be the attempt to rationalize biochemistry. Just as physical chemistry is what makes sense in chemistry, so this 'universal biochemistry" would represent the effort to learn what makes sense in biochemistry.

3. The Metabolisms of Stars and of Living Organisms

Biochemistry begins in the stars. Stars are born, mature, grow old and die as do living organisms. Stars too have a metabolism. Main Sequence stars, which alone might have planets that support life, exist by "burning" – i.e., fusing – hydrogen to helium, as organisms live by burning hydrogen to water.

A star is born by the gravitational collapse of a mass of hydrogen. As such a mass condenses, it heats up. When its core temperature reaches about 5 million degrees, the hydrogen begins to 'burn'. Four hydrogen nuclei – protons – each of mass 1 fuse to form a helium nucleus of mass about 4. But in this transaction a little mass (*m*) is lost, and this, multiplied by c^2, the square of the speed of light, 9×10^{20}, pours out the vast quantities of heat and radiation that keep the star from collapsing further, and hold it on the Main Sequence.

Inevitably the time comes in the life of every star when its core begins to run out of hydrogen. With that it begins to collapse again, and so to heat further. When the temperature in its deep interior reaches about 100 million degrees, the helium begins to 'burn'. Two He^4 fuse to yield Be^8, a nucleus so unstable that it disintegrates within 10^{-16} s. Yet in these enormous masses of material there are always a few Be^8 nuclei, and here and there one captures another helium: $Be^8 + He^4 = C^{12} + \gamma$.* That is how carbon appears in our universe. Then carbon can add another helium: $C^{12} + He^4 = O^{16} + \gamma$; and that is how oxygen enters our universe. Also in outer, cooler regions of the star, at 10 to 20 million degrees, carbon can capture protons, one by one: $C^{12} + 2H^1 = N^{14} + 2\gamma +$ positron and neutrino; and that is the source of nitrogen. By now the star has blown up to enormous size, and for a time these new sources of radiation hold it there. It has become a red giant – a dying star.

Red giants are turbulent structures. Through continuous distillation from their surfaces, through such occasional upheavals as flares and novae, or the ultimate castastrophe, a supernova, they spew much of their substance back into space to become part of the great masses of gases and dust that fill interstellar space. In the

* At 100 million degrees, Be^8 constitutes about 1 part in 10^{10} of the total mass. At that temperature the condensation of Be^8 with He^4 to form C^{12} becomes important (Strömgren, 1953; Fowler, 1964, 1967).

course of time, chance eddies in the gases and dust nucleate new condensations that become new stars; but such later-generation stars, unlike the first generation made wholly of hydrogen and helium, contain carbon, nitrogen and oxygen. In them, hydrogen fuses to helium not only by the direct union of protons, but in a new way catalyzed by carbon, the C—N—O cycle. In summary: $C^{12}+2H^1=N^{14}$; $N^{14}+2H^1=$ $=C^{12}+He^4$. This catalytic fusion yields exactly the same products and as much energy as the proton-proton chain. It accounts for about 15% of the burning of H to He on the Sun. Because they play so great a role in the metabolism of stars, H, He, C, N and O are in that order the most abundant elements in the universe.

About 3 billion years ago these same elements bred organic life on the Earth; for 99% of living matter here is made of H, C, N and O. As already said, it *has* to be that way, wherever life arises in the universe; for only these elements possess the properties upon which life mainly depends. It is a strange realization that stars must die before organisms can live.

That is the long-term connection; but there is also an immediate connection between the lives of stars and organisms. For after a time they come to be coupled through light. The nuclear reactions among H, C, N and O that generate the radiation that holds the star on the Main Sequence also come, through photosynthesis, to run the molecular reactions involving the same elements in the living organisms on its planet.

As already said, the special fitness of these elements in organisms lies in their special capacity to gain electrons .The special fitness of their nuclei in the stars lies in their capacity to gain protons. To capture a proton, an atomic nucleus must collide with it; but since both are positively charged, that means overcoming mutual repulsion. The lightest atomic nuclei, since they bear the smallest positive charges, repel protons most weakly, and so can collide with them at relatively low velocities of approach – at relatively low temperatures. Hydrogen nuclei can fuse at about 5 million degrees; carbon and hydrogen nuclei only at about 10 million degrees. To bring helium nuclei together requires 100 million degrees, well beyond the range of temperature of Main Sequence stars.*

Since small size so much favors proton capture, what of helium itself? If helium could add protons, that would produce lithium, beryllium and boron. But helium is blocked by a special circumstance: there is no stable atomic nucleus of mass 5. Hence these are among the rarest elements in the universe.**

* Of course the sizes of atomic nuclei are not the only consideration. Specific resonances – special energies of collision that favor fusion – can greatly alter its kinetics. Such a resonance at 0.460 MeV promotes the reaction, $C^{12}+H^1=N^{13}+\gamma$; and another resonance at 7.656 MeV promotes $Be^8+He^4=C^{12}+\gamma$ – fortunately, since there is so little Be^8 (Fowler, 1967; Bethe, 1968).

** Among the natural elements and their isotopes, every atomic mass is represented from 1 to 238, except for masses 5 and 8. He^5 and Li^5 are estimated to have lifetimes of about 10^{-21} s, Be^8 about 10^{-16} s (Fowler, 1967, p. 5). Hence He^4 is blocked from adding protons to form the stable isotopes of Li, Be and B; and these elements are instead formed in very small amounts by 'spallation' – the disintegration of larger elements in hig-energy collisions (Greenstein, 1961, 1964).

And magnesium for the chlorophylls? That brings us back to red giants. A red giant, if it survives, eventually begins to run short of helium and so starts again to collapse. With that, the temperature in the deep interior rises further, eventually to temperatures at which carbon and oxygen begin to 'burn' by fusing progressively with He^4 nuclei. So they grow in mass by steps of 4: Ne^{20}, Mg^{24}, Si^{28}, S^{32}. Hence, after H, He, C, N and O, these are the most abundant elements in the universe.

Except for iron – the iron that nucleates hemins just as magnesium nucleates chlorophylls. In the cosmic distribution of the elements, iron makes an aberrant peak. That is because iron has the largest packing fraction – the smallest mass per nucleon – of all the elements. Stars can obtain energy by fusing lighter nuclei than iron, or by the fission of heavier nuclei than iron; but iron is the end of that road, the ultimate ash of nuclear reactions. Stars are born in hydrogen, and die an iron death. That is likely to be spectacular; for by the time nuclei in the deep interior of a red giant have built up to that level, its temperature has reached several billion degrees. At such temperatures even large nuclei can collide and fuse; but beyond iron their fusion no longer *yields* energy, but *consumes* energy. Rather than retarding further collapse, these fusions, which yield small amounts of all the heavier elements to the top of the Periodic System, promote the collapse of the star. It can become catastrophic, ending in a supernova. Such supernovae eject into space, eventually to be gathered up by later-generation stars, the iron, copper, manganese, cobalt, zinc, iodine – the entire array of heavier elements traces of which have become indispensable to living organisms on the Earth.

Perhaps enough has been said to make it clear that fitness of the elements is as great a consideration in the lives of stars as of living organisms. Indeed the parallelism of these relationships has a striking outcome: the proportions of the elements in living organisms is much closer to their distribution in later-generation stars than in the planets. The planets are dead residues; the stars are in every way closer to life.

And now a final consideration: Though the lifetimes of individual organisms are trivial compared with those of stars, life itself competes very well in durability with the stars. Our Sun is about halfway through what – barring some catastrophe – should be a total lifetime of about 10 billion years on the Main Sequence. Larger stars stabilize on the Main Sequence at higher temperatures. Hence they use up their hydrogen more quickly, and have shorther lifetimes. A star twice the mass of the Sun would remain on the Main Sequence less than one billion years, no longer than respiring, eucaryotic organisms have existed on the Earth. A star of 10 Sun masses would survive only about 40 million years; horses and camels have existed that long. A star of 30 Sun masses – and there are such stars – would last only about 3 million years; such stars have come and gone since manlike creatures began to walk the Earth. Conversely a star half the mass of the Sun could have about 40 billion years on the Main Sequence instead of the paltry 10 billion years our Sun affords.

These relationships have an important consequence. It took about two billion years of the solar system before life arose on the Earth. If that is any indication of what to expect in general, no star much more than 1.6 times the mass of the Sun, hence remaining on the Main Sequence less than 1.6 billion years, is likely ever to have a planet that bears life, however fortunate its other circumstances.

4. Electrons and Protons

Some time ago the thought occurred to me, how strange that particles so unlike in every other way as protons and electrons should possess the same numerical charge. These particles are of course opposite in charge, and the proton has about 1840 times the rest mass of the electron; yet their charges are numerically equal.

There are other such pairs of oppositely charged elementary particles that do not raise this problem. – the electron and positron, and the proton and anti-proton – for such pairs of antiparticles on contact mutually annihilate each other to yield one or more photons, and can in turn be generated as pairs of anti-particles from photons. So this behavior is just an aspect of conservation of charge. No such symmetrical relations however exist between protons and electrons.

When this thought first struck me, I did as before on such occasions. I called up Victor Weisskopf, Professor of Theoretical Physics at the Massachusetts Institute of Technology, and an old friend. He heard me out and then said, "That's an interesting question". Then, after a pause, he added, "I suppose that's why it took so long to discover electricity."

For if the proton and electron had unequal charges, almost everything would be, charged. Electricity and the essentially electrical nature of matter would have been apparent to all from the earliest times. As it is, the first substantial work on electrostatics, William Gilbert's *De Magnete*, appeared in 1600; and our knowledge of current electricity began almost two centuries later, with the experiments of Luigi Galvani (1786) and Alessandro Volta (1800). The realization that matter is composed of electrically charged particles emerged gradually throughout the nineteenth century and was given definitive form only in the present century.

As for the problem I raise, it may help one to accept it as real to know that in 1959 Lyttleton and Bondi proposed that the proton and electron differ in charge by about $2 \times 10^{-18}\,e$ (where e is the unit charge, 4.80×10^{-10} esu). That minuscule difference would be enough to account for the observed expansion of the universe.

This proposal impelled a number of investigators to test the equality of charge of protons and electrons, and hence the neutrality of neutrons, by determining whether a variety of atoms and molecules possess residual charges. King and his co-workers at the Massachusetts Institute of Technology have pursued such measurements on a wide variety of gases, from such atomic gases as helium and argon to molecular gases ranging from hydrogen to SF_6 (King, 1960; Dylla and King, 1973). It came out that any possible inequality of charge between the proton and electron, hence charge on the neutron, cannot be larger than 10^{-20} to $10^{-21}\,e$.

We tend to take too much for granted, particularly when it is so fundamental that it surrounds us.* The properties of protons and electrons have determined the nature of our universe. If protons and neutrons had not so much greater mass than electrons, all motions involving these particles would be mutual, and nothing would stay put. As it is, the nucleons are so much heavier that the motions of electrons about atomic nuclei hardly perturb them. It is the heavy nuclei that fix the positions of atoms. Without this great difference in mass there could be no solids; all matter would be fluid.

Similarly if the proton and electron differed in charge, almost everything would be charged. Neutrality could be achieved only through a relatively few, highly limited combinations. If the present rules still held for forming elements, all of them would be charged, and in the same sense. Hence all atoms would repel one another.

That is just the point of the suggestion of Lyttleton and Bondi. If the proton and electron differed in charge by 2×10^{-18} e, that would explain the observed expansion of the universe; but would such a universe do anything other than expand? A charge of about $10^{-18}e$ on each hydrogen atom would negate their gravitational attraction for one another. How then achieve condensations? How form stars and galaxies?

Lyttleton and Bondi suggest local regions of ionization to solve this problem. Such ionized regions would be conducting, and could achieve neutrality by expelling excess charge in the form of free electrons or protons. But we generally rely upon condensations and the high temperatures and radiation that they generate to produce ionizations. How initiate ionization prior to, indeed as a condition for, condensation? It seems likely that a universe of hydrogen atoms each bearing a charge of 2×10^{-18} e would expand, and nothing more: no further elements, no stars or galaxies, no life.

From our parochial point of view, it is fortunate that protons and electrons have what Aristotle would have called the 'accidental' properties they do. Much of the capacity for variety and development in our universe depends upon those properties. One could tinker with the difference in mass between electrons and nucleons without changing matters very much, just so the nucleons remained very much the heavier. In the matter of charge, however, the limitation is extreme. Only precise equality of charge would seem to yield the universe that we know.**

So – as we see it – this may be the best way to build a universe. What I want to know is: How did the universe find that out? How could it have arrived at those properties

* There is a Hindu story of a fish that asked the queen fish, "What is the sea? I have always heard about it, but what is it?" The queen fish answered, "You live and move in the sea. It flows through you and around you. You are made of sea. You are born in it and of it, and will become it again when you die." The Hindu story ends there; but I like to think that the fish promptly asked again: "What is the sea?"

** Quarks, with sub-charges $\frac{1}{3}$ e or $\frac{2}{3}$ e, would just carry this problem back a step. To achieve the equality of charge exhibited by protons and electrons with sub-charges, the latter would have either to be equal or simple sub-multiples of each other, or to vary continuously.

in its elementary particles without trials? – and what could have been the nature of such trials?*

5. Natural Selection in the Universe

We are used to thinking of natural selection in connection with living organisms and their evolution; and of these events occurring in a relatively stable physical context – a planet attached to a relatively stable star; a stable Periodic System of elements composed of stable elementary particles. But if the attention shifts from the planets to the stars, now the elements become unstable. Neutrons have only to be expelled from atomic nuclei to become unstable, disintegrating with a half-life of about 11 s, to protons electrons and antineutrinos. Pairs of antiparticles annihilate each other to become photons; photons may conversely breed pairs of antiparticles. Are there circumstances in which protons and electrons become unstable?*

As it operates among living organisms, natural selection involves three components: mutation as the source of continuous variation; genetic mechanisms of inheritance; and competition, the selective factor – what Darwin called the 'struggle for existence', with its outcome, the 'survival of the fittest'.

To extend a principle of natural selection to molecules and elements, one must seek physical analogues to these factors. That is a task too large to undertake here; but I should like at least to comment upon it.

I think that what replaces mutation in the physical realm, as already noted above, are the instabilities induced mainly by high temperatures, radiation, electric discharges, and high energy collisions. Taking the place of mechanisms of inheritance are the laws of quantum mechanics. They set the patterns, divide the stable from the unstable species, the permitted from the forbidden. Finally, the element of competition enters through the conservation laws. In a finite universe, the modes of physical organization must divide up a limited sum of energy and mass. More of this inevitably means less of that.

However one draws such analogies, there remains a big difference. The natural selection of elements can operate only within sharply proscribed limits. Various conditions can change the proportions of elements; they do not create new elements beyond those we know or can anticipate. Also in principle quantum mechanics encompasses all molecular possibilities. We can anticipate no surprises; really not

* It is harder to convince biologists than physicists of the reality of such problems; for biologists tend to impute to physicists a degree of assurance involving elementary particles that physicists know to be groundless. Note a comment by Fowler (1967, p. 9): "– in low energy interactions, in a world of our kind of matter, protons and neutrons, collectively as nucleons, are stable and immutable.... This in spite of the fact that no previously known laws prevent them from transforming entirely to electrons and other lighter particles with the disappearance of mass and the release of large amounts of energy. For this reason we have to accept another law of nature: Nucleons at low energy and well separated from antinucleons are immutable. If this law did not hold, then the universe as we know it would not exist.

"Here I will not attempt to account for the origin or *nucleogenesis* of the protons and neutrons. Some cosmologists believe that nucleons have always existed and that the question of their origin is not a scientific problem. I must say that I have little sympathy with this point of view."

even in living organisms, where though the macromolecules present virtually endless possibilities for further variation, their variations tend to be no more than changes in the numbers, proportions and sequences of a few simple units.

Living organisms however add another dimension to variation: individuality. Every sample of an isotope is identical with every other; the very definition of a substance is that all its molecules are identical; but no two living organisms are or ever have been identical. Each living organism is unique, different from every other, different even from itself from moment to moment. That is part of its dynamism: not only does its form change constantly, maintaining only a gross consistency; but so does its composition. Its form is only the locus through which flows a constant interchange of matter and energy with the environment. With this almost infinite capacity for variation go particularly stringent conditions for survival. The life of organisms is uniquely competitive. Natural selection operates upon them with particular subtlety and force. Hence they evolve so quickly as occasionally to excite doubts that natural selection by itself could account for their rate of evolution. It takes them into every cranny of the environment; wherever there is a niche, a process that could be made to yield free energy, living organisms occupy and exploit it. Unlike the evolution of elements and molecules, that of living organisms is wholly open-ended and unpredictable.

In this strange paper I have ventured to suggest that natural selection of a sort has extended even beyond the elements, to determine the properties of protons and electrons. Curious as that seems, it is a possibility worth weighing against the only alternative I can imagine, Eddington's suggestion that God is a mathematical physicist. It is the old biological problem of supernatural creation as against 'spontaneous' –i.e., natural – generation, carried back somewhat. Back of the spontaneous generation of life under other conditions than now obtain upon this planet, there occurred a spontaneous generation of elements of the kind that still goes on in the stars; and back of that I suppose a spontaneous generation of elementary particles under circumstances still to be fathomed, that ended in giving them the properties that alone make possible the universe we know.

References

Bara, M. and Galston, A. W.: 1968, *Physiol. Plant. (Copenhagen)* **21**, 109.
Bethe, H. A.: 1968, in *Les Prix Nobel 1967*, Stockholm. Reprinted in *Phys. Today* (Sept. 1968), 36.
Clayton, R. K.: 1971, *Light and Living Matter*, vol. 2: *The Biological Part*, McGraw-Hill, N.Y.
Curry, G. M. and Gruen, H. E.: 1959, *Proc. (U.S.) Nat. Acad. Sci.* **45**, 797.
Dylla, H. F. and King, J. G.: 1973, *Phys. Rev.* **A7**, 1224.
Fowler, W. A.: 1964, *Proc. Nat. Acad. Sci. (U.S.)* **52**, 524.
Fowler, W. A.: 1967, *Nuclear Astrophysics*, Am. Philos. Soc., Philadelphia.
Greenstein, J. L.: 1961, *Am. Scientist* **49**, 449.
Greenstein, J. L.: 1964, *Proc. Nat. Acad. Sci. (U.S.)* **52**, 549.
Hager, A.: 1970, *Planta* **91**, 38.
Henderson, L. J.: 1913, *Fitness of the Environment*, Macmillan. Reprinted Beacon Press, Boston, 1958.
Hubbard, R. and Kropf, A.: 1958, *Proc. Nat. Acad. Sci. (U.S.)* **44**, 130.
King, J. G.: 1960, *Phys. Rev. Letters* **5**, 562.

Lyttleton, R. A. and Bondi, H.: 1959, *Proc. Roy. Soc. London* **A252**, 313.
Oesterhelt, D. and Stoeckenius, W.: 1971, *Nature New Biol.* **233**, 149.
Prout, W.: 1834, *Chemical Meteorology and the Function of Digestion*, Bridgewater Treatise, London, pp. 249–250.
Shropshire, W., Jr.: 1963, *Physiol. Rev.* **43**, 38.
Strömgren, B.: 1953, in G. P. Kuiper (ed.), *The Sun*, Univ. Chicago Press, pp. 36–87.
Thimann, K. V. and Curry, G. M.: 1960, in M. Florkin and H. S. Mason (eds.), *Comparative Biochemistry*, vol. 1, Academic Press, N.Y., p. 243.
Treibs, A.: 1936, *Angew. Chemie* **49**, 682. For a recent review of the geochemistry and paleochemistry of these and other pigments see Dunning, H. N.: 1963, in I. A. Breger (ed.), *Organic Geochemistry*, Macmillan, N.Y. p. 367.
Urey, H. C.: 1952, *Proc. Nat. Acad. Sci. (U.S.)* **38**, 351.
Wald, G.: 1943, *Vitamins Hormones* **1**, 195.
Wald, G.: 1959, *Sci. Am.*
Wald, G.: 1962, in M. Kasha and B. Pullman (eds), *Horizons in Biochemistry*, Academic Press, N.Y., pp. 127–142.
Wald, G.: 1964, *Proc. Nat. Acad. Sci. (U.S.)* **52**, 595.
Wald, G.: 1965, in E. J. Bowen (ed.), *Recent Progress in Photobiology*, Blackwell's, Oxford, pp. 333–350.
Wald, G.: 1968a, *Science* **162**, 230.
Wald, G.: 1968b, in A. Allison (ed.), *Stars and Living Organisms: a Metabolic Connection, in Photobiology, Penguin Science Survey*, Penguin Books, Baltimore p. 33.
Wald, G. and DuBuy, H. G.: 1936, *Science* **84**, 247.
Zechmeister, L.: 1962, *Vitamins A and Arylpolyenes*, Academic Press, N.Y.

GALACTIC CLOUDS OF ORGANIC MOLECULES

DAVID BUHL

National Radio Astronomy Observatory, Greenbank, W.V., U.S.A.*

Abstract. The discovery of immense organic clouds of gas embedded in the dusty regions of our Galaxy is of tremendous importance to the origin of life. It is within these clouds that the formation of stars and planetary systems is believed to take place. The collapse of these clouds to form stars takes place in a very small fraction of the astronomical lifetime of a star. It is precisely at this moment that the clouds of organic molecules appear.

1. Introduction

The presence of enormous clouds of organic molecules in the spiral arms of our Galaxy has only become evident with recent discoveries in radio astronomy (Buhl and Ponnamperuma, 1971). These tenuous clouds are permeated with dust grains. Hence they are completely opaque to optical observations. Radio and infrared radiation can penetrate the clouds, providing us with the only means of studying the physics and chemistry of a molecular cloud. The characteristic emission or absorption frequency of simple molecules can be measured in these clouds to a few parts per million. The radio spectrum is relatively uncrowded, allowing a 99% certainty of identification of a particular molecule on the basis of one spectral line. Most of the species detected are organic, indicating the importance of carbon in interstellar chemistry. Molecules without carbon are relatively simple (H_2, NH_3, H_2O, OH, H_2S and SiO). The molecular cloud has densities and temperatures much lower than terrestial conditions ($n<1$ molecule cc^{-1} and $T<100K$). These clouds are a peculiar laboratory whose dimensions are several light years and whose products include such exotic species as CH, OH, CS, CN and HNC. CO is a major constituent, being second only to H_2 in abundance. Although we have no direct measure of the molecular hydrogen in dense molecular clouds, its presence is implied by the collisional excitation of other molecules. From observations of a number of different molecules the density of H_2 is estimated to run from 100 molecules cc^{-1} at the edges of a molecular cloud to at least 10^{10} molecules cc^{-1} in the central core. At these densities the clouds are rapidly fragmenting and collapsing to form stars and planetary systems in a time scale of 10^5 yr. It would seem most portentous that the clouds of organic molecules appear precisely at the instant that a star is being born in our Galaxy. This certainty suggests that the evolution of a molecular cloud is a significant prelude to the origin of life.

2. Optical Observations

Observations with the Mt. Wilson 100 in. telescope during the late 1930's showed narrow absorption lines in the spectra of several stars. These lines were found to be

* Operated by Associated Universities, Inc., under contract with the National Science Foundation.

due to the diatomic molecules CN, CH and CH^+ located in the tenuous interstellar clouds. At the same time several unidentified diffuse bands were discovered, broad lines which have not been positively connected with any known molecule. Prophyrins have recently been suggested as a possible source of these diffuse bands.

Recently through the use of rockets and satellites the UV portion of the spectrum has become available to astronomers. Using a rocket camera with an electron image intensifier Carruthers (1970) was able to detect molecular hydrogen (H_2) in the 1000 Å spectra of a star. Now a new satellite called Copernicus has been able to confirm the H_2 discovery and also was able to detect the deuterium species HD. Molecular hydrogen (H_2) is an important molecule since it is the major constituent of the denser molecular clouds (>100 molecules cc^{-1}). Due to its symmetry it does not have any microwave lines and is therefore very difficult to detect. This is because the denser regions where the molecular hydrogen is located also contain a large amount of dust grains which strongly attenuate and scatter the UV radiation. Atomic hydrogen (H) fortunately has a 21 cm line which allows the low density regions to be mapped in detail.

3. Radio Observations

In 1963 the 18 cm line of hydroxyl OH was discovered by Weinreb *et al.* (1963) using the Lincoln Laboratories 84 ft antenna. The line had been searched for over a number of years and was first found as an absorption line in the spectrum of the strong radio source Cassopeia A. In 1965 several dense galactic clouds were discovered to have very narrow intense masering lines at 18 cm.

Up till 1968 only diatomic molecules had been found in interstellar clouds. Most astronomers considered it unlikely that anything more complex could be formed and survive under the low density conditions and high energy particles and UV rays that pervade the interstellar medium. What was overlooked was the important role of dust grains in producing and protecting larger molecules. Then, a group at Berkeley (Cheung *et al.*, 1968 and 1969) found both NH_3 and H_2O, with the water line showing spectacular maser emission stronger than the OH molecule.

Encouraged by the success of these experiments Snyder and Buhl teamed up with Zuckerman and Palmer in 1969 to hunt for the 6 cm line of formaldehyde (H_2CO). At the time this was considered a fairly large organic molecule by astronomical standards. The fact that it was discovered in a very large number of galactic sources meant that it was a common constituent of the galactic spiral arms (Snyder *et al.*, 1969; Zuckerman *et al.*, 1970). The 6 cm line always appears in absorption, even in the very cold nearby dust clouds, where the only background to be absorbed is the 3 degree radiation (Palmer *et al.*, 1969). This is most likely the remnant from the fire ball in which the universe originated (the Big Bang) which has now cooled to the temperature and spectrum of a 3 degree Kelvin black body. This was thought to be the lower limit to the temperature in the universe; however, it is possible to cool the formaldehyde transition below this temperature by collisions with hydrogen molecules. In order to understand some of the peculiar properties of molecules in the interstellar environment it is necessary to explain the excitation of a molecular transition.

4. Molecular Energy Levels

Molecules, like everything else in nature have certain resonances, and in particular these resonances are quantized or occur at specific frequencies. The frequencies of rotational transitions are in the radio range of the spectrum, vibrational transitions in the infra-red and electronic transitions in the optical. At this point we must also realize that the resonant frequency of a molecule which is observed is really a transition from one energy level to another. This is generally represented by an energy level diagram (Figure 1) in which the quantized energies of the molecule are plotted in reciprocal wavelength ($1/\lambda$ in cm^{-1}). The line which we observe being produced by a molecule is actually the difference between two energy levels of the molecule and the energy level diagram for each individual molecule is unique. The net result is a specific frequency by which we can identify a particular molecule and this is a very valuable tool for the astronomer.

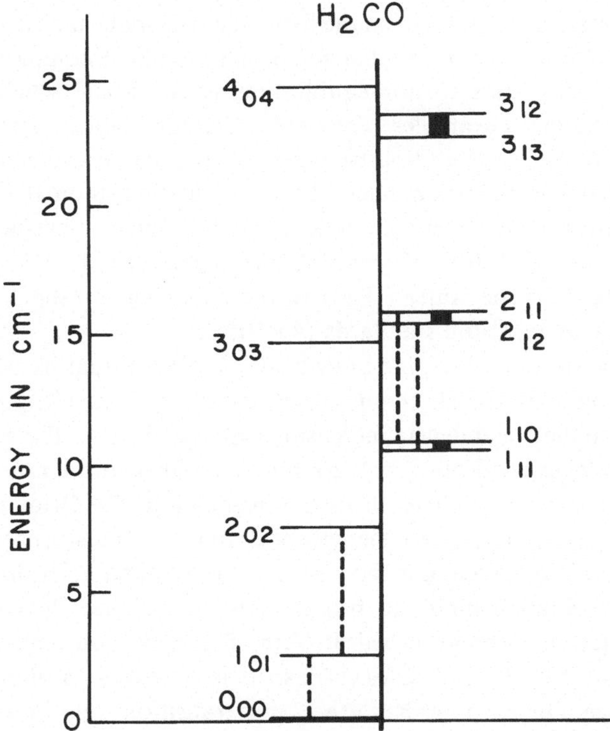

Fig. 1. Energy level diagram for the formaldehyde molecule. The energy scale represents the amount of energy (~ temperature) required to populate a given state. Quantum numbers are indicated for each state ($J_{K^-K^+}$) with the microwave lines indicated as transitions between states. The solid bars indicate absorption and the dotted lines emission transitions observed in molecular clouds.

5. The Excitation Environment

The shorter wavelengths always require more energy and another critical factor in making a molecule resonate is the question of excitation. The higher the energy level the more excitation is required: the number of wave numbers (cm^{-1}) above the ground state is approximately equal to the temperature required to excite the transition. A temperature of 100 K will excite all the levels up to 100 cm^{-1}. In the formaldehyde (H_2CO) energy level diagram (Figure 1) it can be seen that 25 K is sufficient to excite all the transitions that have been found in the radio range of the spectrum. Alternately transitions in the infra-red and optical range require much higher temperatures or excitation energies.

From this discussion of quantum levels two basic principles emerge: molecules have distinct characteristic frequencies by which they can be identified and for temperatures normally encountered in molecular clouds (3 K to 200 K) the main excitation of the molecule is in rotation which corresponds to frequencies in the radio range of the spectrum. A third and more subtle point is that the overlap of lines in the radio range is very small compared with the infra-red and optical regions. What this means is that a molecule like formaldehyde was initially identified by one radio line with 99% certainty. Presently ten lines from the formaldehyde molecule have been identified.

The very low densities of interstellar molecular clouds in comparison to standard laboratory conditions leads to non-equilibrium situations of which the maser and refrigerator are the main examples. They are really complementary processes: in the maser the upper level of a transition becomes overpopulated while in the refrigerator the lower level becomes overpopulated. These population anomalies are most likely produced by collisions with hydrogen molecules (H_2). The H_2 densities required range from 10^3 molecules cc^{-1} for refrigerating the formaldehyde (H_2CO) up to 10^{10} molecules cc^{-1} to heat the water (H_2O) to the masering threshold. These densities are still quite low by terrestial standards (4×10^{19}).

The formaldehyde molecule illustrates some of the unusual excitation conditions which are common in molecular clouds. The 6, 2 and 1 cm wavelength lines all appear in absorption with the line intensity increasing with wavelength. The cooling below 3 K by collisions produces absorption in the 6 and 2 cm lines in the near by dust clouds. However, the mm wavelength lines all appear in emission. The Orion nebulae provides an interesting case where at a wavelength of 2 mm it is the strongest formaldehyde source in the sky while at 6 cm it is the weakest. This is apparently due to the location of the molecular cloud behind the bright emission nebulae, leaving only the 3 K background radiation to be absorbed at 6 cm. The excitation required for the 6 cm transition is about 10^3 H_2 molecules cc^{-1} so it is observed in almost every region where there is dust. On the other hand the 2 mm transitions require 10^7 H_2 molecules cc^{-1} and they are observed only in the most dense and rapidly collapsing molecular clouds. Because of the wide distribution of formaldehyde in the Galaxy and its unusual energy level structure this one molecule is enormously useful in studying our Galaxy.

6. Millimeter Wave Observations

In 1970 a pioneering development of millimeter wave mixers by Burrus, Jefferts, Penzias and Wilson of the Bell Telephone Laboratories (BTL) opened up the mm spectrum for molecular line observations. This resulted in a spectral line receiver which was used at the NRAO 36 ft telescope to discover CO and a new transition of CN.

The CO molecule detected by the BTL group is a very important one from a number of points of view (Penzias *et al.*, 1971). It requires a relatively low H_2 density to excite (10^4 molecules cc^{-1}) and is very strongly bonded. As a result it can wander outside of the protected environment of dense clouds and can be seen almost anywhere in the galactic plane. In this respect it becomes useful for mapping structure in the Galaxy and with a resolution at least a factor of 10 better than the 21 cm hydrogen line. Formaldehyde (H_2CO) on the other hand appears to be confined to the dust clouds despite the low excitation requirements of the 6 cm transition. As such it is a very useful tool for mapping the dust clouds in the Galaxy. CO is also important chemically as a final end product of a number of reactions. The molecules based on the CO bond (H_2CO and H_3COH) appear to predominate.

Subsequently the NRAO 36 ft telescope was used by Snyder and Buhl (1971) to discover the first line of HCN. During a search for the carbon-13 isotope of HCN, an unidentified line appeared at 3.4 mm (Buhl and Snyder, 1973). The line was named X-ogen and appeared about as strong as HCN. There was no detectable splitting of the line profile to aid in the identification and our best guess at present is that it is a simple radical or ion such as HCO^+ or CCH. Another line at a wavelength of 3.3 mm is probably due to an isomer of HCN which has never been identified in a terrestrial laboratory gas cell. This molecule is HNC and its line, like that of X-ogen, is strong and appears in a number of galactic molecular clouds (Snyder and Buhl, 1972). If laboratory measurements were possible on some of these peculiar molecular structures we could identify these lines with a very high degree of certainty since the frequency of a line in a molecular cloud can be measured to a few parts per million.

7. The Isotope Problem

The HCN molecule shows clear evidence of hyperfine splitting of the 3.4 mm line (Figure 2). This is due to the quadrupole interaction of the nitrogen atom and results in three components at an intensity ratio of 3:5:1. The fact that the line observed in the molecular cloud shows approximately this ratio is strong evidence that this HCN emission line is not saturated. This has important implications for interpreting the observations as being a direct measurement of the C^{12}/C^{13} for HCN. Similarly the recent detection by the BTL group of DCN shows a 3:5:1 hyperfine ratio and therefore indicates normal excitation and no saturation. The reason that all this is important is that molecular line observations on a number of different molecules have shown drastically enhanced isotope lines. The terrestrial ratio is $C^{12}/C^{13}=89$ while some clouds have molecules with $C^{12}/C^{13} = 10$. Similarly the DCN lines as well as the UV lines of HD show peculiar H/D ratios. The explanation may be in formation of the molecules where the carbon-13 and deuterium species form at a greater rate than the normal isotope. By no means has the last word been said on the isotope abundance question.

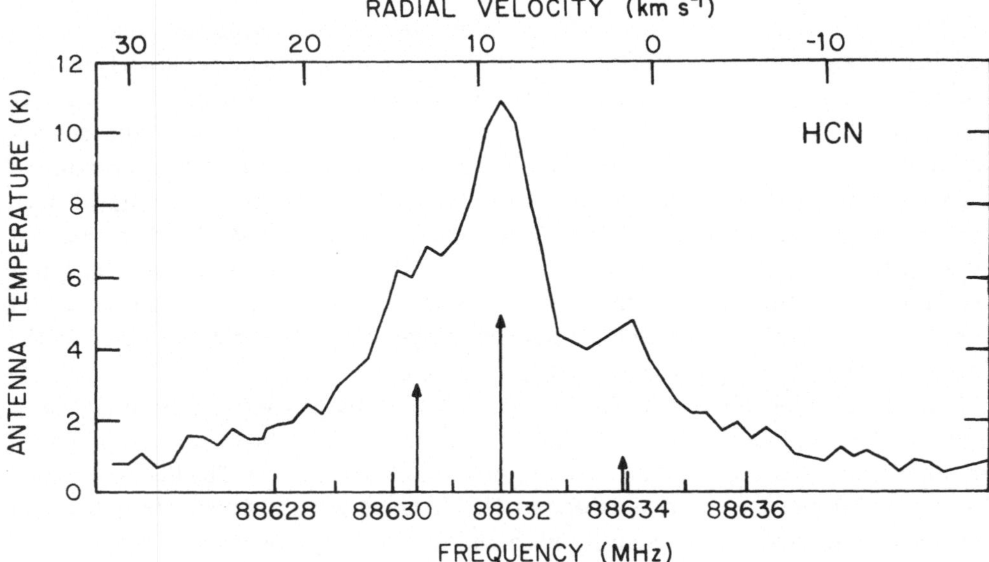

Fig. 2. The 3.4 mm line of HCN in Orion showing the three hyperfine components. The velocity scale applies to the largest component. The observed lines have approximately the expected intensity ratio indicated by the arrows.

8. The Birth of a Star

One of the most fascinating processes in nature is the birth of a star (Bok, 1972). It is suspected that dust clouds in the Galaxy are prime sites for the formation of stars and planets. But precisely because there is a great amount of dust in these clouds, we are unable to see this process take place. However mm wavelength line observations along with infrared measurements are beginning to give us a picture of how a star is formed. The map of Orion in HCN and H_2CO shows several protostars in the process of condensing (Figure 3). The molecular hydrogen density necessary to excite these lines is 10^6 molecules cc^{-1}. The time scale for the collapse of these protostars under gravitational free fall is about 10^5 yr (Larson 1972). Since the lifetime of a star is several billion years this is a very short fraction of the stars life. It is intriguing that the molecules in the dust cloud appear precisely at the instant when the star is born. The H_2 density in the cloud runs from 10^3 molecules cc^{-1} at the edge to over 10^{10} near the core of the protostar where the water masers are excited. This can be seen by comparing the table of molecular excitation parameters (Table I) with the cloud

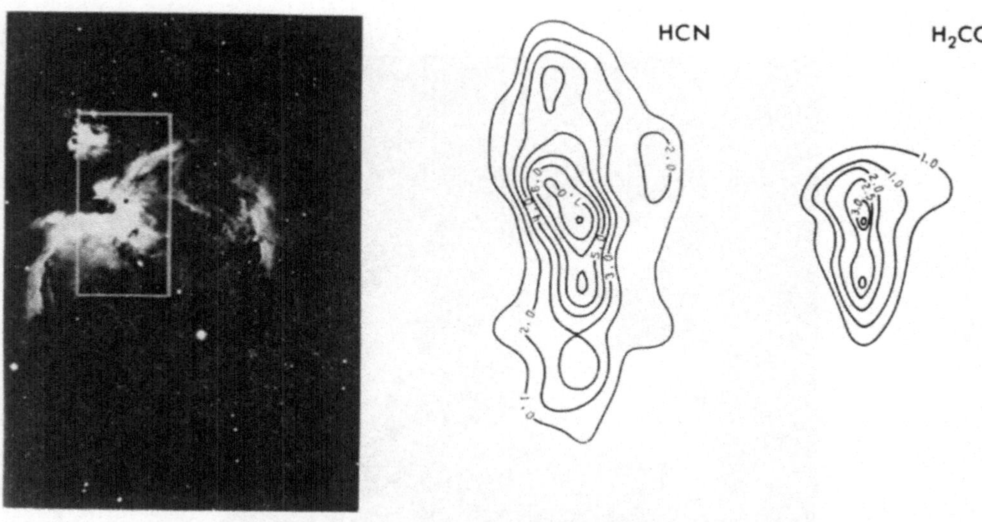

Fig. 3. A contour map of HCN (3.4 mm) and H_2CO (2.1 mm) in Orion. The contour maps have been expanded in size by a factor of 3.4 from the photograph. The box indicates the $8' \times 16'$ of arc region of the nebulae over which the HCN was mapped. (H_2CO: Thaddeus *et al.*, 1971.)

TABLE I

Temperature and density required to excite a molecule

Molecule	Temperature (K)	H_2 density (molecules cc^{-1})
H_2CO (6 cm)	10	10^3
CO	6	10^4
HCN	4	10^6
H_2CO (2 mm)	7	10^7
OH	100	10^8
H_2O	600	10^{10}

extents illustrated for the Orion molecular cloud (Figure 4). It shows that the molecules which require higher excitation densities show emission lines in a smaller region closer to the center of the protostar condensations.

What we see in the Orion nebulae is several prestellar condensations imbedded in the gas and dust of nebulae. In the region where the H_2O emission is occurring the temperature is about 600K and the H_2 density is 10^{10} molecules cc^{-1}. This is a relatively small region, a few astronomical units in radius ($1 \text{ AU}=1.5\times10^8$ km). It probably is a spherical shell surrounding the central core of the condensation (Figure 5). The H_2O molecules in this shell are providing part of the cooling for the gas as it collapses to form the star. The amounts of energy involved are quite large: 10^{30} ergs s^{-1} in the case of the water emission from the central region of the Orion molecular cloud. This is about 0.1% of the power which would be released during the collapse and formation of a star of about one solar mass. Essentially some of the

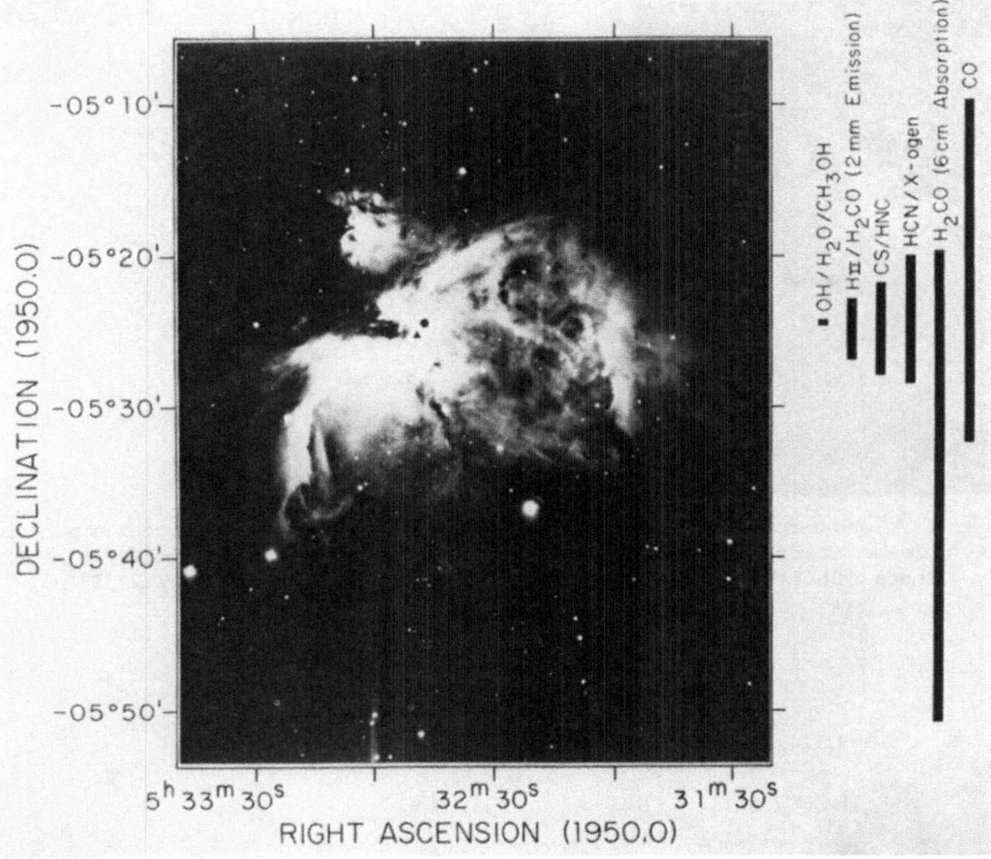

Fig. 4. The approximate N–S extent of the molecular cloud in Orion as seen on various molecular lines. The triangle marks the center of the ionized hydrogen (H II region) and the circle the position of the OH and H_2O masers. The size of these masers is unresolved on this scale being ~ 0.002″. Methyl Alcohol (CH_3OH) also appears to maser in Orion.

velocity built up in the gas as it collapses is converted into microwave energy by the maser action of the water molecules and then radiated out of the condensing cloud.

It is interesting to compare this picture of the action of H_2O molecules with a theoretical model of a collapsing protostar (Larson, 1972). After 10^5 yr the star which will eventually be $1 M_\odot$ has accumulated about half of this mass in the central core (Figure 6). At a distance of 1 AU from the center the environment is about right for exciting the H_2O molecules. The model gives a density of 10^{10} molecules cc^{-1} and a temperature of 500 K, almost identical to the conditions given in Table I which were obtained from analyzing the characteristics of the H_2O molecule. Another interesting comparison is in the predicted and observed velocities for the H_2O emission lines (Figure 5). One of the puzzles in studying the H_2O spectra is the large number of

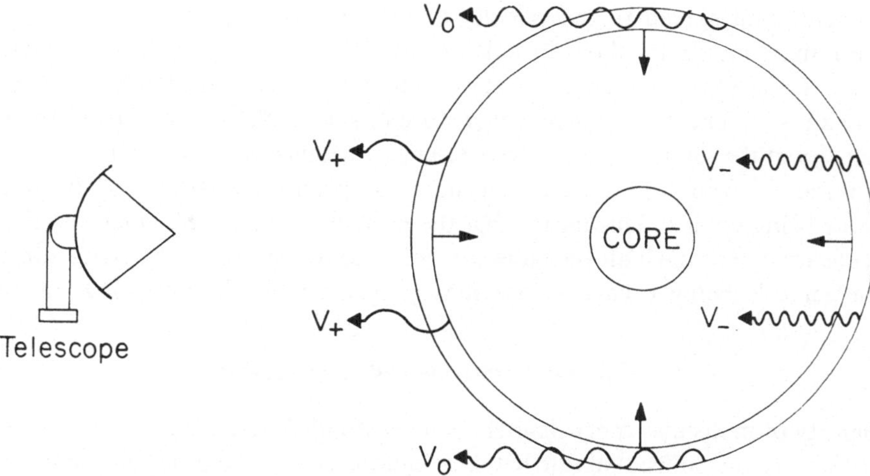

Fig. 5. A diagram of the collapse of a protostar showing the core and the spherical shell where the water maser originates. Gas and dust from the molecular cloud are continuously collapsing into the core at high velocities. As the water and H_2 pass through the spherical shell the critical density for excitation of the water maser is reached. Photons emitted from the front and back of the shell are doppler shifted due to the high infall velocites and form the weak high velocity lines (V_+ and V_-) observed in the water spectrum. The photons emitted from the sides of the shell form the strong low velocity lines (V_0).

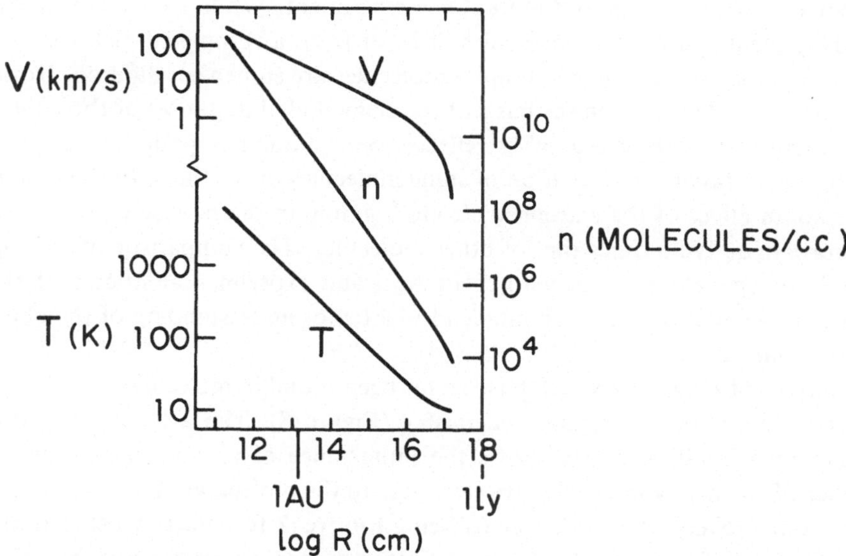

Fig. 6. The temperature, density and velocity inside a protostar as a function of distance from the center. This is taken from the studies of Larson (1972) and represents a protostar of 1 $M_\odot$ at a time when half of the infalling material has formed the central core of the star. The protostar consists mostly of hydrogen molecules and dust grains with smaller amounts of other interstellar molecules.

components and spread in the velocity of the various features (Buhl *et al.*, 1969). In Orion most molecules show a single line at a velocity of 9 km s^{-1} with a width of 5 km s^{-1} while the water emission shows about 10 lines spread over -10 km s^{-1} to $+30$ km s^{-1}. The protostar model (Figure 6) shows that at 1 AU from the center the velocity of the infalling gas is about 30 km s^{-1}. Hence velocities from -20 km s^{-1} to $+40$ km s^{-1} would be expected depending on precisely what part of the spherical shell was being cooled. This means that the peculiar velocities observed for the H_2O molecule are due to the high velocities present close to the core of the protostar, where the molecule is being excited into a maser emission by the collapsing gas cloud.

9. Molecule Formation and Astrochemistry

The density of molecules other than H_2 in most clouds is less than one molecule cc^{-1}. At this density the collisional rate between molecules is of the order of one per 100 yr which is probably too low for any significant amounts of chemical reactions in the gas phase. In addition, two molecule collisions seldom work because the energy arising from the chemical formation of a new molecule cannot be liberated. One therefore looks for chemical reactions on the surface of the dust grains since the energy of collision and formation can be easily absorbed by the grain. The grain is assumed to consist of a core of carbon atoms 1/10 μ in diameter covered by a mantle of organic molecules which have collected one atom at a time. The molecules are then released into the molecular cloud by high energy cosmic rays or UV light. In this way numerous radicals and ions will be produced which will persist for a much longer period of time than they would in a terrestrial laboratory experiment. While this is considered the most likely means of making molecules, it is far from a completed picture.

The abundance of some of the simpler molecules are somewhat different from what would be expected strictly on the basis of the elemental abundances of the constituent atoms (Figure 7). This presumably reflects some dominant paths in the chemical reactions which favor the formation of some molecules over others. In the case of CO this is also an effect of the extremely strong bonding in the molecule giving it a lifetime approximately 30 times that of other molecules. The comparison of these molecular and isotope abundances with theoretical and experimental studies of the formation of molecules on grains should lead to a better understanding of the chemistry of a molecular cloud.

The pattern of molecules which have so far been found in molecular clouds strongly suggests a dominance of organic chemistry (Figure 8). The attempt to group the molecules into families clearly shows the importance of carbon in astrochemistry. A number of the heavy molecules are seen only in the molecular clouds located in the center of our Galaxy, their line signals being too weak to detect in other molecular clouds. The next few years should see a dramatic increase in the number of heavy molecules detected throughout the Galaxy. A number of combinations of the most abundant elements of H, O, C, N and S have been found with up to 7 atoms. With the probable increase in receiver sensitivity by a factor of 10 at mm wavelengths, the

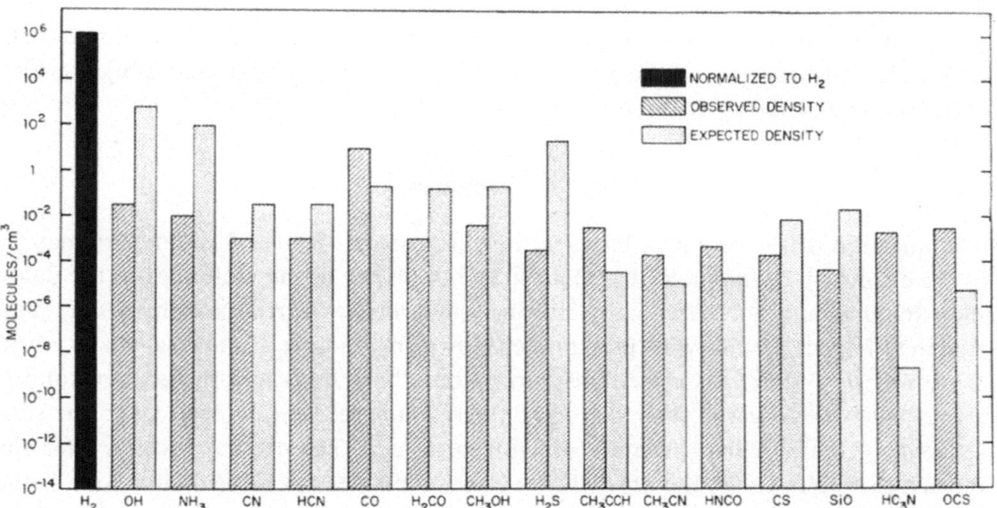

Fig. 7. The expected and observed abundances for several important molecules. The expected abundances are based only on the atomic element abundance of the constituent atoms. Note the predominance of organic molecules both in abundance and variety. The molecules are ordered by molecular weight.

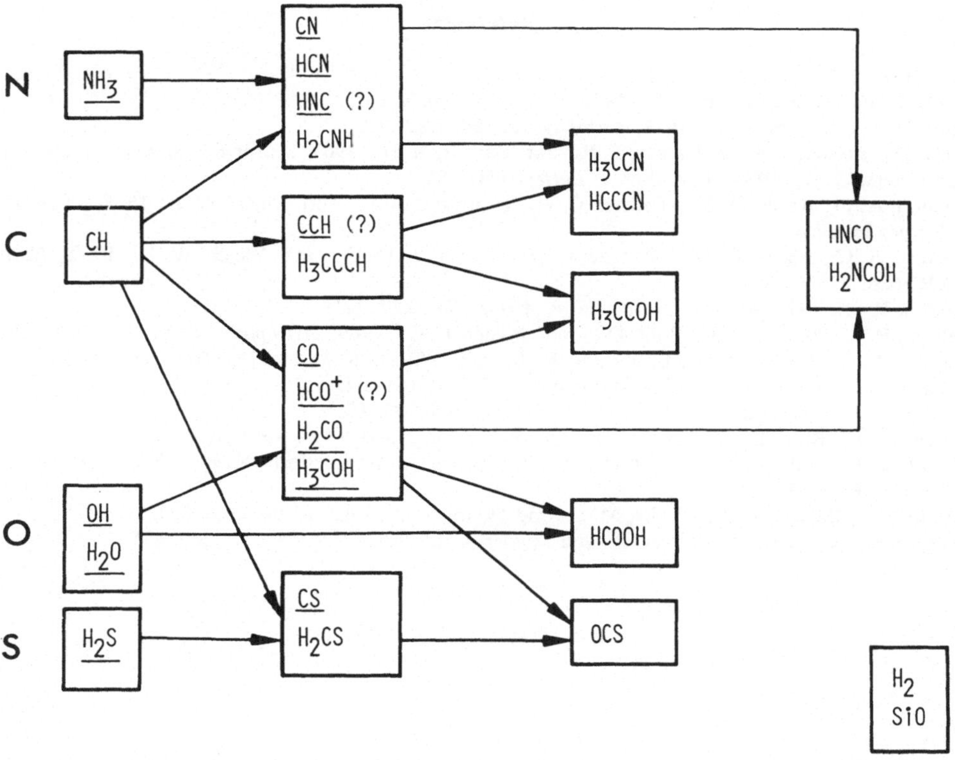

Fig. 8. Molecular chart indicating the relation between the molecules found in galactic molecular clouds. The molecules are grouped into families based on the heavy atom composition. Organic molecules based on carbon are the most prolific. The underlined molecules have been detected in a number of molecular clouds and have relatively large abundances. The (?) are three possible identifications for the 3.4 mm and 3.3 mm X-ogen and Y-ogen lines.

possibility of finding even more complex molecules is immense. This would then help to establish the direction of the organic chemistry in molecular clouds, a subject which is in complete disarray at the moment.

10. Conclusion

The question of the origin of life is a difficult problem. Some of the answer may lie in the chemistry of the molecular clouds. The similarity of the molecules in the dense interstellar clouds to the reaction products in primitive earth experiments is quite striking. This close link with prebiotic chemistry may imply a universal direction in the synthesis. While it is clear that the primordial synthesis in a molecular cloud is a preview of the organic chemistry to come, it is not necessarily identical to the subsequent synthesis which produces a living organism. The crucial point is that the molecules are present at the time when a star and planets are being formed out of the dust and gas in the molecular clouds. An evolutionary sequence appears to be established in which the condensation of a star the accumulation of the dust and molecules into planets and atmospheres and the subsequent evolution of life are all part of an astronomical evolutionary cycle stretching over billions of years.

References

Bok, B. J.: 1972, *Sci. Am.* **227**, 48.
Buhl, D. and Ponnamperuma, C.: 1971, *Space Life Sci.* **3**, 157.
Buhl, D. and Snyder, L. E.: 1973, *Astrophys. J.* **180**, 791.
Buhl, D., Snyder, L. E., Schwartz, P. R., and Barrett, A. H.: 1969, *Astrophys. J. Letters* **158**, L97.
Carruthers, G. R.: 1970, *Astrophys. J. Letters* **161**, L81.
Cheung, A. C., Rank, D. M., Townes, C. H., Thornton, D. D., and Welch, W. J.: 1968, *Phys. Rev. Letters* **21**, 1701.
Cheung, A. C., Rank, D. M., Townes, C. H., Thornton, D. D., and Welch, W. J.: 1969, *Nature* **221**, 626.
Larson, R. B.: 1972, *Monthly Notices Roy. Astron. Soc.* **157**, 121.
Palmer, P., Zuckerman, B., Buhl, D., and Snyder, L. E.: 1969, *Astrophys. J. Letters* **156**, L147.
Penzias, A. A., Solomon, P. M., Jefferts, K. B., and Wilson, R. W.: 1971, *Astrophys. J.* **165**, 229.
Snyder, L. E. and Buhl, D.: 1971, *Astrophys. J. Letters* **163**, L47.
Snyder, L. E. and Buhl, D.: 1972, *Ann. N.Y. Acad. Sci.* **194**, 17.
Snyder, L. E., Buhl, D., Zuckerman, B., and Palmer, P.: 1969, *Phys. Rev. Letters* **22**, 679.
Thaddeus, P., Wilson, R. W., Kutner, M., Penzias, A. A., and Jefferts, K. B.: 1971, *Astrophys. J. Letters* **168**, L59.
Weinreb, S., Barett, A. H., Meeks, M. L., and Henry, J. C.: 1963, *Nature* **200**, 829.
Zuckerman, B., Buhl, D., Palmer, P., and Snyder, L. E.: 1970, *Astrophys. J.* **160**, 485.

THE OUTER SOLAR SYSTEM: PERSPECTIVES FOR EXOBIOLOGY

TOBIAS OWEN

Dept. of Earth and Space Sciences, State University of New York, Stony Brook, N.Y. 11790, U.S.A.

Abstract. The outer solar system contains many environments of interest for studies of the origin of life. Recent observations support the idea that Jupiter and Saturn have retained the mixture of elements originally present in the solar nebula. Subsequent low temperature chemistry has produced the expected array of simple molecules giving characteristic absorption bands in the spectra of these planets. Microwave and infrared observations show that the lower atmospheres are at temperatures above 300 K. Sources of energy for non-equilibrium chemistry seem available at least on Jupiter and the presence of an array of colored materials in the Jovian cloud belts has often been cited as evidence for the existence of complex abiogenic organic molecules. Further study of both planets in an exobiological context seems well worthwhile; potentially productive methods of investigation (including planned space missions) can be described and evaluated from this point of view. Uranus and Neptune are clearly deficient in light gases, but otherwise little is known with certainty about these distant planets. Again unusually high temperatures have been reported, but not above 273 K.

Pluto and many of the outer planet satellites appear to represent a class of small bodies very unlike our neighbors in the inner solar system. Titan, Saturn's largest satellite, is especially interesting for our purposes because of its atmosphere. Methane and hydrogen are both present, and Titan's unusually reddish color again suggests the presence of organic compounds. The hydrogen-methane ratio is likely to be more similar to that of a primitive reducing terrestrial atmosphere than the ratios for Jupiter and Saturn, suggesting that in some respects this satellite may provide an even better model for early organic synthesis on the Earth. The problem of Titan's heat balance and atmospheric composition are currently under active investigation.

1. Introduction

One does not commonly associate the bodies in the outer solar system with the subject of exobiology. The great distances at which these objects move around the sun suggest that their temperatures will be well below the freezing point of water. Rudimentary optical observations confirm the notion that the outer planets and their satellites represent extremely hostile environments to life as we customarily conceive of it. However, closer examination indicates that this impression is not correct. As we shall see, there are regions in the atmospheres of Jupiter and Saturn where conditions are reasonably clement, even by limited terrestrial standards. Furthermore, the atmospheres of these giant planets have a composition that seems well-suited to the abiogenic production of organic molecules, a possibility that was first elaborated by Urey (1959). It has been less widely recognized that the atmosphere of Titan, Saturn's largest satellite, is also a likely place for such chemical reactions to occur.

In this review, I shall attempt to summarize our current knowledge about the composition and structure of outer planet atmospheres with special emphasis on Jupiter, Saturn and Titan. The reader should remember that in addition to the major planets, the outer solar system contains 29 known satellites, most of which exhibit remarkably low densities and hence are probably composed mainly of ices. This frozen material must be made up primarily of compounds whose elements have relatively high cosmic abundances; one suspects in fact that these objects are very similar to the nuclei of comets, already discussed at this meeting by Dr Biermann. Both the comets and this collection of satellites appear to form a vital link between the organic chemistry in the interstellar medium and the processes that formed the planets and satellites.

Two recent reviews should be consulted for additional background material: a general treatment of the outer solar system by Newburn and Gulkis (1973) and an approach oriented toward exobiology offered by Sagan (1971). The present paper will emphasize only the most recent developments in this field.

2. Composition

The present status of our information about atmospheric composition in the outer solar system is indicated in Table I. It should be obvious that this table is woefully incomplete; we expect several new entries to be made in the next few years. The available data already exhibit a significant trend, however, and this is unlikely to change dramatically as new results come in: Jupiter and Saturn exhibit a value of H/C that is much more similar to the solar value than that found for Uranus, Neptune, or Titan. These atmospheric abundance determinations are supported by efforts to

TABLE I

Abundances in the outer solar system

Object	H_2 (km atm)	NH_3 (m atm)	CH_4 (m atm)	H/C
Jupiter	75 ± 15	12 ± 5	50 ± 15	3000 ± 300
Saturn	75 ± 20	?	60 ± 12	2500 ± 400
Uranus	450 ± 100	–	3500 ± 1500	250 ± 50
Neptune	(750 ± 250)	–	(6000 ± 2500)	(250)
Titan	5 ± 2.5	–	(200 to 1600)	(6 to 50)
			Sun	2700 ± 300

Notes:

Jupiter: Abundances from Owen, T.: 1970, *Science* **167**, 1675.

Saturn: H_2 abundance from Encrenaz, Th. and Owen, T.: 1973, *Astron. Astrophys.* (in press). CH_4 abundance from Trafton, L.: 1973, *Astrophys. J.* **182**, 615.

Uranus: H_2 abundance from Encrenaz, Th. and Owen, T.: 1973, *Astron. Astrophys.* (in press). CH_4 abundance from Owen, T.: 1967, *Icarus* **6**, 108.

Neptune: H_2 and CH_4 abundances are estimates developed from Owen, T.: 1967, *Icarus* **6**, 108.

Titan: H_2 abundance from Trafton, L.: 1972, *Astrophys. J.*, **175**, 285. CH_4 abundance from Trafton, L.: 1972, *Astrophys. J.* **175**, 295.

Sun: H to C ratio from Cameron, A. G. W.: 'Abundances of the Elements in the Solar System' (to be published).

build model planets; the bulk compositions of Jupiter and Saturn again require a near-solar mixture of the elements whereas hydrogen and helium must be seriously deficient in the other bodies (Smoluchowski, 1972).

What other compounds may we reasonably expect to find in these atmospheres? In the case of Jupiter and Saturn, one approach to this question is to consult a table of solar abundances and then carry out some chemical equilibrium calculations (Greenspan and Owen, 1967; Lippincott, *et al.*, 1967; Lewis, 1969a). Thus we expect helium to be present at about 20% of the molecular hydrogen abundance, neon to be comparable to ammonia and both water and hydrogen sulfide to be prominent at levels in these atmospheres where temperatures are high enough to permit substantial vapor pressures. With careful work, it should be possible to detect both H_2O and H_2S on Jupiter from ground-based observations. The additional simple, fully reduced compounds that we expect will be much more difficult to observe, as may be seen from the careful study by Lewis (1969a).

The atmospheric composition for the other objects in Table I is more difficult to assay from *a priori* arguments. Solar abundances are no longer a reliable guide and the fractionation processes that have been active are still not understood. Nevertheless, the same basic idea should work in these cases too, once appropriate allowances are made for hydrogen and helium deficiencies. Another approach is simply to examine the spectra of these objects to look for absorptions from likely gases. An example of the kind of observations needed to add both objects and constituents to the table is given in Figure 1, which illustrates low resolution spectra of Pluto, Triton (Neptune's largest satellite) and Titan. Only Titan shows evidence for the presence of methane, leading to model-dependent upper limits of 20 m atm for the abundance of this gas on each of the other two objects.

2.1. Chromophores

Inspection of a good color photograph of Jupiter (Figure 2) shows immediately that substances other than those given in Table I must be present in the planet's atmosphere. Hydrogen, methane, and ammonia are not capable of producing the observed coloration in either the liquid or solid state. However, even under equilibrium conditions at ambient temperature, ammonium monosulfide and hydrosulfide probably form (Owen and Mason, 1969; Lewis, 1969b). These compounds may contribute to the yellow coloration observed on both Jupiter and Saturn. Higher polysulfides may also be produced under the stimulus of UV radiation and some of these may be yellow-orange in color (Lewis and Prinn, 1970; Sill, 1972).

In the case of Jupiter, we must clearly do more than account for a yellowish tint. The colors on this planet have often been attributed to the presence of organic polymers, following Urey's early suggestion (Urey, 1959). The same may well be true for colors on Saturn and on some of the satellites. Indeed, once the basic 'cosmic' mixture of CH_4, NH_3, H_2O, and H_2S is assembled, a variety of colored compounds can be produced, depending on the relative proportions of the constituents and the energy source used to make the reactions go (Woeller and Ponnamperuma, 1969; Chada

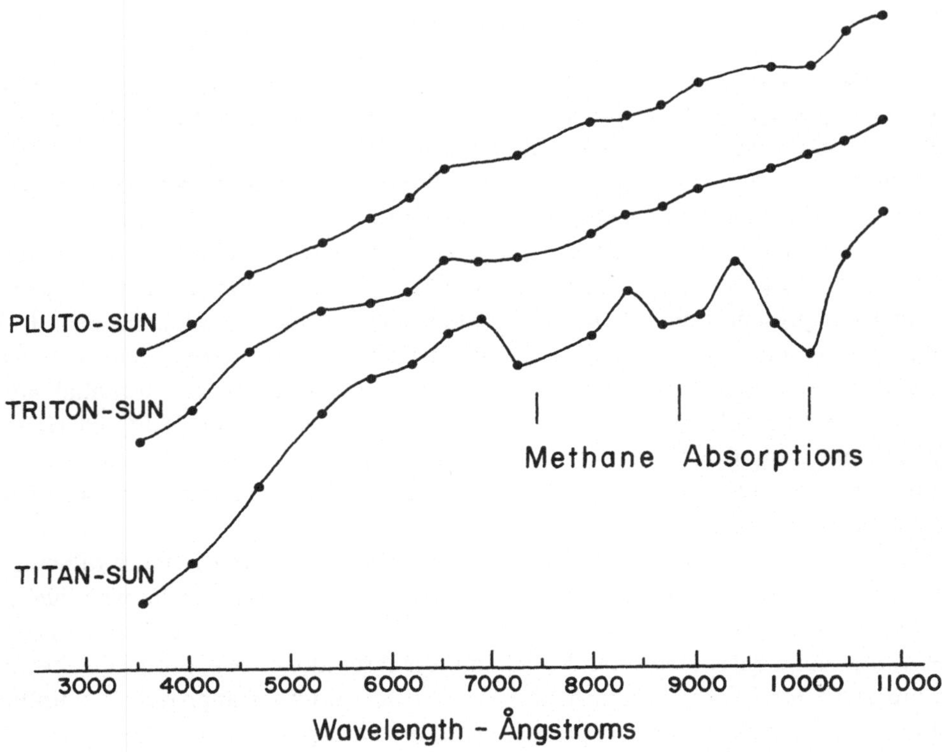

Fig. 1. Low-resolution spectrophotometry of Titan, Triton and Pluto (200-in. telescope plus Oke multichannel spectrophotometer). Note the absence of methane on Pluto and Triton and the redder color (steeper slope at short wavelengths) of Titan. (The vertical scale is displaced for each object; these are only *relative* observations.)

Fig. 2. Jupiter and Io photographed in color with the LPL 61-in. telescope on 25 January, 1968. Note the Great Red Spot at the limb of the planet with the shadow of the satellite just below it. (Lunar and Planetary Laboratory photograph, University of Arizona.)

et al., 1971; Sagan and Khare, 1971). There is no problem with energy sources for Jupiter; in addition to solar UV, we know the planet has an internal heat source (Aumann *et al.*, 1969), trapped radiation belts which may precipitate high energy particles into the upper atmosphere (Carr and Gulkis, 1969), and the extraordinary convective activity observed in the upper clouds even at Earth-based resolution strongly suggests charge separation and hence lightning within the lower, droplet-containing cloud decks.

Despite all this inferential reasoning, there is still a lively debate over whether or not the Jovian chromophores are necessarily organic, where in the atmosphere they are formed, how they continue to survive, and how closely they actually correspond to various laboratory simulacra. It is extremely frustrating to be forced to report that the arguments by analogy remain the best method we have for identifying the substances responsible for the colors on Jupiter. Careful searches have been made for gases likely to be produced in the postulated reactions such as ethane, methylamine, and hydrogen cyanide, but only upper limits have resulted from such investigations (Owen, 1967; Cruikshank and Binder, 1969; Gillett *et al.*, 1969). The one slender piece of spectroscopic evidence that we have is a weak, broad absorption at 4.7 μ first reported by Münch and Neugebauer (1971). Beer and Taylor (1973) were able to show that most of this absorption is caused by the *P*-branch of the ν_2 fundamental of CH_3D, but a definite depression of the continuum at this wavelength remained after the CH_3D lines were accounted for. Münch and Neugebauer (1971) pointed out that this feature would be consistent with the absorption observed by Woeller and Ponnamperuma (1969) in the red polymer produced by irradiation of a mixture of CH_4 and NH_3. It should be stressed that the planetary spectra in both cases correspond to a full-disk average, i.e., colored regions on the planet were blended with the much more extensive regions that show no special colors. Thus spectra obtained with high angular resolution, confined to studies of the Great Red Spot or some other specific feature may be expected to be much more instructive. If this spectral identification holds up, it would lend very strong support to the suggestion that organic compounds make an important contribution to the colors on this planet.

Saturn is much more uniform in appearance than Jupiter (Figure 3), a condition that is customarily ascribed to the planet's greater distance from the Sun, hence lower temperature. The observational evidence for an internal heat source is somewhat weaker than for Jupiter (Aumann *et al.*, 1969; Armstrong *et al.*, 1972) and small scale activity in the visible cloud layers is very much less frequent or widespread. Yet from an observational viewpoint, the main difference between the two planets may result from the greater opacity of the uppermost cloud layer on Saturn compared with Jupiter. It is perhaps noteworthy that both the Great Red Spot on Jupiter and the Equatorial Belt on Saturn appear to extend relatively high into the atmosphere, since each of these regions represents the most colorful area on the respective planets (Owen, 1969).

Titan is especially interesting in this context because the hydrogen-methane mixing ratio in its atmosphere must be very low (Table I). This atmosphere appears to be

Fig. 3. Saturn as seen on 19 January, 1973 with the opening of the rings near maximum. This print is a three-color reconstruction from individual black and white images exposed through red, green, and blue filters with the NMSU 24-in. telescope. (New Mexico State University Observatory photograph).

unstable (Sagan, 1973; Gross, 1973) because hydrogen cannot be retained by the satellite's weak gravitational field and thus we must either be witnessing a transient phenomenon or else some kind of dynamic equilibrium must exist between hydrogen escape and outgassing and/or the outgassing and photochemical breakdown of methane. However, it should be emphasized that the observational data we have at our disposal are still in a very rudimentary state: methane was discovered on Titan 30 yr ago by Kuiper (1944) but the quite unexpected presence of hydrogen was discovered by Trafton only two yr ago (Trafton, 1972a). As in the case of Jupiter, we again have a red color to explain, but this time the coloration is sufficiently extensive that it affects the albedo of the entire object (Harris, 1961). Only Mars and Io, the innermost of the Galilean satellites of Jupiter, are as red as Titan, and reflectance studies indicate that the source of coloration of these three bodies can not be the same (Johnson and McCord, 1970; McCord *et al.*, 1971 – see Figure 4). Titan shows no variation in reflectance as it moves in its orbit about Saturn and there is no indication of negative polarization near zero phase. Both of these observations suggest that we are seeing a thick cloud deck with a thin atmosphere above it, rather than

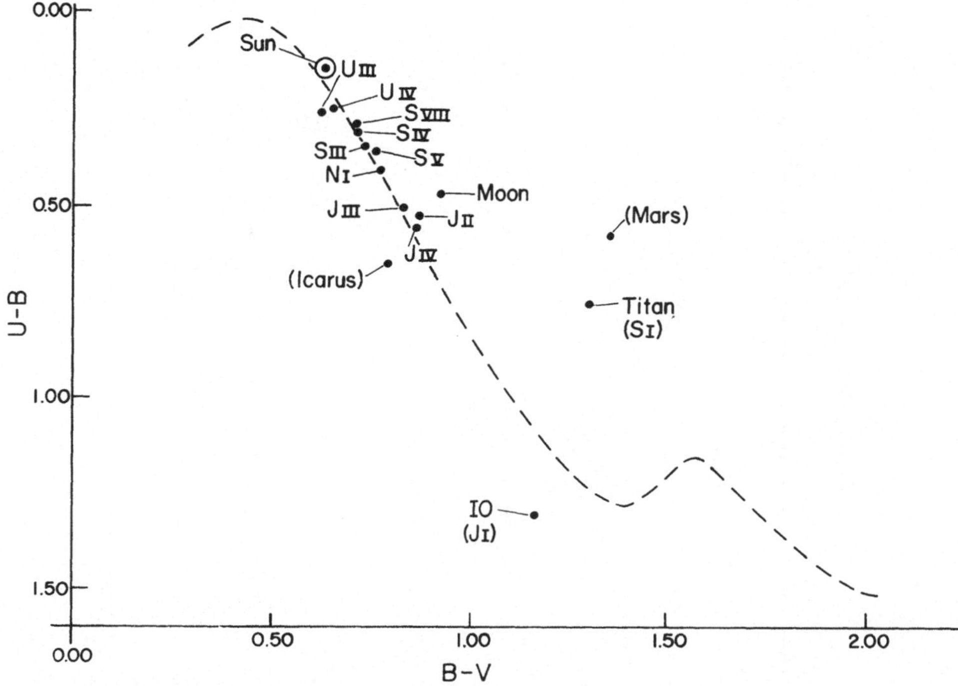

Fig. 4. The colors of the satellites displayed in a conventional astronomical color index plot. The ordinate represents the difference between an object's brightness in Visible and Blue light, the abcissa the difference in Blue and UV light. The dashed line shows the position of so-called main sequence stars on such a plot, the position of the Sun is also shown. Red, M-type stars are in the lower right, as are Io, Titan, and Mars. The asteriod Icarus is also shown as representative of one of the redder asteroids.

the solid surface we view in the case of Io (Veverka, 1973; Zellner, 1973; Sagan, 1973). As in the case of Jupiter, the chromophores are presumably in the cloud deck, but once again we have no unambiguous spectroscopic evidence to use for identification purposes.

Our group at Stony Brook is pursuing the study of these satellites in several ways. From spectroscopic observations of Titan, we have been able to set some preliminary upper limits on two constituents that have often been postulated as intermediaries in the formation of more complex organic substances, viz., ethane and hydrogen cyanide. We find $C_2H_6/CH_4 < 1/25$ and $NCH/CH_4 < 1/100$ (Encrenaz *et al.*, 1973). This result is particularly interesting in the case of ethane, since mixing ratios as high as unity have been predicted by some photochemical calculations (Danielson, 1973) and this gas is also produced in relatively high yield in laboratory experiments set up to simulate reactions in such an atmosphere (Sagan and Miller, 1960). We have also determined an upper limit on the flux from Titan at 4.9 μ which corresponds to a reflectivity of less than 50% (Figure 5 – Joyce *et al.*, 1973). This result provides another means of distinguishing Titan from Io, since the latter exhibits a very high reflectivity

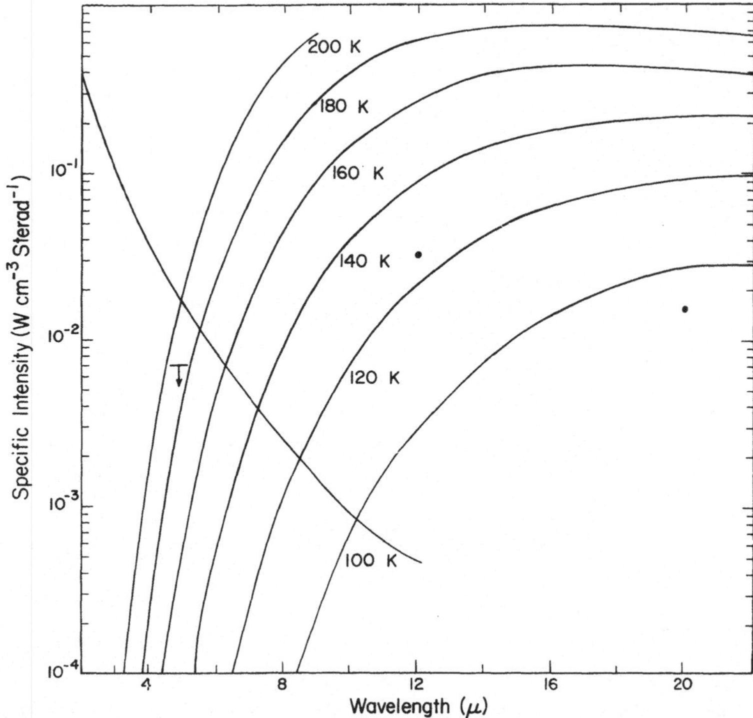

Fig. 5. Titan: a plot of surface brightness *vs* wavelength. The set of curves on the right show the energy distribution for black bodies at the indicated temperatures. The line sloping down from the left indicates the surface brightness of a perfectly diffuse reflecting sheet at Titan's distance from the Sun (cp. Figure 6). The point at 4.9 μ from Joyce *et al.* (1973), at 12 μ from Allen and Murdock(1971), at 20 μ from Morrison *et al.* (1972).

at this wavelength (Figure 6). On the other hand, Jupiter, as we saw above, shows evidence for substances having an absorption in this general spectral region.

I would like to suggest that this rather crude spectroscopy can provide a useful discriminant in deciding among various alternative candidates for chromophores in the outer solar system. It shows, for example, that the Woeller-Ponnamperuma (1969) polymer is more appropriate for Titan and at least some areas on Jupiter while the Sagan-Khare (1973) polymer would be more suitable for Io. Such judgements are *very* qualitative of course, and one hopes for the discovery of more specific spectral signatures.

Finally, experimental analogs to possible conditions on Io and Titan are being studied by T. Scattergood who has irradiated mixtures of CH_4 and NH_3 with a proton beam at energies comparable to those postulated in the Jovian radiation belts. He finds a yellow-brown oil is produced, similar in appearance to descriptions of end products from the more conventional UV irradiation techniques. This work is just in its initial stages and will be extended to include other gases and other experimental conditions. Reflection spectra will be obtained for comparison with observations of the planets and satellites.

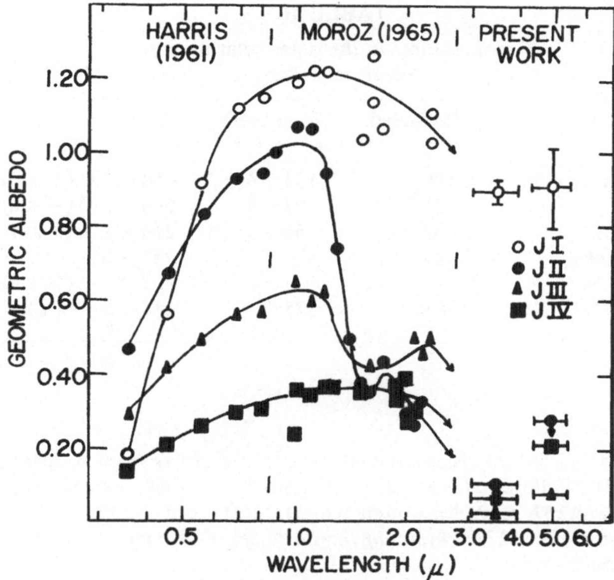

Fig. 6. Geometric albedos (reflectivities) of the Galilean satellites (JI-JIV) as a function of wavelength Smooth curves are qualitative, but note that all are plotted on the same scale. The high reflectivity of Io (JI) beyond 0.7 μ and its red color (cp. Titan in Figure 1) are especially remarkable. Data for 0.35–0.85 μ are from Harris (1961), for 0.85–2.5 μ from Moroz (1967) and for 3–5.4 μ (labelled 'present work') from Gillett *et al.* (1970).

3. Atmospheric Structure

A summary of temperatures for the bodies we have been considering is given in Table II. The 'predicted' temperatures are those that would be assumed by a blackbody with the appropriate rotation rate and reflectivity in equilibrium with incoming solar radiation. The 'measured' values refer to effective temperatures – the actual equivalent blackbody temperatures exhibited by the objects as measured over the largest possible spectral range. Finally, the 'maximum' temperature is the highest value measured anywhere in the spectrum, with the corresponding wavelength given in brackets. The discrepancy between the predicted and maximum temperatures is especially striking and supports the contention made earlier that conditions in the outer solar system may be less extreme than one might suppose *a priori*.

In the case of the major planets, the maximum temperatures are those measured at the longest useable wavelengths in the radio region of the spectrum. There is a clear tendency for brightness temperature to increase with wavelength for all of the major planets as may be seen in more detail by inspecting Figures 7 and 8. This trend is readily understood in terms of model atmosphere calculations which show that the opacity of these atmospheres diminishes with increasing wavelength (just as it does in our atmosphere) and one thus receives radiation from progressively deeper layers (Gulkis and Poynter, 1972; Lewis and Prinn, 1973). In the case of Jupiter and Saturn, we are probing regions in which the temperature is well above the boiling point of

TABLE II

Temperatures in the outer solar system (degrees Kelvin)

Object	Predicted	Measured	Maximum
Jupiter	105	134 ± 4	450 ± 50 (21 cm)
Saturn	71	97 ± 4	540 ± 110 (94 cm)
Uranus	57	55 ± 3	210 ± 17 (6 cm)
Neptune	45	–	227 ± 23 (6 cm)
Pluto	42	–	< 162 (11 cm)
Titan	88	125 (?)	157 ± 5 (8 μ)

Notes:

Jupiter and Saturn: Measured effective temperatures from Aumann, H. H., Gillespie, C. M., Jr., and Low, F. J.: 1969, *Astrophys. J. Letters* **157**, L69. Maximum temperatures from Gulkis, S., and Poynter, R. 1972, *Physics of the Earth and Planetary Interiors* (to be published).

Uranus, Neptune, and Pluto: Measured temperature for Uranus from Low, F. J.: 1966, *Astrophys. J.* **146**, 326. Maximum temperatures (and Pluto upper limit) from Webster, W. J., Jr., Webster, A. C., and Webster, G. T.: 1972, *Astrophys. J.* **174**, 679.

Titan: Measured effective temperature and maximum temperature estimated from Gillett, F. C., Forrest, W. J., and Merrill, K. M.: 1973, *Astrophys. J.* (in press).

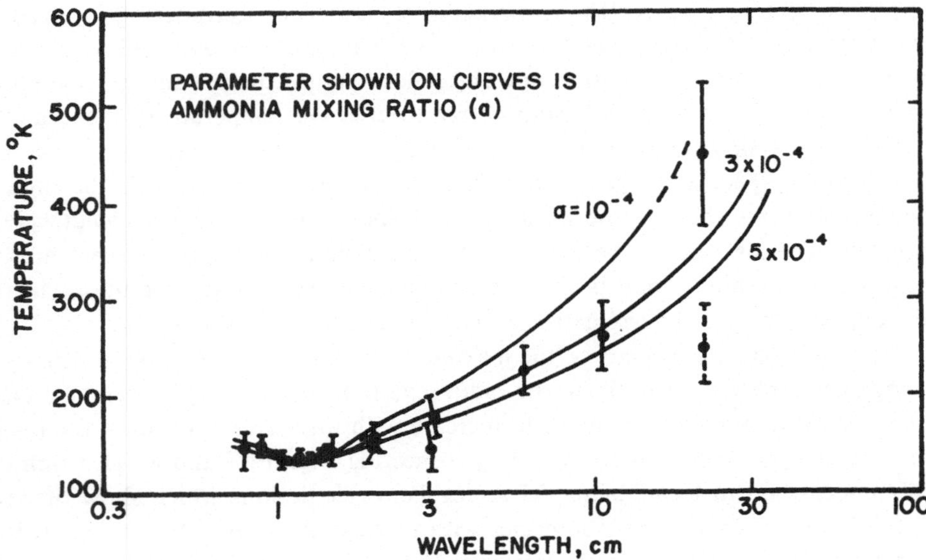

Fig. 7. The radio spectrum of Jupiter, corrected for non-thermal emission. The lines shown correspond to fluxes predicted by model atmospheres with various ammonia-hydrogen mixing ratios. Note the rise in temperature with increasing wavelength (Gulkis and Poynter, 1972).

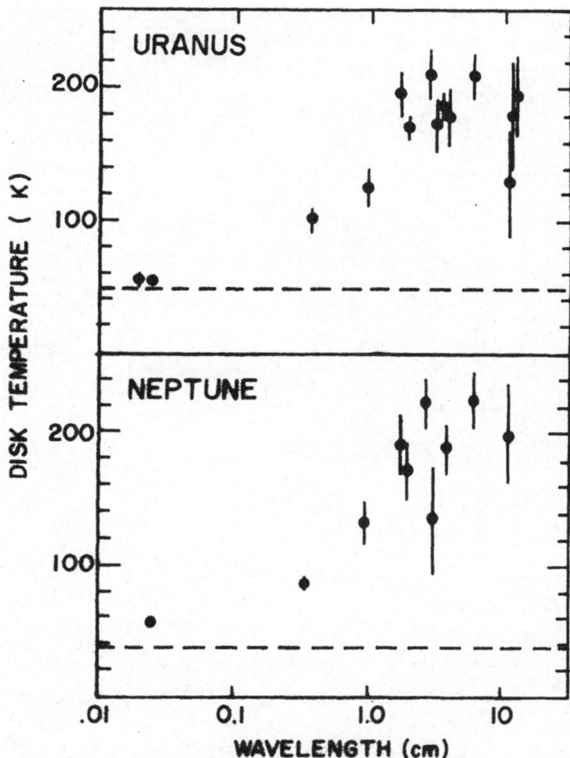

Fig. 8. Brightness temperature as a function of wavelength for Uranus and Neptune. Note the similarity of the data for the two planets. The dashed horizontal lines show the predicted temperatures for equilibrium with the solar flux (Morrison and Cruikshank, 1973; see note to Table I).

water at one atmosphere. It is possible that such regions also exist on Uranus and Neptune and that they will be found once sufficiently long wavelengths are sampled. On the other hand, the spectra of both of these objects seem remarkably flat from 3 to 11 cm, and hence one should keep open the possibility that the atmospheric structure may be rather different from that of Jupiter and Saturn.

Using results from near IR spectroscopy which offers the advantage of permitting both temperature and pressure determinations from studies of absorption lines of the constituent gases, one finds that the pressure on Jupiter is about one atmosphere at the point where $T = 180$ K (Owen and Mason, 1968). Both these numbers are uncertain by about 20% and are somewhat model dependent. However, extrapolation higher or lower in the atmosphere by means of the hydrostatic equation and the assumption of an adiabatic lapse rate for the temperature leads to values of T and P that are consistent with other observations. In fact, it now seems possible to fit a simple model of this type to observations ranging from the near ultra-violet to the radio region, corresponding to an altitude difference of about 125 km. The freezing point of water is reached at a pressure of about 20 atm, and this is likely to be a region in which

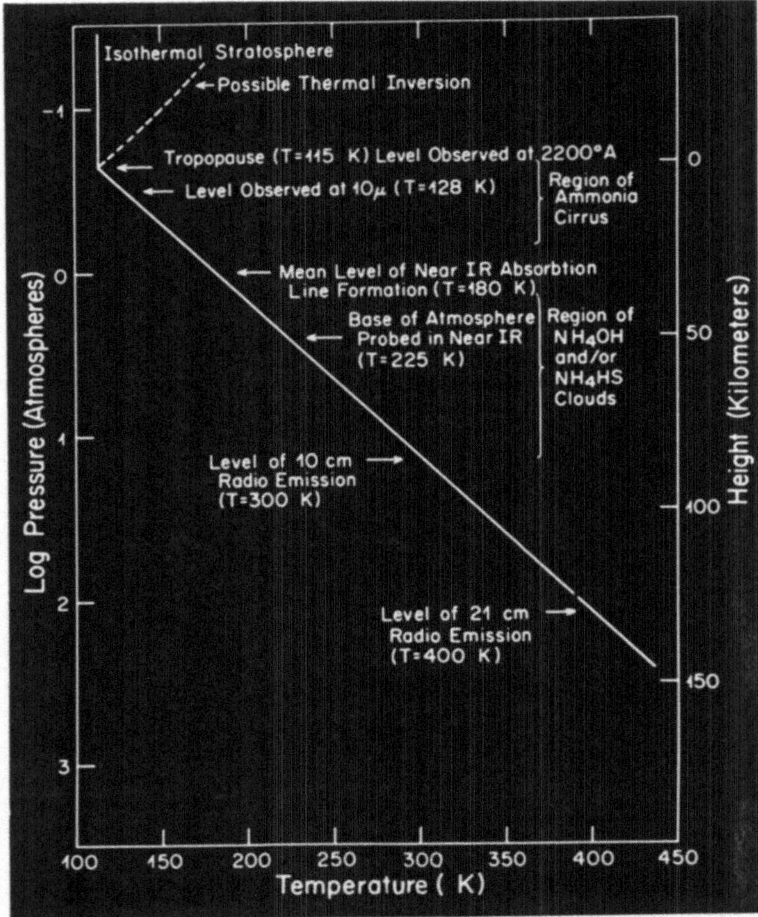

Fig. 9. A Schematic model of pressure and temperature in the atmosphere of Jupiter with locations of major cloud systems indicated. For more detail, see Lewis (1969b).

aqueous ammonia clouds will form (Figure 9 – Lewis, 1969b; Gulkis and Poynter, 1972).

The situation for Saturn is much more poorly defined owing to a less complete set of observations. Models based on cosmic abundances seem capable of explaining the radio observations, so it is now a question of improving the results from IR and UV spectroscopy. Present indications are that pressures and temperatures in the lower atmosphere are similar to those on Jupiter (Gulkis and Poynter, 1972). The smaller mass of Saturn affects both the local pressure and the temperature lapse rate in exactly compensating ways, so to first order one simply travels farther vertically in the atmosphere of Saturn (by the ratio g_J/g_S) to move from (P_1, T_1) to (P_2, T_2). Once again aqueous ammonia clouds are indicated.

The lower atmospheres of Uranus and Neptune have been only marginally probed by microwave observations. If internal heat sources are present, one may expect a similar result to that obtained for Jupiter and Saturn, although now a different (and presently unknown) mixture of gases must be used. An analysis by Lewis and Prinn (1973) suggests layers of clouds analogous to those expected on Jupiter and Saturn. We may expect more work on this problem in the immediate future as new radio observations become available.

The temperatures quoted for Titan have a different significance. This object has not yet been observed at radio frequencies and not enough points have been sampled to lead to a reliable effective temperature. Measurements in the 7 to 25 μ region have revealed an unexpectedly complex spectrum that must be determined by various sources of atmospheric opacity (Morrison *et al.*, 1972). The maximum value occurs in a region of the spectrum where there is an extremely strong methane absorption and hence the atmospheric opacity is very great. One must thus be sampling radiation from an inversion layer high in the planet's atmosphere (Gillett *et al.*, 1973). The lowest temperature occurs in the 20 μ region, where a value of 90 K has been obtained, presumably referring to some level intermediate between the inversion and the lower atmosphere (Morrison *et al.*, 1972). As can be seen from Table II, this would be an appropriate temperature for the surface if no atmosphere were present. The actual surface temperature remains unknown, although a value on the order of 155 K is appropriate for some current models (Pollack, 1973; Cess and Owen, 1973). Much higher temperatures have been suggested (Sagan, 1973), and a completely different interpretation has also been offered, in which the 7–14 μ spectrum is viewed entirely in terms of upper atmosphere emission features, leading to a very low surface temperature (Danielson *et al.*, 1973). Pressures are correspondingly uncertain, making efforts to determine an accurate methane abundance particularly frustrating (Trafton 1972b – cf. Table I). The observation at 4.9 μ already referred to (Figure 5) leads to an upper limit of 190 K for an equivalent temperature, but the effect of the clouds at this wavelength is not known (Joyce *et al.*, 1973). Hydrogen is commonly assumed to be the basic source of opacity required to produce a greenhouse effect that would lead to appropriately high temperatures, but no model successfully explains all the observations as yet (Pollack, 1973; Sagan, 1973; Cess and Owen 1973). This object clearly will be the focus of much additional research during the next few years.

In the case of Pluto, we have only an upper limit on the temperature set from radio observations. The absence of noticeable methane absorptions in the near IR spectrum (see above) can also be used to set a model-dependent limit on the surface temperature. If one assumes that methane is present in saturation equilibrium with the solid and that the total atmospheric pressure (from neon, argon, helium) is on the order of one atmosphere, then $T_S < 60$ K. It is an open question at present why Pluto and Triton do not have warm atmospheres and Titan does. Perhaps these two bodies have always been sufficiently cold that they never evolved enough gas to develop self-sustaining atmospheres of the type presently exhibited by Titan. Be that as it may, they do not appear to constitute very interesting environments for our present purposes.

4. Conclusions

It should be apparent from this discussion that the outer solar system offers us a variety of environments in which natural experiments in pre-biotic organic synthesis must be taking place at the present time. These environments comprise a broad range of temperatures, pressures, and compositions, high and low humidities of both H_2O and NH_3, and various types of energy sources. Furthermore, the resulting chemical reactions have presumably been going on throughout the lifetime of the solar system; it will be most interesting to determine the level of complexity that has been achieved.

We are witnessing a sudden growth in our knowledge about these distant places as a result of the application of several new sophisticated observational techniques. Among these are Fourier transform spectroscopy, high-resolution spectrophotometry, and the use of very sensitive bolometers for photometry and low resolution spectroscopy in the infrared. We can expect a further harvest as these are employed with greater intensity and finesse during the next few years. But it is the missions to comets and the outer planets planned for the end of this decade that should really bring us the information we seek, especially if it is possible to send suitably instrumented probes into the atmospheres of these mysterious objects. It will be extremely surprising if such experiments fail to reveal the chemical precursors of life as we find it on Earth. We can only guess at what else may be revealed.

Acknowledgements

I am grateful to Drs F. C. Gillett, S. Gross, and S. Gulkis for supplying results prior to publication. Figures 2 and 3 were furnished through the courtesy of Mr J. Fountain and Dr B. A. Smith, respectively. The other figures are reproduced through the courtesy of the University of Chicago Press (Figures 5 and 8, copyright by the University of Chicago, all rights reserved), Gordon and Breach (Figure 6), and Dr S. Gulkis (Figure 7). The research at Stony Brook reported here has been supported by NASA under grants NGR-33-015-141, 33-015-169, and 33-015-165.

Note added in Proof. S. Ridgway (*Astrophys. J.*, 1974) has reported the detection of ethane and acetylene in the spectrum of Jupiter. Abundances are still uncertain but the mere presence of these non-equilibrium products greatly strengthens the argument for organic chromophores presented above. B. L. Ulich and B. K. Conklin (*IAU Circular*, 2607) have discovered methyl cyanide emission from Comet Kohoutek (1973f), the first direct evidence for an association between the parent molecules in comet nuclei and the interstellar molecules (discussed in these Proceedings by D. Buhl). We have re-examined the problem of the CH_4 abundance in the atmospheres of Uranus and Neptune and concluded that H/C is even lower than the value of 250 given in Table I (Owen, Lutz, Porco and Woodman – *Astrophys. J.*, 1974). We have also detected Titan at 4.9 μm and found that the albedo may be as low as 0.05 (Joyce, Knacke, and Owen – in preparation for *Astrophys. J.*, 1974).

References

Allen, D. A. and Murdock, T. L.: 1971, *Icarus* **14**, 1.
Armstrong, K. R., Harper, D. A., Jr., and Low, F. J.: 1972, *Astrophys. J. Letters* **178**, L89.
Aumann, H. H., Gillespie, C. M., Jr., and Low, F. J.: 1969, *Astrophys. J. Letters* **157**, L69.
Beer, R. and Taylor, F. W.: 1973, *Astrophys. J.* **179**, 309.
Carr, T. D. and Gulkis, S.: 1969, *Ann. Rev. Astron. Astrophys.* **7**, 577.
Cess, R. and Owen, T.: 1973, *Nature* **244**, 272.
Chada, M. S., Flores, J. J., Lawless, J. G., and Ponnamperuma, C.: 1971, *Icarus* **15**, 39.
Cruikshank, D. P. and Binder, A. B.: 1969, *Astrophys. Space Sci.* **3**, 347.
Danielson, R. E.: 1973, private communication.
Danielson, R. E., Caldwell, J. J., and Larach, D. R.: 1973, *Astrophys. J.* (in press).
Encrenaz, Th., Hardorp, J., and Owen, T.: 1973, in preparation.
Gillett, F. C., Forrest, W. J., and Merrill, K. M.: 1973, *Astrophys. J. Letters* **184**, L93.
Gillett, F. C., Low, F. J., and Stein, W. A.: 1969, *Astrophys. J.* **157**, 925.
Gillett, F. C., Merrill, K. M., and Stein, W. A.: 1970, *Astrophys. Letters* **6**, 247.
Greenspan, J. A. and Owen, T.: 1967, *Science* **156**, 1489.
Gross, S.: 1973, to be published.
Gulkis, S. and Poynter, R.: 1972, *Physics of the Earth and Planetary Interiors* **6**, 36.
Harris, D. L.: 1961, in G. P. Kuiper and B. M. Middlehurst (eds.), *The Solar System*, Vol. III: *Planets and Satellites*, U. of Chicago Press, Chicago, Chap. 8.
Johnson, T. V. and McCord, T. B.: 1970, *Icarus* **13**, 37.
Johnson, T. V. and McCord, T. B.: 1971, *Astrophys. J.* **169**, 589.
Joyce, R. R., Knacke, R. F., and Owen, T.: 1973, *Astrophys. J. Letters* **183**, L31.
Kuiper, G. P.: 1944, *Astrophys. J.* **100**, 378.
Lewis, J. S.: 1969a, *Icarus* **10**, 393.
Lewis, J. S.: 1969b, *Icarus* **10**, 365.
Lewis, J. S. and Prinn, R. G.: 1970, *Science* **169**, 472.
Lewis, J. S. and Prinn, R. G.: 1973, *Astrophys. J.* **179**, 333.
Lippincott, E. R., Eck, R., Dayhoff, M. O., and Sagan, C.: 1967, *Astrophys. J.* **147**, 753.
Moroz, V. I.: 1967, *Physics of Planets*, NASA Technical Translation F-515 of 'Fizika Planet', Clearinghouse for Federal Scientific and Technical Information, Springfield, Va., p. 366.
Morrison, D., Cruikshank, D. P., and Murphy, R. E.: 1972, *Astrophys. J. Letters* **173**, L143.
Münch, G. and Neugebauer, G.: 1971, *Science* **174**, 940.
Newburn, R. L., Jr. and Gulkis, S.: 1973, *Space Sci. Rev.* **13**, 179.
Owen, T.: 1967, *Icarus* **6**, 138.
Owen, T.: 1969, *Icarus* **10**, 355.
Owen, T. and Mason, H. P.: 1968, *Astrophys. J.* **154**, 317.
Owen, T. and Mason, H. P.: 1969, *J. Atmospheric Sci.* **26**, 870.
Pollack, J. B.: 1973, *Icarus* **19**, 43.
Sagan, C.: 1971, *Space Sci. Rev.* **11**, 827.
Sagan, C.: 1973, *Icarus* **18**, 649.
Sagan, C. and Khare, B. N.: 1971, *Astrophys. J.* **168**, 563.
Sagan, C. and Khare, B. N.: 1973, these Proceedings.
Sagan, C. and Miller, S. L.: 1960, *Astron. J.* **65**, 499.
Sill, G.: 1972, *Communications Lunar Planet. Lab.*, in press.
Smoluchowski, R.: 1972, in S. I. Rasool (ed.), *Physics of the Solar System*, NASA SP300, Washington, D.C, Chap. 8.
Trafton, L.: 1972a, *Astrophys. J.* **175**, 285.
Trafton, L.: 1972b, *Astrophys. J.* **175**, 295.
Urey, H. C.: 1959, in S. Fluegge (ed.), *Handbuch der Physik*, Springer, Berlin, Vol. 52, p. 409.
Veverka, J.: 1973, *Icarus* **18**, 657.
Woeller, F. and Ponnamperuma, C.: 1969, *Icarus* **10**, 386.
Zellner, B.: 1973, *Icarus* **18**, 661.

CATALYTIC REACTIONS IN THE SOLAR NEBULA: IMPLICATIONS FOR INTERSTELLAR MOLECULES AND ORGANIC COMPOUNDS IN METEORITES*

EDWARD ANDERS and RYOICHI HAYATSU

Enrico Fermi Institute and Dept. of Chemistry, University of Chicago, Chicago, Ill. 60637, U.S.A.

and

MARTIN H. STUDIER

Argonne National Laboratory, Argonne, Ill. 60439, U.S.A.

Abstract. Organic compounds in meteorites seem to have formed by Fischer-Tropsch-type, catalytic reactions of CO, H_2, and NH_3 in the solar nebula, at 360–400K and (4–10) $\times 10^{-6}$ atm. The onset of these reactions was triggered by the formation of catalytically active grains of magnetite and serpentine at these temperatures.

Laboratory experiments show that the Fischer-Tropsch reaction gives a large kinetic *isotope fractionation* of C^{12}/C^{13}, duplicating the hitherto unexplained fractionation in meteorites. All of the principal compound classes in meteorites are produced by this reaction, or a variant involving a brief excursion to higher temperatures. (1) normal, mono-, and dimethyl*alkanes*; (2) *arenes* and *alkylarenes*; (3) dimeric *isoprenoids* from C_9 to C_{14}; (4) *purines* and *pyrimidines*, such as adenine, guanine, uracil, thymine, xanthine, etc.; (5) *amino acids*, including tyrosine and histidine; (6) *porphyrin*-like pigments; (7) aromatic *polymer* with –OH and –COOH groups.

These reactions may also have played a major role in the evolution of life: first, by converting carbon to a sufficiently non-volatile form to permit its accretion by the inner planets; second, by synthesizing organic compounds on the primitive planets whenever CO, H_2, NH_3, and clay or magnetite particles came together at the right temperature. Similar reactions in other solar nebulae may be the source of interstellar molecules, as first suggested by G. H. Herbig. Ten of the twelve polyatomic interstellar molecules have in fact been seen in these syntheses or in meteorites.

1. Introduction

It is generally agreed that type 1 carbonaceous chondrites (C1) are a primitive condensate from the solar nebula, only a step or two away from solar composition (Anders, 1971a; Cameron, 1973). Thus it seems surprising that they contain a rich variety of complex organic compounds. Let us examine the behavior of carbon in the solar nebula, and see what compounds are produced in the primary condensation process. We can then decide what additional processes must be invoked to produce the observed array of compounds in meteorites.

Only a brief summary of the evidence can be given here. The reader is referred to more detailed reviews on the origin of meteorites (Anders, 1971b, 1972), organic compounds in meteorites (Anders *et al.*, 1973; Oró, 1972; Hayes, 1967; Vdovykin, 1967), and to the experimental papers cited in the text.

* This paper is a revised and abridged version of the authors' article in *Science* **182** (1973), 781.

2. Formation Conditions of Carbonaceous Chondrites

Three 'cosmothermometers', based on equilibrium or kinetic isotope fractionations of O and C, consistently give formation temperatures near 360 K for C1 chondrites (Anders *et al.*, 1973). This is the temperature at which isotopic fractionations were frozen in, owing to sluggish reaction rates or physical isolation of the meteoritic dust by accretion to larger bodies.

The pressure has been estimated as $\simeq 4 \times 10^{-6}$ atm, from the Ar^{36} content of meteoritic magnetite (Lancet and Anders, 1973). This value is tentative, but happens to be close to Cameron and Pine's (1973) theoretical estimate of 8×10^{-6} atm at 360 K, for their model of the solar nebula.

3. Behavior of Carbon in a Solar Gas

With the physical conditions thus defined, it is instructive to consider what happens to carbon in a cooling solar gas (Figure 1). CO is the stable form at high tempera-

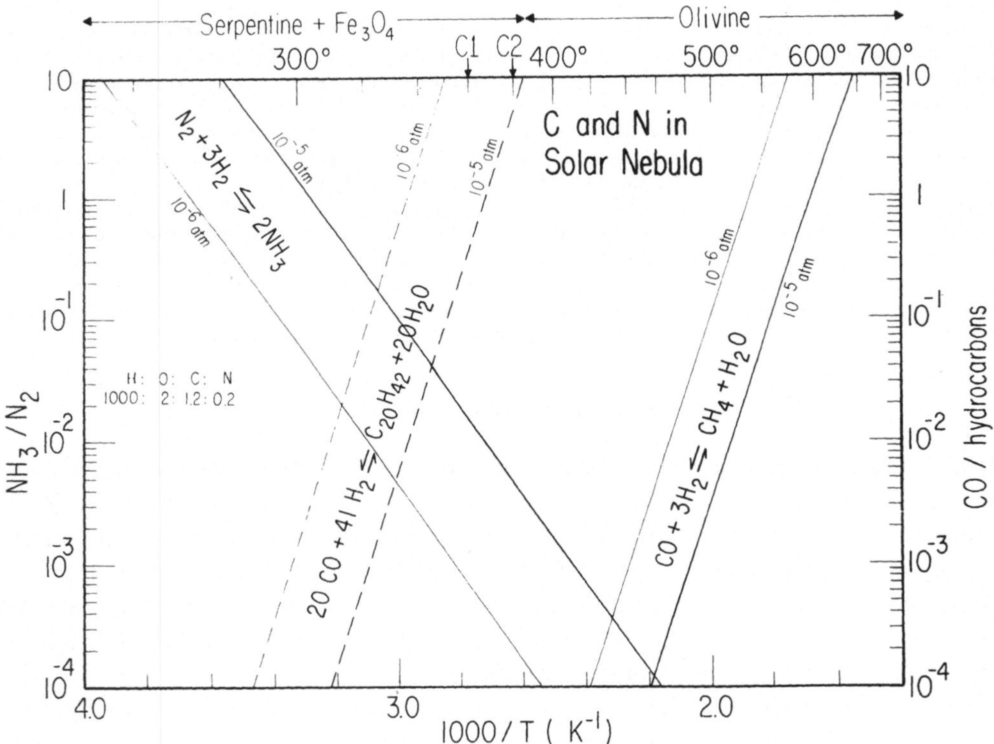

Fig. 1. If equilibrium is maintained on cooling, CO will be converted largely to CH_4 (solid lines) before metastable formation of more complex hydrocarbons by the Fischer-Tropsch reaction becomes possible (dashed lines). However, the reaction is very slow in the absence of catalysts, and may not have begun until ~ 400 K when catalysts such as serpentine and magnetite became available through the hydration of olivine.

tures, but becomes less stable on cooling and should transform to CH_4 below 600 K, as shown by the solid lines in the right-hand portion of Figure 1. However, CH_4 has a condensation temperature of less than 100 K, and if this reaction had gone as written, there should be no carbon and no life anywhere in the inner solar system. Thus events must have taken a different course.

Urey (1953) first noted this paradox in a classic paper. He suggested that the hydrogenation of CO may not have gone smoothly to CH_4 in the absence of man-made catalysts, but stopped at 'complex tarry carbon compounds', representing intermediate stages of hydrogenation. Such compounds, typified by $C_{20}H_{42}$ in Figure 1, are less stable than methane, but could form metastably below $\simeq$400 K.

4. Model Experiments: The Fischer-Tropsch Synthesis

Urey's idea soon fell into oblivion, because of the phenomenal success of the Miller-Urey synthesis. We began to investigate this neglected idea in 1964, when it first became apparent that Miller-Urey reactions could not account for certain features of meteorite organic matter. Our approach was to see how CO and H_2 behaved in the presence of some *natural* catalysts expected in the solar nebula: nickel-iron, magnetite, hydrated silicates. We found (Studier *et al.*, 1965a, 1968) that the reaction indeed tends to stop at intermediate stages of hydrogenation, giving metastable products of $H/C \approx 2$ (e.g. $C_{20}H_{42}$) rather than stable methane of $H/C = 4$. In fact, this process (the Fischer-Tropsch synthesis, discovered in 1923) has been used industrially for the production of gasoline.

It is not feasible to conduct such model experiments at the pressure and H_2/CO ratio of the solar nebula – the total amount of carbon in a 1-liter vessel would be only $\simeq 5 \times 10^{-9}$ g. Accordingly, we used higher pressures (0.1–10 atm), and lower H_2/CO ratios (generally 1, sometimes as high as 120). We shall show later on that these results are nonetheless applicable to the solar nebula.

As a safeguard against contamination, we used deuterium rather than light hydrogen in our syntheses, and identified reaction products by mass spectrometry. In general, products were separated by gas chromatography or other techniques prior to mass spectrometry.

The reaction was always carried out under static conditions in a closed vessel, not in a flow system as in the industrial synthesis. We also broadened the range of experimental conditions beyond those of the classical Fischer-Tropsch synthesis. Reaction times ranged from a fraction of an hour to a few months, and the temperature was sometimes raised briefly from 150–250 to 500–700 °C. Such thermal pulses were intended to simulate short-term heating events in the nebula, such as collisions, shock waves, or the chondrule-forming process. In runs where nitrogen compounds were sought we added ND_3 to the reactant gases, and occasionally used other types of catalysts, such as montmorillonite clay, Al_2O_3, or SiO_2. For want of a better term, we shall call this class of reactions 'Fischer-Tropsch' Type (= FTT).

5. Comparison of Meteoritic and Synthetic Organic Compounds

5.1. Heavy Alkanes

The same few compounds dominate in meteoritic and FT hydrocarbons (Studier *et al.*, 1968, 1972; Nooner and Oró, 1967; Gelpi and Oró, 1970a). Normal alkanes are most prominent, followed by five slightly branched isomers (2Me, 3Me, 2,3DiMe, 3,4DiMe, and 4,5DiMe). This resemblance gains significance if one considers that some 10^5 structural isomers can exist for alkanes with 16C atoms. Terrestrial hydrocarbons show a similar pattern, and so contamination must be considered. However, this seems to be unlikely at least for the meteorites analyzed by Studier *et al.* (1968, 1972), for reasons given by Anders *et al.* (1973).

It seems that the meteoritic hydrocarbons were made by FT reactions, or a process of the same extraordinary selectivity. The Miller-Urey reaction, incidentally, shows no such selectivity. Gas chromatograms of spark discharge hydrocarbons show no structure (Calvin, 1969). Apparently all 10^5 possible isomers are made in comparable yield, as expected for random recombination of free radicals.

5.2. Carbon isotope fractionations

Meteorites show a very large difference in C^{12}/C^{13} ratio between carbonate and organic carbon: 60 to 80 permil (Clayton, 1963; Smith and Kaplan, 1970). This trend remained unexplained for a number of years, because coexisting carbonate and organic matter on Earth shows a much smaller difference, typically 25–30 permil. It probably is a primary feature, unaffected by the later thermal history of the meteorite. Terrestrial calcium carbonate is not known to equilibrate with coexisting organic matter in sediments (Smith and Kaplan, 1970).

It turns out that the Fischer-Tropsch reaction gives an isotopic fractionation of just the right sign and magnitude, owing to a kinetic isotopic effect (Lancet and Anders, 1970). From the temperature dependence of the fractionation between 375 and 550 K, the observed fractionations in C1 and C2 chondrites correspond to about 360 and 400 K (Lancet, 1972). These values agree rather well with the O^{18}-based formation temperatures of carbonates and silicates, 360 K for C1's and 380 K for C2's (Onuma *et al.*, 1972). The Miller-Urey reaction gives a fractionation of only -0.4 ± 0.2 permil (Lancet, 1972).

5.3. Aromatic hydrocarbons; light alkanes

Carbonaceous chondrites contain a wide range of aromatic hydrocarbons, from benzene through alkylbenzenes and -naphthalenes to polynuclear hydrocarbons of up to six fused benzene rings (see the reviews cited above). At higher carbon numbers, aromatics tend to be less abundant than normal alkanes, but below about C_{11}, the reverse is true (Studier *et al.*, 1965b, 1968, 1972). In fact, virtually no normal alkanes at all are found between C_2 and C_8, their place having been taken largely by benzene, toluene, xylene, and various alkenes or branched alkanes.

A pattern of this sort does not form directly in the primary Fischer-Tropsch reaction. It does, however, develop when a primary Fischer-Tropsch mixture remains in contact with the catalyst, for a day or so at 350–400 °C, or longer times at lower temperatures (Studier *et al.*, 1968, 1972). Under such conditions, a metastable equilibrium is approached, with methane and aromatic hydrocarbons forming at the expense of ethane and heavier alkanes (Dayhoff *et al.*, 1964). The kinetics and mechanism of such aromatization on the catalyst surface has been discussed by Galwey (1968, 1972). When the heating is prolonged or carried out at higher temperatures, polynuclear aromatics with up to 7 rings are obtained (Studier *et al.*, 1965a, 1968; Oró and Han, 1966). Opportunities for such secondary reactions certainly existed in the history of meteorites, and rates seem to be adequate even at 360 K (Anders *et al.*, 1973).

5.4. Isoprenoid alkanes

Tri- and tetrameric isoprenoids from C_{14} to C_{21} were found in carbonaceous chondrites by Gelpi and Oró (1970a) but not by Studier *et al.* (1968, 1972). The latter workers found only some dimeric isoprenoids (C_9–C_{11}, C_{13}), and the current consensus seems to be that the heavier isoprenoids were terrestrial contaminants (Gelpi and Oró, 1970b; Gelpi *et al.*, 1970; Oró, 1972; Studier *et al.*, 1972).

The lighter isoprenoids C_9 to C_{14} can be produced in the FT synthesis. Contamination was precluded by use of deuterium (see Studier *et al.*, 1968, for mass spectra).

5.5. Fatty acids

Nagy and Bitz (1963) reported fatty acids from C_{14} to C_{28} in Orgueil. This work was substantially confirmed by Hayatsu (1965) and Smith and Kaplan (1970). The latter authors found 3 to 91 ppm fatty acids from C_{12} to C_{20} in 7 carbonaceous chondrites. Smith and Kaplan believe that these acids are largely or entirely terrestrial contaminants, because unstable unsaturated acids comprise about 30% of the total, and $C_{16}+C_{18}$ acids are predominant. No attempts have yet been made to look for fatty acids in the model experiments described here, but it is well known that they can be produced in the Fischer-Tropsch and related syntheses (Storch *et al.*, 1951).

5.6. Purines, pyrimidines, and other nitrogen bases

Several nitrogen heterocyclics (underlined in Figure 2) have been found in the Orgueil meteorite (Hayatsu, 1964; Hayatsu *et al.*, 1968). All these compounds and several others (Figure 2) are made in an FTT reaction in the presence of NH_3 (Hayatsu *et al.*, 1968, 1972). Pyrroles and nitriles were also observed.

A somewhat different set of N-heterocyclic compounds, including 4-hydroxypyrimidines but no biological heterocyclics, was reported by Folsome *et al.* (1973) in three carbonaceous chondrites. However, this discrepancy seems to reflect problems of technique. With suitable extraction methods, biological heterocyclics such as adenine and guanine are readily identified (see Figure 6 of Anders *et al.*, 1973).

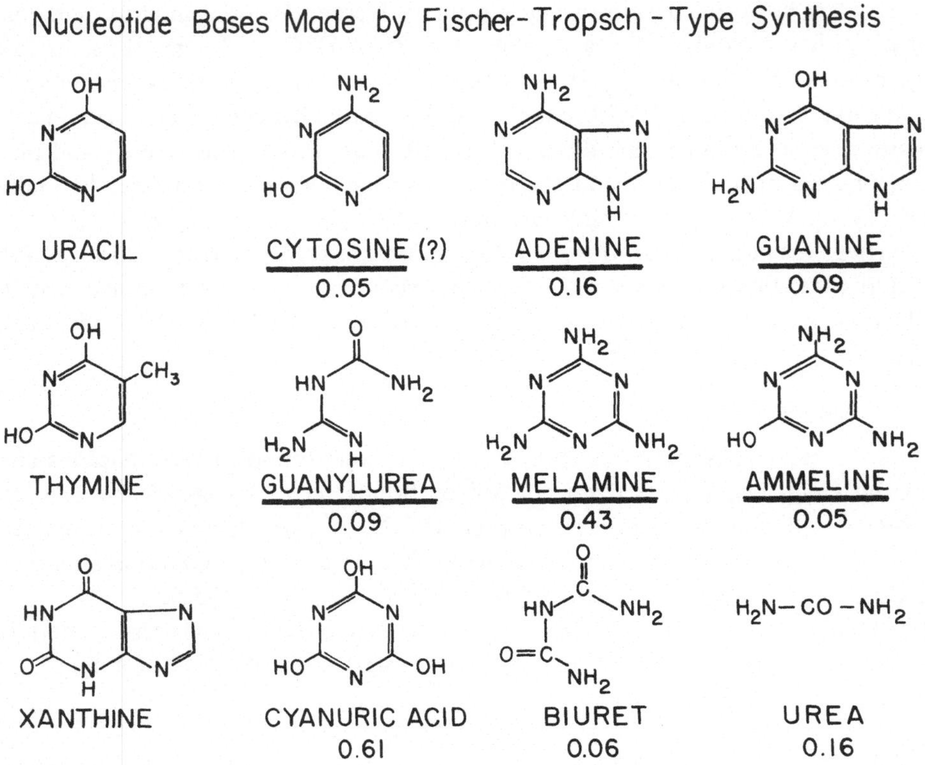

Nucleotide Bases Made by Fischer-Tropsch-Type Synthesis

URACIL

CYTOSINE (?) 0.05

ADENINE 0.16

GUANINE 0.09

THYMINE

GUANYLUREA 0.09

MELAMINE 0.43

AMMELINE 0.05

XANTHINE

CYANURIC ACID 0.61

BIURET 0.06

UREA 0.16

Fig. 2. Underlined compounds have been identified in meteorites. Numbers indicate yields in percent (Hayatsu *et al.*, 1968, 1972).

5.7. Amino Acids

Kvenvolden *et al.* (1970, 1971) have shown that the Murchison meteorite contains at least 17 indigenous amino acids, or compounds hydrolyzable thereto. Ten of these are not commonly found in terrestrial proteins. These findings were substantially confirmed by Oró *et al.* (1971) and Cronin and Moore (1971).

Many of the same amino acids, plus several others (Table I) are produced in FTT syntheses involving a brief initial heating to 500–700 °C (Yoshino *et al.*, 1971; Hayatsu *et al.*, 1971). Yields are low (0.01–0.1%), but the products include structurally complex aromatic or heterocyclic amino acids such as tyrosine and histidine that cannot be made by conventional Miller-Urey syntheses. Regrettably, this work was completed before the meteorite analyses became available, and so the non-protein amino acids were not systematically looked for. Some were tentatively identified when seen in large amounts; others may have comprised the up to 6 unidentified compounds that were observed in most of these syntheses.

A directed search for these amino acids in the Miller-Urey synthesis has been largely successful, yielding all those seen in Murchison (Ring *et al.*, 1972; Wolman *et al.*, 1972). Interestingly, the proportions of amino acids in the synthesis agree to

TABLE I

Amino acids in meteorites and the modified Fischer-Tropsch synthesis

(× × = definite; × = tentative; ○ = not sought)

	Meteorites (Kvenvolden *et al.*, 1970, 1971)	FTT Synthesis (Yoshino *et al.*, 1971; Hayatsu *et al.*, 1971)
Glycine	× ×	× ×
Alanine	× ×	× ×
Valine	× ×	×
Leucine		×
Isoleucine		×
Aspartic Acid	× ×	× ×
Glutamic Acid	× ×	× ×
Tyrosine		× ×
Proline	× ×	×
Ornithine		× ×
Lysine		× ×
Histidine		× ×
Arginine		× ×
N-methylglycine	× ×	× ×
N-ethylglycine	× ×	○
β-alanine	× ×	× ×
N-methylalanine	× ×	○
Isovaline	× ×	○
Norvaline	× ×	○
α-aminoisobutyric Acid	× ×	×
α-amino-n-butyric Acid	× ×	×
β-aminoisobutyric Acid	× ×	×
β-amino-n-butyric Acid	× ×	○
γ-aminobutyric Acid	× ×	×
Pipecolic Acid	× ×	○

within 1–2 orders of magnitude with those in the meteorite. To the extent that such a comparison can be made from the more limited data for the FTT synthesis, the agreement is of the same order.

5.8. Porphyrins

Hodgson and Baker (1969) have detected pigments resembling porphyrins in several carbonaceous chondrites. It is not clear whether these were true (=cyclic) porphyrins or linear pyrrole polymers, which mimic porphyrins in many respects. Both kinds have been seen in FTT syntheses (Hayatsu *et al.*, 1972). One pigment whose cyclic nature was confirmed by mass spectrometry had a major peak at mass 580, as expected for an alkyl-substituted porphin $C_{20}D_{14}N_4+16CD_2$ (Anders *et al.*, 1973). Though it did not show a doubly-charged ion at mass 290, it displayed other characteristics of porphyrins: strong absorption at 394 nm (in the Soret band range), formation of a red copper complex, and chromatographic and solvent extraction behavior similar to that of porphyrins (Hayatsu *et al.*, 1972).

Porphyrin-like pigments of similar properties have been made by the Miller-Urey synthesis (Hodgson and Ponnamperuma, 1969), but their cyclic nature has not yet been verified by mass spectrometry.

5.9. Chloro and thio compounds

Mono- and dichlorobenzenes have been found in several carbonaceous chondrites (Studier *et al.*, 1968, 1972), but in view of the widespread human use of such compounds, it is not at all certain that they are indigenous. Benzothiophenes, first reported by Hayes and Biemann (1968) in pyrolysis experiments, have also been seen in room-temperature solvent extracts (Studier *et al.*, 1972) and hence must be original constituents of the meteorites, not thermal degradation products. It is not known whether these compounds can be made in FTT syntheses, because the necessary experiments have not yet been attempted.

5.10. Macromolecular material

At least 70% of the organic matter in meteorites consists of an ill-defined, insoluble solid, said to resemble humic acids in soils (Briggs and Mamikunian, 1963). It has an aromatic skeleton bearing functional groups such as –COOH and –OH (Bitz and Nagy, 1966; Hayes, 1967). A similar material was obtained in an FTT synthesis extended over 6 months. The mass spectra of the meteoritic and synthetic polymer are similar (Anders *et al.*, 1973), and show mainly benzene, naphthalene, and their alkyl derivatives, as well as alkylindanes, fluorene, anthracene/phenanthrene, alkenes, alkanes, and alkylphenols (Hayatsu *et al.*, manuscript in preparation).

A Miller-Urey reaction in the presence of H_2S (Khare and Sagan, 1973) also gave a polymer, but of aliphatic rather than aromatic structure.

6. Relevance to the Solar Nebula

It appears that FTT reactions can account reasonably well for all features of organic matter in meteorites. The only alternative process, the Miller-Urey synthesis, fails to account for the *n* and *iso*alkanes, arenes, isoprenoids, most *N*-heterocyclics, the polymer, and carbon isotope fractionations, though it remains a possible and perhaps superior source of amino acids.

However, it is not immediately obvious that the FTT model experiments are relevant to the solar nebula. First and foremost, it must be shown that CO was still present at the time the nebula had cooled to 360 K. How did CO traverse the no-man's land between 600 K where it becomes unstable with respect to CH_4, and $\sim$400 K where formation of heavy hydrocarbons first becomes thermodynamically feasible (Figure 1)? Second, was the reaction rate fast enough at the extremely low pressures in the nebula?

6.1. Survival of CO

An answer to the first question was suggested by Lancet and Anders (1970), and is shown at the top of Figure 1. The high-temperature meteoritic phases stable above $\simeq$350–400 K (olivine, pyroxene, Fe, FeS) are not effective catalysts for the Fischer-Tropsch reaction, while the low-temperature phases forming below $\simeq$400 K (hydrated silicates, magnetite) are. (Though metallic iron is often regarded as a catalyst for this synthesis, the catalytically active phase actually is a thin coating of Fe_3O_4 formed on the surface of the metal; Anderson, 1956). Laboratory data show that catalysis by the high-temperature phases is slower by at least a factor of 10^8. Thus CO may have survived metastably between 600 and $\simeq$400 K until catalysts became available by reactions such as:

$$12\,(Mg, Fe)_2SiO_4 + 14H_2O \rightarrow 2Fe_3O_4 + 2H_2 + \\ + 3\,(Mg, Fe)_6\,(OH)_8Si_4O_{10}$$

$$4\,(Mg, Fe)_2SiO_4 + 4H_2O + 2CO_2 \rightarrow 2\,(Mg, Fe)\,CO_3 + \\ + (Mg, Fe)_6\,(OH)_8Si_4O_{10}.$$

This would also explain why the hydrated silicates, carbonates, and organic compounds all have the same formation temperature of 360 K.

6.2. Rate of reaction in the nebula

The second question cannot be answered unequivocally because the kinetics of the Fischer-Tropsch reaction are not well enough understood to permit reliable extrapolations to very low pressures. Still, a tentative analysis of available rate data suggests that the reaction half-time may be no greater than a few centuries, even at 4×10^{-6} atm (Lancet, 1972; see Anders *et al.*, 1973 for a summary).

7. Interstellar Molecules

It seems that FTT reactions in solar nebulae may also be responsible for interstellar molecules, as first suggested by Herbig (1970). Main sequence stars show a marked discontinuity of rotation rates at 1.5 $M_\odot$, suggestive of angular momentum transfer to extrastellar material. Thus solar nebulae may be a common byproduct of star formation. Such nebulae, embedded in interstellar clouds, provide a high-density environment ($\sim 10^{15}$ molecules cm^{-3}) in which matter can be transformed to grains and molecules which are then returned to the interstellar cloud when the nebula is dissipated. The denser parts of such clouds are well-shielded from UV radiation, and may serve as a long-term source of interstellar molecules.

In our own solar system, nearly all volatiles complementary to the inner planets ($3 \times 10^{-3}\, M_\odot$) were so lost. Earth and Venus contain only about 10^{-4} their complement of C, and even lesser amounts of H_2O, N, and noble gases. Because the retained C appears to show the imprint of the Fischer-Tropsch reaction, it seems likely that the lost C, too, had been involved in this process. Of the 12 known interstellar molecules with 3 or more atoms, at least 10 are found in meteorites or in FTT syntheses (Anders, 1973): H_2O, HCHO, CH_3OH, HCOOH, HCN, HC≡CCN, HNOC, $CH_3C{\equiv}CH$, CH_3CN, and COS. This is not a compelling argument in favor of the MFT synthesis, because these structurally simple molecules can also be made by the Miller-Urey synthesis (Sagan, 1973). A choice may become possible when more complex molecules are discovered in interstellar space.

Acknowledgements

This work was supported in part by NASA Grant NGR 14-001-203 and the U.S. Atomic Energy Commission.

References

Anders, E.: 1971a, *Geochim. Cosmochim. Acta* **35**, 516.
Anders, E.: 1971b, *Ann. Rev. Astron. Astrophys.* **9**, 1.
Anders, E.: 1972, in H. Reeves (ed.), *L'Origine du Système Solaire*, CNRS, Paris, 179.
Anders, E.: 1973, in M. A. Gordon and L. Snyder (eds.), *Molecules in the Galactic Environment*, John Wiley and Sons, New York, 429.
Anders, E., Hayatsu, R., and Studier, M. H.: 1973, *Science* **182**, 721.
Anderson, R. B.: 1956, in P. H. Emmett (ed.), *Catalysis. IV: Hydrocarbon Synthesis, Hydrogenation and Cyclization*, Reinhold, New York, p. 29.
Bitz, M. C. and Nagy, B.: 1966, *Proc. Nat. Acad. Sci.* **56**, 1383.
Briggs, M. H. and Mamikunian, G.: 1963, *Space Sci. Rev.* **1**, 647.
Calvin, M.: 1969, *Chemical Evolution*, Clarendon, Oxford, Chapters 4–6.
Cameron, A. G. W.: 1973, in D. N. Schramm and W. D. Arnett (eds.), *Explosive Nucleosynthesis*, Univ. of Texas Press, Austin, p. 3.
Cameron, A. G. W. and Pine, M. R.: 1973, *Icarus* **18**, 377.
Clayton, R. N.: 1963, *Science* **140**, 192.
Cronin, J. R. and Moore, C. B.: 1971, *Science* **172**, 1327.
Dayhoff, M. O., Lippincott, E. R., and Eck, R. V.: 1964, *Science* **146**, 1461.
Emmett, P. H. (ed.): 1956, *Catalysis. IV: Hydrocarbon Synthesis, Hydrogenation and Cyclization*, Reinhold, New York.
Folsome, C. E., Lawless, J. G., Romiez, M., and Ponnamperuma, C.: 1973, *Geochim. Cosmochim. Acta* **37**, 455.
Galwey, A.: 1968, *J. Catalysis* **12**, 352.
Galwey, A.: 1972, *Geochim. Cosmochim. Acta* **36**, 1115.
Gelpi, E. and Oró, J.: 1970a, *Geochim. Cosmochim. Acta* **34**, 981.
Gelpi, E. and Oró, J.: 1970b, *Geochim. Cosmochim. Acta* **34**, 995.
Gelpi, E., Nooner, D. W., and Oró, J.; 1970, *Geochim. Cosmochim. Acta* **34**, 421.
Hayatsu, R.: 1964, *Science* **146**, 1291.
Hayatsu, R.: 1965, *Science* **149**, 443.
Hayatsu, R., Studier, M. H., Oda, A., Fuse, K., and Anders, E.: 1968, *Geochim. Cosmochim. Acta* **32**, 175.
Hayatsu, R., Studier, M. H., and Anders, E.: 1971, *Geochim. Cosmochim. Acta* **35**, 939.
Hayatsu, R., Studier, M. H., Matsuoka, S., and Anders, E.: 1972, *Geochim. Cosmochim. Acta* **36**, 555.
Hayes, J. M.: 1967, *Geochim. Cosmochim. Acta* **31**, 1395.
Hayes, J. M. and Biemann, K.: 1968, *Geochim. Cosmochim. Acta* **32**, 239.
Herbig, G. H.: 1970, *Mém. Soc. Roy. Sci. Liège* **XIX**, 13.
Hodgson, G. W. and Ponnamperuma, C.: 1968, *Proc. Nat. Acad. Sci.* **59**, 22.
Hodgson, G. W. and Baker, B. L.: 1969, *Geochim. Cosmochim. Acta* **33**, 943.
Khare, B. N. and Sagan, C.: 1973, *Icarus*, in press.

Kvenvolden, K., Lawless, J., Pering, K., Peterson, E., Flores, J., Ponnamperuma, C., Kaplan, I. R., and Moore, C.: 1970, *Nature* **228**, 923.
Kvenvolden, K., Lawless, J., and Ponnamperuma, C.: 1971, *Proc. Nat. Acad. Sci.* **68**, 486.
Lancet, M. S.: 1972, Ph. D. thesis, University of Chicago.
Lancet, M. S. and Anders, E.: 1970, *Science* **170**, 980.
Lancet, M. S. and Anders, E.: 1973, *Geochim. Cosmochim. Acta* **37**, 1371.
Nagy, B. and Bitz, M. C.: 1963, *Arch. Biochem. Biophys.* **101**, 240.
Nooner, D. W. and Oró, J.: 1967, *Geochim. Cosmochim, Acta* **31**, 1359.
Onuma, N., Clayton, R. N., and Mayeda, T. K.: 1972, *Geochim. Cosmochim. Acta* **36**, 169.
Oró, J.: 1972, *Space Life Sci.* **3**, 507.
Oró, J. and Han, J.: 1966, *Science* **153**, 1393.
Oró, J., Gibert, J., Lichtenstein, H., Wikstrom, S., and Flory, D. A.: 1971, *Nature* **230**, 105.
Ring, D., Wolman, Y., Friedmann, N., and Miller, S. L.: 1972, *Proc. Nat. Acad. Sci.* **69**, 765.
Sagan, C.: 1973, in M. A. Gordon and L. Snyder (eds.), *Molecules in the Galactic Environment*, John Wiley, New York, p. 451.
Smith, J. W. and Kaplan, I. R.: 1970, *Science* **167**, 1367.
Storch, H. H., Golumbic, N., and Anderson, R. B.: 1951, in *The Fischer-Tropsch and Related Syntheses*, John Wiley and Sons, New York, p. 9.
Studier, M. H., Hayatsu, R., and Anders, E.: 1965a, Enrico Fermi Institute preprint No. 65-115.
Studier, M. H., Hayatsu, R., and Anders, E.: 1965b, *Science* **149**, 1455.
Studier, M. H., Hayatsu, R., and Anders, E.: 1968, *Geochim. Cosmochim. Acta* **32**, 151.
Studier, M. H., Hayatsu, R., and Anders, E.: 1972, *Geochim. Cosmochim. Acta* **36**, 189.
Urey, H. C.: 1953, in *XIIIth Intern. Congr. Pure Applied Chem.* (Plenary Lectures), IUPAC, London, 188.
Vdovykin, G. P.: 1967, *Carbonaceous Matter of Meteorites (Organic Compounds, Diamonds, Graphite)*, Nauka Publishing Office, Moscow. English Translation NASA TT F-582.
Wolman, Y., Haverland, W. J., and Miller, S. L.: 1972, *Proc. Nat. Acad. Sci.* **69**, 809.
Yoshino, D., Hayatsu, R. and Anders, E.: 1971, *Geochim. Cosmochim. Acta* **35**, 972.

PART II

PALEOBIOLOGY

NATURAL EVIDENCE FOR CHEMICAL AND EARLY BIOLOGICAL EVOLUTION

KEITH A. KVENVOLDEN

Planetary Biology Division, Ames Research Center, NASA, Moffett Field, Calif. 94035, U.S.A.

Abstract. Meteorites, particularly type II carbonaceous chondrites, provide natural, tangible evidence for chemical evolution, but they do not appear to contain any evidence for biological evolution. On the other hand, some of the oldest sedimentary rocks of the earth have yielded good evidence for early biological evolution; whatever evidence there may be for chemical evolution in these old rocks is generally obscure.

Carbonaceous chondrites (types I, II, and III) have been examined for their content of various kinds of organic compounds. Amino acids have been reported to be present in the three types, but only in type II carbonaceous chondrites (Murray and Murchison) has an indigenous suite of amino acids been found which is apparently free of most terrestrial contaminations. These indigenous compounds are thought to have resulted from extraterrestrial, abiotic, chemical syntheses, and the presence of the amino acids in meteorites provides strong support for the theory of chemical evolution.

The geological record of the Swaziland Sequence and Bulawayan System of southern Africa contains morphological and chemical fossils which indicate that early biological evolution was taking place at least 3.0 to 3.3 aeons ago. Interpretation of the significance of the chemical fossil record has proven to be difficult. At present the occurrence of simple compounds in these very ancient rocks is believed to have little or nothing to do with biochemical processes three aeons ago. The bulk of the reduced carbonaceous material in these rocks, however, probably represents the residue of three billion years old and older organic matter. Isotopic studies of this carbonaceous material may provide chemical evidence for early biological evolution.

Modern concepts of chemical evolution were formulated in the 1920's when Oparin (1924) and Haldane (1929) independently hypothesized that life arose under reducing conditions through an evolutionary sequence of events involving increasingly complex organic substances which were synthesized and accumulated over long periods of time. Tests of these concepts generally have been confined to laboratory experiments wherein attempts have been made to simulate primitive, prebiotic environmental conditions. The first really successful test of the concept was reported by Miller (1953), who demonstrated the generation of amino acids resulting from the interaction of an electric discharge on a mixture of gases which was presumed to simulate the atmosphere of the primitive Earth. Numerous subsequent experiments over the last 20 yr have followed this same theme, and these have been succinctly reviewed by Lemmon (1970).

Besides the experimental results from laboratory simulation studies, naturally occurring evidence also has been observed which relates to concepts of chemical evolution. Except for that evidence found in meteorites, lunar samples and ancient rocks of the Earth, the majority of the evidence has come about through astronomical observations of the Sun, interstellar media, comets, and the planets (Oró, 1972). Although the astronomical information is important in any consideration of chemical evolution, this paper will focus on the natural evidence found in tangible samples which have been chemically analyzed in earth-bound laboratories. These samples include the extraterrestrial materials, meteorites, and the terrestrially occurring early Precambrian rocks.

1. Carbonaceous Meteorites

Before 1969, the only tangible samples available of extraterrestrial origin which could potentially relate to chemical evolution were meteorites, particularly carbonaceous chondrites. These substances, at first thought to represent some of the well-mixed rubble from the inner solar system, are now generally believed to be relatively unaltered condensate from the solar nebula (Anders, 1971). After 1969, extraterrestrial samples from several localities on the earth's Moon became available through the space programs of the United States and the Soviet Union. Whether or not any of these samples contain evidence related directly to the theory of chemical evolution is not yet clear. Although the number of extraterrestrial samples available for study has increased markedly as a result of the Apollo Program, it is still the carbonaceous meteorites that provide the best evidence with regard to chemical evolution.

That carbonaceous meteorites contain organic substances has been known for more than a century. Berzelius (1834) extracted complex organic substances from the Alais meteorite and wondered about the significance of his findings and their relationships to the possibility of extraterrestrial life. Since that time a number of studies of the organic chemistry of meteorites have been undertaken sporadically. Starting with the work of Mueller (1953) and Nagy *et al.* (1961) the field has gained momentum, and during the last twelve years significant advances have been made. Review articles and summaries of the work have become common (Briggs and Mamikunian, 1963; Mason, 1962–1963; Urey, 1966; Nagy, 1966; Hayes, 1967; Nagy, 1968; Baker, 1971; Ponnamperuma, 1972; Lawless *et al.*, 1972a; Oró, 1972). This paper will focus on the discoveries of amino acids in carbonaceous meteorites.

Amino acids, being components of proteins, seem to have received the most attention in considerations of chemical evolution. Ease of both synthesis and analysis probably has promoted this attention. Amino acids were first reported to be present in meteorites (Murray and Bruderheim) by Degens and Bajor (1962). The following amino acids were observed either free and/or combined: serine, glycine, alanine, leucine, aspartic acid glutamic acid, ornithine/arginine, threonine, lysine, histidine, valine, proline, tyrosine and phenylalanine. Most of these amino acids are known to be in proteins of living systems, but the authors cautiously avoided the interpretation that the source of these compounds was extraterrestrial life. Rather they suggested that the compounds were abiotic in origin or were terrestrial contamination or both.

1.1. Orgueil (a type I carbonaceous chondrite)

In a more detailed paper, Kaplan *et al.* (1963) assessed the amino acid content of eight carbonaceous chondrites (including Orgueil) and five non-carbonaceous chondrites. They reported, in addition to the compounds just listed, methionine and β-alanine. Anders *et al.* (1964) found that their analysis of amino acids in water extracts of Orgueil were in good agreement with the results of Kaplan *et al.* (1963). These compounds were interpreted to have resulted from abiotic chemical synthesis with a possible small overprint of terrestrial contamination. Certainly if these compounds did result from extraterrestrial chemical processes, their presence in meteorites would afford strong support for the theory of chemical evolution because most of the amino acid building blocks of life could be accounted for in this extraterrestrial synthesis. Detailed considerations, however, of the amino acid results by Hamilton (1965), Oró and Skewes (1965) and Hayes (1967) clearly showed that most of the amino acids in Orgueil were probably not indigenous but rather resulted from terrestrial contamination, particularly fingerprints. Only three compounds, glycine, phenylalanine, and β-alanine were in excess of amounts attributable to fingerprints.

Two recent analyses have begun to clarify the picture with regard to the amino acid content of Orgueil. By means of sophisticated gas chromatographic techniques, Oró *et al.* (1971b) generally confirmed and extended previous work by showing that an acid hydrolysate of Orgueil contained L-serine, L-leucine, β-alanine, L-isoleucine, L-threonine, glycine, L-valine, L-alanine and 22.5% D-alanine. The amino acids were thought to be a result of terrestrial contamination by microorganisms, with the D-alanine coming from bacterial cell walls. That the D-alanine could represent part of an abiotically-formed racemix mixture was not suggested because, if true, other amino acids should also have been present in the D isomeric configuration.

Additional work on another sample of Orgueil has shown that a hydrolyzed water extract contains both contaminating and indigenous, possibly abiotically synthesized amino acids (Lawless *et al.*, 1972b). In this report five protein amino acids were confirmed by gas chromatographic-mass spectrometric techniques: L-valine, L-alanine, glycine, L-proline, and L-aspartic acid. Also identified by the same techniques were α-aminoisobutyric acid, D-alanine, N-methylglycine, D- and L-β-aminoisobutyric acids, β-amino-*n*-butyric acid, β-alanine and γ-aminobutyric acid. D-proline and D-aspartic acid were indicated by gas chromatography, but were not confirmed by mass spectrometry, and β-aminoisobutyric acid appeared to be present as a racemic mixture. Clearly, β-alanine and glycine dominated the population of amino acids recovered. Contamination alone cannot likely account for the dominance of just two individual compounds. Therefore, some of the glycine and β-alanine appears to be indigenous to the Orgueil, a predication made earlier by Hayes (1967). Although most of the L-protein amino acids are likely contaminants, the other compounds probably are indigenous. The finding of D-alanine confirmed the earlier report of Oró *et al.* (1971b); but, in contrast to that report where D-alanine was considered a bacterial contaminant, the D-alanine was interpreted to be indigenous and present along with an equal amount of L-alanine as a racemix mixture.

Finding indigenous amino acids in Orgueil certainly supports ideas of chemical evolution, but the population of amino acids that can be recovered from this stone is limited. The amino acids that both can be identified with confidence and apparently are indigenous are, to a large extent, not the common building blocks of proteins.

1.2. Murray (a type II carbonaceous chondrite)

The history of analyses of Murray meteorite for amino acids parallels that of the Orgueil except the final outcome has been different. As mentioned earlier, Degens and Bajor (1962) found a suite of protein amino acids in Murray. This first report was followed up by Kaplan *et al.* (1963), who showed that the Murray contained a population of amino acids distributed similarly to those in Orgueil. Sixteen protein amino acids plus β-alanine were reported. As with the Orgueil, amino acids in Murray were thought to have been synthesized by extraterrestrial abiogenic processes. Analyses of Murray meteorite conducted eight years later showed that specimens of this chondrite contain a suite of amino acids very much different from that described earlier. Oró *et al.* (1971b), using gas chromatographic techniques, first reported that acid hydrolysates of Murray contain both D and L isomers of alanine, valine, isoleucine, proline and aspartic acid with the L-isomers dominant. Also reported were L-serine, L-leucine, β-alanine and glycine. A water extract of Murray contained a racemic mixture of alanine. The conclusion reached was that about 90% of the amino acids in Murray result from biological contamination and the remainder (both D and L configurations) are probably of chemical origin.

By ion exchange procedures, Cronin and Moore (1971) were able to show that Murray meteorite contains six protein amino acids (glycine, alanine, valine, proline, aspartic and glutamic acids) and at least five nonprotein amino acids (α-aminoisobutyric, β-amino-*n*-butyric, γ-aminobutyric acids and β-alanine and possibly isovaline). The suite of protein amino acids differs markedly from that reported earlier. The probable reason for the difference is that the samples used by Kaplan *et al.* (1963) were received as a crushed fine black powder which could easily have been contaminated, while the sample analyzed by Cronin and Moore (1971) was carefully removed from the interior of a large specimen.

Later the same year, Lawless *et al.* (1971), using mainly gas chromatography-mass spectrometry, found in Murray the same six protein amino acids reported by Cronin and Moore (1971) but also twelve non-protein amino acids including the five mentioned above. The seven additional non-protein amino acids were N-methylalanine, N-methylglycine, N-ethylglycine, norvaline, β-amino-*n*-butyric acid, pipecolic acid, and γ-aminobutyric acid. At least seven amino acids were present as racemic mixtures (alanine, valine, glutamic acid, proline, pipecolic acid, and α-amino-*n*-butyric and β-aminoisobutyric acids). The results on Murray are qualitatively similar to those reported earlier for the Murchison meteorite, another Type II carbonaceous chondrite (Kvenvolden *et al.*, 1970). Because of the presence of several amino acids not used in proteins, and because of the racemic nature of seven protein and non-protein amino acids, the conclusion was reached that these compounds are indigenous to the meteorite and may have been formed by an extraterrestrial chemical process.

1.3. Murchison (a Type II Carbonaceous Chondrite)

The fall of the Murchison meteorite in September of 1969 has provided extraterrestrial material which has characteristics consistent with an abiotic synthesis. A number of organic compounds have been found including the amino acids (Kvenvolden *et al.*, 1970). This meteorite fell at a particularly opportune time, for many geochemical laboratories were in the process of preparing for the study of another kind of extraterrestrial sample from the Moon. Contamination from sample handling and from laboratory procedures had been reduced considerably. Not only were laboratories well prepared for 'extraterrestrial geochemistry', but conditions at the time and place of the fall were conducive to minimize contamination (Lovering *et al.*, 1971). More than 80 kg of fragments have been collected; many specimens were picked up very soon after the fall. The short interval between time of fall and time of analysis also minimized the chance of contamination. Many fragments were of sufficient size so that interior pieces could be recovered for analysis. The presence of amino acids in the Murchison meteorite has been confirmed independently in three different laboratories (Kvenvolden *et al.*, 1970; Kvenvolden *et al.*, 1971; Cronin and Moore, 1971; Oró *et al.*, 1971a). Ion exchange, gas chromatographic and mass spectrometric techniques have been used and the results are in close agreement. These results "represent, taken together, the first solid evidence for the presence of amino acids in meteorites which are probably indigenous and abiogenic" (Oró, 1972).

Eighteen amino acids were first identified in Murchison (Kvenvolden *et al.*, 1970; Kvenvolden *et al.*, 1971), and these same compounds were found later in Murray (Lawless *et al.*, 1971) as mentioned before. Figure 1 shows an approximate distribution of amino acids. Although protein amino acids generally dominate the distribution

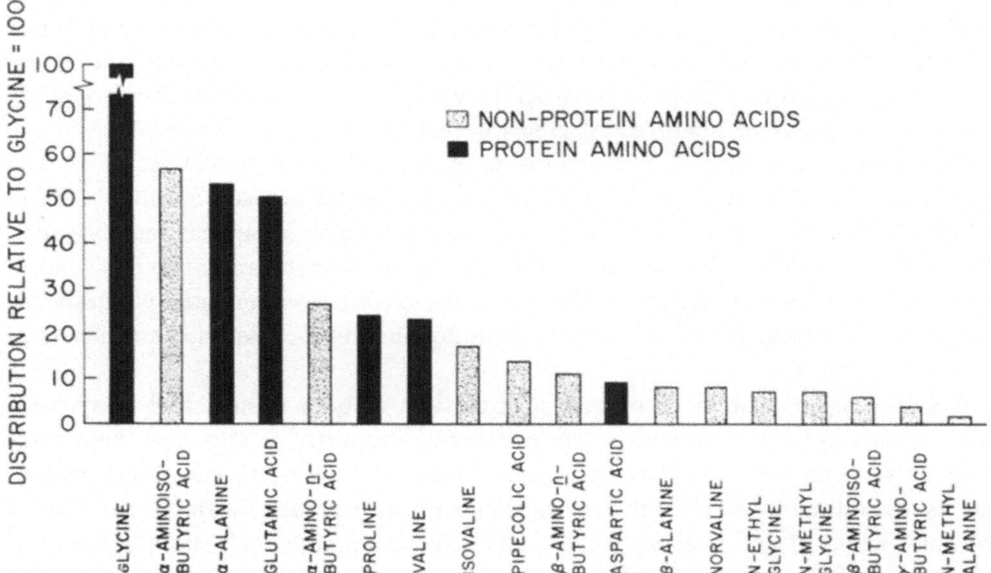

Fig. 1. Approximate distribution of amino acids in Murchison meteorite. Distribution is shown relative to glycine = 100.

in terms of concentrations, non-protein amino acids dominate in the number of different species present.

Continuing studies of the Murchison have shown that at least 35 amino acids are present (Lawless, 1973). Only six protein amino acids have been identified thus far: they are glycine (C_2), alanine (C_3), aspartic acid (C_4), and valine, proline, and glutamic acid (C_5). Although three isomers of leucine (C_6) have been characterized, mass spectrometric and gas chromatographic information to date do not permit any of these compounds be assigned the structures of the common protein amino acids, leucine and isoleucine. Examination of the structures of twenty-six naturally occurring protein amino acids (Morrison and Boyd, 1969) shows that there are one C_2, three C_3, three C_4, six C_5, seven C_6, four C_9, one C_{11}, and one C_{15} amino acids. Twenty of these compounds differ in structural detail and number of carbon atoms from the amino acids found in Murchison. For example, there are at least eight C_4 amino acids in this meteorite while in modern proteins there are only three. Table I compares by carbon number to C_6 the amino acids in Murchison and in modern proteins. If these meteorite amino acids represent products of an abiotic synthesis, and if a similar synthesis took place to create the original suite of amino acids from which life eventually evolved, then proto-proteinaceous material may have been characterized by compositions different from those found in today's proteins. Early evolution may have required that organisms synthesize many of the fundamental amino acid building blocks used now in proteins.

Amino acids in proteinaceous material of living organisms not only differ in certain structural details from amino acids in meteorites, but also differ in their stereochemistry. While protein amino acids are dominantly in the L configuration, meteorite amino acids appear to be present as racemic mixtures (equal concentrations of D and L forms). Evidence for racemic mixtures of most asymmetric amino acids in meteorites was provided by the gas chromatographic response of diastereomeric derivatives which show peaks of about equal area for each racemic pair (Kvenvolden *et al.*, 1971; Lawless *et al.*, 1971). The separation of diastereomeric derivatives of D- and L-α-amino-*n*-butyric acid and D- and L-α-alanine from Murchison are illustrated in Figure 2. The peak areas of the two diastereomeric pairs are approximately equal. This kind of chromatographic evidence helped assure that the amino acids found in Murchison were not terrestrial contaminants. A meteorite contaminated with modern terrestrial biological products would contain dominantly L-amino acids rather than racemic mixtures.

The finding in Murchison of such a large suite of both protein and nonprotein amino acids present as racemic mixtures strongly supports the idea that these compounds were created by abiotic processes. Since this discovery, laboratory experiments attempting to simulate the environments of the primitive Earth have produced complex suites of amino acids, including the eighteen amino acids initially found in Murchison (Ring *et al.*, 1972; Wolman *et al.*, 1972). Even the relative distribution of the amino acids from the simulation experiment and from the meteorite are strikingly similar. Indeed the amino acids found both in the Murchison and Murray meteorites seem to provide naturally occurring evidence in support of the theory of chemical evolution.

1.4. Allende (a type III carbonaceous chondrite)

The year of 1969 was a good one for falls of carbonaceous chondrites. Besides the

TABLE I

Comparison of amino acids to C_6 in the Murchison meteorite and in modern proteins

CARBON NUMBER		2	3	4	5	6
MURCHISON METEORITE	NEUTRAL AMINO ACIDS	GLYCINE	α–ALANINE	α–AMINOISOBUTYRIC ACID	VALINE	PIPECOLIC ACID
			β–ALANINE	α–AMINO–n–BUTYRIC ACID	ISOVALINE	2 ISOMERS OF PIPECOLIC ACID
			N–METHYLGLYCINE	β–AMINOISOBUTYRIC ACID	NORVALINE	
				β–AMINO–n–BUTYRIC ACID	5 OR 6 ISOMERS OF VALINE	
				γ–AMINOBUTYRIC ACID		3 ISOMERS OF LEUCINE
				N–METHYLALANINE	PROLINE	
				N–ETHYLGLYCINE	ISOMER OF PROLINE	
	ACIDIC AMINO ACIDS			ASPARTIC ACID	GLUTAMIC ACID	6 ACIDIC AMINO ACIDS
MODERN PROTEIN	NEUTRAL AMINO ACIDS	GLYCINE	α–ALANINE	THREONINE	PROLINE	CYSTINE
			CYSTEINE		HYDROXY-PROLINE	ISOLEUCINE
			SERINE		METHIONINE	LEUCINE
					VALINE	
	ACIDIC AND BASIC AMINO ACIDS			ASPARAGINE	GLUTAMINE	ARGININE
				ASPARTIC ACID	GLUTAMIC ACID	HISTIDINE
						HYDROXYLYSINE
						LYSINE

Murchison in September, Allende fell in February. The organic chemistry of Allende has received modest attention because early results suggested that the indigenous concentration of organic compounds within most of the specimens of this meteorite is very small. Han *et al.* (1969), collected and analyzed fragments of this meteorite immediately after the fall. They reported that hydrocarbons and fatty acids were confined to the exterior portions of the stone. Terrestrial biological contamination could readily account for the results they obtained. We also examined Allende for amino acids and found that samples containing surface materials showed amino acids (Figure 3), but interior samples yielded negative results, a finding equivalent to that of Cronin and Moore (1971). The amino acids found in surface material are common to proteins and, undoubtedly, represent terrestrial contamination. From the point of view of chemical evolution, Allende offers at present little in the way of support of the theory. Recently, Breger *et al.* (1972), reported finding formaldehyde in Allende. This result certainly has implication for chemical evolution; however, to date no confirmation of this work has been obtained by other investigators, and only with future studies will the significance of this discovery become clear.

Of the three types of carbonaceous chondrites that have been studied in detail, it is the type II specimens that seem to be richest in their content of amino acids,

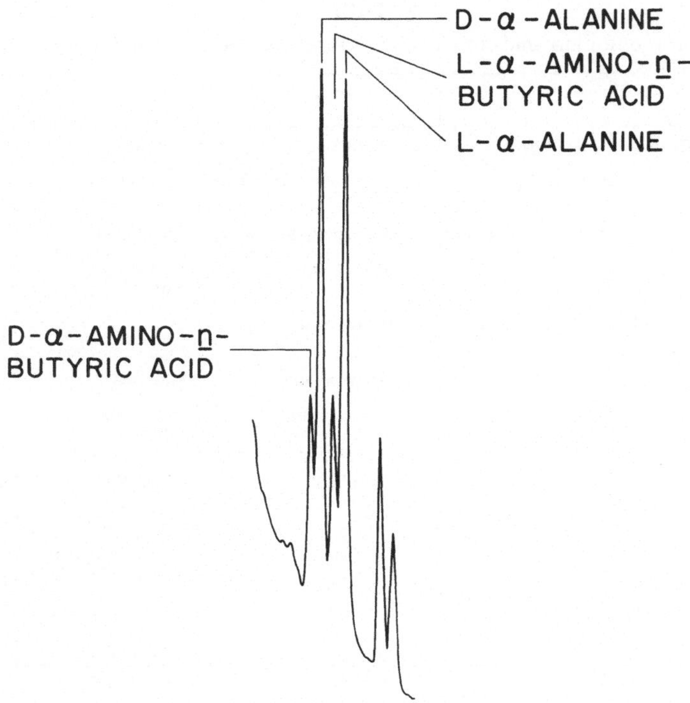

Fig. 2. Gas chromatogram showing the separation and resolution of N-trifluoroacetyl-(+)-2-butyl esters of α-amino-*n*-butyric acid and α-alanine from a hydrolyzed water extract of Murchison meteorite.

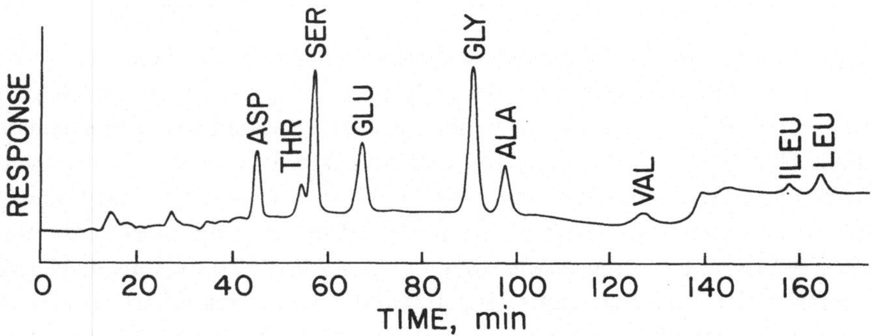

Fig. 3. Amino acid analyzer (ion exchange) chromatogram of amino acids in surficial material from Allende meteorite.

although type I carbonaceous chondrites contain the greatest concentration of organic carbon. This observation does not necessarily mean that the physical and chemical conditions experienced by type II chondrites is most conducive to the formation of specific molecules such as amino acids. Only the type I Orgueil has been examined in detail. This meteorite has been contaminated by terrestrial substances and has not been preserved under optimum conditions since the time of its fall in 1864.

Certainly alteration of the indigenous individual organic compounds could easily have taken place. Not until a new type I carbonaceous chondrite falls and is examined immediately will it be possible to compare the indigenous compounds in type I and type II specimens.

Although other classes of organic compounds, such as hydrocarbons, nucleic acid bases, fatty acids, porphyrins, etc., are found in meteorites, this discussion has been limited to only amino acids. This kind of molecule, being so dominant in living systems, is of great significance to the theory of chemical evolution as envisioned by Oparin and Haldane. The confirmed presence of a complex suite of amino acids in meteorites seems to show that extraterrestrial, abiotic chemical processes have been active sometime in the past. There seems to be no evidence that the amino acids found have anything to do with the occurrence of an extraterrestrial life. Consequently, meteorites provide naturally occurring evidence in support of chemical evolution, but not of biological evolution.

2. The Early Earth

Evidence supporting the concepts of chemical evolution is not readily deciphered from the rocks of the Earth. The inorganic chemistry of some of the early sedimentary rocks, however, support the idea that the atmosphere at the time these sediments were deposited was severely depleted in oxygen; that is, this early atmosphere was reduced and, therefore, in accord with the concepts of chemical evolution. For review of the environment of the primitive Earth, see Rutten (1971) and Cloud (1972). On the other hand, organic geochemical studies of early sedimentary rocks have not produced any positive evidence for the primitive organic materials that would have accumulated during the early history of the Earth as a consequence of chemical evolution. These studies, particularly on the early Precambrian Swaziland Sequence and Bulawayan System of southern Africa, have, however, provided evidence for very early biological evolution (for a review see Kvenvolden, 1972).

2.1. Carbon and Carbon Isotopes

Studies focusing on carbon and its isotopic abundance have given some possible clues to early Earth history. The organic carbon content of these ancient Precambrian sediments is variable. Schopf and Barghoorn (1967) pointed out that carbonaceous cherts and shales from the Fig Tree Group (including samples of the Swartkoppie Formation) contain about 0.5% organic matter. Elemental analyses show that samples from the Theespruit and Kromberg Formations contain about 1% total carbon (Scott *et al.*, 1970). Organic carbon determined by dry combustion methods on six samples including Fig Tree shale, Swartkoppie chert, Kromberg chert and Theespruit chert range from 0.2 to 1.02% (Kvenvolden, 1972). The organic carbon content of a sample of Bulawayan stromatolitic limestone is about 0.5% (Hoering, 1964).

Systematic determinations of the organic content of a number of Swaziland Sequence sediments have been determined by Dungworth and Schwartz (personal communication) and by Moore and Lewis (personal communication). The preliminary values, listed in order of increasing age, from these studies are given in Table II, along with the carbon isotopic composition of total carbon as determined by Oehler *et al.* (1972). Examination of the organic carbon content shows a few discrepancies between measurements on the same sample probably due to the use of samples from different individual specimens from the same localities. The most obvious conclusion from these data is that the lowermost and oldest specimens from the Onverwacht Group contain significantly higher concentrations of organic carbon than do the younger specimens in the Sequence except for the graphitic shale in the Swartkoppie Formation.

Before the systematic study by Oehler *et al.* (1972) on the carbon isotopic composition of organic matter in Precambrian rocks only a limited number of isotopic measurements had been made on the Swaziland Sequence and Bulawayan System samples (Kvenvolden, 1972). Hoering (1965) determined that the $\delta\,^{13}C_{PDB}$ for insoluble organic matter in Fig Tree shale was -28.3 per mil and for soluble organic matter -28.9 per mil. Later Brooks and Shaw (1972) obtained $\delta\,^{13}C$ values of -26.9 and -27.5 per mil for the insoluble and soluble organic material, respectively, in Fig Tree chert. A $\delta\,^{13}C$ value of -15 per mil for a sample from the Theespruit Formation had been determined by Silverman (personal communication) and recently, Brooks and Shaw (1972) reported $\delta\,^{13}C$ values for Onverwacht chert (presumably Theespruit Formation) of -15.8 and -24.2 per mil for insoluble and soluble organic matter, respectively. Carbon isotopic compositions of coexisting carbonates and reduced organic carbon in samples of stromatolitic limestones from the Bulawayan System were first determined by Hoering (1967) and later by Schopf *et al.* (1971).

In the study by Oehler *et al.* (1972) samples from sixteen different localities of the Swaziland Sequence were examined for their carbon isotopic compositions. The values obtained for total organic carbon fall into two distinct groups (Figure 4). Thirteen samples which are all stratigraphically above the Middle Marker ranged in $\delta\,^{13}C$ values from -26.1 to -33.0 per mil. Hoering's value of -28.3 per mil and Brooks and Shaw's value of -26.9 per mil for insoluble organic carbon in the Fig Tree Formation fall within this range. Three samples, all stratigraphically below the Middle Marker and confined to the Theespruit Formation, have organic carbon with $\delta\,^{13}C$ values ranging from -15.0 to -19.5 per mil. Silverman's determination of -15 per mil (personal communication) and Brooks and Shaw's value of 15.8 per mil fall in this same grouping. Although the $\delta\,^{13}C$ values of the soluble organic fractions in the three samples ranged from -25.5 to -26.2 per mil, the values for the kerogen fractions ranged from -14.3 to -18.9 per mil. The difference in isotopic compositions between the soluble material and the kerogen suggests the soluble organic substances were probably not derived from the kerogen which is believed to be syngenous. A possible conclusion is that the soluble material is younger than the kerogen and the rocks themselves.

Therefore, a carbon isotopic discontinuity of about 12 per mil seems to exist between organic matter in the Theespruit Formation and younger organic matter in the Swaziland Sequence. In fact, the organic matter in the Theespruit Formation is heavier isotopically than most organic matter in all younger sedimentary rocks up

TABLE II

Carbon content of the Swaziland Sequence

SAMPLE NO.	ORGANIC[1] CARBON %	ORGANIC[2] CARBON %	$\delta^{13}C_{PDB}$[3] (ORGANIC CARBON) ‰	FORMATION	DESCRIPTION
28	0.20	0.14	-28.0	FIG TREE GROUP	BLACK CHERT
21	0.17	0.04	-28.7	FIG TREE GROUP	FERRUGINOUS CHERT
27	0.28	0.21	-31.4	SWARTKOPPIE	BLACK CHERT
26	1.27	—	-26.9	SWARTKOPPIE	GRAPHITIC SHALE
25	0.12	—	—	SWARTKOPPIE	BANDED CHERT
24	0.18	—	—	SWARTKOPPIE	BLACK CHERT
23	0.45	0.65	—	SWARTKOPPIE	BLACK CHERT
22	0.04	—	-24.9	SWARTKOPPIE	OOLITIC CHERT
20	0.17	0.10	-30.6	KROMBERG	BANDED CHERT
19	0.62	0.74	-26.2	KROMBERG	BLACK CHERT
18	0.33	0.20	-26.1	KROMBERG	MASSIVE BLACK CHERT
16	0.12	—	—	KROMBERG	COARSE GRAINED CARBONATE
15	0.27	—	—	KROMBERG	BLACK CHERT
14	0.12	0.02	—	KROMBERG	BLACK CHERT
13	0.22	—	-27.5	KROMBERG	BLACK CHERT
12	0.61	0.23	-33.0	KROMBERG	BLACK CHERT
11	0.44	—	—	KROMBERG	SHALY BLACK CHERT
10	0.35	0.36	-32.5 (av)	HOOGGENOEG	BLACK CHERT
8	0.22	—	-28.4	HOOGGENOEG	THIN-BEDDED CARBONATE
7	0.23	0.17	-28.8	HOOGGENOEG	FRACTURED BLACK CHERT
6	1.16	1.15	-15.1	THEESPRUIT	MEDIUM-GRAINED BLACK CHERT
3	1.95	1.55	-15.0 (av)	THEESPRUIT	COARSE GRAINED BLACK CHERT
5	0.49	—	-19.5	THEESPRUIT	COARSE GRAINED BLACK CHERT

SAMPLES 1, 2, 9 and 29 ARE GRANITES; SAMPLES 4 AND 17 ARE TUFFS AND LAVAS RESPECTIVELY.

[1]MOORE AND LEWIS

[2]DUNGWORTH AND SWARTZ

[3]OEHLER ET AL, 1972

to recent time (Degens, 1969). Clearly, the carbon isotopic values of the Theespruit sample are anomalous. Brooks and Shaw (1972) suggested that the $\delta^{13}C$ value of -15.8 from the Onverwacht chert was heavy because of thermal alteration. Oehler *et al.* (1972) also considered the possibility that thermal metamorphism could account for a loss of ^{12}C thus producing an increase in $\delta^{13}C$. As a test of this idea, carbon

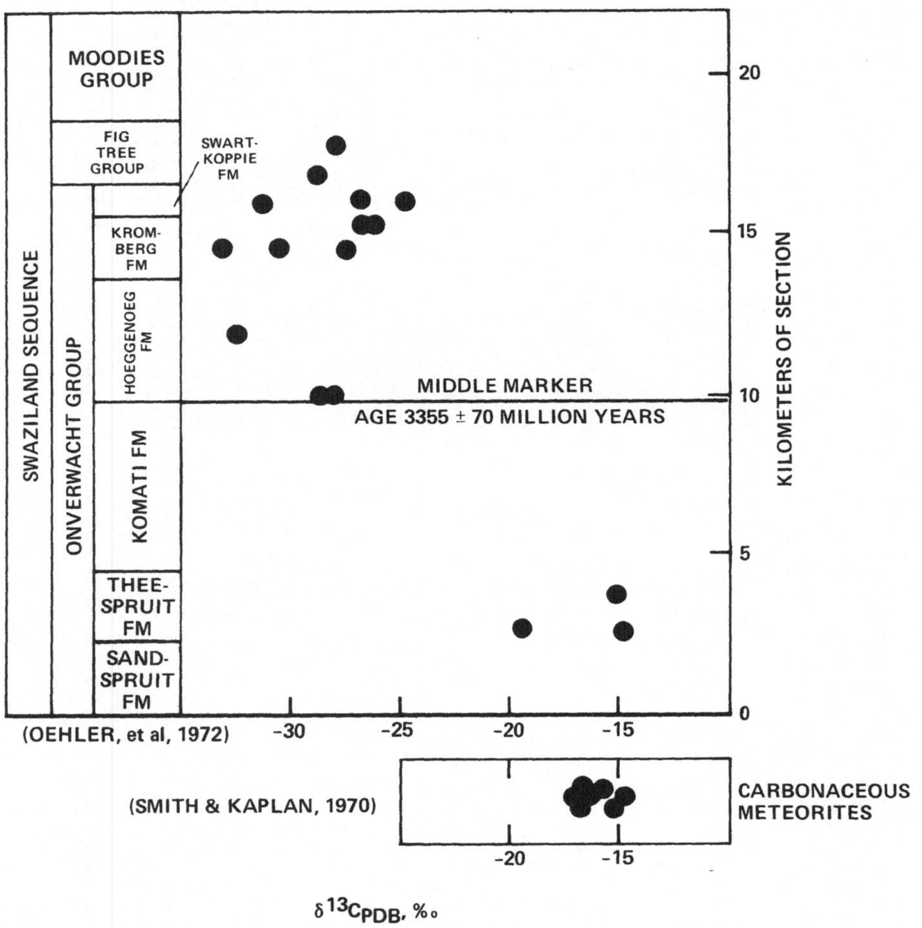

Fig. 4. Carbon isotopic values ($\delta^{13}C_{PDB}$) of organic carbon in Swaziland Sequence and carbonaceous meteorites.

isotopic composition of a sample from the Hooggenoeg Formation, which was collected adjacent to the Dalmein Spruit intrusive body, was measured. The $\delta^{13}C$ value was found to be -32.5 per mil, which does not differ significantly from values obtained from other samples collected above the Middle Marker. Although this sample does not test a regional metamorphic effect, it does provide a test case for local thermal metamorphism due to intrusion. The results suggest that the anomalously heavy carbon isotopic compositions of Theespruit samples are not necessarily caused by thermal effects.

Attention has been called (Oehler *et al.*, 1972) to the similarity between the isotopic composition of the organic carbon in the Theespruit Formation and in carbonaceous chondrites (Smith and Kaplan, 1970). Figure 4 displays this correlation. Such similarity could be coincidental or it could indicate that the Theespruit organic carbon may in part be the remains of abiologically-produced organic material. If this supposition is correct, then the Theespruit chert contains the first organic geochemical evidence yet found on Earth for the primitive organic materials which should have accumulated during the early history of the Earth as a consequence of chemical evolution. It was suggested (Oehler *et al.*, 1972) that the discontinuity in isotopic values may reflect some kind of break in the early carbon cycle; it may mark the time of origin of biochemical mechanisms capable of fractionating carbon isotopes in a manner similar to that of modern autotrophs. Isotopic values of biogenic organic matter in most sedimentary rocks of the entire known geologic record fall within the range of -20 to -40 per mil (Degens, 1969), and fossil organic material with values in this range have usually been interpreted as having been derived from photosynthetic or chemosynthetic organisms.

More confidence could be placed in the interpretation of the carbon isotopic abundances of the Swaziland Sequence sediments, if the lithologies of the cherts were similar throughout the entire section. Unfortunately, the cherts of the Theespruit Formation are distinctly different in appearance from cherts in the overlying formations. In the Theespruit Formation the chert is rather coarse grained and appears to be recrystallized. It occurs in pods and lenses, measuring up to 18 in. thick, in the tuffaceous beds. Barghoorn (personal communication) has pointed out that the Theespruit rocks should be called phyllites and that the carbon in these rocks is detrital. In the lowest Hooggenoeg rocks the chert is interbedded with carbonate and is microcrystalline and badly fractured. Localities higher in the section have medium to thick-bedded deposits of microcrystalline, black chert.

The $\delta^{13}C$ values of the organic carbon in Bulawayan System stromatolites also fall in the range that is generally attributed to products from photosynthesis. The average $\delta^{13}C$ value of the organic carbon is -32.1 per mil and the carbonate carbon has an average value of -0.1 per mil. Perry and Tan (1972) show a range of $\delta^{13}C$ values of -0.3 to $+0.7$ per mil for five carbonate samples from the Bulawayan Group. The average of $+0.1$ per mil is not greatly different from the average value of -0.1 per mil reported by Schopf *et al.* (1971). For the Bulawayan stromatolites the carbon isotopic fractionation between carbonate and organic carbon has been suggested to result from the photosynthetic processes of primitive blue-green algae (Schopf *et al.*, 1971).

Interpretation of the carbon isotopic record of the Precambrian is difficult. The isotopic composition of organic material buried in most Precambrian sediments (except the Theespruit Formation) is in the same range as organic material found in sedimentary rocks which were deposited during the entire Phanerozoic aeon. Because higher plant photosynthesis fractionates carbon isotopes (Park and Epstein, 1960) to a greater extent than does algal photosynthesis (Behrens and Frishman, 1971; Calder and Parker, 1973) the carbon isotopic record of the Precambrian, where algae and bacteria dominated, should be slightly different from the post-Precambrian record where contributions from algae, bacteria, and higher plants would be expected.

Studies of the modern geologic record where blue-green algae contribute significantly to the organic material in sediments have shown that the difference between the $\delta^{13}C$ values of coexisting organic carbon and carbonate carbon range between -16 and -14 per mil (Behrens and Frishman, 1971). For the Precambrian record this difference between reduced and oxidized carbon is much greater. For example, Hoering (1967) reports differences of -21 to -31 per mil for Precambrian stromatolites; Schopf *et al.* (1971) showed an average difference of about -32 per mil for Bulawayan stromatolites. With the exception of Theespruit rocks differences between reduced and oxidized carbon in the Fig Tree and Onverwacht sediments are more negative than about -21 per mil. The observed enrichment in ^{12}C is in accord with the enrichment observed in modern, higher plant photosynthesis not as observed in algal photosynthesis. Only the carbon in the Theespruit Formation has isotopic compositions in the same range as carbon found in modern algal mats. Clearly, the $\delta^{13}C$ values of algal-derived organic material from modern times are significantly different from values obtained from Precambrian organic materials presumably derived from algal precursors. Calder and Parker (1973) have attempted to explain this apparent inconsistency in carbon isotopic data by suggesting that special environmental factors in the Precambrian may have resulted in the greater enrichment of ^{12}C in reduced carbon. They have shown that the amount of photosynthetic fractionation of carbon isotopes by blue-green algae in laboratory cultures depends upon the concentration of CO_2 in the atmosphere. Minimum fractionation occurred at a CO_2 concentration of 0.2%. At higher and lower CO_2 concentrations the fractionation increased. Maximum fractionation of 18 per mil was observed in the laboratory cultures at a CO_2 concentration of 3%, but these experiments did not yield isotopic composition comparable to those in Precambrian sediments. Earlier, Abelson and Hoering (1961) observed, however, that a culture of the green alga, *Chlorella pyrenoidosa*, grown with vigorous aeration of a 5% CO_2 mixture exhibited a fractionation of -25.8 per mil. Indeed, increased CO_2 pressures appear to cause algae to fractionate carbon isotopes to an extent at least approaching values observed in the Precambrian record. Possibly the carbon isotopic record of the Precambrian is providing clues to evolutionary changes in the composition of the early atmosphere of earth. Perhaps the CO_2 concentration was very much higher in Precambrian times than at present. Because of this higher CO_2 concentration the primitive algae may have fractionated carbon isotopes to a much greater extent than they do today.

The idea of higher CO_2 concentrations in the early Precambrian atmosphere is certainly not new. For example, Rutten (1971) suggested, based on various geologic considerations, that the early atmosphere had a CO_2 concentration about 10 times the present atmospheric level or about 0.3% by volume. The consequences of such a CO_2 enriched atmosphere would have led to a lack of carbonate deposition (Mason, 1966), and, interestingly, carbonate deposits are rare in the early Precambrian record. The first distinguished carbonate bed known is the 20-ft thick Middle Marker at the base of the Hooggenoeg Formation which has an age of about 3,360 m.y. (Hurley *et al.*, 1972). Higher in the Swaziland Sequence section, carbonates form only a minor part of the lithologies present.

In this discussion it has been assumed that the biochemical mechanisms for algal photosynthesis have not changed or evolved greatly since Precambrian times. That is, it has been assumed that if Precambrian algae were living today they would fractionate carbon isotopes to the same extent as today's algae given the same conditions. That the biochemical mechanisms then and now were different is certainly a possibility, but there is no obvious direct or indirect test for such an hypothesis.

An alternative explanation for the isotopic compositions of early Precambrian organic material involves the possibility of enrichment of ^{12}C or depletion of ^{13}C due to diagenetic processes operating on the algal and bacterial debris. Indeed, preliminary work by Oehler and Schopf (personal communication) suggest that the isotopic composition of blue-green algae can be altered during laboratory simulations of diagenesis, but this work is as yet incomplete.

It seems certain now that within the Swaziland Sequence and Bulawayan System is a record of very early life on Earth. Clues to this early life have been provided by morphological and chemical fossils. Although the significance of soluble organic materials in these rocks is yet uncertain (Kvenvolden, 1972), the insoluble organic carbon appears to be potentially valuable in the interpretation of the early carbon cycle. Carbon isotopic studies in particular have yielded data for a number of speculations regarding earliest life, the origin of autotrophy and the composition of the early atmosphere.

References

Abelson, P. H. and Hoering, T. C.: 1961, *Proc. Nat. Acad. Sci. (U.S.)* **47**, 623.

Anders, E.: 1971, *Ann. Rev. Astron. Astrophys.* **9**, 1.

Anders, E., DuFresne, E. R., Hayatsu, R., Cavaille, A., DuFresne, A., and Fitch, F. W.: 1964, *Science* **146**, 1157.

Baker, B. L.: 1971, *Space Life Sci.* **2**, 472.

Behrens, E. S. and Frishman, S. A.: 1971, *J. Geol.* **79**, 94.

Berzelius, J. J.: 1834, *Ann. Phys. Chem.* **33**, 113.

Breger, I. A., Zubovic, P., and Chandler, J. C.: 1972; *Nature* **236**, 155.

Briggs, M. H. and Mamikunian, G.: 1963, *Space Sci. Rev.* **1**, 647.

Brooks, J. and Shaw, G.: 1972, *Chemical Geol.* **10**, 69.

Calder, J. A. and Parker, P. L.: 1973, *Geochim. Cosmochim. Acta* **37**, 133.

Cloud, P. E., Jr.: 1972, *Am. J. Sci.* **272**, 537.

Cronin, J. R. and Moore, C. B.: 1971, *Science* **172**, 1327.

Degens, E. T.: 1969, in G. Eglinton and M. T. J. Murphy (eds.), *Organic Geochemistry, Methods and Results*, Springer-Verlag, Berlin, pp. 304–329.

Degens, E. T. and Bajor, M.: 1962, *Naturwissenschaften* **49**, 605.

Haldane, J. B. S.: 1929, *Rationalist Annual; (1933) Science and Human Life*, Harper Bros., N.Y., p. 149.

Hamilton, P. B.: 1965, *Nature* **205**, 284.

Han, J., Simoneit, B., Burlingame, A. L., and Calvin, M.: 1969, *Nature* **222**, 364.
Hayes, J. M.: 1967, *Geochim. Cosmochim. Acta* **31**, 1395.
Hoering, T. C.: 1964, *Carnegie Inst. Yearbook* **63**, 258.
Hoering, T. C.: 1965, *Carnegie Inst. Yearbook* **64**, 215.
Hoering, T. C.: 1967, in P. H. Abelson (ed.), *Researches in Geochemistry 2*, John Wiley and Sons, Inc., N.Y., pp. 87–111.
Hurley, P. M., Pinson, W. H., Jr., Nagy, B., and Teska, T. M.: 1972, *Earth Planetary Sci. Letters* **14**, 360.
Kaplan, I. R., Degens, E. T., and Reuter, J. H.: 1963, *Geochim. Cosmochim. Acta.* **27**, 805.
Kvenvolden, K. A.: 1972, 'Organic Geochemistry of Early Precambrian Sediments', presented at *24th International Geological Congress*, Section 1, pp. 31–41.
Kvenvolden, K., Lawless, J., Pering, K., Peterson, E., Flores, J., Ponnamperuma, C., Kaplan, I. R., and Moore, C.: 1970, *Nature* **228**, 923.
Kvenvolden, K. A., Lawless, J. G., and Ponnamperuma, C.: 1971, *Proc. Nat. Acad. Sci. (U.S.)* **68**, 486.
Lawless, J. G., Kvenvolden, K. A., Peterson, E., Ponnamperuma, C., and Moore, C.: 1971, *Science* **173**, 626.
Lawless, J. G., Folsome, C. E., and Kvenvolden, K. A.: 1972a, *Sci. Am.* **226**, 38.
Lawless, J. G., Kvenvolden, K. A., Peterson, E., Ponnamperuma, C., and Jarosewich, E.: 1972b, *Nature* **236**, 66.
Lawless, J. G.: 1973, *Geochim. Cosmochim. Acta* **37**, 2207.
Lemmon, R. M.: 1970, *Chem. Rev. Am. Chem. Soc.* **70**, 95.
Lovering, J. F., Le Maitre, R. W., and Chappell, B. W.: 1971, *Nature Phys. Sci.* **230**, 18.
Mason, B.: 1962–1963, *Space Sci. Rev.* **1**, 621.
Mason, B.: 1966, *Principles of Geochemistry* (3rd. ed.), John Wiley and Sons, N.Y., p. 329.
Miller, S. L.: 1953, *Science* **117**, 528.
Morrison, R. T. and Boyd, R. N.: 1969, *Organic Chemistry* (2nd. ed.), Allyn and Bacon, Inc., Boston, 1204 p.
Mueller, G.: 1953, *Geochim. Cosmochim. Acta* **4**, 1.
Nagy, B., Meinschein, W. G., and Hennessy, D. J.: 1961, *Ann. N.Y. Acad. Sci.* **93**, 25.
Nagy, B.: 1966, *Geol. Foren. Stockholm Forhand.* **88**, 235.
Nagy, B.: 1968, *Endeavour* **27**, 81.
Oehler, D. Z., Schopf, J. W., and Kvenvolden, K. A.: 1972, *Science* **175**, 1246.
Oparin, A. I.: 1924, *Proiskhozhdenie zhizni*, Izd. Moskovskii Rabochii, Moscow.
Oró, J.: 1972, *Space Life Sci.* **3**, 507.
Oró, J. and Skewes, H. B.: 1965, *Nature* **207**, 1042.
Oró, J., Gilbert, J., Lichtenstein, H., Wikstrom, S., and Flory, D. A.: 1971a, *Nature* **230**, 105.
Oró, J., Nakaparksin, S., Lichenstein, H., and Gil-Av, E.: 1971b, *Nature* **230**, 107.
Park, R. and Epstein, S.: 1960, *Geochim. Cosmochim. Acta* **21**, 110.
Perry, E. C., Jr. and Tan, F. C.: 1972, *Geol. Soc. Am. Bull.* **83**, 647.
Ponnamperuma, C.: 1972, *Ann. N.Y. Acad. Sci.* **194**, 56.
Ring, D., Wolman, Y., Friedmann, N., and Miller, S. L.: 1972, *Proc. Nat. Acad. Sci. (U.S.)* **69**, 765.
Rutten, M. G.: 1971, *The Origin of Life by Natural Causes*, Elsevier, Amsterdam, 420 p.
Schopf, J. W. and Barghoorn, E. S.: 1967, *Science* **156**, 508.
Schopf, J. W., Oehler, D. Z., Horodyski, R. J., and Kvenvolden, K. A.: 1971, *J. Paleont.* **45**, 477.
Scott, W. M., Modzeleski, V. E., and Nagy, B.: 1970, *Nature* **225**, 1129.
Smith, J. W. and Kaplan, I. R.: 1970, *Science* **167**, 1367.
Urey, H. C.: 1966, *Science* **151**, 157.
Wolman, Y., Haverland, W. J., and Miller, S. L.: 1972, *Proc. Nat. Acad. Sci. (U.S.)* **69**, 809.

ASPECTS OF THE GEOLOGIC HISTORY OF SEAWATER

HEINRICH D. HOLLAND
Dept. of Geology, Harvard University, Cambridge, Mass., U.S.A.

It would be marvelous if I could unfold before you a grand panorama of the chemistry of the earliest oceans; unfortunately I cannot. We now know a good deal about the limits to the chemistry of the oceans during the last 600 m.y., and about the mechanisms that have kept the composition of ocean water within its rather narrow bounds. We also know something, but a good deal less, about ocean water chemistry during the preceding 2500 m.y.; but there are still no compelling geologic data bearing on the chemistry of ocean water during the first 1000 m.y. of Earth history. There are, however, some interesting alternatives; I would like to explore these, because they do set at least some limits on the chemistry of the most likely incubating medium for the earliest organisms.

The residence times of the major cations and anions in modern seawater are all shorter than 100 m.y.; most of the residence times are shorter than 10 m.y., and one of the major anions: bicarbonate, has a residence time of only 100000 yr. The composition of seawater is therefore potentially quite variable on a geologically short time scale.

However, all the current evidence indicates that the composition of ocean water has been surprisingly constant during the Phanerozoic Era, that is, during the past 600 m.y. This constancy was implicit in Sillén's (1961) equilibrium model for ocean water and in the numerous subsequent models which amplified Sillén's original theme. More recent studies of the rate of reaction between seawater and silicate minerals have cast considerable doubt on the efficacy of his original buffer mechanisms. There is still uncertainty even regarding the actual reactions by which seawater manages to dispose of the river flux of dissolved salts. The Sillén model per se can therefore contribute only modestly to defining the history of seawater.

At present, marine evaporites probably supply the most compelling evidence regarding the past chemistry of the oceans. Marine evaporite sequences typically start with a carbonate section, pass into a gypsum-anhydrite section, and thence into a halite section; in those marine evaporites which contain the last salts to crystallize from seawater, chlorides and sulfates of potassium and magnesium predominate. Many of the common minerals in non-marine evaporites are conspicuously absent in marine evaporites. This is particularly true of the carbonates, bicarbonates and sulfates of sodium. The mineralogy of all Phanerozoic marine evaporites can be explained in terms of the evaporation of present-day seawater; they therefore demand no change in the composition of seawater during the Phanerozoic, and set limits on possible excursions of seawater chemistry during this Era (Holland, 1972). Whether these limits are considered severe or not depends somewhat on one's perspective.

Variations of a factor of three in either direction for the concentration of the major cations and anions are allowed. The compositional range covered by the permitted variability is therefore quite considerable, but it is small compared to the compositional range permitted by the rather short residence time of the major cations and anions in modern seawater.

Unfortunately, evaporites older than 1000 m.y. are not known, and the record between 600 m.y. and 1000 m.y. ago is very incomplete. It seems likely that this is due to preferential solution of marine evaporites rather than to their scarcity or absence in Precambrian sediments at their time of deposition. In the absence of older marine evaporites we have to fall back on evidence that can be garnered from the mineralogy and chemistry of other marine sediments, and on inferences from the behavior of the ocean-atmosphere system during the Phanerozoic Era. Differences between Phanerozoic marine sediments and those of the preceding 2500 m.y. are rather slight; they can be explained by a lower oxygen pressure and a somewhat higher CO_2 pressure in the atmosphere, and by a higher concentration of SiO_2 in seawater. A lower oxygen pressure in the atmosphere would account for the apparent persistence of minerals that are rapidly destroyed by oxidation during weathering; it would also help to explain the origin of the enormous Precambrian banded iron formations which seem to demand that the fluviatile and marine geochemistry of iron during the early and mid-Precambrian was different from that of iron today (Holland, 1973; Garrels *et al.*, 1973). In Precambrian banded iron formations iron minerals alternate with chert; the presence of the chert is most easily accounted for by a concentration of dissolved silica near 100 ppm in Precambrian ocean water. Such a higher silica concentration is a reasonable consequence of the absence of marine organisms with siliceous shells in the Precambrian. Perhaps it would be more logical to say that the low concentration of dissolved silica in seawater today is a reasonable consequence of the efficient removal of silica by marine organisms, largely by radiolarians and diatoms.

If the current view of silicate-carbonate relationships (Holland, 1965; 1968) is correct, a high silica concentration in seawater demands a somewhat higher atmospheric CO_2 pressure. A higher CO_2 pressure would also help to explain the efficiency of chemical weathering during the Precambrian despite the absence of a cover of higher land plants, and may help to explain the absence of an abnormal number of ice ages despite a proposed lower solar constant.

Just how low the pressure of atmospheric oxygen and how high the pressure of atmospheric CO_2 was during early to mid-Precambrian time is not certain. Recent experiments by Grandstaff (in preparation) indicate that an oxygen pressure less than 0.02 P.A.L., that is less than 4 mbar, was probably required to allow the mineral uraninite to survive weathering, transport, and incorporation into Precambrian sediments.

CO_2 pressures may have been considerably higher than at present, although values in excess of 0.1 atm seem unlikely. Limestones are found in some of the earliest sedimentary rocks, notably those in the Bulawayan series, which is thought to be slightly more than 3000 m.y. old. Near-surface seawater must have been saturated or somewhat supersaturated with respect to calcite at that time and ever since. The pH of seawater has therefore been determined at least roughly by the reaction

$$CaCO_3 + 2H^+ \rightarrow Ca^{+2} + CO_2 + H_2O.$$

Even if the course of the CO_2 pressure in the Precambrian atmosphere were well known, the pH of seawater could be calculated only if the activity of Ca^{+2} or of HCO_3^- were known. They are not; all we can do for the present is to surmise that the pH of early Precambrian seawater was probably somewhat but not very much lower than its present value near 8.0.

Finally we have now worked ourselves back to the oceans at the time that is of particular interest to this conference. We still have no sedimentary rocks older than approximately 3400 m.y. and are therefore only lightly restrained by the burden of facts. The major variables in designing reasonable alternatives for the nature of the earliest oceans are surely the thermal history of the Earth and the oxidation state of the atmosphere at the end of the accretion process. It still seems possible to propose an Earth that was either cold, cool, warm, or hot at the end of the accretion process; a case can be made for a highly reduced initial oxidation state for the atmosphere-ocean system but also for one that is not very different from that of Bulawayan time some 3100 m.y. ago. On the whole, however, the evidence presently speaks for high temperatures and highly reducing conditions at the end of the accretion process. Early melting in the parent bodies of meteorites, and early extensive differentiation of the Moon are strong arguments in favor of melting temperatures at and close to the surface of the Earth during and shortly after accretion. The highly reduced state of the lunar basalts suggests a similar oxidation state for the earliest terrestrial basalts; such basalts would have been accompanied by much more reducing volcanic gases than those that have accompanied basaltic volcanism during the past 2500 m.y.

With the data of Speidel and Nafziger (1968) on the oxygen fugacity in gases equilibrated with mineral assemblages and melts in the system FeO–MgO–SiO_2, one can calculate the composition of volcanic gases at pressures up to 15 kbar in equilibrium with highly reduced basalts. These calculations show that in equilibrium with olivine, silica, pyroxene, and iron the H_2/H_2O ratio in volcanic gases is on the order of unity, and that CH_4, CO, and CO_2 will all be present. On cooling to 25 °C H_2 will tend to react with CO and CO_2, and the final mixture will tend to consist of H_2 and CH_4 equilibrated with liquid water. It is hard to say whether these reactions will go to completion or whether CO and CO_2 would be removed from the atmosphere by one or more of many possible alternative reactions. It is also hard to say how rapidly H_2 would have escaped from such a reducing atmosphere. Exospheric temperatures in CH_4–H_2 atmospheres may have been sufficiently low to stabilize a reducing atmosphere for a period of several hundred million years. If so, the atmosphere could have evolved at a geologically slow rate from a highly reducing state to the CO_2-rich, near-neutral state that we have inferred for the period between 2000 and 3400 m.y. ago.*

The consequences of a highly reducing atmosphere for the chemistry of an asso-

* See note added in proof (p. 91).

ciated ocean are rather interesting. The major weathering acids were probably volcanic HCl and H_2S. Neutralization of volcanic HCl with surface rocks should have been geologically rapid, and ocean water was probably never more than slightly acid.

In many ways the role of H_2S as a weathering agent would have been similar to that of CO_2 today. The first ionization constant of H_2S is only slightly smaller than that of carbonic acid, the solubility of H_2S in water at 25°C somewhat greater than that of CO_2. River waters would have been dilute bisulfide solutions of the same major cations as those in modern streams: calcium, magnesium, sodium, and potassium. The iron sulfides are so insoluble that the concentration of iron in rivers would have been very small. The oceans would have tended toward a steady state in which the rate of influx of dissolved salts was equal to their rate of removal. Clay minerals and zeolites are the most likely authigenic minerals forming in such an ocean.

A large fraction of the H_2S was probably recycled from the ocean to the atmosphere much as CO_2 is recycled today. The pressure of H_2S in the atmosphere and the concentration of H_2S and HS^- in ocean water are hard to estimate. P_{H_2S} could have been as large as several tenths of an atmosphere; but it was probably much smaller, both because sulfur must have been removed continuously by the formation of iron sulfides, and because at a steady state oceanic pH between 8 and 10 much of the sulfur in the ocean-atmosphere system was probably present as HS^- dissolved in ocean water.

Nitrogen was probably present in large part as molecular nitrogen in the atmosphere and as ammonium ion in seawater. Again, convincing estimates of the NH_4^+ concentration in seawater are impossible to make, but the permitted range includes the concentrations demanded by the stability requirements of essential amino acids.

The availability of phosphorus in such an ocean is severely limited by the low solubility of calcium phosphates. The presently available thermochemical data for phosphine and for the phosphorus halides suggest that none of these compounds were present in volcanic gases in more than trace concentrations. On the other hand the ablation of phosphides in iron meteorites during passage through a reducing atmosphere could have produced enough phosphine to serve as an important starting material in the synthesis of organic phosphorus compounds.

At least some of these organic compounds were probably synthesized in the atmosphere. A methane-hydrogen atmosphere could have been polymerized by solar ultraviolet radiation in times on the order of 10^7 to 10^8 yr (Lasaga *et al.*, 1971). The products would have been largely saturated hydrocarbons unless the hydrogen/methane ratio was less than 10^{-4}, and it is possible that the high molecular weight fraction of such polymerization processes formed what might be termed a primordial 'oil slick' on the early oceans.

Sillén (1965) and others have pointed out that during the waning stages of the reducing stage the replacement of methane by CO_2 in the atmosphere and the replacement of bisulfide by sulfate and bicarbonate in the oceans takes place almost simultaneously when the hydrogen pressure has dropped to about 10^{-5} atm. The development of limestones in sedimentary rocks somewhat more than 3000 m.y. old is a strong indication that the reducing stage had passed by that time. It is perhaps not unreasonable to hope for the discovery of sedimentary rocks from an earlier period of Earth history and for the discovery of really substantial evidence regarding the chemistry of the atmophere-ocean-system before life began.

Note added in proof. Recent calculations by D. Hunden suggest that H_2 escape was probably geologically rapid.

References

Garrels, R. M., Perry, E. A., Jr., and Mackenzie, F. T.: 1973, *Econ. Geol.*

Holland, H. D.: 1965, *Proc. Nat. Acad. Sci.* **53**, 1173.

Holland, H. D.: 1968, in L. H. Ahrens (ed.), *Origin and Distribution of the Elements*, Pergamon Press, Oxford and New York, p. 949.

Holland, H. D.: 1972, *Geochim. Cosmochim. Acta* **36**, 637.

Holland, H. D.: 1973, *Econ. Geol.* **68**, 1169.

Lasaga, A. C., Holland, H. D., and Dwyer, M. J.: 1971, *Science* **174**, 53.

Sillén, L. G.: 1961, in M. Sears (ed.), *Oceanography*, A.A.A.S. Publ. 67, 549.

Sillén, L. G.: 1965, *Arkiv Chemi* **25**, 159.

Speidel, D. H. and Nafziger, R. H.: 1968, *Am. J. Sci.* **266**, 361.

HOMEOSTATIC TENDENCIES OF THE EARTH'S ATMOSPHERE

JAMES E. LOVELOCK
University of Reading, Dept. of Applied Physical Science, Reading, England

and

LYNN MARGULIS
Dept. of Biology, Boston University, Boston, Mass., U.S.A.

Abstract. The atmosphere of the earth differs greatly from that of the other terrestrial planets with respect to composition, acidity, redox potential and temperature history predicted from solar luminosity. From the fossil record it can be deduced that stable optimal conditions for the biosphere have prevailed for thousands of millions of years. We believe that these properties of the terrestrial atmosphere are best interpreted as evidence of homeostasis on a planetary scale maintained by life on the surface. Some possible mechanisms of this biological homeostasis have been noted and the implications of this concept for experimental studies pointed out.

The purpose of this paper is to develop the concept that the atmosphere of the Earth flows in a closed system controlled by and for the biosphere. We view the Earth as a planet whose surface physical and chemical state is in homeostasis at an optimum set by the contemporary biota and reexamine in this new context some questions on the past and present condition of the Earth. These questions arise from the observation that the Earth, unlike the other terrestrial planets, has been evading the laws of equilibrium thermodynamics for millions of years. For example, given the temperature and pressure and the amounts of oxygen in the Earth's atmosphere "one can calculate what the thermodynamic equilibrium abundance of methane ought to be.... The answer turns out to be less than one part in 10^{36}. This is then a discrepancy between theory and observation of at least 30 orders of magnitude and cannot be dismissed lightly." (Sagan, 1970)

Surface life on the planet as a whole has certain environmental requirements of temperature, acidity and nutrient elements. The atmosphere of the Earth is anomalous with respect to its predicted temperature, acidity and the presence of certain elements in gaseous form; conditions on the Earth are not those expected for a planet interpolated between Venus and Mars. Conditions, at least at the 'core' (the tropical and temperate regions) are skewed from their predicted values in directions favored by most species of organisms. Such anomalies in the Earth's atmosphere have persisted for times that are very long relative to the residence times of non-noble gases in the atmosphere. We argue that it is unlikely that chance alone accounts for the fact that temperature, pH and the presence of compounds of nutrient elements have been, for immense periods of time, just those optimal for surface life. Rather we present the 'Gaia hypothesis'* the idea that energy is expended by the biota to actively maintain

* 'Gaia' is taken from the Greek for 'Mother Earth', see Lovelock and Lodge (1972) and Margulis and Lovelock (1974).

these optima. This hypothesis generates numerous experiments in the search for biological mechanisms to maintain homeostasis, as noted.

1. Environmental Factors Delimiting the Biosphere

Ultimately, the organic compounds of life depend on the absorption of visible solar radiation for their production. Life thrives in the temperature range of 20–28 °C, under neutral or slightly alkaline conditions, and only with an assured supply of the elements carbon, nitrogen, oxygen, hydrogen, phosphorus, sulfur, calcium, potassium, sodium, magnesium, chlorine, and about a dozen others required in trace quantities. Active growing replicating organisms always require water. The external environment may vary drastically from optimal conditions of temperature and ionic concentrations in aqueous solution, yet optimal conditions are maintained within the interiors of cells. Desert organisms do not live in the absence of water, they conserve water. Temperate and arctic organisms conserve heat rather than metabolize at subzero temperatures. Aridity and low temperatures are significant limiting variables on earth: no large organisms and only a limited microbiota are associated with the Anarctica desert, an environment still more permissive than that expected on Mars (Horowitz *et al.*, 1972). It is true that some organisms are able to survive and grow in extreme environments such as boiling water (Bott and Brock, 1969), subzero temperatures, pressures exceeding 10^3 atm, or pH 1 (Kushner, 1971). Yet if such conditions became world-wide, almost all life would instantly perish, and presumably the survivors could not persist indefinitely. Furthermore life is a surface phenomenon and most abundant at the interface between water, solid substratum and air.

Although some oxygen respiring organisms require a greater quantity of O_2 in solution than that in equilibrium with 20% in the gas phase, many grow only in the total absence of free oxygen. Nearly 10^3 species of anaerobic organisms, varied with respect to their oxygen tolerances, have been described (Prevot and Fredette, 1966). From the alpine to the ocean abyss, living organisms can be found; life can survive over a total pressure variation from about 0.3 to 10^3 atm. Neither oxygen nor pressure per se limit the distribution of life as a whole. Rather the major physical variables determining the distribution of organisms are: solar radiation, temperature, water abundance, and the concentrations of hydrogen and other ions and elements.

Life has been present at interfaces (intertidal, lake edge and riverbank environments) throughout most of the Precambrian and Phanerozoic – that is, over 3 b.y. (Kvenvolden, 1972 and Schopf, 1972). Therefore, these stable temperature, pH and element cycling requirements for life must have been met on this planet consistently for the entire 'recorded' history of the Earth. Organisms found in extreme environments are highly specialized. If the Earth had frozen out for even a few tens of thousands of years, or if hot acid springs had been widely distributed for even a single epoch, these occurrences would have been discerned form the fossil record.

It doesn't seem possible to us that this veneer of living slime, composed of several million species, all within a few thousand feet of the surface of the Earth could possibly affect the solar system; thus the environment of the biosphere should be considered to be outer space. The atmosphere, far from being an inert sink, we regard as a regulated fluid component of the biosphere, a contrived circulatory system to assure the perpetuation of conditions optimal to the whole of the interconnected living organisms. We do not mean to imply any mystical vital force. Our point is that it is not merely a coincidence that the conditions from which the Earth has not deviated are those optimal for life.

Our assumption that the atmosphere-biosphere system is actively controlled clarifies some problems in atmospheric science and generates useful experiments.

2. Were there no Life on Earth

To model the temperature and atmospheric composition of a lifeless Earth has been very difficult; the factors determining the mean surface temperature of the Earth as a physical system are still only poorly understood. For a recent review of this complex and controversial subject see the SMIC report* and (Robinson and Robins, 1971). The responses of the atmosphere-ocean to small changes, for example, in solar output are very non-linear. A comparatively small decrease should lead (assuming the absence of perturbing influences of life) to an accelerating decline in temperature. A lower surface temperature implies more snow cover which in turn reduces the quantity of sunlight retained by the surface, which leads to further cooling, and so on. Similarly, it has been calculated (Rasool and De Bergh, 1970) that if the Earth had formed only 6 million miles closer to the Sun the positive feedback of temperature increase, through the greenhouse properties of water vapor and CO_2, would have led to Venus-like conditions with surface temperatures above the boiling point of water. Thus the atmosphere-hydrosphere temperature seems to be dynamically maintained between the two precarious and stable extremes.

Although the details are obscure there is general agreement that from a physical viewpoint the climate of the Earth is not at a stable equilibrium. That it should have persisted in a state of disequilibrium for billions of years strikes us as highly improbable. Were life obliterated now, the nitrogen and oxygen of the atmosphere would react to form nitric acid which would soon dissolve in the seas and the nitrogen would revert to its stable chemical form, the nitrate ion (Sillen, 1966). The atmosphere would then be a reasonable interpolation between Mars and Venus with a redox potential $pE = 5-7$, rather than $pE = 13$**. Also like Venus, and possibly also Mars, the Earth would then have acidic land surfaces. Because the climate is strongly dependent on the chemical composition of the atmosphere, such chemical changes would profoundly affect the climate.

The sources, sinks, and residence times for the major nonnoble gaseous components

* *Inadvertent climate modification: Report of the Study of Man's Impact on Climate, 1971*, MIT Press, Cambridge, Mass. (multi-authored).

** pE, a concept analogous to pH, measures electron concentration and equals $-\log [e^-]$.

TABLE I

Principal sources and sinks

Gas	Concentration		Production			
	(parts per million)	Inorganic sources [a]	Emissions ($\times 10^9$ tons yr^{-1}) Biological nonanthropogenic		Anthropogenic	
			Quantity	Source	Quantity	Source
Nitrogen, N_2	7.9×10^5	< 0.001	1	denitrifying bacteria, from nitrate and nitrite	0	—
Oxygen, O_2	2.1×10^5	1.6×10^{-4}	110	blue green algal, nucleated algal, green plant photosynthesis	0	—
Carbon dioxide, CO_2	320	0.01	140	waste product of aerobic respiration, bacteria, animals and plants	16	waste product of fuel oxidation
Methane, CH_4	1.5	0	2	product of bacterial fermentation	0	—
Nitrous oxide, N_2O	0.35	C	1	produced from nitrate by denitrifying, heterotrophic; ammonia oxidizing bacteria	0	—

monoxide, CO				marine, soil bacteria		of fuel oxidation
Ammonia, NH_3	0.01	0	1–2	excretory product of nearly all organisms from amino acids, nucleotides via urea or uric acid	0	—
Hydrocarbons $(CH_2)_n$	0.001	0	0.2	plant emissions; released from organisms buried under anaerobic conditions	0.2	waste product of fuel oxidation
Oxides of nitrogen, NO_x	0.001	(total, 0.7)		?	0.16	waste products of fuel oxidation
Hydrogen sulfide, H_2S	$< 10^{-4}$	?	?	produced by sulfate-reducing bacteria; waste product of sulfur-containing amino acids	0.001	waste product of chemical industry
Sulfur dioxide, SO_2	2×10^4	variable	0	—	0.16	waste product of fuel oxidation
Dimethyl-sulfide, CH_3SCH_3	2×10^{-4}	0	0.2	marine algal emissions	0	—

[a] Abiologic sources: includes volcanic, tectonic emissions and upper atmosphere photolysis.

Table I (Continued)

Gas	Removal	Residence time	References
N_2	Nitrogen fixing bacteria and blue green algae	10^6–10^7 yr	See Brock (1966, 1970) for general discussion of all microbial activities in this table and further references
O_2	Respiration: aerobic bacteria, fungi, animals, plants	10^3 yr	Robinson and Robins (1971); Donahue, 1966
CO_2	Fixation by heterotrophic and photosynthetic bacteria, algae, green plants	2–5 yr	Schutz *et al.* (1970)
CH_4	Methane-oxidizing bacteria, oxidation to water and CO_2 in stratosphere	7 yr	See note on p. 95; Donahue (1966)
N_2O	Photolyzed, reacts with ozone in stratosphere to form higher oxides	10 yr	Schutz *et al.* (1970); Yoshida and Alexander (1970)
CO	Utilized by organisms, reacts with OH radicals in troposphere	months	Weinstock and Niki (1972)
NH_3	Oxidized by aerobic marine and soil bacteria; reactions with acids to form ammonium chloride and sulfate: fixation into amino acids by most bacteria; direct incorporation into leaves of plants	≈ 1 week	See note on p. 95; Hutchinson *et al.* (1972)
$(CH_2)_n$	Atmospheric oxidation, source of food for specialized heterotrophic bacteria	≈ 1 day	Robinson and Robins (1971); Brock (1970)
NO_x	Atmospheric reactions to form nitrate and nitrite salts; major N source for green plants, algae, denitrifying bacteria	< 1 year	Robinson and Robins (1971); Alexander (1961)
H_2S	Atmospheric and hydrospheric oxidation to sulfur or surlfate; to organic sulfur by photosynthetic bacteria	≈ 1 day	Schiff and Hodson (1970); Grey and Jensen (1972)
SO_2	Atmospheric oxidation; sulfate ion reduced by organisms to organic sulfur, H_2S and elemental sulfur	days	Lovelock *et al.* (1972)
$(H_3C)_2S$	Atmospheric oxidation to dimethyl sulfoxide	days	Lovelock *et al.* (1972)

of the atmosphere are shown in Table I. The residence times of these gases are very small fractions of the total history of the Earth, which again suggest an ongoing dynamic process maintains atmospheric stability. This stability is even more impressive if the concept is correct that the Sun's luminosity has increased by several tens of per cent as the Sun proceeds up the main sequence (Sagan and Mullen, 1972; Dilke and Gough, 1972). If the Earth's temperature merely reflected the solar luminosity, a change in luminosity in either direction of more than ten per cent might be all that would be required to freeze the oceans or to set off a runaway to extremes of heat (see note on p. 95; Rasool and De Bergh, 1970). But the fossil evidence very eloquently shows us that throughout periods of billions of years the oceans have neither frozen nor boiled (Schopf, 1972).

3. Questions Generated by the Homeostasis Hypothesis

We assume that the quantities actively modified by the biota are the gas composition, acidity and temperature of the atmosphere. Because the dominant species on the present Earth are obligate aerobes, we assume that oxygen, too, is maintained at its current high level by organisms. What are the mechanisms by which life homeostats the Earth's surface? At this stage in our knowledge we can only list (in Table II) some samples of the many possibilities biology offers. Detailed discussion of these may be found elsewhere (Lovelock and Margulis, 1974). Although the environmental control mechanisms are likely to be subtle and complex, we believe their evolution can be comprehended broadly in terms of Neodarwinian thought (Mayr, 1972). All organisms at any given time are, if circuitously, connected to all others. People are misled by the ease with which 'individuals' can be identified in human and animal populations. (Ambiguity seems to arise only in exceptional cases, such as pregnant women or Siamese twins). However, when considered from the point of view of the survival of the individuals to reproduce and leave offspring to the next generation, the 'individual' is very difficult to delineate from the 'group' or 'population'. Among organisms more distantly related to man the concept of the individual is extremely elusive. Club mosses can be counted as separate plants but when pulled out of the ground it is obvious that hundreds of feet of rhizoid underground connect the countable individuals. Fairy ring mushrooms, social insect and coral colonies are other examples. Passage through the guts of Galapagos tortoises is a requirement for germination of Galapagos solanaceous seeds, thus the tortoise may be considered part of the plant life cycle. Examples like this abound; all species are dependent upon other species for nutrients, the delivery of gases, shelter, support and so forth; none could survive in the total absence of the others.

Presumably the same mechanisms of natural selection that have led to local environmental control have led to near planetarywide control. (For example, humidity and thermoregulation in honeybee hives: the extent of local biological temperature control by these social insects is amazing. Typical hive temperatures of 31 °C have been maintained in −28 °C weather (Wilson, 1971). Analogous with the evolution of local environmental or internal control, in the evolution of atmospheric homeostasis those species of organisms that retain or alter conditions optimizing their fitness (i.e., proportion of offspring left to the subsequent generation) leave more of the same. In this way conditions are retained or altered to their benefit. If the atmosphere is a function-

TABLE II

Possible biological mechanisms to achieve homeostasis of the present terrestrial atmosphere

Quantity controlled	Property changed to achieve control	Biological mechanisms available
Temperature	Surface albedo	Darkening by direct uptake of water: lichen, algal and moss rock cover to retain moisture.
		Physiological control of pigments (carotenoids, xanthophylls, melanins, phycocyanin, phycoerythrin, chlorophylls, hemes, anthocyanins, etc.).
		Shadow casting, soil formation, alteration of surface textures.
		Trapping and precipitation of sediments such as $CaCO_3$, carbon black, iron sulfides.
	Surface IR emissivity	Surface structures and textures, comparable with IR wavelengths (8–14 μm).
	Atmospheric albedo	Emission of dust and aerosol precursors, e.g., terpenes (Rasmussen, 1970), sulfur gases, ammonia Acid and base excretions that react to form precipitating salt particles.
		Plant transpiration, excretion of lipid and detergent surfactants on water surfaces, bacterial and algal scums and slimes, i.e., control of H_2O evaporation and hence of cloud cover (Margulis and Lovelock, 1974).
	Atmospheric emissivity	Emission and removal of IR active gases, e.g., NH_3, CO_2, H_2O (via fermentation, photosynthesis, excretion, etc., see Table I).
	Circulation and heat transfer	Emission of N_2O which could modify ozone layer and hence circulation (Lovelock, 1971) by alteration of the N_2O/N_2 ratio by denitrifying bacteria (Alexander, 1961); evaporation control as above.
pH	Ammonia gas concentration	Control of NH_3 sources so that emission just titrates atmospheric production of H_2SO_4 and HNO_3.
	Removal of carbonic acid	CO_2 removal by blue green algae catalyzes $CaCO_3$ deposition and leads to local alkalinities up to pH 12 (Golubic, 1973).
		Some organisms directly excrete acids and bases (lactic, acetic, uric, nitric, etc.)
pE (See note on p. 95)	Oxygen concentration	O_2 is major excretory product of photosynthesis of algae and green plants.
		Bacterial CH_4 production and transport may be involved in maintaining redox potential, see Lovelock and Lodge (1972).

Table II (Continued)

Quantity controlled	Property changed to achieve control	Biological mechanisms available
		H_2, CH_4, H_2S are produced by photosynthetic bacteria, desulfovibrios, and methane bacteria in anaerobic environments to locally lower redox potentials.
		O_2 is removed by respiration, e.g., nitrogen fixers (*Azotobacter*), animals, chemoautotrophs. Redox potential control via O_2 removal must be good for small increase in per cent O_2 greatly increases the probability of direct combustion by forest fires, see note on p. 95.
pCx (where x is some essential element, see Frieden (1972) for entire list)	Concentration of gas vapor bearing element x	Emissions of gases such as N_2, N_2O, NH_3, CO_2, H_2S, $(CH_3)_2S$, CH_3I, etc. and where necessary control also of sinks so that pCx is kept within acceptable limits. Volatile methylated derivatives of certain essential elements required in trace quantities such as I, Se, and Br produced by marine algae and bacteria.

ing part of the biological cybernetic system which sustains homeostasis, it is appropriate to question the purpose of its various components. Just as it is reasonable to ask of honeybees: By what mechanisms is hive temperature controlled, or of mammals: What is the function of bicarbonate ion in the blood, it becomes reasonable to ask: What is the function of nitrous oxide or methane? Why are these gases released into the atmosphere in quantities of 10^9 tons yr^{-1}? Such questions would be rightfully considered illogical if the atmosphere were an open system, a product of steady state chemistry only; but if we consider the atmosphere to be in homeostasis we raise as critical questions at least the following: 1. Are the limiting elements essential to life (such as nitrogen, phosphorus, sulfur, iodine, bromine and others (Frieden, 1972)) returned through the atmosphere as volatile biological products in quantities that compensate the losses from the land surfaces in the run-off of rivers? 2. Does the large biological production of ammonia, 2×10^9 tons yr^{-1}, act to maintain the pH of the land surfaces close to neutral? 3. Does the comparably large production of methane keep the atmosphere oxidizing by transporting hydrogen to the stratosphere where it ultimately escapes by photolysis? (The maintenance of oxygen in the atmosphere by hydrogen loss via methane, is an example of an explanation generated by the Gaia hypothesis; it solves the issue raised by Gregor (1971) and Van Valen (1971); if O_2 is maintained by H escape from H_2O, the water originates from the oxidation of methane above the cold trap, the paucity of buried carbon from photosynthesis can be explained. See Lovelock and Lodge (1972) for details. 4. What are the sensors, amplifiers and control mechanisms operating to maintain constant the steady state chemical composition of the gases of the atmosphere? What mechanisms are involved in the maintenance of the physical steady state of atmospheric temperature? 5. What are the limits on these control mechanisms? How did they evolve? How did they cope with the two presumed transitions of the atmosphere in the Earth's history (from primordial hydrogen-methane-ammonia to N_2; from N_2 to the present N_2–O_2 mixture (Cloud, 1968))?

Asking these questions has already led to the discovery of the probable balancing factor in the sulfur cycle of the biosphere. Dimethylsulfide, a product of marine biological activity, was sought as an emission from the oceans on the grounds of the Gaia hypothesis. It was found in quantities sufficient to balance the input of inorganic sulfates washed from the land surfaces (Lovelock *et al.*, 1972). Similarly, iodine is emitted in substantial quantities as methyl iodide from the oceans. We expect to find analogous volatile bromine, phosphorus and selenium compounds that should be emitted by marine organisms.

The temperature of both the hives of honeybees and the blood of mammals is maintained constant by means of highly complex systems. These involve several different control loops to facilitate heat generation and heat loss (Wilson, 1971 and Myers, 1969). Similarly, if the biospheric thermostat is responsible for the constancy of the Earth's surface temperature over the last 3 thousand million years, then it is probable that it also requires a number of different control systems. The concentration of carbon dioxide in the atmosphere has increased during the past 25 yr*. Given this information on a physical model (because of its role as a greenhouse gas), one would expect an increase in the mean temperature over the same period. However, contrary to expectations, there has been a significant decrease in Northern Latitude mean temperatures during the past 25 yr*. On the basis of the Gaia hypothesis it may be worth seeking biological regulatory mechanisms that could overcompensate for the CO_2 increase. In general we are beginning to investigate our prediction that the annual biological production and removal of vast quantities of reactive gases (Table I) will be understood in the context of complex atmospheric control mechanisms involving many species of organisms (Table II). One corollary of 'Gaia' is that air pollution has an ancient history: oxygen itself was one of the first major pollutants. Life, although not necessarily man, has a remarkable ability to adapt to the 'pollutants' produced by the biosphere itself. Life as a whole has survived through the mechanism of the evolution and persistence of new species (and eventually higher taxa) better able to survive and reproduce under changed atmospheric conditions. There is no reason to believe this mechanism will change in the near future, on a geological time scale.

Acknowledgements

We gratefully acknowledge the support by NASA NGR 22-004-025 (to LM) and of Shell Research Ltd. (to JEL). We thank Mark Winston, S. Riggs, J. K. Kelleher, J. G. Schaadt, L. Machta, B. Cameron, E. S. Barghoorn, S. Golubic, D. Hall, W. Kap-

* See note on p. 95.

lan, Philip Morrison and C. Junge. We are also indebted to Prof. G. E. Hutchinson, whose life work has surmised rather than explicitly acknowledged the existence of Gaia. For a complete list of his contributions, including those relevant to this work, see G. A. Riley, *Limnol. Oceanography* **16**, (1971), 157.

References

Alexander, M.: 1961, *Introduction to Soil Microbiology*, John Wiley and Sons, New York.
Bott, T. L. and Brock, T. D.: 1969, *Science* **164**, 411.
Brock, T. D.: 1966, *Principles of Microbial Ecology*, Prentice-Hall, Englewood Cliffs, New Jersey.
Brock, T. D.: 1970, *Biology of Microorganisms*, Prentice-Hall, Englewood Cliffs, New Jersey.
Cloud, P. E., Jr.: 1968, *Science* **160**, 719.
Dilke, F. W. W. and Gough, D. O.: 1972, *Nature* **240**, 262.
Donahue, T. M.: 1966, *Ann. Geophys.* **22**, 175.
Frieden, E.: 1972, *Sci. Am.* **227**, 52.
Golubic, S.: 1973, in N. G. Carr and B. A. Whitton (eds.), *Biology of the Blue-Green Algae*, Blackwell, Oxford, pp. 434–472.
Gregor, B. and Van Valen, L.: 1971, *Science* **174**, 316.
Grey, D. C. and Jensen, M. L.: 1972, *Science* **177**, 1099.
Horowitz, N. H., Cameron, R. E., and Hubbard, J. S.: 1972, *Science* **176**, 242.
Hutchinson, G. L., Millington, R. J., and Peters, D.: 1972, *Science News* **101**, 363.
Kushner, D. J.: 1971, in R. Buvet and C. Ponnamperuma (eds.), *Chemical Evolution and the Origin of Life*, North Holland Publ. Co., Amsterdam, pp. 485–491.
Kvenvolden, K.: 1972, *Proc. of the 24th International Geological Congress*, Sec. 1, p. 31–41.
Lovelock, J. E.: 1971, *Atmospheric Environ.* **5**, 403.
Lovelock, J. E. and Lodge, J. P., Jr.: 1972, *J. Atmospheric Environ.* **6**, 575–579.
Lovelock, J. E., Maggs, R. J., and Rasmussen, R. A.: 1972, *Nature* **237**, 452–453.
Margulis, L. and Lovelock, J. E.: 1974, *Icarus*, in press.
Mayr, E.: 1972, *Science* **176**, 981.
Myers, R. D.: 1969, in *The Hypothalamus: Anatomical, Functional and Clinical Aspects*, Charles Thomas, Springfield, Illinois, pp. 406–522.
Prevot, A. R. and Fredette, V.: 1966, *Manual for the Classification and the Determination of the Anaerobic Bacteria*, Lea and Febiger, Philadelphia.
Rasmussen, R. A.: 1970, *Environ. Sci. Technol.* **4**, 667.
Rasool, S. and DeBergh, C.: 1970, *Nature* **226**, 1037.
Robinson, E. and Robins, R. C.: 1971, in S. F. Singer (ed.), *Global Effects of Environmental Pollution*, D. Reidel Publ. Co., Dordrecht and Springer-Verlag, New York.
Sagan, C.: 1970, *Planetary Exploration*, Condon Lectures, Oregon State System of Higher Education, Eugene, Oregon, p. 31.
Sagan, C. and Mullen, G.: 1972, *Science* **177**, 52.
Schiff, J. A. and Hodson, R. C.: 1970, *Ann. N.Y. Acad. Sci.* **175**, 555.
Schopf, J. W.: 1972, in C. Ponnamperuma (ed.), *Exobiology*, North Holland Publ. Co., Amsterdam, pp. 6–61.
Schutz, K., Junge, C., Beck, R., and Albrecht, B.: 1970, *J. Geophys. Res.* **75**, 2230.
Sillen, L. G.: 1966, *Tellus* **18**, 198.
Van Valen, L.: 1971, *Science* **171**, 439.
Weinstock, B. and Niki, H.: 1972, *Science* **176**, 290.
Wilson, E. O.: 1971, *The Insect Societies*, Harvard University Press, Cambridge, Mass.
Yoshida, T. and Alexander, M.: 1970, *Proc. Soil Sci. Soc. Am.* **34**, 880.

MICROFOSSILS FROM THE MIDDLE PRECAMBRIAN McARTHUR GROUP, NORTHERN TERRITORY, AUSTRALIA

M. D. MUIR

Geology Dept., Royal School of Mines, Prince Consort Road, London SW7 2 BP, England

Abstract. Microfossils have been detected in several formations of the McArthur Group (about 1600 m.y. old), in petrological thin sections and in macerations. Some of these occur associated with a lead-zinc ore body; others have been found in well-preserved dolomites and cherts. The variety of microfossils observed is considerable, and their state of preservation is good. Large numbers of singlecelled, colonial and multicellular organisms occur; in the latter, at least one clear case of cell differentiation can be demonstrated. Some of the organisms are morphologically comparable with blue-green algae such as the Chroococaceae, but, unlike the well-known Bitter Springs microflora, the assemblage is notably poor in filamentous algae. The filaments that do occur are not septate, and may represent discarded blue-green algal sheaths. Many of the microorganisms are extremely small in size, and in some cases, colonial structures composed of large numbers of 1 μm diameter cells are present, that may represent bacterial remains. Most of the microfossils occur in stromatolitic cherts, but the lead-zinc orebody from which some were obtained is a fine-grained dolomitic shale.

Stratigraphically, this new assemblage occurs in sediments of age intermediate between the well-known Gunflint Chert assemblage and the equally well-known Bitter Springs flora. The level of organization of the microfossils represents a great advance on that of the Gunflint Chert microfossils, in that demonstrably colonial and large multicellular microorganisms occur, as well as cells of a relatively large size. No convincing evidence for the presence of nuclei or nuclear membranes has yet been found in McArthur Group microorganisms, but the large size and organizational complexity of some of the structures suggests that the origin of the eukaryotic cell may occur rather earlier in geologic times than previous indications have suggested.

1. Introduction

The microfossils described here occur within the Carpentarian carbonate rocks of the McArthur Group which crop out in the Northern Territory of Australia, around the southwestern and western shores of the Gulf of Carpentaria from the border with Queensland to Arnhem Land. The Carpentarian succession (Dunn *et al.*, 1966) straddles the division between the Middle and Late Precambrian as defined by an orogenic event in areas outside Australia, and ranges in age from 1800×10^6 to 1400×10^6 yr. The age of McArthur Group rocks appears most likely on present evidence (Croxford *et al.*, 1973) to be about 1600×10^6 yr old. An extensive age determination study is in progress on the 'type' Carpentarian succession; the results of this will undoubtedly help in making more accurate age assessments.

The McArthur Group is 5.4 km thick and conformably overlies a dominantly arenite succession, the Tawallah Group, which is 4.0 km thick. It is unconformably succeeded by the 3.0 km thick Roper Group which is a sequence of arenites and lutites.

The two major divisions of the McArthur Group are the Batten Subgroup and Umbolooga Subgroup. Microfossils have so far been found in two subdivisions of the Umbolooga Subgroup; the HYC Pyritic Shale Member of the Barney Creek Formation (Hamilton and Muir, 1974) and the Amelia Dolomite (Croxford *et al.*, 1973) which overlies the basal Mallapunyah Formation (see simplified stratigraphic section in Table I).

While the rocks of the McArthur Group as a whole are dominantly carbonates, that

TABLE I

Schematic diagram of the stratigraphy of the McArthur Group at MucArthur River. After Croxford *et al.* (1973).

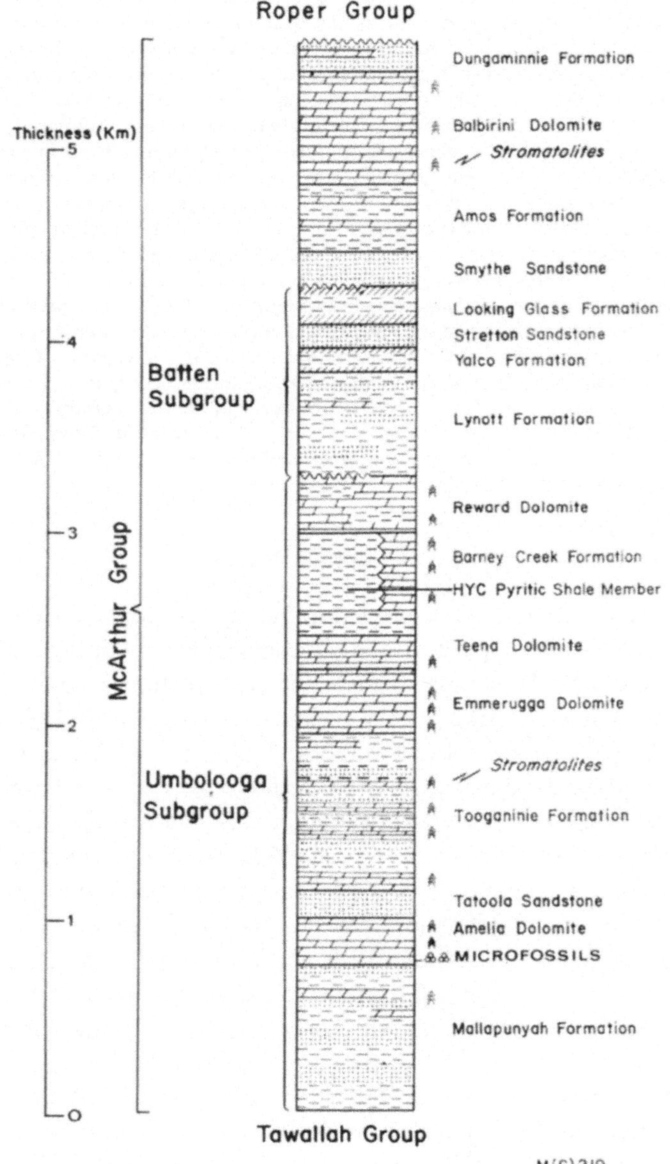

may or may not have suffered silicification, there are several arenite horizons occurring mainly in the lower part of the succession (the Mallapunyah and Tatoola) and in the upper, Batten Subgroup (Stretton Sandstone). There are minor sandstones in the cyclic Tooganinie Fm. [see Plumb and Brown (1973) for the detailed stratigraphy of the McArthur Group].

2. Preservation of the Microfossils

The Amelia Dolomite assemblage occurs within finely laminated silicified carbonate rocks (Croxford *et al.*, 1973) and were discovered by Croxford during petrological examination of the core from the diamond drill hole Tawallah Pocket No. 1. This hole was drilled by Carpentaria Exploration Co. Pty. Ltd. (a subsidiary of Mount Isa Mines Holdings Ltd.) in the course of base metal exploration in the area. In outcrop, the Amelia Dolomite is commonly stromatolitic and the fine, often wavy, laminations seen in thin section represent bedding traces made visible by finely divided organic matter, probably the structureless remains of algal mucilage. The microfossils described here occur only in the silicified parts of the core rocks. Under crossed polars, it is clear that microcrystallites of the silica frequently pass through the walls of the microfossils, suggesting that the silica originated or at least crystallized at some time after the incorporation of the microfossils in the sediment which may have originally been carbonate. There are patches of coarsely crystalline dolomite within the chert, but this occurs in veins and cracks and invariably has a cross-cutting relationship with the bedding. Microfossils are never present in this dolomite, and it is probably much later in origin than the fossiliferous chert. Thus, the microfossils may have been first enclosed in carbonates which became diagenetically silicified. The resulting cherts were subsequently invaded by late-stage dolomite.

The HYC Pyritic Shale Member of the Barney Creek Formation is a bituminuous tuffaceous shale which contains the orebody, mainly sulphides of lead, zinc, and iron, with minor associated silver. In petrological thin section, because of the abundance of sulphides (Croxford and Jephcott, 1972; Hamilton and Muir, 1974) the rock is full of opaque minerals, and while bituminous matter is clearly visible, it is impossible to determine its structure. Maceration of the shale, following the technique described by Saxby (1970), gave rise to abundant organic matter, most of which was structurally preserved. This has been interpreted (Hamilton and Muir, 1974) as the remains of cells in a gelatinous matrix of organic matter representing the fossilized remains of microbial slime or mucilage. Some of the cells are deformed by development of sulphide minerals, but most retain a spheroidal shape.

3. Description of the Microfossils

cf. *Huroniospora microreticulata*, Barghoorn, 1965 (Figures 1, 2 and 3)

Description: Cells solitary, occasionally in pairs or fours; wall thick with reticulate or punctate structure, outline circular to ellipsoidal. Diameter ranges from 4.0–12.0 μm with an average of 6.9 μm (13 specimens).

Discussion: These cells are very similar to those described by Barghoorn (1965) in the Gunflint Chert, excepting that no pairs or fours have so far been described from the Gunflint Chert.

The decussate tetrad is in accordance with that described by Schopf (1968) for *Myxococoides* from the Bitter Springs Formation; a unicellular form of this genus has

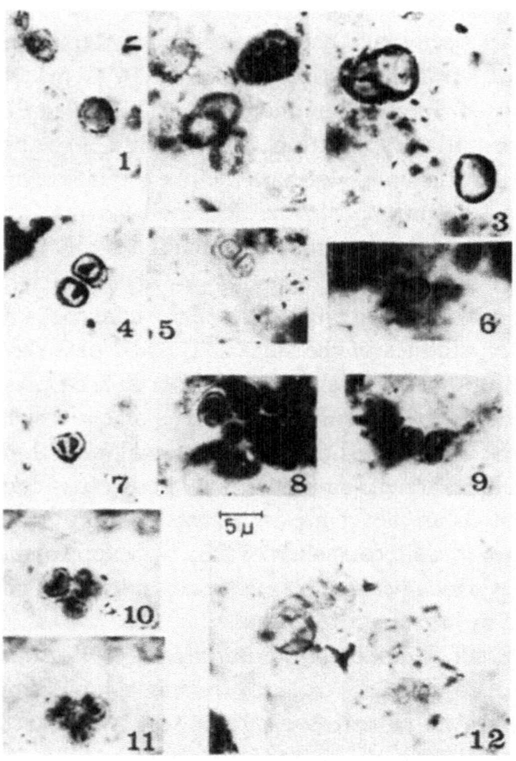

Figs. 1, 2, and 3. Cf. *Huroniospora microreticulata*, Barghoorn (1965). Amelia Dolomite, slides TSA and Amelia 3. Figure 1 is MCA1/5 (103.6:3.7), Figure 2, MCA5/12 (93.0:13.0), Figure 3, is MCA10/6 (97.7:13.1). Single cells, and pairs, and threes can be seen.

Figs. 4–9. *Sphaerophycus parvum*, Schopf (1968). Amelia Dolomite, slides TSA and Amelia 3. Figure 4 is MCA15/1 (108.7:14.5), Figure 5, MCA6/11 (94.9:9.8, Figures 6, MCA11/1 (99.2:8.5), Figure 7, MCA15/3 (108.7:145), Figure 8, MCA6/8 (98.4:10.1), Figure 9, MCA4/15 (92.0:8.9). These figures illustrate the range of morphological variation in this species.

Figs. 10 and 11. Form A. Amelia Dolomite, slide Amelia 3. MCA5/15 and 16, (92.9:11.0). Six cells in different focal planes each with its included pyrite crystal.

been described by Schopf and Barghoorn (1969) from the Skillogalee Dolomite of South Australia as *Myxococcoides muricata*. The difference between *H. reticulata* and *M. muricata* was not discussed by the authors of the latter species and appears to be very slight. It is possible that *M. muricata* is a junior synonym of *H. reticulata*, but I do not propose the formal synonymy here.

Sphaerophycus parvum, Schopf, 1968, (Figures 4–9)

Description: Cells generally spherical, occasionally elliptical; may be solitary or, more commonly, in pairs, or larger loosely aggregated groupings; surface texture smooth. Cell diameters range from 2.0–5.0 μm, average 3.9 μm (58 specimens measured).

Discussion: The size range of these specimens fits well within that described by Schopf (1968) for his material from the Bitter Springs Formation, and the morphology of the specimens appears to be identical. Schopf regards the species as being of blue-green algal origin, possibly a member of the Chroococcaceae, and he believes that reproduction is by fission.

The cells are frequently set in an organic matrix, but equally as often they are not. A relationship between the cells and matrix is difficult to determine.

Form A (Figures 10 and 11)

Description: Cells spherical to slightly ellipsoidal, usually in pairs or fours, occasionally solitary. Wall coarsely reticulate or punctate, encompassed in a granular organic matrix. Size range from 2.0–4.0 μm with an average of 2.0 μm (42 specimens), outer layer (sheath) 1 μm wider.

Discussion: These small cells are distinguished by the coarse nature of the wall structure from any other of the species of *Sphaerophycus*, Schopf (1968). They are further characterized by the invariable presence in the centre of each cell of a single cube of a sulphide mineral, probably pyrite, although the exact nature of the mineral has yet to be determined. At the present time the relationship between the cell and sulphide mineral is unclear.

Form B (Figures 12, 14 and 15)

Description: Cells usually ellipsoidal, sometimes spherical, arranged in fours with a common, very distinct, sheath. Surface texture of cells smooth to slightly granular, minimum diameter 1.1–2.0 μm (average 1.5 μm, 48 specimens), maximum diameter 1.8–2.7 μm (average 2.2 μm). Smooth hyaline sheath encompassing four cells always present, usually circular in outline, and disc shaped in section. Diameter of sheath 6.5–10.8 μm (average 7.4 μm, 35 specimens measured). Large agglomeration of sheath-surrounded cells frequently observed.

Discussion: This form differs from previously described species of *Sphaerophycus* in that four cells are consistently associated within a mutual sheath. It cannot be

Fig. 13.

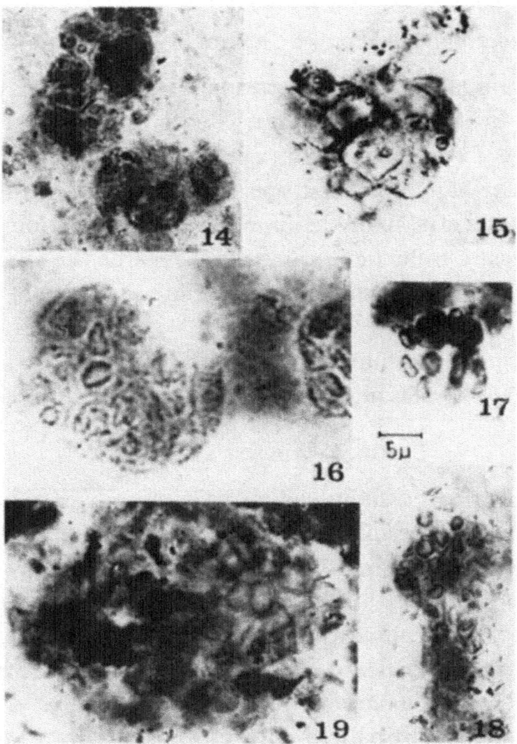

Figs. 12, 14 and 15. Form B. Amelia Dolomite, slides TSA and Amelia 3. Figure 12 is MCA10/11 (98.6:13.1), Figure 14, MCA1/3 (98.3:10.1) and Figure 15, MCA15/3 (111.1:14.2). The range of The range of variation of the Form is shown in these illustrations.

Fig. 16. Form C. Amelia Dolomite, slide TSA MCA1/9 (101.1:13.2).

Figs. 17 and 18. Form D. Amelia Dolomite, slide Amelia 3. Figure 17 is MCA4/9 (91.9:10.8), and Figure 18, MCA5/10 (92.9:13.1).

Fig. 19. Form E. Amelia Dolomite, slide Amelia 3 MCA4/8 (91.9:10.9).

confused with *S. parvum* (see Figure 13) because the four cells are not separated by any kind of internal wall between the cells of the sheath. Figure 12 shows both Form B and *S. parvum*, and the dividing sheath wall between the two cells of *S. parvum* can clearly be seen.

Affinity: A habit of this kind is not infrequently found in the blue-green algal family of the Chroococcaceae which form colonial structures within multilayered sheaths. In the present-day algae, cells reproduce by fission, but when more than four cells are present within a single sheath, the sheath breaks up to form a number of smaller, four membered colonies which remain in association.

Form C (Figure 16)

Description: Cells ellipsoidal to angular, each with an encompassing sheath. Cell outlines apparently controlled by their relationship with other cells in the colony. Surface texture of cells and sheaths coarsely granular. Average diameter of cells 4.2 μm (range

2.8–7.8 μm); distance between cells about 1.7 μm (63 cells measured). Colonies frequently kidney shaped, sometimes ellipsoidal. Average maximum diameter 18.6 μm (range 15.3–20.4 μm, 14 colonies measured). Colonies frequently occur in clusters of five and six.

Discussion: The cells of this Form are associated together within a gelatinous-looking matrix (? originally algal mucilage) and are kept apart by a thin layer of this sheath material. They resemble the colonial *Myxococcoides* Schopf (1968) described from the Bitter Springs Formation. In the Amelia Dolomite assemblage, they are very abundant and both the individual cells and the colonies possess very distinctive shapes.

Form D (Figures 17 and 18)

Description: Cells usually ellipsoidal, some sub-angular, each within its own sheath. Cell wall thick, smooth, sheath smooth, hyaline. Cell diameters range from 1.0–3.2 μm, (average 1.7 μm, 30 specimens). Cells form colonies which are enveloped in originally gelatinous organic material, probably mucilaginous. Colonial outline irregular.

Discussion: The well-separated but obviously associated small cells make this a particularly distinctive form.

Form E (Figures 19–23)

Description: Cells more or less spherical, shape sometimes altered by relationship with neighboring cells; cells sometimes in pairs. Indistinct organic matrix present throughout colony, sheaths occasionally present. Cell surface smooth, sometimes slightly granular, sheath, where present, smooth. Diameter of cells ranges from 2.7–5.0 μm (average 3.9 μm, 107 specimens measured). Outline of colony indistinct, but colonies up to 55 μm in diameter have been observed.

Discussion: These large colonies form a distinctive component of the Amelia Dolomite assemblage. They are characterised by smooth walled globular cells, and irregular outlines of the colonies. Their form and organization is very similar to the cells extracted from the HYC Pyritic Shale Member, and they may be conspecific. However, further work remains to be done to establish such a relationship.

Form F (Figure 24)

Description: Cells spherical, but outlines affected by contact with adjoining cells. No sheath or organic matrix determinable. Cell surface smooth, hyaline. Average diameter of cells 3.8 μm, size range from 3.0–4.5 μm (28 specimens). Colonies have irregular outlines.

Discussion: In this form, cells in colonies appear to have walls in common between adjacent cells. Rather similar forms have been described by Konzalova (1972) from the Upper Proterozoic of Bohemia, although she preferred to make no definite assignment.

Myxococcoides minor, Schopf, 1968 (Figures 25 and 26)

Discussion: Colonies of this type are common in the Amelia Dolomite assemblage and characteristically possess a rather definite outline which varies from ellipsoidal to

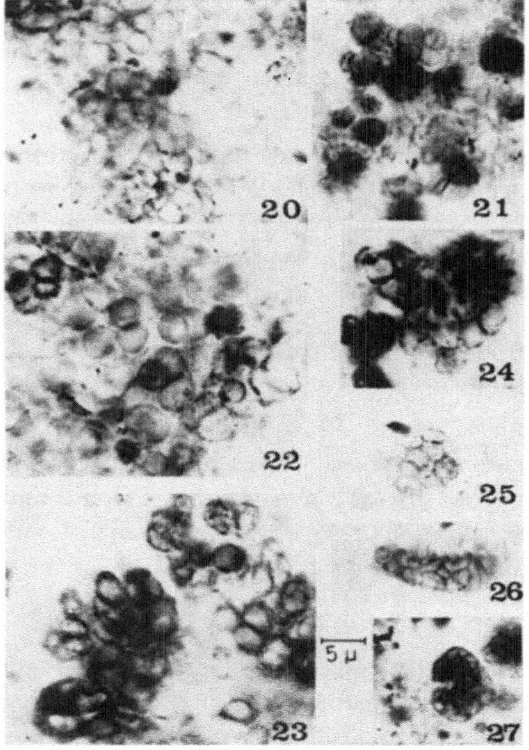

Figs. 20–23. Form E. Amelia Dolomite, slide Amelia 3, Figure 20, MCA10/3 (95.9:13.4), Figure 21, MCA4/19(92.3:12.8), Figure 22, MCA5/8 (92.8:12.9), Figure 23, MCA10/9 (98.1:13.4). These micrographs illustrate the range of variation of these colonial bodies.

Fig. 24. Form F. Amelia Dolomite, slide Amelia 3, Figure 24, MCA4/17 (92.0:8.1), showing the features of this Form.

Figs. 25 and 26. *Myxococcoides minor*, Schopf (1968). Amelia Dolomite, slide Amelia 3. Figure 25, MCA10/1 (95.7:13.4), Figure 26, MCA4/14 (92.0:9.1). These show the characteristics of the Form.

Fig. 27. *Palaeoanacystis vulgaris*, Schopf, 1968. Amelia Dolomite, slide Amelia 3, MCA4/14 (92.2:9.7).

reniform. The Amelia Dolomite specimens are smaller than those described by Schopf, but otherwise are identical.

Palaeoanacystis vulgaris, Schopf, 1968 (Figures 27 and 28)

Discussion: These small colonial organisms resemble the Bitter Springs Formation specimens very closely, forming a tightly packed, almost gelatinous mass of cells with a dense brown colour. Cells have walls in common.

Form G (Figures 29 and 30)

Description: Cells spherical to ellipsoidal, shape occasionally affected by contact with walls of adjoining cells. Surface distinctly granulose. A general organic matrix is present, but individual cells are not surrounded by a sheath. Average cell diameter

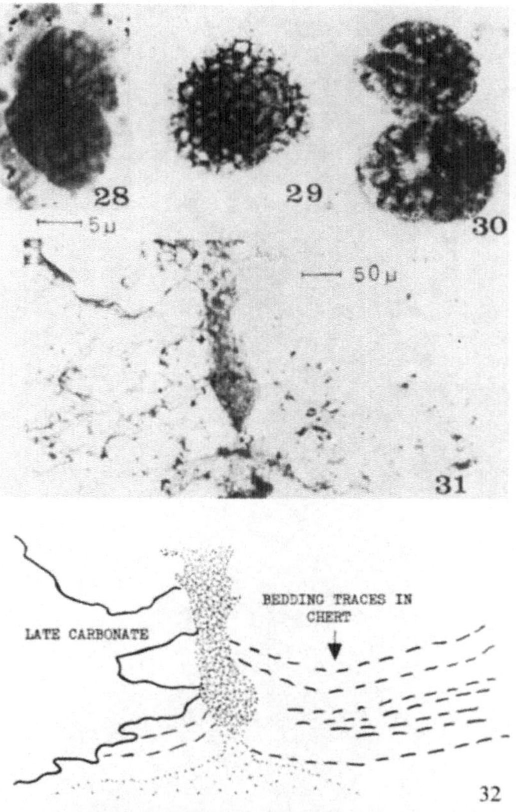

Fig. 28. *Palaeoanacystis vulgaris*, Schopf, 1968. Amelia Dolomite, slide Amelia 3. MCA6/10 (94.9:12.4). This specimen shows the salient features of the Form.

Figs. 29 and 30. Form G. Amelia Dolomite, slide TSA, Figure 29, MCA15/13 (94.8:12.2), Figure 30, MCA15/12 (94.8:12.2). These micrographs show the general shape of the colony and the wall structure.

Figs. 31–33. Form H. Amelia Dolomite, slide Amelia 3. MCA6/19–23 and MCA7/9 (95.7:11.1). Figure 31 shows the relationship of the organism to the surrounding sediment, Figure 32 is a diagrammatic representation of the same, and Figure 33 is a composite at the same magnification as all the other illustrations (save Figures 31and 32) showing details of cells and organization.

2.5 μm (range 1.0–3.4 μm, 111 specimens). Colonies spherical to ellipsoidal, frequently occurring in pairs or small clusters.

Discussion: This Form differs from *Palaeoanacystis vulgaris* by its very distinctive wall structure.

Form H (Figures 31–33)

Description: Multicellular structure of microscopic, but relatively large size. Individual cells are polygonal as a result of compression from adjoining cells. All cells have walls in common with neighbouring cells; the common walls are smooth. Cells average 6.8 μm in diameter and range from 4.0–8.0 μm in size (250 measured). The overall shape of the colony is cylindrical (25–30 μm in diameter), broader in the lower

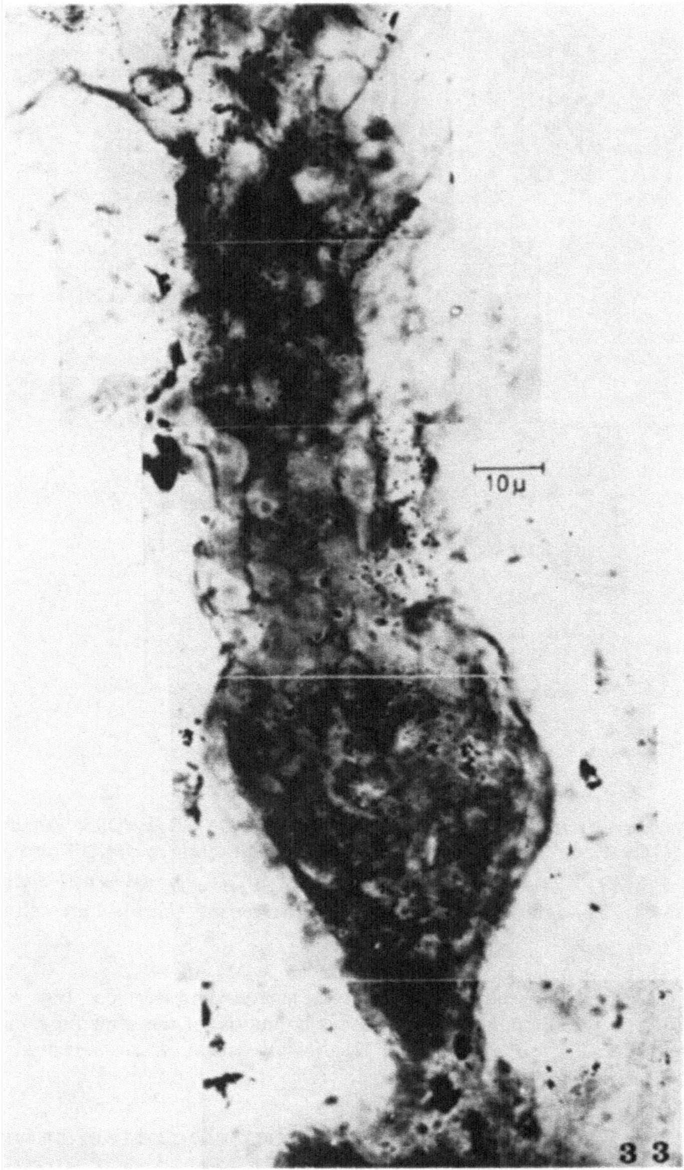

third of its length (50 μm in diameter). From the broad section it tapers towards its base. The basal part is noncellular and after reaching a diameter of about 12 μm at its narrowest part, it spreads out laterally.

Discussion: The organism appears to have been fossilized in the position of growth. Traces of bedding can be seen in the surrounding chert (Figure 32 shows this diagrammatically) and the broadening of the basal non-cellular part of the organism suggests that it may have been where the organism was attached to the substrate – possibly a kind of hold-fast.

TABLE II

Dimensions of the forms described in the text, except for Form H and Form I

Cell Type	Number off cells measured	Cell diam. average	(μm) Range	Surface texture	Organic matrix	Sheath	Other comments
Huroniospora microreticulata	13	7.4	4.0–12.0	Thick wall coarsely punctate	None	None	Thick walled spheroidal cells. Usually solitary but may occur in associated pairs, threes or fours.
Sphaerphycus parvum	58	2.5 (inner) 3.7 (outer)	1.0–3.0 2.0–5.0	Both bodies smooth and hyaline	None	Yes	Usually paired may be solitary, sheath may be multilayered.
Form A	42	2.7 (inner) 3.6 (outer)	2.0–4.0	Both layers coarsely reticulate	Yes	Yes	May be in pairs or fours micron-size sulphide cube in centre of cells.
Form B	35	1.5 min 2.2 max 7.4 (sheath)	1.1– 2.0 1.8– 2.7 6.5–10.8	Psilate Hyaline	Yes	Yes	4 cells invariably present in sheath, agglomeration into large colonies.
Form C	63 cells 14 colonies	4.2 18.6 min 27.5 max	2.8–7.8	Granular Granular	Yes	Yes	Cells angular colonies reniform.
Form E	30	1.7 2.1	1.0–3.2 1.3–3.4	Smooth smooth	Yes	Yes	Cells ellipsoidal, colonies irregula
Form F	28	3.8	3.0–4.5	Smooth	None	None	Colonies irregular in shape.
Myxococcoides minor	133	2.6	1.5–4.5	Smooth	Yes	None	Colonial, reniform, ellipsoidal
Palaeoanacystis vulgaris	62	1.9	1.6–4.6	Smooth	?	None	Colonies reniform, ellipsoidal.
Form G	111	2.5	1.0–3.4	Granular	Yes	None	Colonies spherical to ellipsoidal.

In the narrow cylindrical part of the organism there are two lateral 'branches' (Figure 33), and the peripheral cells of the broadest part are more hyaline in appearance and more regular in shape than those of the main bulk. Residues from original cell contents are present in all cells. Thus, the apparent cell differentiation may represent merely preservational variability.

Although only one specimen so far has been found which shows all the features described above, some oblique sections have also been observed with the same general dimensions and cell size and morphology.

Form I (Figures 34–36)

Description: Multicellular structure, with cells of two distinct morphological types. Some cells are elongate, average width ranging from 2.8–5.3 μm, average length from 10.1–18.6 μm (35 measured). These cells are aggregated together and form sub-parallel to radiating groups. The ends of these cells are diamond shaped and new cells are inserted between two earlier ones. The other type of cell is spheroidal with a reticulate or punctate cell wall, ranging in diameter from 1.5–6.0 μm (average 3.9 μm, 26 specimens). The spheroidal cells occur at the periphery of the organism and are

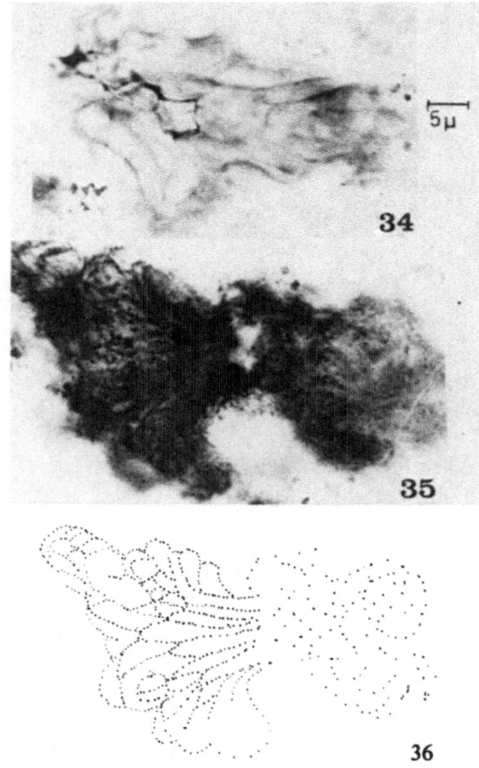

Figs. 34–36. Form I. Amelia Dolomite, slide Amelia 3. Figure 34, MCA5/21 (93.5:11.2), Figures 35 MCA3/8 (116.0:12.2). Figure 36 is a diagrammatic representation of Figure 35, showing the relationship between cells.

terminal to the elongated cells. Overall shape of the organism is triangular to subtriangular.

Discussion: This Form appears to show clear evidence of cell differentiation, with two distinct types of cell present. Eight complete or partial specimens have been found. In the specimen figured as Figure 35, there is a mass of structureless organic matter at the narrowest part of the elongated cell bundle. Although it is not evident that this mass of organic material was ever attached to the cell bundle, the overall appearance of this organism is rather similar to *Kakabekia* Barghoorn (1965). However, the greater size of the Amelia Dolomite microfossil, together with its complex detail distinguish it from this genus.

The affinities of this organism are not known, although it possesses some of the morphological characteristics of Myxobacterial fruiting bodies.

4. Discussion of the Microfossil Assemblage

The assemblage differs from both the Gunflint Chert and Bitter Springs Assemblages in that it is noticeably poor in filamentous remains. There are some filaments, but so far those that have been found have been nonseptate, and may represent the discarded sheaths of Oscillatorian-type blue-green algae. The dominant element of the assemblage are coccoid forms, and these are mostly colonial. Unicellular forms are common, and generally have thick walls, which are in many cases impregnated with finely divided sulphide crystals.

The colonial forms show a wide range of organization from pairs of cells, to fours, and then to large numbers of cells set in multilayered sheaths or in more-or-less amorphous organic matter. Some of the colonial forms have walls in common between cells, and thus become multicellular. Two forms have so far been observed where cell differentiation has taken place. One of these, Form H has a massive structure with a large number of cells, most of which posses the remains of the original cell contents. Whether or not these could be nuclear residues has yet to be demonstrated, but this combined with the large size and complex organization of the body would tend to suggest a level of development higher than that of the prokaryotes. It could therefore indicate the earliest appearance of primitive eukaryotes.

5. Conclusions

Some microfossils are described for the first time from the Carpentarian McArthur Group (1600×10^6 yr old) and show a variety of forms from single celled and colonial to multicellular forms. The assemblage differs from the older Gunflint Chert (1900×10^6 yr old) assemblage in its lack of filaments, and in the possession of stratigraphically early colonial forms, and appears to have more taxa in common with the much younger (800×10^6 yr old) Bitter Springs Formation. The presence of relatively large, complex structures may represent the earliest appearance of primitive cukaryotes.

Note about the figures. All co-ordinates given refer to the mechanical stage co-ordinates of the Zeiss Photomiscroscope in the Geology Department, Royal School of Mines, Imperial College, London.

When the investigation is complete, described material will be lodged in the Museum of the Bureau of Mineral Resources, Canberra. A duplicate collection will be available in the Geology Department, Imperial College.

Acknowledgements

I gratefully acknowledge the help of Dr N. J. W. Croxford and Mr J. Janecek for their help in allowing me access to this material; also the Carpentaria Exploration Co. Pty. Ltd., and Mount Isa Mines Holdings Ltd., for allowing me to examine the drill core. I am indebted to CRA who provided financial support in the form of a grant to the Baas Becking Geobiological Laboratory for me to visit the area and collect further material, and also appreciate the interest and support of Mr Haddon F. King. I am particularly grateful to Mr K. A. Plumb for organising and conducting a collecting trip to many Precambrian microfossil localities in Australia, including the McArthur River area in May–June, 1972.

References

Barghoorn, E. S. and Tyler, S. A.: 1965, *Science* **147**, 3658, 563.
Croxford, N. J. W. and Jephcott, 1972, *Proc. Aust. Inst. Min. Met.* **243**, 1
Croxford, N. J. W., Jahecer, J., Muir, M. D., and Plumb, K. A.: 1973, *Nature* **245**, 28–30.
Dunn, P. R., Plumb, K. A., and Roberts, H. G.: 1966, *J. Geol. Soc. Aust.* **13**, 593.
Hamilton, L. H. and Muir, M. D.: 1974, *Miner. Deposita* (in press).
Konzalova, M.: 1973, *Casopis pro Miner. Geol.* **17**, 267.
Plumb, K. A. and Brown, M. C.: 1973, *Bur. Miner. Resour. Aust. Bull.* **139**, 101.
Saxby, J. D.: 1970, *Geochim. Acta.* **34**, 1317.
Schopf, J. W.: 1968, *J. Palcont.* **42**, 651.
Schopf, J. W. and Barghoorn, E. S.: 1969, *J. Paleont.* **43**, 111.
Schopf, J. W. and Blacic, J. M.: 1971, *J. Paleont.* **45**, 925.

THE DEVELOPMENT AND DIVERSIFICATION OF PRECAMBRIAN LIFE

J. WILLIAM SCHOPF

Dept. of Geology, University of California, Los Angeles, Calif. 90024, U.S.A.

Abstract. During the past decade, important strides have been made toward deciphering the paleobiology of the Precambrian Eon, the earliest seven-eighths of Earth history. This progress has accrued chiefly from micropaleontological and organic geochemical studies of fine-grained, ancient cherts. Although understanding of the early biota – of its composition, diversity, paleoecology and evolution – still remains far from adequate, three particularly significant generalizations have emerged: (i) Living systems were extant earlier than about 3000 m.y. ago; (ii) between about 3000 and 1000 m.y. ago, the Earth's biota was dominated by prokaryotic blue-green algae; and (iii) the development of the nucleated, eukaryotic cell type somewhat earlier than 1000 m.y. ago led to a stage of rapid diversification that culminated with the appearance of megascopic life near the close of the Precambrian. Consideration of these generalizations, and of the evidence bearing on them provides a 'state-of-the-art' assessment of the current status of Precambrian paleobiology.

In the introduction to a recent (and assertedly speculative) essay on the origins and early development of multicellular, megascopic organisms, my colleagues and I suggest that "when one surveys the panorama of innovations in morphology and biochemistry that has appeared during the long course of biological evolution, one is struck by the realization that a surprisingly small number of these developments – probably fewer than a dozen – stand out as having had a profound and lasting impact on subsequent evolutionary history. These signal innovations are of the type that Stebbins (1969) has characterized as leading to the emergence of 'new levels of organizational complexity', innovations resulting in such major events as the development of autotrophy, eukaryotic organization, sexuality, multicellularity, homoiothermy, angiospermy, and so forth. Interestingly, almost any such list of major evolutionary events could also serve as an inventory of 'major unsolved problems in paleobiology', for the fossil record has provided only limited insight into the nature of these important evolutionary transitions" (Schopf *et al.*, 1973). Extending this theme a bit further, it is, I believe, at least equally true that until quite recently – prior to the past decade or so – the known fossil record afforded virtually *no* insight into several of these transitions, especially those occurring early in Earth history. Indeed, it is only during the past few years, as a direct result of discoveries of cellularly preserved Precambrian microorganisms and the application of organic geochemistry to problems of Precambrian evolution, that paleobiologists have addressed themselves in earnest to a consideration of the early development and diversification of life. Paleobiologic study of the Precambrian is thus relatively novel; there are comparatively few workers in the field and, at present, a paucity of firm data. In view of this novelty, it can be safely assumed that some, and perhaps much of what is now 'known' about Precambrian life could well be inaccurate – today's 'facts' might readily be supplanted by tomorrow's discoveries, just as current understanding has replaced that of a decade ago.

What concepts, then, appear to be relatively well established regarding early evolution? What aspects of current thought hold promise of having lasting validity? An ongoing assessment and reassessment of data, interpretations and working hypotheses are, of course, typical of the formative stages of any newly and rapidly developing science; Precambrian paleobiology is no exception. For example, recent studies have shown that 'early' (circa 1965) inferences based on the types of organic components extracted from Precambrian sediments may have been at least partially in error (see, for example, Schopf, 1970; Smith *et al.*, 1970; Kvenvolden, 1972). Similarly, the biological affinities originally suggested for several types of Precambrian microfossils (and, in a minor number of cases, their putative biogenicity) have been subject to considerable debate and, in some cases, reinterpretation. Thus, various 'facts' have been modified or rejected, a process that will no doubt continue as new data accumulate. At the same time, however, it is possible, and I tend to believe rather likely, that certain of the *generalizations* that have emerged in recent years – especially those based on several, distinct, complementary lines of evidence – will be reinforced, rather than rejected, as the early fossil record becomes better known. In short, although the data currently available are very notably limited, and although a great deal remains to be learned regarding the course of early evolution, important strides have been made toward understanding the timing and nature of major evolutionary events occurring during the Precambrian.

The following is a brief discussion of three major generalizations that appear to me to be among the most important conclusions to be drawn, albeit tentatively, from the studies of the past decade. [For recent, detailed reviews of data bearing on this discussion, see Glaessner (1962, 1966), Cloud (1968) and Schopf (1970)]. Taken together, these generalizations represent a 'state-of-the-art' assessment of Precambrian paleobiology. In addition to summarizing, in a cursory fashion, the current status of important aspects of the field, the following discussion points out several 'problem areas' in need of further, detailed investigation.

GENERALIZATION ONE: *Living systems were probably extant as early as* 3300 m.y. *ago; photoautotrophs, apparently including blue-green algae, originated earlier than* 3000 m.y. *ago.*

As is summarized in Table I, the Precambrian fossil record is limited by a variety of factors. In general, the importance of these factors increases markedly as a function of geologic age – the older the sedimentary unit, the greater the probability that it will have been eroded and recycled, severely altered by metamorphism, or be simply buried and thus unavailable for study. It is perhaps not surprising, therefore, that the known early Precambrian fossil record is extremely limited.

The oldest generally accepted microfossils now known (*Archaeosphaeroides barbertonensis* Schopf and Barghoorn; and *Eobacterium isolatum* Barghoorn and

TABLE I

Limitations of the Precambrian fossil record

Fossiliferous units eroded and recycled:	Record lost
Fossiliferous units metamorphosed:	Record destroyed
Fossiliferous units not exposed:	Record unknown
Few fossiliferous units currently known:	Record incomplete
Known fossiliferous units predominantly cherts:	Record unrepresentative

Schopf) occur in cherts and shales of the upper Onverwacht and Fig Tree Groups of the Swaziland Supergroup in the eastern Transvaal, South Africa. These sediments are apparently on the order of 3200 m.y. in age. The most common of these taxa, *A. barbertonensis*, has been detected by several workers at several stratigraphic horizons in the sequence (reviewed by Schopf, 1970). In general, these are rather unimpressive microfossils – spheroids about 20 μm in diameter, commonly exhibiting a granular, reticulate-like surface texture. They are, in fact, of such simple morphology that their biogenicity might be questioned; might these forms be the result of solely nonbiologic processes? Probably not. The spheroids assigned to *A. barbertonensis* fall within a relatively restricted range of size, shape and surface texture, a characteristic of modern unicellular microorganisms; in addition, they bear very substantial resemblance to algal unicells (viz., chroococcacean blue-green algae) of younger geologic age, both Precambrian and Phanerozoic, and to unicellular algae of the modern biota. Moreover, and perhaps most importantly, they are of demonstrable organic composition – they can be freed from the encompassing mineral matrix by palynological maceration; they are thus easily distinguishable from mineral grains and inorganic microspheroids such as those recently described by Merek (1973). Their abundance and their occurrence in both cherts and shales of the sequence are further consistent with a biogenic interpretation, suggesting that the spheroids probably represent unicellular microorganisms of planktonic habit. Finally, organic geochemical data – especially results of recent studies of the carbon isotopic composition of insoluble organic matter ('kerogen') occurring in these strata (Oehler *et al.*, 1972) – indicate that the carbonaceous material in these sediments was probably produced by photoautotrophs, such as photosynthetic bacteria or blue-green algae [other organic geochemical data from the sequence have been summarized by Kvenvolden (1972) and by Oehler *et al.* (1972)]. Thus, several lines of evidence indicate that living systems were probably extant as early as 3200 m.y. ago, and that photoautotrophs may have been represented in this early biota.

It seems significant that several of the lines of evidence bearing on the interpretation of *A. barbertonensis* can be applied, if perhaps with somewhat less force, to microscopic spheroids detected in underlying sediments of the middle and lower Onverwacht Group by Nagy and Nagy (1969) and by Brooks and Muir (in press). These spheroids, apparently more than 3300 m.y. in age, are reported to be of somewhat more variable morphology than those of the upper Onverwacht and Fig Tree Groups; furthermore, certain of these older spheroids are of mineralic, rather than organic composition. Thus, the biogenicity of these spheroids, especially those of inorganic composition, seems open to question; and, it remains to be established whether the *organic* members of the group fall within a range of size and shape comparable to that exhibited by modern (or fossil) taxa. Organic geochemical data may prove relevant to this problem. Several workers (Brooks, 1971; Oehler *et al.*, 1972; Silverman, quoted in Kvenvolden, 1972) have detected a distinct discontinuity in carbon isotopic composition between organic matter of the lower Onverwacht and that occuring higher in the sequence. The implications of this discontinuity are difficult to assess, but it seems possible that it could mark the time of origin of autotrophic life (Oehler *et al.*, 1972). If so, this break could be correlative with the appearance of organic spheroids (algal microfossils?) in the sequence; unfortunately, this possibility has yet to be investigated. Nevertheless, based on the limited data at hand, it seems to me probable that some, and perhaps all of the organic spheroids detected in the middle and lower Onverwacht sediments are biogenic; together with other factors discussed above, the coincidence of organic composition and biologic morphology, even including such simple morphology as a relatively unornamented spheroid (a morphology reasonably expected for primitive microorganisms), cannot be easily attributed to nonbiologic processes.

Additional evidence of the very early existence of biological organization is afforded by the well known stromatolitic structures of the Bulawayan Group of Rhodesia, currently regarded as having an age of between about 3000 and 3300 m.y. (Vail and Dodson, 1969). Although long a subject of controversy, the biological origin of these calcareous deposits now seems firmly established (Schopf *et al.*, 1971). And, by comparison with younger stromatolitic communities, both fossil and modern, it seems evident that the Bulawayan structures were produced by filamentous, phototactic microorganisms, presumably blue-green algae. [Although a photobacterial origin for these structures is conceivable, assuming the atmosphere was virtually anoxic, there is no obvious evidence to support this contention; 'bacterial stromatolites' reported from siliceous hot springs by Walter *et al.* (1972) differ from the Bulawayan structures in mineralogy, morphology and ecologic setting.]

Thus, a rather broad spectrum of data – including geologic evidence, the occurrence of stromatolites, microfossils and probable microfossils in early Precambrian sediments, and the results of carbon isotopic and other organic geochemical analyses of ancient sediments – seem to indicate that living systems were extant earlier than 3000 to 3300 m.y. ago. Aspects of the foregoing interpretations may change as new data accumulate; the basic generalization, however, appears to be on solid footing.

GENERALIZATION TWO: *The Precambrian Eon of Earth history can be properly termed 'The Age of Blue-Green Algae'; these primitive, prokaryotic microorganisms, characterized by a marked degree of evolutionary conservatism, were the dominant components of Earth's biota for the period extending from about* 3000 *to* 1000 m.y. *ago.*

In a manuscript currently in press, I have reviewed in detail the geologic distribution of blue-green algae and the phylogenetic implications of the known cyanophytic fossil record. The following is an abridged version of the conclusions presented in that paper (to which the reader is referred for a thorough discussion of the subject; Schopf, in press), together with several additional observations.

(i) During the past two decades, impressive progress has been made toward documenting the geologic distribution and diversity of fossil blue-green algae. Much of this progress has accrued from studies of fine-grained, Precambrian chert deposits in which cyanophytes are preserved as three-dimensional, structurally intact, organic residues. In such deposits and in similar chemical sediments of the Phanerozoic, evidence of ecologic setting, growth habit, general morphology, detailed cellular anatomy and mode of reproduction commonly can be discerned, features that provide a sound basis for comparison of extant and fossil taxa.

(ii) The known cyanophytic fossil record is far from complete. Nevertheless, the most active phase of blue-green algal evolution, that occurring during the middle and late Precambrian, is relatively well documented; overall, the quantity, quality and geologic distribution of available fossil evidence seem sufficient for tentative identification both of major events in blue-green algal evolution and of phylogenetic relationships among certain of the principal cyanophytic families.

(iii) The earliest blue-green algae apparently were unicellular, noncolonial coccoid forms ('Coccogoneae') of the Chroococcales. By middle Precambrian time (2.5 to 1.7×10^9 yr ago), 'coccogoneans' had become moderately diversified; by the late Precambrian (extending from 1.7 to about 0.6×10^9 yr ago) the 'tribe' was highly diverse and quite 'modern' in character, represented by members of at least three extant families (Chroococcaceae, Entophysalidaceae and Pleurocapsaceae) and a large variety of growth-forms (unicells, sheath-enclosed pairs, decussate quartets, irregular colonies of a few to many cells, and highly ordered cuboidal and spheroidal colonies).

(iv) The occurence of algal stromatolites in early Precambrian sediments suggests that the filamentous 'tribe' ('Hormogoneae'), derived from chroococcacean progenitors, was extant prior to about 3000 m.y. ago and that the earliest 'hormogoneans' were relatively simple forms of the oscillatoriacean type. At least two, and probably three, 'hormogonean' families (Oscillatoriaceae, Nostocaceae and, apparently, Rivulariaceae) evolved earlier than 2000 m.y. ago; the middle Precambrian biota appears to have been dominated by heterocystous, presumably nitrogen-fixing, nostocaceans. By late Precambrian time, the Nostocaceae had been supplanted by the Oscillatoriaceae as the dominant cyanophytic family and most, and possibly all, major 'hormogonean' families had become established. The fossil evidence, therefore, strongly supports the 'progressive', rather than the 'retrogressive' concept of cyanophytic phylogeny (cf. Desikachary and Padmaja, 1970).

(v) The course of cyanophytic evolution has been influenced markedly by the evolution of the Earth's atmosphere from an essentially anoxic to a relatively oxic state; this atmospheric evolution, in turn, has been strongly influenced by the evolution of the Cyanophyta. Thus, the development of an oxic environment was almost entirely the result of cyanophytic photosynthesis; and, as oxygen concentrations increased, cyanophytes evolved mechanisms for both coping with (e.g., heterocysts to 'protect' the oxygen-sensitive nitrogenase enzyme complex) and utilizing (e.g., aerobic respiration) the newly available free oxygen. The reported occurrence of heterocysts in middle Precambrian cyanophytes provides strong evidence for the existence of relatively oxic conditions; it seems possible that the earliest widespread occurrence in the geologic record of algal stromatolites, near the beginning of the middle Precambrian, may reflect the origin of heterocystous nostocaceans capable fixing atmospheric nitrogen in an aerobic environment.

(vi) The evolutionary history of the 'Coccogoneae' and 'Hormogoneae' seem notably similar; both cyanophycean 'tribes' orginated during the early Precambrian, had become moderately diversified by the middle Precambrian, and were highly diversified and quite 'modern' in character by the late Precambrian, reaching their zenith in distribution and ecologic importance about 1000 m.y. ago. Subsequently, members of both 'tribes' have exhibited an extreme degree of evolutionary conservatism – as evidenced by detailed similarities in morphology, ecology, growth habit and mode of reproduction between Precambrian, Phanerozoic, and extant taxa (compare Figures 1, 2 and 3, 4 and 5, 6 and 7). This feature seems primarily attributable to the wide ecologic tolerance and versatile physiology characteristic of the class (summarized in Table II). The Cyanophyceae represents perhaps the most impressive example of bradytelic evolution now known in any biologic group.

(vii) It has long been assumed by phycologists that fossil evidence can make no useful contribution to understanding cyanophytic evolution. For example, Fritsch (1965, p. 859) has asserted that "even should some of the fossil types referred to Blue-Green Algae actually belong to this class, as well they may, they afford no morphological data that might help in the elucidation of structural features or of the evolutionary sequence;" this theme has often been repeated (e.g., Desikachary, 1959; Desikachary and Padmaja, 1970). Indeed, it is believed by some, and seems widely accepted, that a fossil record of thallophytic evolution is 'lacking' and that this is "a state of affairs which is unlikely to be remedied" (Klein and Cronquist, 1967, p. 256). Such views are no longer tenable. It is now evident that the fossil record represents, both actually and potentially, a rich source of significant data that can make a major contribution toward deciphering the course of thallophytic phylogeny.

GENERALIZATION THREE: *The nucleated, eukaryotic cell type had become established at least as early as* 900, *and possibly prior to* 1300 m.y. *ago*; *a series of significant advances, apparently 'triggered', in part, by the development of meiosis and eukaryotic sexuality, occurred between* 900 *and* 700 m.y. *ago and culminated in the emergence of megascopic, multicellular organisms near the close of the Precambrian.*

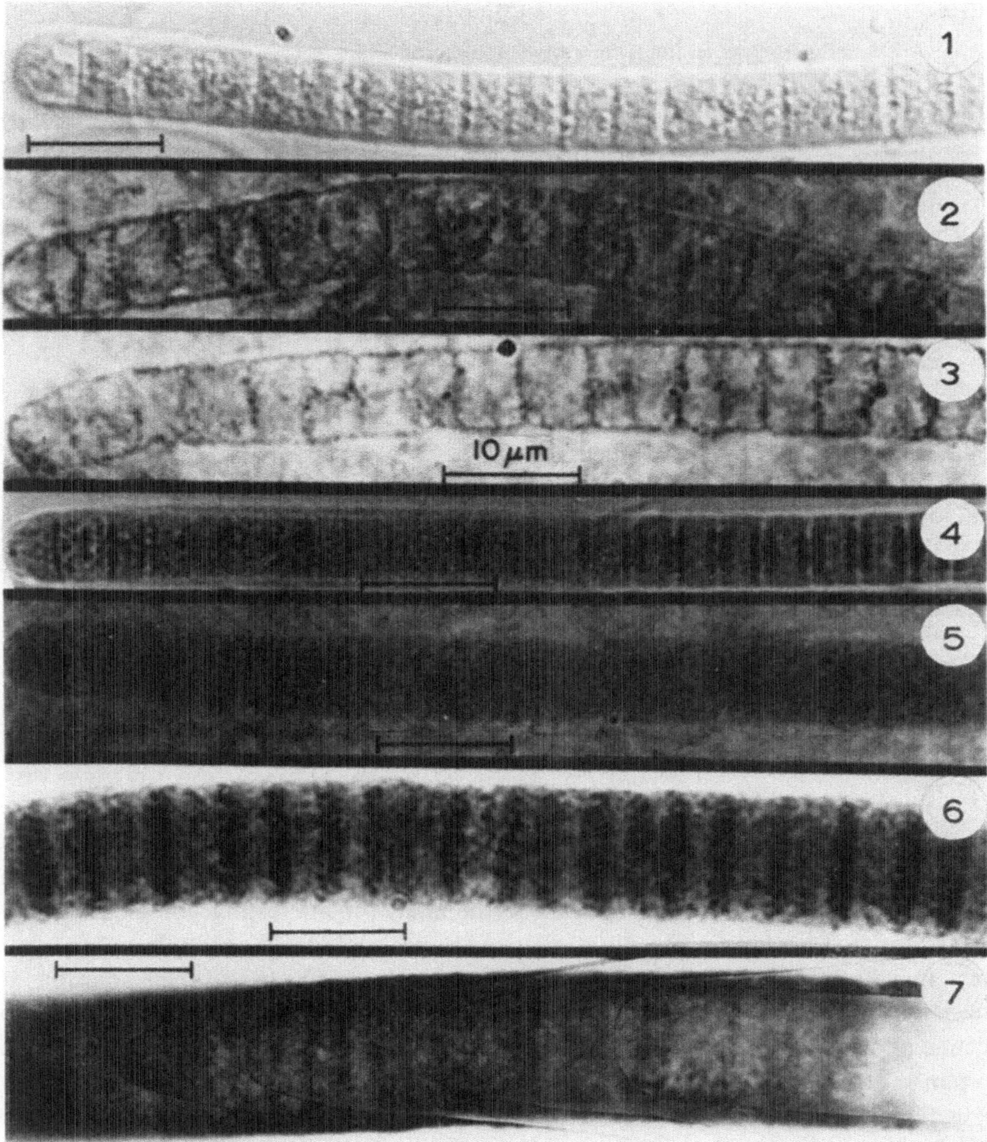

Figs. 1–7. Optical photomicrographs showing modern oscillatoriacean blue-green algae (Figures 1, 4, 6) for comparison with fossil oscillatoriaceans in petrographic thin sections of the Middle Devonian (ca. 370 m.y. old) Rhynie Chert of Aberdeenshire, Scotland (*Archaeothrix* spp.; Figures 2, 3, 5) and of black chert from the late Precambrian (ca. 900 m.y. old) Bitter Springs Formation of central Australia (unnamed taxon; Figure 7). Lines for scale represent 10 μm.

Perhaps foremost among the various aspects of Precambrian paleobiology that are being actively investigated at the present time are problems relating to the development, during the late Precambrian, of multicellular, megascopic life. This interest in late Precambrian evolution is neither new nor surprising – 'the origin of the

TABLE II

Environmental limits for growth and reproduction of blue-green algae

	Lower limit	Upper limit
Temperature, Survival	−269 °C (liquid helium)	+112 °C (dry soil) + 90 °C (wet soil)
Temperature, Photosynthesis	− 30 °C (lichens) − 9 °C (Saline Lake, U.S.S.R.)	+ 75 °C (*Synechococcus*, hot springs)
Hydrogen-ion concentration	pH: 3.0–3.5 (hot springs, U.S.A., Japan, Iceland)	pH: 10–11 (*Arthrospira* in alkaline lakes, Kenya) pH: 13 (*Plectonema* on silica)
Oxidation potential	Eh: −200 mv (at pH 6)	Eh: +700 mv (at pH 4)
Total salinity	freshwater (0.001–0.1%) brackish (1.0%) marine (3.5%)	Great Salt Lake (27.5%) Dead Sea (31.5%)
Altitude	396 m below sea level (Dead Sea, Palestine)	> 5200 m above sea level (Himalaya Mts., Kashmir)
Resistance to aridity and desiccation	*Schizothrix* and *Anacystis* at the Atacama Desert, Chile, perhaps the most arid locale known. *Nostoc* revived after 107 yr of storage as a dried herbarium specimen	
Resistance to ionizing radiation	*Microcoleus* survived 2560 kr γ-irradiation form Co^{60} source. (25 kr lethal to the chlorophycean alga *Eudorina*) *Oscillatoria* survived 200 kr X-irradiation. (20 kr lethal to the chlorophycean alga *Spirogyra*)	
Data from:	Ahmadjian, 1962; Brock, 1969; Brock and Brock, 1970; Cameron, 1963; Davis, 1972; Drouet, 1963; Forest and Weston, 1966; Fuhs, 1968; Godward, 1962; Lund, 1962; Shields and Drouet, 1962; Vallentyne, 1963.	

Metazoa' and its putative stratigraphic corollary, 'the Precambrian-Cambrian boundary problem', have long proved fertile areas for interesting, imaginative, but inconclusive speculation. The discoveries of the past decade, however, have provided new insight into these problems and suggested new approaches to their solution. It has been only in recent years, for example, that paleobiologists have attempted to define and to place in geochronologic perspective those major innovations in biologic organization involved in the evolutionary transition from microscopic prokaryote to megascopic, multicellular eukaryote. Thus, such matters as the origin of the eukaryotic cell type, the development of eukaryotic sexuality, the evolution of life cycles among the algae, and the temporal and ecological interrelatedness of early metaphytic and metazoan evolution, are now being actively investigated by paleobiologists for the first time. What follows here is a discussion of new data bearing on the timing of one of the most critical stages in the evolutionary sequence – the origin of the nucleated, eukaryotic cell type – followed by a short summary of those events that may been particularly important in subsequent Precambrian evolution.

All higher organisms are composed of nucleated, eukaryotic cells. The evolutionary progression leading to the development of metazoans and metaphytic algae therefore had its beginning in the Precambrian derivation of primitive eukaryotes from prokaryotic (anucleate, non-mitotic) ancestors. When did this event occur? To answer this question, criteria must first be established by which microfossils of primitive, eukaryotic microorganisms can be differentiated from those of prokaryotes. Two types of criteria can be applied: (i) morphology – eukaryotes are of more complex morphology and are generally of larger cell size than prokaryotes; and (ii) intracellular organization – unlike prokaryotic bacteria and blue-green algae, the intracellular 'machinery' of the eukaryotic cell is compartmentalized into membrane-bound organelles such as nuclei, chloroplasts, mitochondria, etc. Unfortunately, the former of these criteria is of rather limited applicability; tissues, complex branching patterns, differentiated sporangia, etc., while assured evidence of eukaryotic organization, are relatively advanced features and would not have been exhibited by primitive unicellular members of the lineage. Cell size might be of somewhat more general use; prokaryotes tend to be relatively small (ca. 1–10 μm) and eukaryotic cells, including those of unicellular forms, relatively large (commonly 10–100 μm). There are, however, many known exceptions. Cell size, by itself, cannot be considered compelling evidence of affinity.

What of the other criterion? Can such presumably delicate structures as nuclei, chloroplasts, etc., be preserved in fossils? Although it may seem somewhat surprising, the answer to this question is 'yes'. Over the years, a variety of organelles and structurally preserved intracellular bodies has been reported from fossil plants: nuclei, some containing chromatin-like granules, have been detected in calcified sphenopsid spores from the Carboniferous of Indiana (Baxter, 1950); starch granules occur in Carboniferous conifer seeds preserved in Kansas coal balls (Baxter, 1964); nuclei and chromosome-like structures have been reported from silicified fern sporangia from the Jurassic of India (Vishnu-Mittre, 1967); nuclei and nucleoli occur in permineralized female gametophytes of a Carboniferous club moss (Darrah, 1938); pyrenoids, nuclei and degraded chloroplasts have been detected in silicified algal unicells from the Eocene of Wyoming (Bradley, 1946); nuclei, some with nucleoli, occur in silicified conifer pollen from the Carboniferous of France (Florin, 1936); and nuclei and nucleoli have been detected in cortical cells of a silicified fossil cycad from the Triassic of New Mexico (Gould, 1971). In all of these examples, the apparent organelles are preserved in permineralized plant parts, the cells being three-dimensionally embedded in a fine-grained, chemically deposited, calcareous or siliceous matrix. This type of preservation ('permineralization' or 'petrifaction') is exhibited by virtually all of the important Precambrian microbiotas discovered in recent years. Thus, it is perhaps not surprising that organelle-like bodies, apparently indicative of eukaryotic organization, have been detected in the Precambrian.

The oldest known unicells exhibiting what might be termed 'fairly convincing' organelle-like bodies (Figures 8 and 9) are those occurring in black stromatolitic cherts of the Beck Spring Dolomite of eastern California, thought to be on the order of 1300 m.y. in age (Cloud *et al.*, 1969). Comparable, but apparently somewhat younger microfossils (Figures 11–13), perhaps 1000 to 1100 m.y. old, have recently been detected in black, stromatolitic cherts near Boorthanna, South Australia

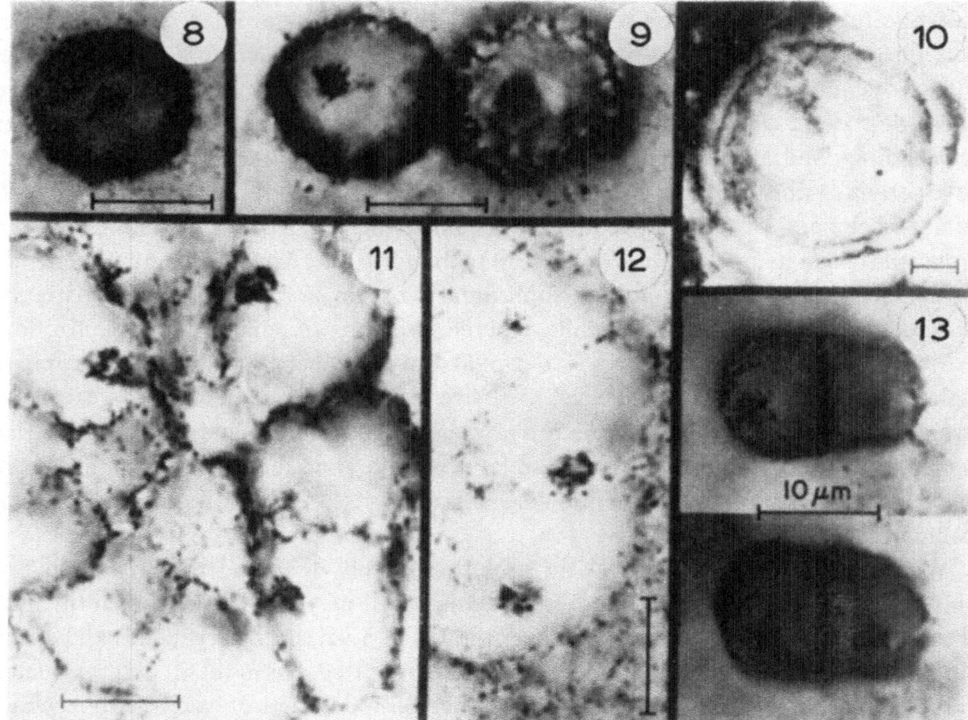

Figs. 8–13. Optical photomicrographs showing possible eukaryotes in petrographic thin sections of black cherts from the late Precambrian (ca. 1300 m.y. old) Beck Spring Dolomite of California (Figures 8, 9) and from a late Precambrian (ca. 1000 m.y. old) dolomitic formation occurring near Boorthanna, South Australia (Figures 10–13; also see Schopf and Fairchild, 1973). Lines for scale represent 10 μm.

(Schopf and Fairchild, 1973). In addition to solitary and colonial cells containing possible organelles, the Boorthanna assemblage includes unusually large unicells (Figure 10), 50–70 μm in diameter, that similarly might be of eukaryotic affinities. Although these data are highly suggestive – and may ultimately be accepted as establishing the existence of the nucleated cell type – additional evidence is needed to 'prove' the point. Specifically, many of the possible organelles in both deposits could be interpreted with equal force as remnants of collapsed, 'coalified' cytoplasm; as such, the cells could be of either pro- or eukaryotic affinity.

The earliest assured eukaryotes now known are those occuring in bedded black cherts of the approximately 900 m.y.-old Bitter Springs Formation of central Australia; nine taxa of eukaryotes, including possible fungi and several types of unicellular algae, have been described from this extraordinary well preserved assemblage (Schopf, 1968; Schopf and Blacic, 1971). Two principal categories of intracellular bodies occur in the Bitter Springs unicells: dense, micrometer-sized, organic 'spots' (Figure 14), and somewhat larger, spheroidal, 'granular' bodies (Figure 15). As is shown in Figure 16, both types of bodies occur in cells of about the same diameter; however, the bodies

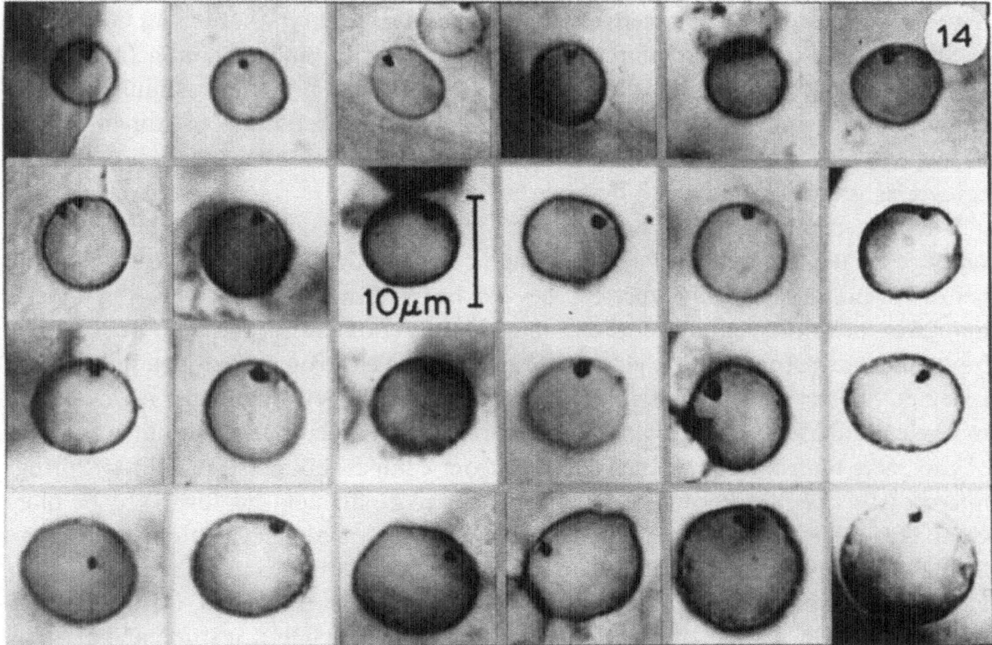

Fig. 14. Optical photomicrographs showing unicells containing intracellular bodies ('spots') in petrographic thin sections of black chert from the late Precambrian (ca. 900 m.y. old) Bitter Springs Formation of central Australia. Magnification is indicated by the line representing 10 μm.

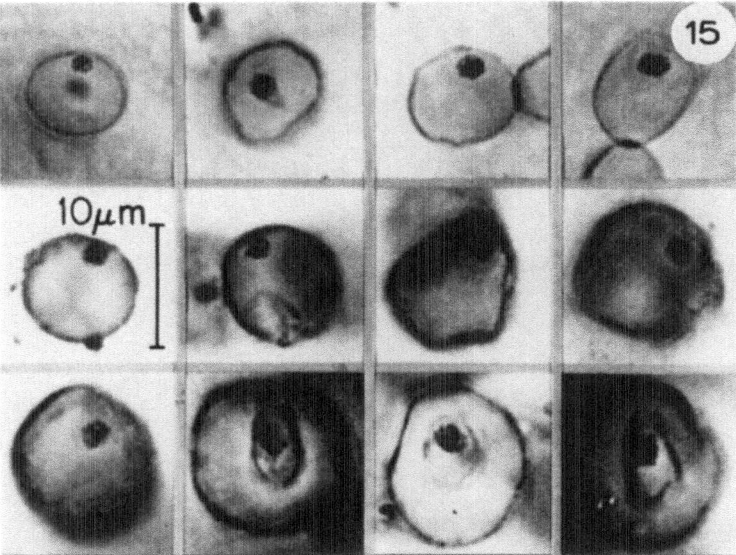

Fig. 15. Optical photomicrographs showing unicells containing intracellular bodies ('granulars') in petrographic thin sections of black chert from the late Precambrian (ca. 900 m.y. old) Bitter Springs Formation of central Australia. Magnification is indicated by the line representing 10 μm.

always occur singly, one per cell, and in no case have both types of bodies been found in the same cell. Cells containing 'spots' are decidedly more common (amounting to roughly 90% of the 573 cells measured) than those containing 'granulars'. As is shown in Figure 17, the 'granulars', about 2 μm in diameter, are approximately twice as large as the 'spots'; although the size ranges are somewhat overlapping, the two categories are roughly defined by both texture and size.

What interpretation can be given these two classes of intracellular bodies? Both types of bodies are demonstrably organic in composition – they have the color and texture of preserved organic matter; they can be freed from the encompassing mineral matrix by palynological maceration; embedded in epoxy resin, they can be sectioned on an ultramicrotome with a glass knife; and, in the electron microscope they exhibit a degree of electron transmissibility typical of carbonaceous materials. As evidenced by their regularity in size, shape and texture, by their relationship to mineral grains of the chert matrix, and by their absence from comparably preserved Bitter Springs microfossils that are known on other grounds (e.g., filamentous forms) to be blue-green algae, they appear to be original, or only slightly altered structures, rather than artifacts of preservation. They are, in general, substantially too small and too regular to be reasonably interpreted as cytoplasmic remnants (degraded cytoplasm also occurs in certain of the Bitter Springs unicells; it differs in color, texture, and electron microscopic ultrastructure from the organelle-like intracellular bodies). In short, the 'spots' and 'granulars' appear to be reasonably well preserved, original organic structures. Possibly the two types of bodies represent two different types of organelles.

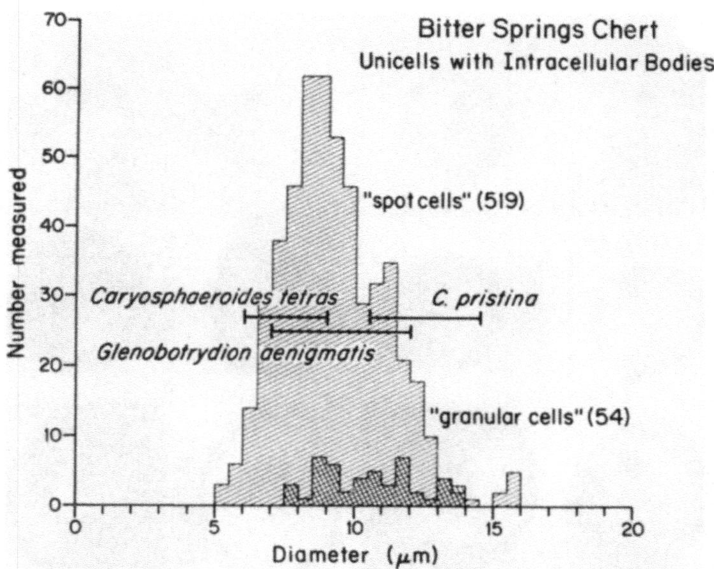

Fig. 16. Histogram showing size distribution of Bitter Springs unicells containing organelle-like 'spots' (diagonally lined area) and 'granulars' (cross-hatched area). Size ranges of comparable unicellular taxa, previously described from the Bitter Springs cherts, are indicated (see Schopf, 1968; Schopf and Blacic, 1971).

This, however, seems rather unlikely – if they represent, for example, preserved nuclei and pyrenoids, why are not both types of organelles preserved within a single cell? It is perhaps more likely that the two types of bodies represent differing degrees or stages of preservation of a single type of organelle. Specifically, as shown in Table III,

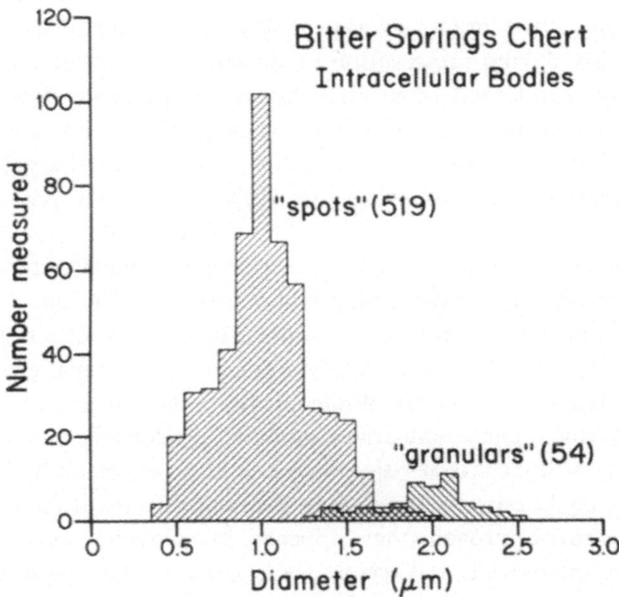

Fig. 17. Histogram showing size distribution of organelle-like 'spots' and 'granulars' occurring in unicells preserved in the late Precambrian Bitter Springs cherts.

TABLE III

Comparison of fossil and modern unicells

	Measured diameter (μm)			
	Cell	'Spot'	'Granular'	Nucleus
Bitter Springs unicells				
Number Measured	573	519	54	
Range	5.0–16.0	0.4–2.0	1.3–2.5	
Most (> 85%)	7.0–12.0	0.7–1.3	1.6–2.2	
Average	9.5	1.0	1.9	
Modern eukaryotic unicells (Chlorophyceae)				
Chlorococcum polymorphum [a]	8.0			1.5
Chlorosarcinopsis eremi [b]	8.3			1.6
Chlorosarcinopsis gelatinosa [b]	9.0			1.8
Chlorosarcina brevispinosa [b]	9.0			2.0
Spongiochloris incrassata [b]	9.2			2.0
Friedmannia israeliensis [b]	9.7			2.2

[a] Bischoff and Bold, 1963.

[b] Chantanachat and Bold, 1962.

the 'granulars' are of about the same size as the nuclei of morphologically comparable modern eukaryotic algae; further, 'granular cells' seem generally somewhat better preserved than 'spot cells', not uncommonly containing additional intracellular structures such as preserved cellular membranes (Figure 15), and the 'granulars' themselves exhibit a chromatinlike, reticulate substructure (Figure 15). Thus, I regard the majority of the 'granulars' as probably representing relatively well preserved nuclei; presumably during preservation or diagenesis, many of these structures became altered and condensed to produce 'spots'. Regardless, however, of whether this suggested relationship is correct, the important generalization seems clear – microorganisms containing some type of organelle(s), and thus of eukaryotic affinities, were represented in the biota at least as early as 900 m.y. ago and, possibly, earlier than 1300 m.y. ago.

In a recent paper, my colleagues and I have reviewed much of the available paleobiologic evidence bearing on the timing and nature of evolutionary events involved in the late Precambrian transition from unicellular to multicellular organization (Schopf *et al.*, 1973). The following is a synopsis of the principal evolutionary events tentatively postulated in that paper. While highly speculative and likely to be in error, at least in detail, this scenario illustrates results of a relatively new type of approach to the problem, one that stresses the interrelatedness of metaphytic and metazoan evolution and relies heavily on the known Precambrian fossil record, a paleobiological-paleoecological approach that appears to provide considerable promise.

(i) Unicellular, planktonic eukaryotes were derived from prokaryotic ancestors probably prior to 1300 m.y. ago; these earliest eukaryotes were haploid, asexual algae that reproduced exclusively by mitotic cell division.

(ii) Sexual, eukaryotic algae, having an alternation of diploid and (meiotically produced) haploid generations, apparently originated about 1000 m.y. ago. The advent of this 'haploid-dominant' sexual life cycle resulted in a marked increase in both diversity and evolutionary rate among late Precambrian eukaryotes; the development of eukaryotic sexuality may have served as an 'evolutionary trigger' leading to the emergence of megascopic organization.

(iii) Between about 900 and 700 m.y. ago, algal life cycles evolved through a series of stages from 'haploid-dominant' to 'haploid-diploid coequal' and 'diploid-dominant'; this sequence was paralleled by the development of multicellular organization, megascopic size and a benthic habit (thus resulting in the origin of metaphytes).

(iv) The late Precambrian diversification of eukaryotic microorganisms led to the emergence of pelagic, heterotrophic protists from which were derived, perhaps as early as 750 m.y. ago, primitive eumetazoans. The earliest eumetazoans were small, soft-bodied and pelagic, primarily relying on phytoplankton as a food source; subsequently, and prior to about 650 m.y. ago, eumetazoans 'discovered the bottom,' a shallow, shelf-like environment replete with a diverse assemblage of encrusting and erect metaphytes.

(v) Near the close of the Precambrian, 'hard parts', of diverse types and compositions, appeared more or less concurrently in various groups of benthic metaphytes and metazoans. The origin of this skeletal material, which served supportive and protective functions, seems interrelated with the development of relatively large body size and with the appearance of increasingly efficient food gathering capability in mobile, bilaterally symmetrical eumetazoans.

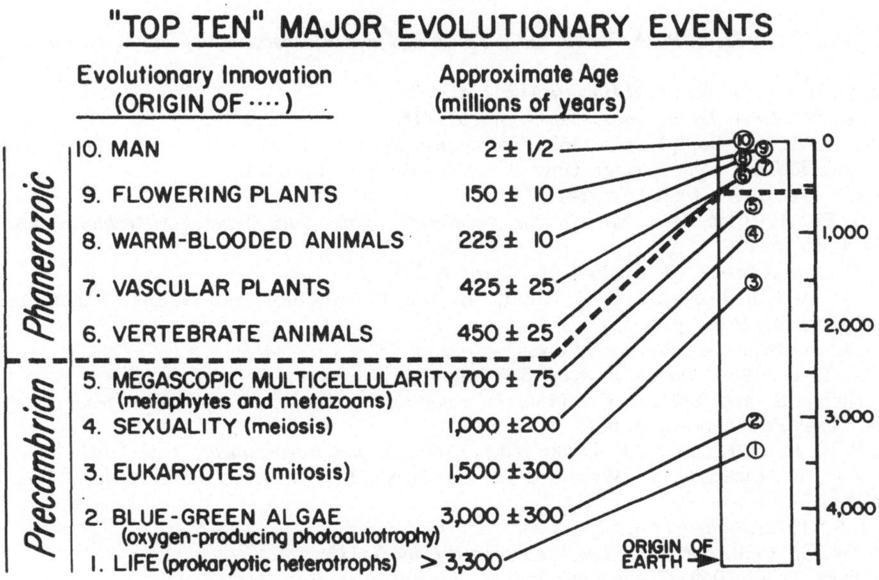

"TOP TEN" MAJOR EVOLUTIONARY EVENTS

	Evolutionary Innovation (ORIGIN OF ····)	Approximate Age (millions of years)
Phanerozoic	10. MAN	2 ± 1/2
	9. FLOWERING PLANTS	150 ± 10
	8. WARM-BLOODED ANIMALS	225 ± 10
	7. VASCULAR PLANTS	425 ± 25
	6. VERTEBRATE ANIMALS	450 ± 25
Precambrian	5. MEGASCOPIC MULTICELLULARITY (metaphytes and metazoans)	700 ± 75
	4. SEXUALITY (meiosis)	1,000 ± 200
	3. EUKARYOTES (mitosis)	1,500 ± 300
	2. BLUE-GREEN ALGAE (oxygen-producing photoautotrophy)	3,000 ± 300
	1. LIFE (prokaryotic heterotrophs)	> 3,300

ORIGIN OF EARTH

0 · 1,000 · 2,000 · 3,000 · 4,000

Fig. 18. Chart illustrating the temporal distribution of important innovations that have occurred during the course of biological evolution.

Conclusion

Figure 18 summarizes the temporal relationships among the various evolutionary events here discussed and between them and major events of the Phanerozoic. It is evident that the Precambrian encompasses an enormous segment of geologic time and includes the vast majority, more than 80%, of the history of life on this planet. In recent years, substantial progress has been made toward deciphering the course of this early biological development. The three generalizations presented and discussed in the foregoing paragraphs summarize important results of the active investigations of the past decade. The task of the next decade will be to evaluate these generalizations and to add depth and detail to the current broad brush picture of the early evolutionary progression.

Acknowledgments

This study has been supported by Grant GB-37257 from the U.S. National Science Foundation, Systematic Biology Program, and by Grant NGR 05-007-407 from the National Aeronautics and Space Administration. This is the published version of a paper delivered at the 4th International Conference on the Origin of Life held in Barcelona, Spain, in June, 1973. Funds enabling my participation in this conference were provided by a Fellowship from the John Simon Guggenheim Memorial Foundation.

References

Ahmadjian, V.: 1962, in R. A. Lewin (ed.), *Physiology and Biochemistry of Algae*, Academic Press, p. 817.

Baxter, R. W.: 1950, *Botanical Gazette* **112**, 174.

Baxter, R. W.: 1964, *Trans. Kansas Acad. Sci.* **67**, 418.

Bischoff, H. W. and Bold, H. C.: 1963, *Phycological Studies. IV: Some Soil Algae From Enchanted Rock and Related Algal Species*, Univ. Texas Publication No. 6318.

Bradley, W. H.: 1946, *Am. J. Sci.* **244**, 215.

Brock T. D.: 1969, in *Microbial Growth, Nineteenth Symp. Soc. General Microbiol.*, Cambridge Univ. Press, p. 15.

Brock, T. D. and Brock, M. L.: 1970, *J. Phycol.* **6**, 371.

Brooks, J.: 1971, in J. Brooks, P. R. Grant, M. Muir, P. van Gijzel, and G. Shaw (eds.), *Sporopollenin*, Academic Press, p. 351.

Brooks, J. and Muir, M.: *Nature*, in press.

Cameron, R. E.: 1963, *Ann. N.Y. Acad. Sci.* **108**, 412.

Chantanachat, S. and Bold, H. C.: 1962, *Phycological Studies. II: Some Algae from Arid Soils*, Univ. Texas Publication No. 6218.

Cloud, P. E., Jr.: 1968, in E. T. Drake (ed.), *Evolution and Environment*, Yale Univ. Press, p. 1.

Cloud, P. E., Jr., Licari, G. R., Wright, L. A., and Troxel, B. W.: 1969, *Proc. Nat. Acad. Sci. (U.S.)* **62**, 623.

Davis, J. S.: 1972, *Biologist* **54**, 52.

Darrah, W. C.: 1938, *Bot. Mus. Leaflets Harvard Univ.* **6**, 113.

Desikachary, T. V.: 1959, *Cyanophyta*, Indian Council Agric. Res., New Delhi.

Desikachary, T. V. and Padmaja, T. D.: 1970, *Rev. Algol.* (Paris) **10**, 8.

Drouet, F.: 1963, *Proc. Acad. Nat. Sci. Philadelphia* **115**, 261.

Florin, R.: 1936, *Svensk Botanisk Tidskrift.* **30**, 624.

Forest, S. and Weston, C. R.: 1966, *J. Phycol.* **2**, 163.

Fritsch, F. E.: 1965, *The Structure and Reproduction of the Algae*, Vol. 2, Cambridge Univ. Press.

Fuhs, G. W.: 1968, in D. F. Jackson (ed.), *Algae, Man and Environment*, Syracuse Univ. Press, p. 213.

Glaessner, M. F.: 1962, *Biol. Rev. Cambridge Phil. Soc.* **37**, 467.

Glaessner, M. F.: 1966, *Earth Sci. Rev.* **1**, 29.

Godward, M. B. E.: 1962, in R. A. Lewin (ed.), *Physiology and Biochemistry of Algae*, Academic Press, p. 551.

Gould, R. E.: 1971, *Am. J. Bot.* **58**, 239.

Klein, R. M. and Cronquist, A.: 1967, *Quart. Rev. Biol.* **42**, 105.

Kvenvolden, K. A.: 1972, *Proc. 24th Internat. Geol. Cong.*, Sect. 1, Precambrian Geology, Montreal, p. 31.

Lund, J. W. G.: 1962, in R. A. Lewin (ed.), *Physiology and Biochemistry of Algae*, Academic Press, p. 759.

Merek, E. L.: 1973, *Bioscience* **23**, 153.

Nagy, B. and Nagy, L. A.: 1969, *Nature* **223**, 1226.

Oehler, D. Z., Schopf, J. W., and Kvenvolden, K. A.: 1972, *Science* **175**, 1246.

Schopf, J. W.: 1968, *J. Paleontol.* **42**, 651.

Schopf, J. W.: 1970, *Biol. Rev. Cambridge Phil. Soc.* **45**, 319.

Schopf, J. W.: in T. Dobzhansky, M. K. Hecht, and W. C. Steere (eds.), *Evolutionary Biology*, Vol. 7, Appleton-Century-Crofts, in press.

Schopf, J. W. and Blacic, J. M.: 1971, *J. Paleontol.* **45**, 925.

Schopf, J. W., Oehler, D. Z., Horodyski, R. J., and Kvenvolden, K. A.: 1971, *J. Paleontol.* **45**, 477.

Schopf, J. W., Haugh, B. N., Molnar, R. E., and Satterthwait, D. F.: 1973, *J. Paleontol.* **47**, 1.

Schopf, J. W. and Fairchild, T. R.: 1973, *Nature* **242**, 537.

Shields, L. M. and Drouet, F.: 1962, *Am. J. Bot.* **49**, 547.

Smith, J. W., Schopf, J. W., and Kaplan, I. R.: 1970, *Geochim. Cosmochim. Acta* **34**, 659.
Stebbins, G. L.: 1969, *The Basis of Progressive Evolution*, Univ. North Carolina Press, Chapel Hill.
Vail, J. R. and Dodson, M. H.: 1969, *Trans. Geol. Soc. Africa* **72**, 79.
Vallentyne, J. R.: 1963, *Ann. N.Y. Acad. Sci.* **108**, 342.
Vishnu-Mittre: 1967, in C. D. Darlington and K. R. Lewis (eds.), *Chromosomes Today*, Oliver and Boyd, p. 250.
Walter, M. R., Bauld, J., and Brock, T. D.: 1972, *Science* **178**, 402.

PART III

PRIMORDIAL ORGANIC CHEMISTRY

THE ATMOSPHERE OF THE PRIMITIVE EARTH AND THE PREBIOTIC SYNTHESIS OF AMINO ACIDS

STANLEY L. MILLER

Dept. of Chemistry, University of California, San Diego, La Jolla, Calif. 92037, U.S.A.

Abstract. The atmosphere of the Earth at the time of its formation is now generally believed to have been reducing, an idea proposed by Oparin and extensively discussed by Urey. This atmosphere would have contained CH_4, N_2 with traces of NH_3, water and hydrogen. Only traces of NH_3 would have been present because of its solubility in water. UV light and electric discharges were the major sources of energy for amino acid synthesis, with electric discharges being the most efficient, although most other sources of energy also give amino acids.

The first prebiotic electric discharge synthesis of amino acids showed that surprisingly high yields of amino acids were synthesized. Eleven amino acids were identified, four of which occur in proteins. Hydroxy acids, simple aliphatic acids and urea were also identified. These experiments have been repeated recently, and 33 amino acids were identified, ten of which occur in proteins, including all of the hydrophobic amino acids.

Methionine can be synthesized by electric discharges if H_2S or CH_3SH is added to the reduced gases. The prebiotic synthesis of phenylalanine, tyrosine and tryptophan involves pyrolysis reactions combined with plausible solution reactions.

Eighteen amino acids have been identified in the Murchison meteorite, a type II carbonaceous chondrite, of which six occur in proteins. All of the amino acids found in the Murchison meteorite have been found among the electric discharge products. Furthermore, the ratios of amino acids in the meteorite show a close correspondence to the ratios from the electric discharge synthesis, indicating that the amino acids on the parent body of the carbonaceous chondrites were synthesized by electric discharges or by an analogous process.

1. Introduction

The atmosphere of the primitive Earth at the time of its formation is now generally believed to have been reducing, an idea proposed by Oparin (1938) and extensively discussed by Urey (1952). This atmosphere is sometimes spoken of as a methane, ammonia, water and hydrogen atmosphere, but most of the nitrogen was probably molecular nitrogen and only traces of ammonia would have been present. Ammonia is very soluble in water and would be largely as NH_4^+ in the primitive ocean, if the pH of the ocean was about 8 as it is at present (Miller and Urey, 1959). The partial pressure of ammonia was not likely to have been much greater than 10^{-5} atm. This is based on an ocean of pH 8 and the fact that the NH_4^+ concentration could not have been much more than 0.01 M because of the absorption of NH_4^+ by clay minerals (Sillén, 1967; Bada and Miller, 1968). Although 10^{-5} atm was a small percentage of the primitive atmosphere, this value corresponds to a significant concentration of NH_4^+ and NH_3 in the ocean, and this dissolved ammonia would have played an important role in the prebiotic synthesis of organic compounds. On the other hand, if the partial pressure of NH_3 in the atmosphere drops much below 10^{-6} atm, the NH_3 and NH_4^+ in the ocean will be to low to obtain amino acids by a Strecker synthesis and aspartic acid will be unstable with respect to deamination to fumaric acid (Bada and Miller, 1968).

An atmosphere of this type is thermodynamically stable in the presence of sufficient hydrogen, and only extremely small amounts of organic compounds would be present at thermodynamic equilibrium. However, various sources of energy can produce activated molecules (e.g. HCN and aldehydes) which would have reacted in the primitive ocean to produce biologically interesting organic compounds. The major source of energy in the atmosphere is UV light, with electric discharges being the next most abundant source. Most prebiotic experiments have used electric discharges because of their convenience, and they may have been the most important source of energy because of their efficiency, especially for the synthesis of hydrogen cyanide.

In the first prebiotic synthesis of amino acids and other organic compounds by electric discharges (Miller, 1953, 1955, 1957a, b), the discharge apparatus was designed to circulate the gases past the electrodes in a five liter flask and to condense the discharge products into a 500 ml flask about half-filled with water. The water was boiled in the 500 ml flask and the spark run for a week. The results of this experiment gave the yields in Table I. The yields of amino acids from this experiment appear to be the highest obtained in any prebiotic experiment of this type, even though no attempt was made to maximize the yields.

A great deal of effort was taken to properly identify the compounds. Most of the compounds in the table were positively identified by obtaining a melting point of a suitable derivative and showing that the mixed melting point with an authentic sample was not depressed.

It was shown that most if not all of the amino and hydroxy acids were produced by the Strecker and cyanohydrin synthesis from the corresponding aldehyde.

$$\mathrm{RCHO + HCN + NH_3 \rightleftarrows \underset{\displaystyle NH_2}{\underset{|}{RCH}}{-}CN + H_2O}$$

$$\mathrm{\underset{\displaystyle NH_2}{\underset{|}{RCN}}{-}CN + 2H_2O \rightarrow \underset{\displaystyle NH_2}{\underset{|}{RCH}}{-}COOH + NH_3}$$

$$\mathrm{RCHO + HCN \rightleftarrows \underset{\displaystyle OH}{\underset{|}{RCH}}{-}CN}$$

$$\mathrm{\underset{\displaystyle OH}{\underset{|}{RCH}}{-}CN \rightarrow \underset{\displaystyle OH}{\underset{|}{RCH}}{-}COOH}$$

The formation of the nitriles is a reversible reaction, while the hydrolysis of the nitrile, first to the amide and then to the acid, is irreversible. It should be noted that no acid hydrolysis step was used in these experiments; the nitriles were hydrolyzed to the amide and acid by the basic conditions of the boiling ammonia solution in electric discharge apparatus.

The Strecker synthesis can account for the synthesis of the α-amino acids and α-

TABLE I

Yields from sparking a micture of CH_4, NH_3, H_2O and H_2: 59 *m*mole (710 mg) of carbon was added as CH_4. The percent yields are based on the carbon

Compound	Yield (μmole)	Yield (%)
Glycine	630	2.1
Glycolic acid	560	1.9
Sarcosine	50	0.25
Alanine	340	1.7
Lactic acid	310	1.6
N-Methylalanine	10	0.07
α-Amino-n-butyric acid	50	0.34
α-Aminoisobutyric acid	1	0.007
α-Hydroxybutyric acid	50	0.34
β-Alanine	150	0.76
Succinic acid	40	0.27
Aspartic acid	4	0.024
Glutamic acid	6	0.051
Iminodiacetic acid	55	0.37
Iminoacetic-propionic acid	15	0.13
Formic acid	2330	4.0
Acetic acid	150	0.51
Propionic acid	130	0.66
Urea	20	0.034
N-Methyl urea	15	0.051
Total		15.2

hydroxy acids. The β-alanine and succinic acids were probably derived from acrylonitrile (Miller, 1957a).

$$NH_3 + CH_2 = CH-CN \rightarrow H_2N-CH_2-CH_2-CN \rightarrow H_2N-CH_2-CH_2-COOH$$

$$NC^- + CH_2 = CH-CN \rightarrow NC-CH_2-CH_2-CN \rightarrow HOOC-CH_2-CH_2-COOH$$

The addition of HCN to acrylonitrile appears to be irreversible, but it is not clear whether this is the case for NH_3.

2. Prebiotic Synthesis of the Hydrophobic Amino Acids

In these first experiments, the only hydrophobic amino acids that could be identified were glycine, alanine, α-amino-n-butyric acid, and α-aminoisobutyric acid. Subsequently, the prebiotic synthesis of the higher aliphatic amino acids was claimed, for example, by the action of electric discharges on $CH_4+NH_3+H_2O$ (Grossenbacher and Knight, 1965; Czuchojowski and Zawadzki, 1968; Oró, 1963; Ponnamperuma *et al.*, 1969; Matthews and Moser, 1966), by heating $CH_4+NH_3+H_2O$ to 900–1200° (Harada and Fox, 1964; Taube *et al.*, 1967), and by the action of shock waves on CH_4, C_2H_6, NH_3, and H_2O (Bar-Nun *et al.*, 1970). The amino acids were identified only by an amino-acid analyzer (Grossenbacher and Knight, 1965; Czuchojowski and Zawadzki, 1968; Oró, 1963; Matthews and Moser, 1966; Harda and Fox, 1964; Bar-Nun *et al.*, 1970), only by paper electrophoresis (Taube *et al.*, 1967), or only by gas chromatography (Ponnamperuma *et al.*, 1969). However, these techniques are not sufficient by themselves to identify an amino acid.

The synthesis under prebiotic conditions of aspartic acid, glutamic acid, serine, threonine, and proline have been reported in a number of prebiotic experiments, but they have not been properly identified (except for aspartic acid (Oró, 1968)). The synthesis of these amino acids (except proline) has also been reported from the polymerization of HCN (reviewed by Lemmon, 1970), but again without proper identification. A prebiotic synthesis of threonine should also yield allothreonine, but this amino acid has never been reported. In addition, several investigators have reported the appearance of a large peak at the isoleucine position on the amino-acid analyzer (Grossenbacher and Knight, 1965; Oró, 1963; Ponnamperuma *et al.*, 1969; Matthews and Moser, 1966; Fox and Windsor, 1970). The identification of this peak as isoleucine has been questioned (Oró, 1963; Fox and Windsor, 1970). It is evident that this compound cannot be isoleucine, since a corresponding peak for alloisoleucine is not observed.

It seemed likely that part of the reason for the low yield of the higher aliphatic amino acids in the first experiment was the high partial pressure of water near the spark. Therefore, the spark discharge experiments were repeated using lower temperatures and also lower pressures of ammonia (Ring *et al.*, 1972; Wolman *et al.*, 1972).

The electric discharge flask was patterned after that of Oró (1963) and is shown in Figure 1. One hundred ml of 0.05 M NH_4Cl was added to the flask, the flask evacuated, and sufficient NH_3 was added to bring the pH to 8.7. The partial pressure of NH_3 was calculated to be 0.1 mm. Methane (200 mm) and N_2 (80 mm) were then added, and the spark discharge was run for 48 hr. The tesla coil was the same kind as previously used (Miller, 1955). The temperature of the flask remained between 20 and 25°. The aqueous solution, presumably containing the amino nitriles rather than the amino acids (Miller, 1957), was hydrolyzed with 3 M HCl for 24 hr, desalted, and evaporated to dryness. The dried sample was hydrolyzed again with 3 M HCl in order to open the lactam rings of glutamic acid, α,γ-diaminobutyric acid, and α-hydroxy-γ-aminobutytic acid that may have been formed during the desalting. Seven similar runs were combined. A sample of the desalted amino acids was then run on the amino-acid analyzer (Dus *et al.*, 1967). The chromatogram is shown in Figure 2.

In order to separate the various amino acids, the combined runs were chromatographed on a column (38.5 × 2.2 cm) of Dowex 50, in the hydrogen form, and eluted

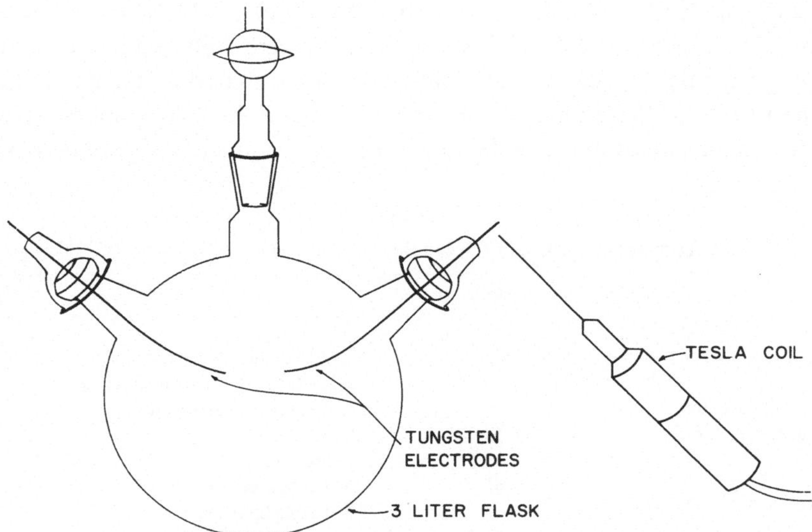

Fig. 1. Electric discharge apparatus used to synthesize amino acids at room temperature (after Oró, 1963).

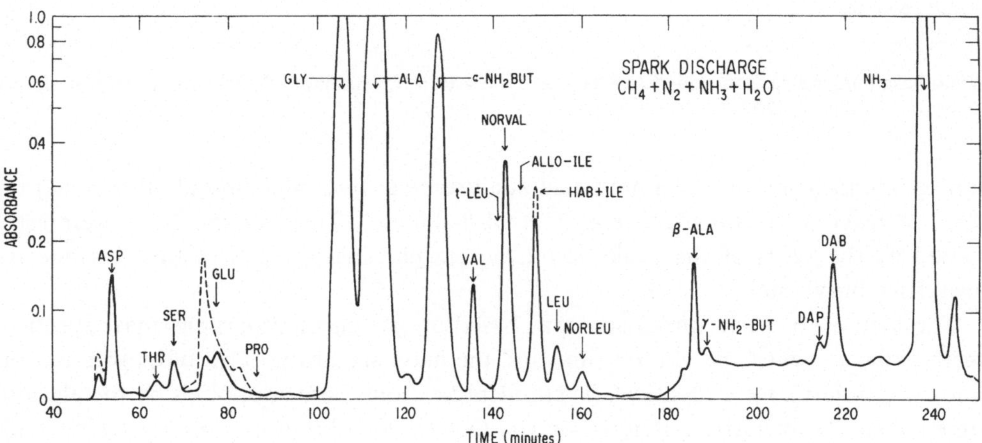

Fig. 2. Amino acid analyzer chromatogram of the desalted amino acids after the electric discharge synthesis. The arrows show the elution time of the indicated amino acids but this is not the basis for their identification. The peak labeled valine, norvaline (norval), allisoleucine, leucine and norleucine (norleu) contain between 50 and 80% of the indicated compound; most of the peaks labeled HAB + + Ile is α-hydroxy-γ-diaminobutyric acid; DAB is α, γ-diaminobutyric acid; DAP is α,β-diamino-propionic acid. The dashed line shows the increase in color on heating the column eluent and ninhydrin for 30 min instead of the usual 8 min; the color yield of proteins amino acids is not changed by the additional heating, but N-substituted amino acids give substantial increases in color yield.

with HCl (Wall, 1953) (400 ml of 1.5 M HCl, 700 ml of 2.5 M HCl, 400 ml of 4.0 M HCl, and 600 ml of 6.0 M HCl). Eighteen fractions were collected and evaporated to dryness in a vacuum desiccator; each fraction was quantitated on the amino-acid analyzer. The results of these analyses are shown in Table II. Those amino acids given only approximate values in the table were either not completely separated from inter-

TABLE II

Yields from sparking CH_4 (336 μmole), N_2, and H_2O with traces of NH_3

	μmole		μmole
Glycine	440	α, γ-Diaminobutyric acid	33
Alanine	790	α-Hydrocy-γ-aminobutyric acid	74
α-Amino-n-butyric acid	270	α, β-Diaminopropionic	6.4
α-Aminoisobutyric acid	~ 30	Isoserine	5.5
Valine	19.5	Sarcosine	55
Norvaline	61	N-Ethylglycine	30
Isovaline	~ 5	N-Propylglycine	~ 2
Leucine	11.3	N-Isopropylglycine	~ 2
Isoleucine	4.8	N-Methylalanine	~ 15
Alloisoleucine	5.1	N-Ethylalanine	< 0.2
Norleucine	6.0	β-Alanine	18.8
tert-Leucine	< 0.02	β-Amino-n-butyric acid	~ 0.3
Proline	1.5	β-Amino-isobutyric acid	~ 0.3
Aspartic acid	34	γ-Aminobutyric acid	2.4
Glutamic acid	7.7	N-Methyl-β-alanine	~ 5
Serine	5.0	N-Ethyl-β-alanine	~ 2
Threonine	~ 0.8	Pipecolic acid	~ 0.05
Allothreonine	~ 0.8		

Yield based on the carbon added as CH_4. Glycine = 0.26%, Alanine = 0.71%, total yield of amino acids in the table = 1.90%.

fering amino acids on the amino-acid analyzer (e.g. threonine and allothreonine) or did not react with ninhydrin (e.g. N-ethyl-β-alanine). These amino acids were estimated by the areas of the peaks found on gas chromatography of the N-trifluoroacetyl-sec-butyl esters.

The identity of each amino acid was based on gas chromatography-mass spectrometry, that is, when the elution time and the mass spectrum of an unknown and of known standards were identical. These identifications were supported by the elution times of the known and unknown on the Dowex 50 column and on the amino-acid analyzer.

In addition to confirming the identity of the unknown, the gas-chromatographic analysis showed that each of the amino acids (except for isovaline, α-hydroxy-γ-aminobutyric acid, α,γ-diaminobutyric acid, and aspartic acid that do not form two peaks on the colums used) were racemic within the experimental error (45–55% D-isomer). This result shows that there was no significant contamination from reagents or dust during the separation process. This conclusion applies particularly to the proline, where the yield was sufficiently low that contamination was a reasonable possibility.

It was particularly interesting to find that the amino acid which has the same elution time as isoleucine on the amino acid analyzer is α-hydroxy-γ-aminobutyric acid. The aldehyde precursor of this is β-aminopropionaldehyde. The Strecker amino acid of this aldehyde is α,γ-diaminobutyric acid, which was found among the electric discharge products. Similarly, both isoserine and α,β-diaminopropionic acid were found, the aldehyde precursor of which is aminoacetaldehyde.

The results in Table II show that there is no selective synthesis of the branched-chain amino acids that occur in proteins. Indeed, the yield of norvaline is three times that of valine, although the yield of norleucine is 50% that of leucine and that of (isoleucine + alloisoleucine). Therefore, the occurrence of glucine, alanine, valine, isoleucine, and leucine in proteins, but the absence of α-amino-n-butyric acid, norvaline, alloisoleusine, and norleucine, cannot be understood on the basis of the yields from this type of synthesis.

The absence of *tert*-leucine in this synthesis may be due to the instability of its amino nitrile or its precursor aldehyde (pivaladehyde). Two aliphatic amino acids with both α hydrogens substituted, α-aminoisobutyric acid and isovaline (α-amino-α-methyl-n-butyric acid) were found. The six-carbon amino acids of this class were not looked for. The relatively low yield of α-aminoisobutyric and isovaline may be due to the instability of the corresponding aminonitrile, as has been discussed elsewhere (Miller, 1957a).

The yield of proline is quite low – seven times lower than leucine. A yield this low suggests that an electric-discharge synthesis of this type was not the only source of proline on the primitive Earth.

The yield of 5- and 6-carbon amino acids is substantially lower than the glycine, alanine, or even the α-amino-n-butyric acid. The mole ratios of glycine: alanine: α-aminobutyric acid: (valine + norvaline): 6-carbon amino acids are 100:180:60:18:6. There is some variation in these ratios in different experiments, with the same conditions and spark source. The reason for the lower yields of the 5- and 6-carbon amino acids is not clear. When the temperature was lowered to 0° during the sparking, the yields of the 5- an d6-carbon amino acids were not increased, nor did the use of ethane instead of methane increase the yields.

It seems unlikely that the amino acids that were important in prebiotic polypeptides were present in the primitive ocean in about equal concentrations. Several mechanisms would have been available to concentrate certain amino acids in prebiotic polypeptides. One possible mechanism would have concentrated the sea water in a lagoon by evaporation until the amino acids were partially precipitated, and then synthesized peptides with the precipitated amino acids. The process would have concentrated the 5- and 6-carbon amino acids, since they are less soluble than glycine, alanine, and α-amino-n-butyric acid. The concentration of the higher aliphatic amino acids could also have occurred by adsorption on suitable mineral surfaces.

Peptide bonds of the 5- and 6-carbon amino acids are more stable to hydrolysis than those of glycine and alanine. This stability has been correlated with the 'rule of 6'

(Whitfield, 1964), and it has been suggested that the relative rates of hydrolysis generated sequences of peptides in the primitive ocean that were hydrolytically stable (Nicholson, 1970). The same considerations would predict that the higher aliphatic amino acids would concentrate in the peptides of a primitive ocean.

These considerations make it plausible that the yields of the 5- and 6-carbon amino acids obtained in this type of experiment would have been adequate for prebiotic peptide synthesis.

3. Prebiotic Synthesis of Methionine

Methionine might be considered 'too complex' an amino acid to be synthesized in significant yield in an electric discharge reaction because of the large number of possible isomers. However, the large yields of α-hydroxy-γ-aminobutyric acid and α,γ-diaminobutyric acid obtained in the above experiments, raised the possibility that acrolein was the precursor. In the presence of CH_3SH, NH_3, and HCN, acrolein might give methionine (Van Trump and Miller, 1972).

We therefore sparked a mixture of CH_4, N_2, H_2S, H_2O, and a trace of NH_3 and found that methionine was synthesized in 0.03% yield based on the H_2S (2×10^{-4}% based on the carbon). The yields of glycine and alanine in the same experiment were 0.068 and 0.104%, respectively based on the carbon.

It seemed likely that the limiting factor in this methionine synthesis was the formation of the thiomethyl group. We therefore sparked a mixture of CH_4, N_2, H_2O, CH_3SH, and a trace of ammonia and obtained 0.23% methionine. In a similar experiment, a mixture of CH_4, N_2, H_2O and a trace of NH_3 was sparked, and the CH_3SH was added at the termination of the sparking. This gave an 0.63% yield of methionine, based on the sulfur.

The methionine in these experiments was separated from the other amino acids by the amino-acid analyzer without the use of ninhydrin. The methionine peak, which contained norvaline and allisoleucine, was desalted and rerun on the analyzer, with the use of only the pH 3.28 citrate buffer. This procedure separated the methionine from the norvaline and alloisoleucine, and allowed quantification. The methionine peak was converted to the N-tri-fluoroacetyl-D-2-butyl ester and chromatographed on a 50-m gas chromatographic capillary column with OV-225 as a stationary phase. The mass spectrum and gas chromatography retention time of this unknown derivative agreed with the mass spectrum and retention time of an authentic sample of DL-methionine.

Acrolein was shown to be a product of the action of a spark discharge on a mixture of CH_4 and H_2O by means of an acrolein-specific fluorescent assay with m-aminophenol (Alarcon, 1968). The yield of acrolein was 0.04%, based on the methane. An alternate spectrophotometric assay with 4-hexylresorcinol (Cohen and Altshuller, 1961), a method which is sensitive to both acrolein and propiolaldehyde, gave a combined yield of 0.11%. A propiolaldehyde yield of 0.07% is in agreement with the results of Dowler *et al.* (1970).

On the basis of these results we propose the following model for the prebiotic synthesis of methionine.

$$\left.\begin{array}{l} CH_4 + H_2O \xrightarrow{\text{spark}} CH_2{=}CH{-}CHO \\ CH_4 + H_2S \xrightarrow[\text{or spark}]{\text{UV}} CH_3SH \end{array}\right\} \longrightarrow CH_3S{-}CH_2{-}CH_2{-}CHO$$

$$\xrightarrow[\text{HCN}]{\text{NH}_3} CH_3S{-}CH_2{-}CH_2{-}\underset{\displaystyle NH_2}{\underset{|}{CH}}{-}CN \xrightarrow{\text{hydrolysis}} CH_3S{-}CH_2{-}CH_2{-}\underset{\displaystyle NH_2}{\underset{|}{CH}}{-}COOH$$

A model experiment was conducted to determine whether methionine could be synthesized from acrolein under the dilute conditions expected in the primitive ocean, a reaction which is effective using high concentrations (Smith, 1962). A mixture of acrolein (8×10^{-4} M), HCN (4×10^{-3} M), NH_3 (2.5×10^{-3} M), and CH_3SH (5×10^{-4} M) was added to a deaerated solution of NH_4Cl (7.5×10^{-3} M, final pH 8.7) and the solution was kept for 28 days. The mixture was hydrolyzed with 3 M HCl, desalted, hydrolyzed again with 3 M HCl, and quantitated on the amino-acid analyzer. The yields were 15% methonine, 0.5% glutamic acid, 0.5% α,γ-diaminobutyric acid, and 13% α-hydroxy-γ-aminobutyric acid, based on the added acrolein. The same experiment omitting the CH_3SH gave 1.5% glutamic acid and 0.8% α,γ-diaminobutyric acid. These results show that CH_3SH adds to acrolein in preference to NH_3 or HCN under the conditions of the experiment. The relative yields of the amino acids in the primitive ocean would depend on the concentrations of CH_3SH, HCN, and NH_3 as well as the temperature and hydrolytic conditions.

It appears likely that acrolein was a key intermediate in prebiotic amino acid synthesis, being a precursor not only of methionine but also of glutamic acid, homocysteine, homoserine, α,γ-diaminobutyric acid, and α-hydroxy-γ-aminobutyric acid (Figure 3).

The results from these low temperature electric discharge experiments are extremely encouraging. Table II shows that 10 of the 20 amino acids in proteins are obtained. Asparagine and glutamine can be included, since they would be obtained as hydrolysis products of the nitriles or from aspartic and glutamic acids under polymerization conditions. Methionine is obtained in the same experiment if H_2S or CH_3SH is added

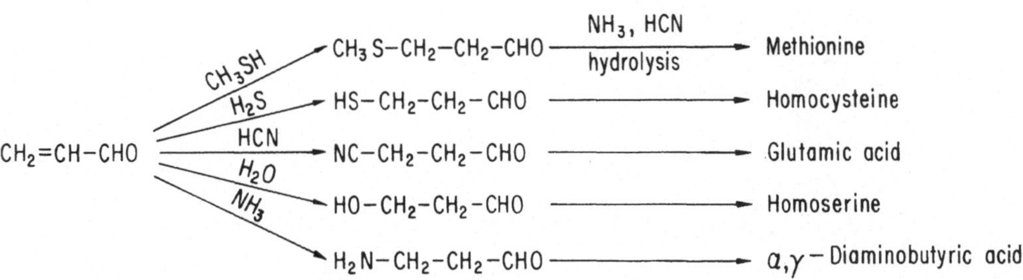

Fig. 3. Previotic synthesis of amino acids from acrolein.

TABLE III

Relative abundances of amino acids in the Murchison meteorite and in an electric discharge synthesis. Mole ratio to glycine (= 100): 0.05–0.5, *; 0.5–5, **; 5–50 ***; > 50, ****.

Amino acid	Murchison meteorite	Electric discharge
Glycine	* * * *	* * * *
Alanine	* * * *	* * * *
α-Amino-n-butyric acid	* * *	* * * *
α-Aminoisobutyric acid	* * * *	* *
Valine	* * *	* *
Norvaline	* * *	* * *
Isovaline	* *	* *
Proline	* * *	*
Pipecolic acid	*	<*
Aspartic acid	* * *	* * *
Glutamic acid	* * *	* *
β-Alanine	* *	* *
β-Amino-n-butyric acid	*	*
β-Aminoisobutyric acid	*	*
γ-Aminobutyric acid	*	* *
Sarcosine	* *	* * *
N-Ethylglycine	* *	* * *
N-Methylalanine	* *	* *

to the gas mixture. Cysteine is obtained by the action of UV on a mixture of methane ethane, ammonia water and hydrogen sulfide (Sagan and Khare, 1971; Khare and Sagan, 1971). The prebiotic synthesis of phenylalanine, tyrosine and tryptophan involves the combination of pyrolysis and plausible solution reactions (Friedmann and Miller, 1969; Friedmann *et al.*, 1971). This adds up to 17 of the 20 amino acids in proteins.

The synthesis of proline and pipecolic acid implies the synthesis of arginine and lysine if they arise from the nitrile,

$$\begin{array}{l} CH_2 - CH_2 \\ | \quad\quad\;\; | \\ CH_2 \quad CH-CN \\ \;\; \diagdown \;\; \diagup \\ \quad\;\; N \\ \quad\;\; H \end{array} + NH_4^+ \rightleftarrows H_3^+N-CH_2-CH_2-CH_2-\underset{\displaystyle NH_2}{\underset{|}{CH}}-CN$$

$$\xrightarrow{H_2N-CN} H_2N-\overset{\displaystyle NH}{\overset{\|}{C}}-NH-CH_2-CH_2-CH_2-\underset{\displaystyle NH_2}{\underset{|}{CH}}-CN \rightarrow \text{Arginine}$$

The nitrile of proline is quite stable relative to ornithine nitrile, but if the nitriles react with cyanamide, the open chain compound should be stable, leading to arginine. Similarly the homolog of proline nitrile, pipecolic acid nitrile, would lead to lysine through homoarginine. However, the details of these reactions need to be worked out.

Only histidine has not been synthesized under prebiotic conditions and properly identified.

4. The Murchison Meteorite

On September 28, 1969 a type II carbonaceous chondrite fell in Murchison, Australia. Surprisingly large amounts of amino acids were found by Kvenvolden *et al.* (1970, 1971). The first report identified seven amino acids (glycine, alanine, valine, proline, glutamic acid, sarcosine and α-aminoisobutyric acid), of which all but valine and proline had been found in the original electric discharge experiments. The most strikking are sarcosine and α-aminoisobutyric acid. The second report identified 18 of the amino acids present of which nine had previously been identified in the original electric discharge experiments, but the remaining nine had not. Oró *et al.* (1971) have confirmed the results of Kvenvolden *et al.* (1970, 1971).

At that time we had identified the hydrophobic amino acids from the low temperature electric discharge experiments described above, and therefore we examined the Dowex 50(H^+) samples for the non-protein amino acids found in Murchison. We are able to find all of them.

There is a striking similarity between the products and relative abundances of the amino acids produced by electric discharge and the meteorite amino acids. Unfortunately, there are only a few quantitative values (Cronin and Moore, 1971) for the meteorite amino acids, but we have estimated their relative abundances from the published gas chromatography data (Kvenvolden *et al.*, 1971). Table III compares the results.

The most notable difference between the meteorite and the electric-discharge amino acids is the pipecolic acid, the yield being extremely low in the electric discharge. Proline is also present in relatively low yield from the electric discharge. The amount of α-aminoisobutyric acid is greater than α-amino-n-butyric acid in the meteorite, but the reverse is the case in the electric discharge. The amounts of aspartic and glutamic acids in the meteorite are comparable, but there is five times as much aspartic acid as glutamic acid in the electric discharge.

We do not believe that reasonable differences in ratios of amino acids detract from the overall picture. Indeed, the ratio of α-aminoisobutyric acid to glycine is quite different in two meteorites of the same type, being 0.4 in Murchison and 3.8 in Murray (Cronin and Moore, 1971).

One would expect quantitative differences between the meteorite composition and the electric-discharge products, even if the mechanism of formation in the two cases were identical. Thus cyanoacetylene, which is synthesized by sparking CH_4 and N_2, is probably the major precursor of aspartic acid, but the yield of cyanoacetylene is decreased by the addition of small amounts of NH_3 to the CH_4+N_2 mixture (Sanchez *et al.*, 1966). Therefore, local differences in the NH_3 partial pressures on the parent body of the carbonaceous chondrites would result in substantial differences in the aspartic acid concentration if the amino acids were not completely mixed on the parent body. Temperature differences can also affect the yields. For example, the stability of α-aminoisobutyronitrile is quite sensitive to temperature and cyanide concentration, while α-amino-n-butyronitrile, being more stable is not particularly sensitive to these factors. Also there is the possibility that the amino acids in Murchison may have been decarboxylated or otherwise destroyed to different extents in the time since they were synthesized.

The very close correspondence between the amino acids found in the Murchison meteorite and those produced by an electric discharge synthesis, both as to the amino acids produced and their relative ratios, suggests that the amino acids in the meteorite were synthesized on the parent body by means of an electric discharge or analogous processes. Electric discharges appear to be the most favored source of energy but sufficient data are not available to make realistic comparison with other energy sources. In any case, it is unlikely that a single source of energy synthesized all of the organic compounds either on the parent body of the carbonaceous chondrites or on the primitive Earth. All sources of energy would have made their contribution, and the problem is to evaluate the relative importance of each source.

Our ideas on the prebiotic synthesis of organic compounds are based largely on the results of experiments in model systems. So it is extremely gratifying to see that such synthesis really did take place on the parent body of the meteorite, and so it becomes quite plausible that they took place on the primitive Earth.

Acknowledgment

This work was supported by National Science Foundation Grant 25048. I am indebted to my collaborators N. Friedmann, W. J. Haverland, D. Ring, J. E. Van Trump and Y. Wolman.

References

Alarcon, R. A.: 1968, *Ann. Chem.* **40**, 1704.
Bada, J. L. and Miller, S. L.: 1968, *Science* **159**, 423.
Bar-Nun, A., Bar-Nun, N., Bauer, S. H., and Sagan, C.: 1970, *Science* **168**, 470.
Cohen, I. R. and Altshuller, A. P.: 1961, *Ann. Chem.* **33**, 726.
Cronin, J. R. and Moore, C. B.: 1971, *Science* **172**, 1327.
Czuchajowski, L. and Zawadzki, W.: 1968, *Rocz. Chem.* **42**, 697.
Dowler, M. J., Fuller, W. D., Orgel, L. E., and Sanchez, R. A.: 1970, *Science* **169**, 1320
Dus, K., Lindroth, S., Pabst, R., and Smith, R. A.: 1967, *Ann. Biochem.* **18**, 532.
Grossenbacher, K. A. and Knight, C. A.: 1965, in S. W. Fox (ed.), *The Origins of Prebiological Systems*, Academic Press, New York, pp. 173–183.
Friedmann, N., Haverland, W. J., and Miller, S. L.: 1971, in R. Buvet and C. Ponnamperuma (eds.), *Chemical Evolution and the Origin of Life*, North Holland Publ. Co., Amsterdam, pp. 123–135.
Friedmann, N. and Miller, S. L.: 1969, *Science*, **166**, 766.
Harada, K. and Fox, S. W.: 1964, *Nature* **201**, 335.
Khare, B. N. and Sagan, C.: 1971, *Nature* **232**, 577.
Kvenvolden, K., Lawless, J., Pering, K., Peterson, E., Flores, J., Ponnamperuma, C., Kaplan, I. R., and Moore, C.: 1970, *Nature* **228**, 923.

Kvenvolden, K. A., Lawless, J. G., and Ponnamperuma, C.: 1971, *Proc. Nat. Acad. Sci. U.S.A.* **68**, 486.
Lemmon, R. H.: 1970, *Chem. Rev.* **70**, 95.
Matthews, C. N. and Moser, R. E.: 1966, *Proc. Nat. Acad. Sci. U.S.A.* **56**, 1087.
Miller, S. L.: 1953, *Science* **117**, 528.
Miller, S. L.: 1955, *J. Am. Chem. Soc.* **77**, 2351.
Miller, S. L.: 1957a, *Biochim. Biophys. Acta* **23**, 480.
Miller, S. L.: 1957b, *Ann. N.Y. Acad. Sci.* **69**, 260.
Miller, S. L. and Urey, H. C.: 1959, *Science* **130**, 245.
Nicholson, I.: 1970, *J. Macromol. Sci. Chem.* **A4**, 1619.
Oparin, A. I.: 1938, *The Origin of Life*, Macmillan, New York.
Oró, J.: 1963, *Nature* **197**, 862.
Oró, J.: 1968, *J. Brit. Interplanetary Soc.* **21**, 12.
Oró, J., Gilbert, J., Lichtenstein, H., Wickstrom, S., and Flory, D. A.: 1971, *Nature* **230**, 105.
Pollock, G. E. and Oyama, V. I.: 1966, *J. Gas Chromatogr.* **4**, 126.
Ponnamperuma, C., Woeller, F., Flores, J., Romiez, M., and Allen, W.: 1969, in R. F. Gould (ed.), *Chemical Reactions in Electric Discharges* (Advances in Chemistry, series no. 80, A.C.S., Washington), pp. 280–288.
Ring, D., Wolman, Y., Friedmann, N., and Miller, S. L.: 1972, *Proc. Nat. Acad. Sci. U.S.A.* **69**, 765.
Sagan, C. and Khare, B. N.: 1971, *Science* **173**, 417.
Sanchez, R. A., Ferris, J. P., and Orgel, L. E.: 1966, *Science* **154**, 784.
Sillén, L. G.: 1967, *Science* **156**, 1189.
Taube, M., Zdrojewski, S. Z., Samochocka, K., and Jezierska, K.: 1967, *Angew. Chem.* **79**, 239.
Urey, H. C.: 1952, *Proc. Nat. Acad. Sci. U.S.A.* **38**, 351.
Van Trump, J. E. and Miller, S. L.: 1972, *Science* **178**, 859.
Wall, J. S.: 1953, *Ann. Chem.* **25**, 950.
Whitfield, R. E.: 1963, *Science* **142**, 577.
Wolman, Y., Haverland, W. J., and Miller, S. L.: 1972, *Proc. Nat. Acad. Sci. U.S.A.* **69**, 809.

BIOMOLECULES FROM HCN*

J. P. FERRIS, J. D. WOS, T. J. RYAN, A. P. LOBO and D. B. DONNER
Dept. of Chemistry, Rensselaer Polytechnic Institute, Troy, N.Y. 12181, U.S.A.

Abstract. The mechanism of the condensation of dilute aqueous solutions of HCN and the products formed by these reactions have been investigated. The initial HCN condensation reactions yield **3**, a compound which is readily oxidized to **4**. A similar oxidation of **5** to **6** was also observed. Urea is formed on hydrolysis of **4**. The oxidation-reduction products formed from HCN may be in part a consequence of the oxidation of **3**. It has been established by combination GC/MS that the amino acids glycine, diaminosuccinic acid, α-amino-isobutyric acid, aspartic acid, alanine and isoleucine are released on acid hydrolysis of the 'HCN polymer'. Hydantoin (**7**), 5,5-dimethylhydantoin (**8**) and 5-carboxymethyldenehydantoin (**10**) are also released on acid hydrolysis of the HCN condensation products. The direct conversion of the dicarbonyl derivative of diaminosuccinic acid to orotic acid via **10** at pH 8 has been observed. This conversion suggests a direct route to pyrimidines from HCN.

HCN is considered to have been one of the more important precursors of biological molecules on the primitive Earth (Sanchez *et al.*, 1967). HCN could have been formed by a number of routes under possible prebiological conditions. Plausible routes to purines and amino acids have been demonstrated starting from HCN (Lowe *et al.*, 1963; Oro and Kimball, 1961).

We have continued to explore the chemistry of HCN with the goals of determining the mechanisms involved in HCN oligomerizations in dilute aqueous solutions and identifying the compounds which are produced in these oligomerization mixtures. Complex mixtures are formed if 0.1–1.0 M aqueous solutions of cyanide are allowed to stand at room temperature (pH 9.2) for one month or longer. Oxalic acid, urea and the so-called 'HCN polymer'** have been isolated from these reaction mixtures. The so-called 'HCN polymer' has been found to have a molecular weight of much less than 1000 so that it does not fall within the accepted definition of a polymer. It has been possible to fractionate this substance(s) into acidic, basic and neutral fractions. Urea is the main constituent of the neutral fraction (Ferris *et al.*, 1973a). Amino acids are released on acid hydrolysis of the acidic and basic fractions, however, neither pronase nor carboxypeptidase catalyzed the release of amino acids from this material. The absence of cleavage by the proteolytic enzymes suggests that peptide links are not present in the 'HCN polymer' (Ferris *et al.*, 1973a; Labadie *et al.*, 1968).

The early stages of the HCN oligomerization reaction are well understood. HCN condenses to form the tetrameric species diaminomaleonitrile (DAMN) (**3**) via a dimer (**1**) and a trimer (**2**) (Sanchez *et al.*, 1967; Ferris *et al.*, 1972a, b). However, the conversion of these oligomers to urea, oxalic acid and the 'HCN polymer' is not understood. The possibility of subsequent nucleophilic condensation to cyanide with these

* Chemical Evolution XV. For the previous paper in the series see Ferris and Ryan (1973).
** The terms 'HCN polymer' and oligomerization mixtures are used interchangeably to describe those products formed by condensation of HCN in aqueous alkaline solution (Ferris *et al.*, 1973a, b)

oligomers was probed by the addition of the nucleophiles to the reaction mixture. There was no observable effect on the rate of the oligomerization, when azide, monomethylamine, trimethylamine, hydroxide or phenoxide were substituted for ammonia in the condensation. The only difference that was observed was the formation of monomethylurea in addition to urea when the oligomerization was carried out in the presence of monomethylamine (Ferris *et al.*, 1973b). This result suggests that cyanate, or a substance similar to cyanate, is formed during the course of the condensation. Cyanogen is a likely source of cyanate since it is hydrolyzed to both oxalic acid derivatives and cyanate at pH 9.

$$\mathrm{HCN} \rightleftarrows \underset{\mathbf{1}}{\mathrm{HN{=}CHCN}} \rightleftarrows \underset{\mathbf{2}}{\mathrm{NH_2CH(CN)_2}} \rightleftarrows \underset{\mathbf{3}}{(\mathrm{H_2N})(\mathrm{NC})\mathrm{C{=}C}(\mathrm{CN})(\mathrm{NH_2})} \xrightarrow{O_2} \underset{\mathbf{4}}{\mathrm{HN{=}C(CN){-}C(CN){=}NH}}$$

The occurrence of oxidation and reduction reactions during the course of the HCN oligomerization is required by the formation of urea and oxalic acid (HCN oxidation products) and, after hydrolysis of the 'HCN polymer', amino acids (HCN reduction products). We investigated the nature of these redox reactions by studying the reactions of diisopropyldiaminomaleonitrile (diisopropylDAMN) **5** in aqueous solution (Ferris and Ryan, 1973). Compound **5** is stable in aqueous solution in the absence of oxygen, however, it is readily oxidized to **6** in the presence of oxygen. No higher molecular weight oligomers were isolated in the reaction of **5** or **6** with cyanide.

$$\underset{\mathbf{5}}{(\mathrm{RHN})(\mathrm{NC})\mathrm{C{=}C}(\mathrm{CN})(\mathrm{NHR})} \xrightarrow{O_2} \underset{\mathbf{6}}{\mathrm{RN{=}C(CN){-}C(CN){=}NR}}$$

$$\mathrm{R} = (\mathrm{CH_3})_2\mathrm{CH{-}}$$

A similar oxidation of **3** to **4** appears to take place in the presence of oxygen. Compound **4** is converted to urea in the presence of ammonia. This suggests a possible pathway for urea formation during the oligomerization of cyanide. The direct oxidation of cyanide to cyanogen and cyanate which in turn react with ammonia to give urea is another possibility (Ferris *et al.*, 1973a).

The facile air oxidation of **3** suggested that the oligomerization of cyanide might follow a different course in the absence of oxygen. The loss of cyanide from an aqueous (pH 9.2) solution proceeded more slowly in the absence of oxygen. However, we did observe the formation of the 'HCN polymer', urea and, after hydrolysis, amino acids when oxygen was excluded from the reaction solution (Ferris and Ryan, 1973). Thus, redox reactions leading to biomolecules from HCN could have taken place on the primitive earth in the absence of oxygen.

The amino acids released on hydrolysis of the 'HCN polymer' have been identified by their retention times on ion exchange columns (amino acid analysis). We confirmed the presence of amino acids in the acid hydrolysate with the same ion exchange retention times as those reported previously (Lowe *et al.*, 1963; Matthews and Moser, 1967). Other ninhydrin positive materials were also observed, one of which was identified as citrulline by comparison of its retention time on ion exchange chromatography with that of an authentic sample and by base hydrolysis to ornithine.

TABLE I

Amino acids released on acid hydrolysis of the 'HCN polymer'[a]

	Use of amino acid analyzer (%)			GC/MS (%)
	A	B	C	
asp	2.4–18	4.2–32	5–15	3–36
thr	0.1–9.3	0.02–0.1	0.1–0.2	0
ser	0.2–6.1	6.2–8.4	0–0.9	0
glu	1.4–6.1	0–0.01	0.9–17	0–50
cit	0.5–1.4	nd	nd	nd
gly	6.7–89	54–85	48–88	30–50
ala	0.8–2.5	2.3–5.3	2.2–5.9	1–10
val	0.2–0.9	0	0.03–0.2	0
ile	tr	0.05–0.1	0–0.13	0–6
leu	0.6	0.05–0.1	0.06–0.13	0
lys	0.2–0.9	0	0.2–0.6	0
arg	0	0	0–1.3	0
his	tr	0	0.9–5.0	0
diaminosuccinic	nd	nd	nd	0–80
α-aminoisobutyric	nd	nd	nd	0–4
α-aminobutyric	nd	0–0.02	nd	0
α, β-diaminopropionic	nd	0–0.6	0	0
β-ala	0	0–1.2	0	0
μ mole mg^{-1}	0.4–1.1	0.34–0.54	nd	1

[a] The assigned amino acid structures are unconfirmed with exception of structures verified by GC/MS or, as in the case of citrulline, by some other method. % = % of the total μ moles of amino acids; μ mole mg^{-1} = total μ moles of animo acids released per mg of 'HCN polymer' hydrolyzed. tr = trace; nd = not determined; A = Ferris *et al.* (1973a); B = Matthews and Moser (1967); C = Lowe *et al.* (1963).

The structures assigned to some of the amino acids using the amino acid analyzer could not be confirmed by combination gas chromatography-mass spectrometry (GC/MS). In studies done collaboratively with Prof. John Oro we found glycine and the *meso-* and *d,l*-isomers of diaminosuccinic acid to be major hydrolytic products (Table I). We also found, in decreasing amounts, α-aminoisobutyric acid, aspartic acid, alanine and isoleucine (citrulline derivatives are not sufficiently volatile to be detected by GC/MS). The *meso-* (Toi *et al.*, 1960) and *d,l*-isomers (McKennis and Yard, 1958) of diaminosuccinic acid were synthesized and were observed to have the same GC/MS retention times and mass spectra as the corresponding amino acids released on acid hydrolysis of the 'HCN polymer'. Furthermore, the retention time on the amino acid analyzer of *meso*-diaminosuccinic acid corresponds closely with that of aspartic acid and the retention of *d,l*-diaminosuccinic acid corresponds closely with that of glutamic acid. These data demonstrate the hazards of identifying amino acids solely by use of ion exchange retention times.

Further evidence for the structures of glycine, α-aminoisobutyric acid and diaminosuccinic acid was obtained when hydantoins corresponding to the cyclization products of the carbonyl derivatives of these amino acids were isolated after acid hydrolysis of the 'HCN polymer'. A 0.5% yield of hydantoin (**7**), a 0.5–1% yield of 5,5-dimethylhydantoin (**8**) and a 2–3% yield of 5-carboxymethylidenehydantoin (**10**) were obtained. Each of the above hydantoins was isolated in pure form and identified by direct comparison with an authentic sample. Compound **10** was established as the hydantoin corresponding to diaminosuccinic acid by acid hydrolysis of the dicarbamyl derivative of diamino succinic acid (**9**).

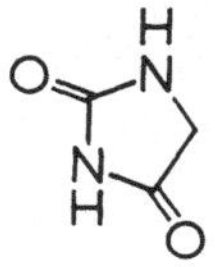

7 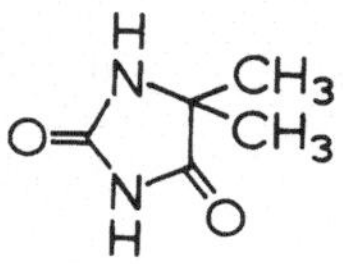 **8**

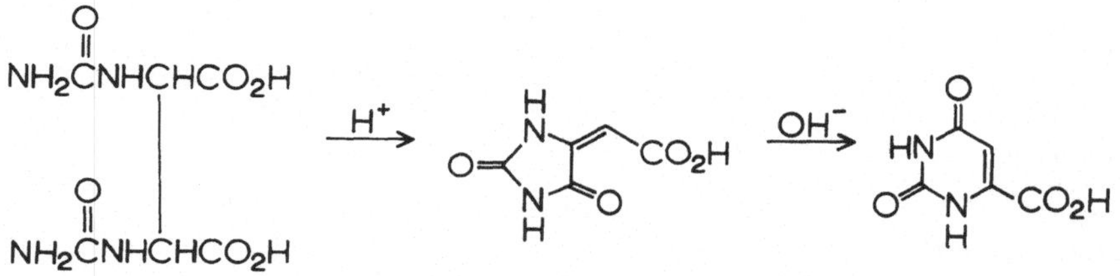

$NH_2C(=O)NHCHCO_2H$ / $NH_2C(=O)NHCHCO_2H$ $\xrightarrow{H^+}$ **10** $\xrightarrow{OH^-}$ **11**

9 **10** **11**

The previously reported conversion of 5-carboxymethylidenehydantoin (**10**) to orotic acid (**11**) (Nyc and Mitchell, 1947) suggests a possible prebiotic route to pyrimidines directly from HCN. We have investigated this possibility using the dicarbamyl derivative of diaminosuccinic acid (**9**) as a model for the diaminosuccinic acid derivative in the oligomerization mixture. The conversion of **9** to **10** was observed to take place in the pH 1–8.5 range in the absence of phosphate buffer. Approximately a 1% yield of **10** is observed at pH 7. The conversion of **10** to **11** does not vary with the buffers used and is observed at pH values greater than 6.5. From these data one would predict that the direct conversion of **9** to **11** would be observed in the pH 7.5–8.5 range and this has been observed. Although the direct formation of **11** from the 'HCN polymer' remains to be carried out, these present experiments strongly suggest the possibility of a direct route from cyanide to the pyrimidines.

The isolation of the pyrimidine orotic acid demonstrates that all three major groups of nitrogen containing biological monomers, purines, pyrimidines and amino acids, could have been formed from cyanide on the primitive Earth. Further, many of the compounds utilized in the biosynthesis of pyrimidines (Chart 1) (Lehninger, 1970) are formed in mildly alkaline aqueous cyanide.

Chart 1

$$\left.\begin{array}{l}\text{Citrulline + phosphate}\searrow\\ \text{Citrulline + ATP}\longrightarrow\\ CO_2 + NH_3 + 2\text{ATP}\nearrow\end{array}\right.\begin{array}{c}\text{carbamyl}\\ \text{phosphate}\end{array}\rightarrow\begin{array}{c}\text{ureidosuccinic}\\ \text{acid}\end{array}\rightarrow$$

$$\begin{array}{c}\text{dihydroorotic}\\ \text{acid}\end{array}\underset{\text{DPN}}{\rightarrow}\begin{array}{c}\text{orotic}\\ \text{acid}\end{array}\twoheadrightarrow\begin{array}{c}\text{pyrimidine}\\ \text{nucleotides}\end{array}$$

Indirect evidence for the formation of cyanate (a precursor of carbamyl phosphate) as well as direct evidence for the formation of citrulline (Ferris *et al.*, 1973a), aspartic acid and orotic acid from aqueous alkaline cyanide has been obtained. It is generally assumed that the first self-replicating system utilized nucleic acid derivatives in the 'prebiotic soup' that had been formed by chemical processes. However, when these sources were exhausted the early life forms may have evolved biosynthetic pathways starting from the other compounds produced by aqueous cyanide. If this hypothesis is correct then the pathway for pyrimidine biosynthesis has not changed significantly over the past 3–4 billion years.

Acknowledgements

This work was supported by grant NGR 30-018-148 from the National Aeronautics and Space Administration, grant GP-19255 from the National Science Foundation and a Career Development Award (G.M. 6380) to J.P.F. from the U.S. Public Health Service. We thank Prof. John Oro for the GC/MS analyses.

References

Ferris, J. P., Donner, D. B., and Lotz, W.: 1972a, *J. Am. Chem. Soc.* **94**, 6968.
Ferris, J. P., Donner, D. B., and Lotz, W.: 1972b, *Bioorganic Chem.* **2**, 95.
Ferris, J. P., Donner, D. B., and Lobo, A. P.: 1973a, *J. Mol. Biol.* **74**, 499.
Ferris, J. P., Donner, D. B., and Lobo, A. P.: 1973b, *J. Mol. Biol.* **74**, 511.
Ferris, J. P. and Ryan, T. J.: 1973, *J. Org. Chem.* **38**, 3302.
Labadie, M., Ducastaing, S., and Breton, J. C.: 1968, *Bull. Soc. Pharm. Bordeaux* **107**, 61.
Lehninger, A. L.: 1970, *Biochemistry*, Worth Publishers, New York, pp. 571–572.
Lowe, C. U., Rees, M. W., and Markham, R.: 1963, *Nature* **199**, 219.
Matthews, C. N. and Moser, R. E. 1967, *Nature* **215**, 1230.
McKennis, H. and Yard, A. S.: 1958, *J. Org. Chem.* **23**, 980.
Nyc, J. F. and Mitchell, H. K.: 1947, *J. Am. Chem. Soc.* **69**, 1382.
Oro, J. and Kimball, A. P.: 1961, *Arch. Biochem. Biophys.* **94**, 293.
Sanchez, R. A., Ferris, J. P., and Orgel, L. E.: 1967, *J. Mol. Biol.* **30**, 223.
Toi, K., Mori K., and Izumi, Y.: 1960, *Bull. Chem. Soc. Japan* **33**, 1529.

THE PREBIOTIC SYNTHESIS OF OLIGONUCLEOTIDES

J. ORÓ and E. STEPHEN-SHERWOOD

Depts. of Biophysical Sciences and Chemistry, University of Houston, Houston, Tex. 77004, U.S.A.

Abstract. This paper is primarily a review of recent developments in the abiotic synthesis of nucleotides, short chain oligonucleotides, and their mode of replication in solution. It also presents preliminary results from this laboratory on the prebiotic synthesis of thymidine oligodeoxynucleotides. A discussion, based on the physicochemical properties of RNA and DNA oligomers, relevant to the molecular evolution of these compounds leads to the tentative hypothesis that oligodeoxyribonucleotides of about 12 units may have been of sufficient length to initiate a self replicating coding system. Two models are suggested to account for the synthesis of high molecular weight oligomers using short chain templates and primers.

1. Introduction

The mode of synthesis of RNA and DNA components, namely the purine and pyrimidine bases, nucleosides and nucleotides under environmental conditions assumed to have prevailed on the primitive earth, has now been reasonably established and was reviewed recently (Stephen-Sherwood and Oró, 1973). The formation of oligonucleotides of sufficient length to carry the necessary information for the coding of small protoenzymes has yet to be accomplished under abiotic conditions.

The problem of nucleic acid synthesis in a primitive earth environment will be discussed under four major headings:

1.1. The formation of nucleotides, since the mode of formation of these compounds is pertinent to their subsequent condensation.

1.2. The synthesis of short-chain oligonucleotides.

1.3. Possible modes of replication and elongation of these oligonucleotides.

1.4. The physicochemical properties of RNA and DNA oligomers relevant to the molecular evolution of these compounds.

1.1. The formation of mononucleotides

The most successful models for the phosphorylation of nucleosides involve reactions using mild temperatures and ambient humidity, simulating the effects resulting from the drying up of primeval lakes and ocean beds. Acceptable temperatures for these abiotic reactions are 60 °C–80 °C (Osterberg and Orgel, 1972), since surface tempertures of up to 80 °C have been recorded in desert areas during present times (Chang, 1958). Reactions carried out at temperatures much greater than 100 °C are of doubtful significance with respect to prebiotic chemistry.

Urea, an important reagent in prebiotic phosphorylation reactions, was probably readily available on the primitive Earth, since it, guanidine, and, to a lesser extent, cyanamide are formed when aqueous solutions of ammonium cyanide are exposed to sunlight or 254 nm UV radiation (Lohrmann, 1972). When a solution of a nucleoside, with urea, ammonium chloride and sodium hydrogen phosphate, is dried and heated to temperatures below 100 °C, extensive phosphorylation occurs. A major product in the case of ribonucleosides is the nucleoside 2′,3′-cyclic phosphate, but considerable amounts of 3′- and 5′-phosphates are also formed. Hydroxylapatite can be substituted for sodium phosphate in this reaction with an only moderate decrease in yields (Lohrmann and Orgel, 1971). The most probable source of phosphate on the primitive Earth was apatite, which is present as an accessory mineral in most igneous rocks; however, it has been suggested that struvite, $MgNH_4PO_4 \cdot 6H_2O$, may also have been a prebiotic source of phosphate. When uridine 5′-phosphate is heated with struvite and urea, uridine 5′-diphosphate and P^1, P^2-diuridine 5′-pyrophosphate are formed in good yield (Handschuh and Orgel, 1973). The phosphorylation of thymidine using mixtures of urea and ammonium chloride with phosphate, pyrophosphate or tripolyphosphate yielded only thymidine 3′-phosphate, thymidine 5′-phosphate and thymidine 3′,5′-diphosphate (Bishop *et al.*, 1972). Urea also promotes transphosphorylation. The heating of pU or Up with urea, or urea and ammonium chloride, produced as a major product uridine 2′,3′-cyclic phosphate (Lohrmann and Orgel, 1971). When an aqueous solution of TMP and urea was dried and heated at 80 °C for 18 hr, transphosphorylation also occurred and 2.3% ppT, 11.7% pTp, and 1.6% of a higher phosphorylation product, probably ppTp or pTpp, were formed. Although 25% thymidine was recovered, the net loss of phosphate with respect to nucleotide components was 10% (Odom *et al.*, 1973), Urea not only facilitates phosphorylation reactions, but appears to aid in the conservation of phosphate. Under similar reaction conditions as those described for the phosphorylation of uridine, urea promotes the formation of polyphosphates from ammonium phosphate. Addition of a nucleoside (thymidine or deoxythymidine) to the system promotes trimetaphosphate synthesis (Osterberg and Orgel, 1972).

Apatite and hydroxylapatite are particularly insoluble phosphates; however, the addition of ammonium oxalate to a suspension of apatite aids in solubilizing the phosphate by rendering the system slightly acidic ($pH \simeq 5$) due to loss of ammonia during evaporation. Cyanamide and its dimer, dicyandiamide, were more effective than urea for the phosphorylation of uridine in an apatite, ammonium oxalate system (Schwartz, 1972). Apatite crystals catalyze the phosphorylation of adenosine and other nucleosides, except guanosine, with pyrophosphate under drying conditions and temperatures below 100 °C (Neuman *et al.*, 1970a). Adenosine 5′-phosphate may be further phosphorylated to ADP and ATP. In an experiment designed to simulate the alternate wetting and drying which would result from tides on a primitive beach, once equilibrium had been established, about 80% adenosine, 20% adenylic acid, and 3% ADP and ATP were found to be present on the apatite surface (Neuman *et al.*, 1970b).

1.2. The synthesis of short-chain oligonucleotides

The majority of the syntheses which have led to the formation of di- and trinucleotides have been conducted under drying conditions. Thus, di- and trinucleotides are formed in yields of 20% and 7% respectively, when uridine and uridine

2′(3′)-phosphate are heated to 160 °C (Moravek *et al.*, 1968a, b). These nucleotides contained only 2′→5′ and 3′→5′ internucleotide bonds, with a preference for the natural 3′→5′ internucleotide linkage. Heating of uridine 2′(3′)-phosphate to 160 °C produces dinucleoside diphosphates and trinucleoside triphosphates of uridine, in yields of 23% and 12%, respectively, containing predominantly 3′→5′ internucleotidic bonds in addition to the unnatural 2′→5′ bond (Moravek *et al.*, 1968c). The polymerization of the triethylammonium salt of cytidine 2′:3′-cyclic phosphate at 138 °C for 48 hr, produced oligomers of cytidylic acid up to the hexamer, consisting of both 3′→5′ and 2′→5′ phosphodiester bonds. Approximately 50% of the dimer had the natural linkage (Tapiero and Nagyvary, 1971).

Oligonucleotides are formed when a uridine, urea, ammonium dihydrogen phosphate system is heated at temperatures between 85 and 100 °C. If the uridine/phosphate molar ratios range from 1.0 to 2.85, UpU and UpUpU are formed in yields of 18.5% and 7.4%, respectively, after 11 days heating at 100 °C in a dry nitrogen atmosphere. Reactions under conditions of ambient humidity produced similar results (Osterberg *et al.*, 1973) These oligomers contained 60% and 40%, respectively, of the natural 3′→5′ phosphodiester bond. Uridine 2′,3′-cyclic phosphate is a major product in this reaction, 38.8% being present after 3 days heating and 44.7% after 11 days. In the presence of excess phosphate, uridine polyphosphates and oligonucleotides having 5′→5′ pyrophosphodiester bonds are the major products of the reaction after 4 days heating at 100 °C and only 11% uridine 2′,3′-cyclic phosphate is formed. Although the mechanism of oligonucleotide synthesis may depend on the formation of polyphosphates and trimetaphosphate, it is more probable that the reactive intermediate is uridine 2′,3′-cyclic phosphate. Polymerization occurs via a transesterification process involving primary hydroxyl groups only (Tapiero and Nagyvary, 1971; Osterberg *et al.*, 1973). Enzymatic synthesis of UpU from cyclic Up and U involves the same transesterification process (Bernfield, 1966; Bernfield and Rottman, 1967).

In this abiotic synthesis of oligouridylates, there is a preference for the natural 3′→5′ phosphodiester bond; however, 2′→5′ bonds are also formed in substantial amounts. It can be anticipated that higher oligomers would possess random bonding with both 3′→5′ and 2′→5′ bonds present in the chain. These oligomers would not necessarily be ineffective in functioning as a primitive template, since in solution both the 3′→5′ and the 2′→5′ dimers of adenyladenine stack with an anti, anti right-handed conformation typical of RNA polymers (Kondo *et al.*, 1970). The selection of the natural linkage may have occurred at some later stage in the development of a self replicating system. It is significant that adenyl(3′→5′) adenosine has a greater selectivity for a right-handed stack conformation than adenyl(2′→5′) adenosine. This suggests that nucleic acids built with a 3′→5′ linkage may have assumed pre-eminence during the evolutionary process because of their superior uniformity of configuration (Tazawa *et al.*, 1970).

The urea-ammonium chloride-phosphate system offers a method for the formation of short chain oligoribonucleotides on the primitive Earth. It does not appear to be applicable to the formation of oligodeoxyribonucleic acids since no oligonucleotide formation was evident after 44 days of heating at 65 °C of a mixture of thymidine, urea, ammonium chloride, and phosphate (Bishop *et al.*, 1972).

Thymidine-oligonucleotides of chain length from 1 to 12 nucleotide units are formed during the solid state condensation of O^2, 5′-cyclothymidine 3′-phosphate at 138 °C for 48 hr (Nagyvary and Nagpal, 1972). The yields of $(Tp)_2$ and $(Tp)_3$ were 13% and 14%, respectively. O^2,5′-cyclothymidine 3′-phosphate has not thus far been produced under prebiotic conditions. Further, this method would not be applicable to the synthesis of purine oligonucleotides.

Two approaches to the synthesis of deoxyoligonucleotides can be envisioned under primitive earth conditions.

(a) A deoxynucleoside 3′,5′-cyclic phosphate could be an intermediate in oligonucleotide formation (Pongs and Ts'o, 1971). Thymidine 3′,5′-cyclic phosphate was reported to be formed on heating an aqueous solution of TMP with cyanamide; and low yields of oligodeoxynucleotides were also obtained (Ibanez *et al.*, 1971b). It is possible that this 3′,5′-cyclic phosphate may have been an intermediate although a more direct cyanamide dehydration reaction cannot be eliminated.

(b) An alternative approach makes use of pre-activated mononucleotides. A number of activating groups could be of importance in prebiotic condensation reactions, namely, polyphosphates, phosphoramidates, acylphosphates and phosphoimidazolides (Figure 1).

In an extensive study of the mechanism of internucleotide bond formation in pyridine using dicyclohexylcarbodiimide as a condensing agent, an alkyl trimetaphosphate has been implicated as one phosphorylating species. The first step in the activation process is the formation of a symmetrical P^1, P^2-dialkyl pyrophosphate (Weimann and Khorana, 1962; Jacob and Khorana, 1964). It is also known that ethyl polyphosphate esters, which are probably mixtures of cyclic metaphosphate and straight-chain pyrophosphate esters (Nooner and Oró, 1973), promote the condensation of

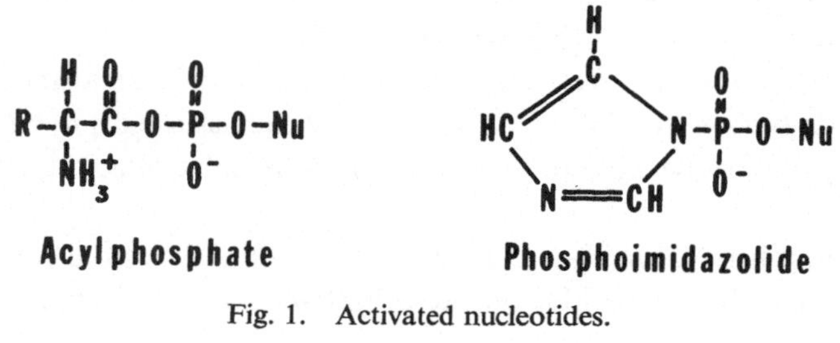

Polyphosphate

Phosphoamidate

Acylphosphate

Phosphoimidazolide

Fig. 1. Activated nucleotides.

nucleotides to high molecular weight polymers (Schramm *et al.*, 1962). Even though subsequent work showed that the polyribonucleotides had branched chains (Gottich and Slutsky, 1964), nevertheless, internucleotide bond formation did occur and it is not surprising that some unnatural bonds were formed. In the polymerization of thymidine 5′-phosphate in boiling N,N-dimethylformamide using β-imidazolyl-4(5)-propionic acid or triethylamine hydrochloride as a catalyst, P^1,P^2-dithymidine 5′-pyrophosphate is an intermediate in the formation of oligonucleotides (Pongs and Ts'o, 1969, 1971). The catalytic function of the β-imidazole-4(5)-propionic acid or the triethylamine hydrochloride is proposed as that of a proton donor.

Under more reasonable conditions for prebiotic synthesis it has been found in our laboratory that imidazole promotes the condensation of thymidylic acid in an aqueous system; and oligomers of up to five units in length were obtained containing mainly $3' \rightarrow 5'$ phosphodiester bonds in yields of less than 1% (Ibanez *et al.*, 1971a).

More recently, the condensation of a preactivated mononucleotide, ^{14}C-labeled thymidine triphosphate with thymidine 5′-phosphate as a primer, in the presence of 4-amino-5-imidazolecarboxamide (AICA) and cyanamide, yielded, as its major products, oligonucleotides of the formula $(pT)_n$, $n = 2 - 4$ (Stephen-Sherwood *et al.*, 1973). The reactants were dissolved in water and evaporated to dryness at 89 °C and allowed to react at this temperature under condition of ambient humidity for 40 hr. On terminating the reaction, the products were subjected to a procedure which degrades pyrophosphate bonds (Moon and Khorana, 1966), and the remaining radioactive products were analyzed by enzymatic degradation. The total yield of oligonucleotide product with at least one terminal phosphate was 32%, and 80% of these oligomers contained natural $3' \rightarrow 5'$ phosphodiester bonds. The yield of $(pT)_2$ and $(pT)_3$ was 9.3% and 5.6%, respectively.

Thymidine 5′-triphosphate was used in this reaction instead of thymidine 5′- phosphate because the latter polymerizes only in very small yields when heated with cyanamide in the presence or absence of AICA. Thymidine 5′-triphosphate has the potential of forming a trimetaphosphate, a molecule already implicated as being a reactive intermediate in the formation of phosphodiester bonds (Figure 2). In the condensation of trialkyl ammonium nucleoside 5′-phosphates with inorganic phosphate in the presence of DCC, tetra- and higher polyphosphates were initially formed; however, nucleoside 5′-triphosphates were formed ultimately at the expense of these higher polyphosphates. The increased yield of nucleoside 5′-triphosphates over other polyphosphates was postulated as being due to stabilization of the molecule by trimetaphosphate formation (Smith and Khorana, 1958; Weimann and Khorana, 1962).

Thymidine 5′-phosphate was chosen in preference to thymidine as a primer in this condensation, since it was considered essential to block the more reactive primary 5′-hydroxyl group in order to avoid appreciable formation of $5' \rightarrow 5'$ phosphodiester bonds. The role of AICA is assumed to be that of a proton donor; the role of cyanamide has not yet been determined. Cyanamide and AICA were chosen because of their similarity in structure to two amino acids, arginine and histidine, which have been shown to be present in the active center of DNA polymerase (Salvo *et al.*, 1973). Although this analogy is somewhat tenuous, it could well be that the guanidine and imidazole groups present in these two amino acids play an important role both in the biological and prebiotic synthesis of oligodeoxynucleotides.

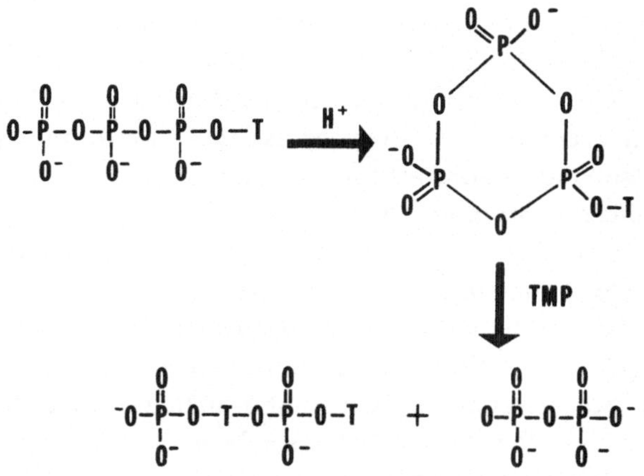

Fig. 2. Possible mechanism for oligonucleotide formation using thymidine-5′-triphosphate.

Though the polymerization of mononucleotides under drying conditions offers a reasonable model for the abiotic synthesis of short chain oligomers, it is doubtful whether this type of condensation is applicable to the formation of long polymers. In the chemical synthesis of homooligomers, the degree of polymerization decreases, according to Poisson's distribution (Cramer, 1966). The longest oligomer so far prepared by chemical means is a polythymidylic acid of 30 chain units (Hayes and Hansbury, 1964).

1.3. Possible modes of replication and elongation of oligonucleotides

As a result of rainstorms, tidal variation, and flooding, compounds produced under dehydrating conditions could ultimately end up in an aqueous medium. The question arises: what are the conditions and catalysts required for the replication and elongation of these short chain oligomers in an aqueous medium? The information available on the chemical replication of oligonucleotides is limited. In aqueous solution, using a water soluble carbodiimide, adenylic acid is preferentially condensed on a double stranded polyuridylic acid template, and guanylic acid on a double stranded polycytidylic acid template (Sulston *et al.*, 1968a, b). The dinucleotides were isolated in 35% yield, but the majority of the isomers contained unnatural $2' \rightarrow 5'$ phosphodiester bonds. These results demonstrate that selectivity with respect to Watson-Crick base pairing does exist at the monomer level and the formation of complementary oligomers would probably have been reasonably accurate with respect to base sequence, if not in respect to linkages. Similar condensation of deoxyadenylic acid with adenosine on a polyuridylic acid template gave dinucleotides containing mainly $5' \rightarrow 5'$

linkages (Schneider-Bernloehr *et al.*, 1968b). Adenyl(5′→5′) adenosine has an anti, anti, right-handed symmetrical conformation as opposed to the anti, anti, righthanded asymmetrical conformation of adenyl(3′→5′) adenosine (Kondo *et al.*, 1970), so it is probable that the formation of a 5′→5′ bond would terminate polymerization in a replicating system.

Adenosine-5′-phosphoimidazolide, an activated nucleotide, reacts on a polyuridylic acid template to form di- and trinucleotides in yields of 43.5% and 2.8%, respectively. The major dinucleotide formed was adenyl(2′→5′) adenine (95.9%) (Weimann *et al.*, 1968). D-adenosine-5′-phosphoimidazolide reacts much more rapidly with D-adenosine than with L-adenosine on a poly-D-uridylic acid template, suggesting that segregation of D- and L-nucleotides may have occurred at an early stage in biochemical evolution (Schneider-Bernloehr *et al.*, 1968a). Deoxyadenosine, 1-β-arabinofuranosyladenine and α-adenosine react less efficiently than β-adenosine with adenosine-5′-phosphoimidazolide on a sodium polyuridylic acid template. Deoxyadenosine-5′-phosphoimidazolide is about as reactive as adenosine-5-phosphoimidazolide (Schneider-Bernloehr *et al.*, 1970). Although adenosine-5′-triphosphate forms a stable helix with polyuridylic acid (Miles *et al.*, 1966), it undergoes hydrolysis without appreciable oligonucleotide formation (Weimann *et al.*, 1968). Deoxyadenyl-(5′→3′)-deoxyadenyl-(5′→N)-phenylalanine condenses on a polyuridylic acid template to yield higher oligomers (Shabarova and Prokofiev, 1970). Polybasic amines and glycine derivatives stabilize the triple helix formed by adenosine-cyclic-2′,3′-phosphate with double-stranded polyuridylic acid and catalyse the formation of dinucleoside diphosphates; the major isomer is 2′→5′ linked. Dinucleotide formation was most efficient in frozen solutions which favor stacking (Renz *et al.*, 1971). The formation of the 2′→5′ internucleotide bond in this reaction is explained on stereochemical grounds (Usher, 1972). Only pyrimidine polynucleotides can serve as templates for chemical replication (Orgel, 1968; Orgel and Sulston, 1971).

The evidence presented suggests that replication of oligonucleotides should have been possible in the pools and oceans of the primitive Earth, particularly in the cold icy conditions existing during winter months, which would favor base stacking and hydrogen bonding. Likely catalysts to promote condensation would be imidazole derivatives, cyanamide, cyanogen, polybasic amines, and glycine derivatives. The length of nucleotide chain required to serve as primers for chemical replication is a matter of speculation, but based on the evidence of enzymatic synthesis using short chain oligonucleotide primers, they could be as short as 9–12 units in length. A deoxyoligonucleotide of six units, $pT(pApT)_2pA$, $[(AT)_3]$, can serve as a primer for the synthesis of a high molecular weight dAdT copolymer using *E. coli* DNA polymerase (Kornberg *et al.*, 1964). The replication was profoundly influenced by temperature, the shorter oligomers requiring lower temperatures in order to maximize hydrogen bonding. For example, $(AT)_4$ replicated best at 10 °C, and less efficiently at higher temperatures due to breakage of hydrogen bonds.

Further studies using DNA-like polymers with repeating trinucleotide sequences (Wells *et al.*, 1967b) and repeating tetranucleotide sequences (Wells *et al.*, 1967a) as primers indicate that a minimum size of 9 to 12 nucleotides is necessary for template activity. An additional requirement for template activity is that both complementary strands be present and be hydrogen bonded to form Watson-Crick base pairs with opposite polarity of the two chains. A slippage mechanism based on a model originally proposed by Chamberlin and Berg (1962) is suggested to account for the synthesis of high molecular weight oligomers using short chain primers. Whether a similar slippage model (Figure 3) could account for replication and elongation of short chain oligomers in primitive pools and oceans has yet to be determined. Temperature fluctuations, as a result of night and day, and also winter and summer cooling and heating could account for the forming and breaking of hydrogen bonds and hence promote both elongation and, of necessity, variety in the long chain oligomers.

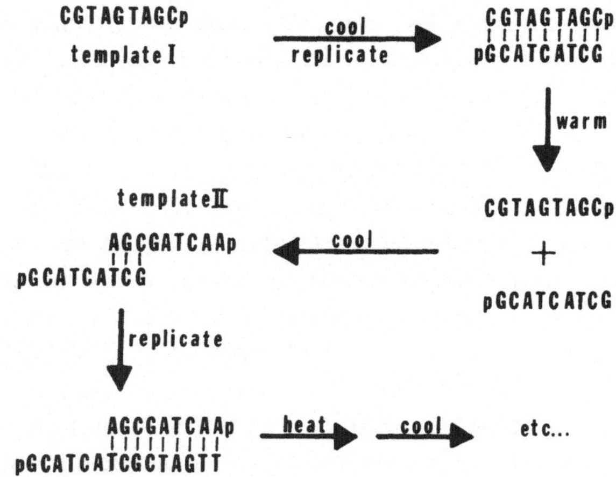

Fig. 3. A slippage model for the replication and elongation of oligonucleotides using short chain primers.

Another possibility for producing long chain oligomers from relatively short chain templates utilizes a mechanism similar to the rolling circle model devised to explain the replication of circular DNA in many prokaryotic cells (Gilbert and Dressler, 1968). Small circles of 12 to 20 units could have a limited degree of stacking of some bases (Gray, 1973); this statement is based largely on the following evidence. RNA oligomers of the form A_6CmU_6 (m=4, 5, 6, or 8) form hairpins with a helix of six AU base pairs and m cytidine residues in the loop. The circular dichroism spectrum of cytosine residues in the C_6 and C_8 loops is very similar to single stranded oligocytidylic acid which is known to stack in solution. However, the C_5 loop has a very different circular dichroism spectrum from the corresponding oligo (C_5) spectrum. It is of interest that the six-membered loop was the most stable of those measured; the smaller loops were strained and the larger showed decreased stability due to decreased probability of the A_6 and U_6 regions meeting. Two of the three loops in the cloverleaf model of t-RNA always have seven residues and may have maximum stability (Uhlenbeck *et al.*, 1973).

The model (Figure 4) proposed to explain replication envisages a circular template with a region of limited stacking on which free nucleotides would stack, hydrogen bond, and polymerize. Further polymerization would occur when some of the bases in the circle unstack as a result of strain and slight fluctuation in temperature, and new bases, ahead of the growing polymer, come into stacked conformation. This model would explain the formation of DNA with a periodic base sequence, from which it has been suggested (Yčas, 1972) that the periodic proteins such as silk fibrin, collagen, keratin, protamines, etc., may have arisen. One of the fundamental questions of the model is the probability of cyclization of short chain oligonucleotides of twelve units or more. Cyclic thymidine polynucleotides of up to five units are formed in the poly-

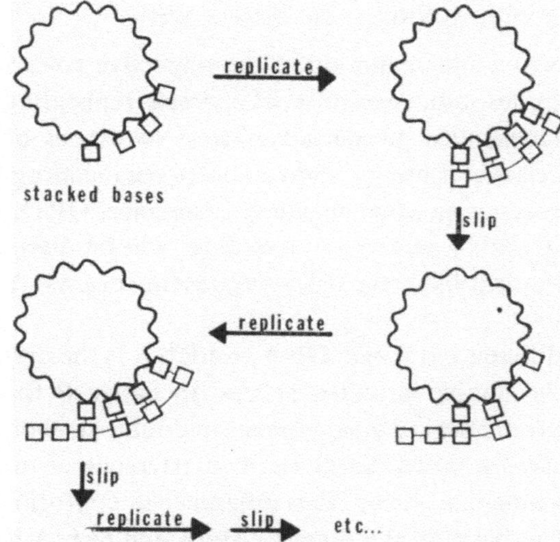

Fig. 4. Rolling circle model for the formation of long chain oligonucleotides.

merization of thymidine-5′-phosphate in anhydrous pyridine using DCC or p-toluene-sulfonyl chloride as a condensing agent (Tener *et al.*, 1958). Thymidine polynucleotides of length greater than 5 units do not cyclize in this system. However, under more natural conditions the polynucleotide joining enzyme from *E. coli* catalyzed the intramolecular joining of linear deoxyadenylatedeoxythymidylate oligomers. The smallest circle which could be synthesized consisted of 17–18 nucleotide units (Olivera *et al.*, 1968). Therefore, at relatively low temperatures it appears quite reasonable that prebiotic cyclization could occur through the mediation of a short chain oligonucleotide containing adjacent sequences which are complementary to the terminal sequences of the longer chain molecule.

Another model devised to explain DNA replication in primitive systems proposes a catalytic role for RNA (Brewin, 1972). In this model, trinucleotides are synthesized on the 3′ end of a primitive RNA adaptor molecule using the anticodon as a template. The 'charged' adaptor then interacts with a parental DNA duplex; the anti-codon bonding to one strand and complementary trinucleotide is excised and bonded to the other. This model would require a more highly evolved system than the simple slippage mechanism. The intermediate oligonucleotides would also have to be of greater length.

Up to this point in the discussion, no mention has been made of the possible catalytic role of oligopeptides and protoenzymes in the chemical evolution of DNA and RNA. Undoubtedly, oligopeptides played an important role in the evolution of the genetic apparatus of present day organisms, but with the information presently available, it is difficult to make any meaningful comments on the stage at which protoenzymes began to play a vital role in the synthesis or replication of DNA and RNA.

1.4. The physiochemical properties of RNA and DNA oligomers relevant to the molecular evolution of these compounds

In order to have a better understanding of the respective roles played by ribo- and deoxyribooligonucleotides in the evolution of a simple replicating system, a brief discussion on the physicochemical properties of these two types of oligonucleotides is advantageous. For the sake of brevity, only a few papers relating to recent studies on the absorption spectroscopy, nuclear magnetic resonance, ORD, CD, and molecular hybridization of RNA, DNA and oligonucleotides will be discussed. Additional information may be obtained from the following recent reviews (Yang and Samejima, 1969; Walker, 1969).

One fundamental difference between DNA and RNA is the difference in conformation of their respective double stranded helices. In native B form DNA, the bases stack perpendicularly to the helical axis, whereas in double stranded RNA or in DNA-RNA hybrids, the bases are tilted due to steric interference of the 2′ hydroxyl group of ribose (Yang and Samejima, 1969). This difference in conformation is also present in short oligomers. The bases in the dimers rApdA and rAprA have an oblique, less stacked, conformation than the dApdA and dAprA dimers, again due to steric interference of the 2′hydroxyl group (Kondo *et al.*, 1972). Thus deoxyribooligonucleotides probably display better stacking than their ribooligonucleotide counterparts. However, the degree of stacking in any short chain oligomer also depends on the bases present. In dinucleoside phosphates, the degree of stacking decreases in the order purine·purine > purine·pyrimidine > pyrimidine·pyrimidine (Warshaw and Tinoco, 1966). Homopolymers of uridylic acid and thymidylic acid exhibit little base stacking at room temperature (Richards *et al.*, 1963). In the case of poly U, stacking is favored by low temperatures and high salt concentrations (Simpkins and Richards, 1967).

Just as stacking interactions between adjacent bases is present in short chain oligomers, so also is complementary Watson-Crick base pairing, by hydrogen bonding, displayed at an elementary level. Even though the free bases do not show any tendency to hydrogen bond, methylation of the glycosyl nitrogens results in the formation of hydrogen bonded complexes between complementary base pairs only. These methyl derivatives of adenine and thymine, adenine and uracil, and guanine and cytosine cocrystallize. The guanine-cytosine mixed crystals have normal Watson-Crick hydrogen bonding (O'Brien, 1963), but in the case of adenine, the pyrimidine is hydrogen bonded to the N-7 instead of the N-1 nitrogen (Hoogstein, 1963; Mathews and Rich, 1964). Complementary base pairing of substituted bases also occurs in nonaqueous solvents (Shoup *et al.*, 1966). As already mentioned, selective Watson-Crick base pairing also occurs between purine mononucleotides and their complementray pyrimidine homopolymers (Sulston *et al.*, 1968a, b).

Molecular hybridization studies on natural polynucleotides of defined length indicate that oligomers of about twelve units are sufficient to form stable duplexes in solutions at relatively low temperatures and high salt concentration. The RNA·RNA duplexes are more stable than DNA·DNA duplexes which in turn are more stable than DNA·RNA hybrids (Walker, 1969; Niyogi and Thomas, 1967; Ruger and Bautz, 1968).

Assuming the premise that short RNA- and DNA-like polymers evolved simultaneously on the primitive Earth, the DNA-like oligomers would have made more suitable templates for the development of a simple replicating system. As the above evidence shows, these linear oligomers (of 12 units or more) probably have a uniform stacked conformation, and the lower stability of the DNA duplexes would have made self replication easier than with their RNA counterparts. In addition, the resistance of DNA to alkaline hydrolysis would have been advantageous since one of the fundamental requirements in any replicating system is chemical stability of the coding template. In contrast, the hairpin like structures found in present day tRNA's would evolve more readily from oligomers which form more stable duplexes. As mentioned earlier, loops of seven bases would probably have maximum stability (Uhlenbeck *et al.*, 1973). It has been suggested that simple hairpin structures may have been important in the development of a system for the controlled synthesis of primitive protoenzymes (Woese, 1972).

In conclusion, it is tempting to speculate whether small DNA molecules containing 12 complementary base pairs could have, in a primitive protein synthesizing system, produced a tetrapeptide. This speculation is not as fanciful as it first appears. It has been suggested that ferredoxin from *Clostridium pasteurianum* has evolved by doubling a shorter protein, which in turn developed from a repeating sequence of four amino acids, alanine, aspartic acid or proline, serine and glycine (Eck and Dayhoff, 1966) which correspond to a dodecanucleotide. Furthermore, such dodecanucleotides form stable duplexes (Walker, 1969; Niyogi and Thomas, 1967; Ruger and Bantz, 1968) and are the smallest molecules to serve as template instructed primers for *E. coli* DNA polymerase (Wells *et al.*, 1967a, b).

The above amino acids are all formed in electric discharge experiments (Ring *et al.*, 1972) and the three major ones (glycine, alanine, aspartic) are easily obtained from HCN in aqueous solutions (Oró and Kamat, 1961) and should therefore have been readily available on the primitive Earth. Repeated duplication of the twelve unit sequence of bases in the DNA template, followed by subsequent modification of these bases by chemical mutagenic agents and ultraviolet light would ultimately produce a template capable of coding for a functional enzyme.

2. Conclusion

The following summary itemizes the main topics covered in this review:

(1) Nucleosides may be phosphorylated under drying conditions using (a) urea, ammonium chloride and hydroxyl apatite or struvite system, or (b) cyanamide or dicyandiamide, and ammonium oxalate and apatite system.

(2) Uridine oligonucleotides are formed when uridine, urea and ammonium dihydrogenphosphate are heated under drying conditions.

(3) Thymidine oligodeoxynucleotides are formed in small yields when thymidine-5′-phosphate is reacted with either cyanamide or imidazole and evaporated to dryness, and in larger yields when thymidine-5′-triphosphate, thymidine-5′-phosphate, cyanamide, and 4-amino-5-imidazole carboxamide are heated under drying conditions.

(4) No satisfactory method has yet been reported for the replication of oligonucleotides under abiotic conditions. In all the methods, the synthesized oligomers contained mainly unnatural 2′→5′ or 5′→5′ bonds with only small amounts of the natural 3′→5′ linkage.

(5) It is suggested on the basis of enzymatic and molecular hydridization studies that oligonucleotides of 12 units in length could serve as templates under abiotic conditions. This seems to be consistent with the postulated biochemical evolution of ferrodoxin.

(6) Two models are suggested to account for the synthesis of high molecular weight oligomers using short chain primers:

(a) A slippage mechanism utilizing linear templates and

(b) A rolling circle model from which DNA with periodic base sequences is formed. It has been suggested that the periodic proteins such as silk fibrin, collagen, protamines, etc. may have arisen from DNA with such repetitive base sequences.

(7) It appears probable that DNA and RNA oligomers could have arisen simultaneously on the primitive Earth. On the basis of physicochemical data, DNA oligomers would have made more suitable templates for a self-replicating system.

Acknowledgements

We wish to thank Miss M. A. Moré for her technical assistance in the preparation of this manuscript. This work was supported by a research grant NGR-44-005-002 from the National Aeronautics and Space Administration and was presented at the 4th International Conference on the Origin of Life, June 25–28, 1973, Universidad Autonoma de Barcelona, Barcelona, Spain.

References

Bernfield, M. R.: 1966, *J. Biol. Chem.* **241**, 2014.
Bernfield, M. R. and Rottman, F. M.: 1967, *J. Biol. Chem.* **242**, 4134.
Bishop, M. J., Lohrmann, R., and Orgel, L. E.: 1972, *Nature* **237**, 162.
Brewin, N.: 1972, *Nature* **236**, 101.

Chamberlin, M. J. and Berg, P.: 1962, *Proc. Nat. Acad. Sci. (U.S.)* **48**, 81.
Chang, H. J.: 1958, *Ground Temperature I*, Blue Hill Meteorological Observatory, Harvard University.
Cramer, F.: 1966, *Angew. Chem. Internat. Edit. Eng.* **5**, 173.
Eck, R. V. and Dayhoff, M. O.: 1966, *Science* **152**, 363.
Gilbert, W. and Dressler, D.: 1968, *Cold Spring Symp. Quant. Biol.* **33**, 473.
Gottich, B. P. and Slutsky, O. I.: 1964, *Biochim. Biophys. Acta* **87**, 163.
Gray, H. B., Jr.: 1973, Private Communication.
Handschuh, G. J. and Orgel, L. E.: 1973, *Science* **179**, 483.
Hayes, F. N. and Hansbury, E.: 1964, *J. Am. Chem. Soc.* **86**, 4172.
Hoogstein, K.: 1963, *Acta. Cryst.* **16**, 907.
Ibanez, J. D., Kimball, A. P., and Oró, J.: 1971a, *J. Mol. Evol.* **1**, 112.
Ibanez, J. D., Kimball, A. P. and Oró, J.: 1971b, *Science* **173**, 444.
Jacob, T. M. and Khorana, H. G.: 1964, *J. Am. Chem. Soc.* **86**, 1630.
Kondo, N. S., Fang, K. N., Miller, P. S., and Ts'o, P. O. P.: 1972, *Biochemistry* **11**, 1991.
Kondo, N. S., Holmes, H. M., Stempel, L. M., and Ts'o, P. O. P.: 1970, *Biochemistry* **9**, 3479.
Kornberg, A., Bertsch, L. L., Jackson, J. F., and Khorana, H. G.: 1964, *Biochemistry* **51**, 315.
Lohrmann, R.: 1972, *J. Mol. Evol.* **1**, 263.
Lohrmann, R. and Orgel, L. E.: 1971, *Science* **171**, 490.
Mathews, F. S. and Rich, A.: 1964, *J. Mol. Biol.* **8**, 89.
Miles, H. T., Howard, F. B., and Frazier, J.: 1966, *Federation Proc.* **25**, 1853 (abstract).
Moon, M. W. and Khorana, H. G.: 1966, *J. Am. Chem. Soc.* **88**, 1798.
Moravek, J., Kopecky, J., and Skoda, J.: 1968a, *Collection Czech. Chem. Commun.* **33**, 960.
Moravek, J., Kopecky, J., and Skoda, J.: 1968b, *Collection Czech. Chem. Commun.* **33**, 4407.
Moravek, J., Kopecky, J., and Skoda, J.: 1968c, *Collection Czech. Chem. Commun.* **33**, 4120.
Nagyvary, J. and Nagpal, K. L.: 1972, *Science* **177**, 272.
Neuman, M. W., Neuman, W. F., and Lane, K.: 1970a, *Curr. Mod. Biol.* **3**, 253.
Neuman, M. W., Neuman, W. F., and Lane, K.: 1970b, *Curr. Mod. Biol.* **3**, 277.
Niyogi, S. K. and Thomas, C. A.: 1967, *Biochem. Biophys. Res. Commun.* **26**, 51.
Nooner, D. W. and Oró, J.: 1973, *J. Mol. Evol.* (in press).
O'Brien, E. J.: 1963, *J. Mol. Biol.* **7**, 107.
Odom, D. G., Ibanez, J. D., Stephen-Sherwood, E., and Oró, J.: 1973, Private Communication.
Olivera, B. M., Scheffler, I. E., and Lehman, I. R.: 1968, *J. Mol. Biol.* **36**, 275.
Orgel, L. E.: 1968, *J. Mol. Biol.* **38**, 381.
Orgel, L. E. and Sulston, J. E.: 1971, in A. P. Kimball and J. Oró (eds.), *Prebiotic and Biochemical Evolution*, North Holland Publ. Co., Amsterdam, p. 89.
Oró, J. and Kamat, S.: 1961, *Nature* **190**, 442.
Osterberg, R. and Orgel, L. E.: 1972, *J. Mol. Evol.* **1**, 241.
Osterberg, R., Orgel, L. E., and Lohrmann, R.: 1973, *J. Mol. Evol.* **2**, 231.
Pongs, O. and Ts'o, P. O. P.: 1969, *Biochem. Biophys. Res. Commun.* **36**, 475.
Pongs, O. and Ts'o, P. O. P.: 1971, *J. Am. Chem. Soc.* **93**, 5241.
Renz, M., Lohrmann, R., and Orgel, L. E.: 1971, *Biochim. Biophys. Acta.* **240**, 463.
Richards, E. G., Flessel, C. P., and Fresco, J. R.: 1963, *Biopolymers* **1**, 431.
Ring, D., Wolman, Y., Friedman, N., and Miller, S. L.: 1972, *Proc. Nat. Acad. Sci. (U.S.)* **69**, 765.
Ruger, W. and Bautz, E. K. F.: 1968, *J. Mol. Biol.* **31**, 83.
Salvo, R., Evans, J. E., and Kimball, A. P.: 1973, in preparation.
Schneider-Bernloehr, H., Lohrmann, R., Orgel, L. E., Sulston, J., and Weimann, B. J.: 1968a, *Science* **162**, 809.
Schneider-Bernloehr, H., Lohrmann, R., Sulston, J., Weimann, B. J., Orgel, L. E., and Miles, H. T.: 1968b, *J. Mol. Biol.* **37**, 151.
Schneider-Bernloehr, H., Lohrmann, R., Sulston, J., Orgel, L. E., and Miles, H. T.: 1970, *J. Mol. Biol.* **47**, 257.
Schwartz, A. W.: 1972, *Biochim. Biophys. Acta.* **281**, 477.
Schramm, G., Grotsch, H., and Polmann, W.: 1962, *Angew. Chem.* **74**, 53.
Shabarova, Z. A. and Prokofiev, M. A.: 1970, *Febs. Letters* **11**, 237.
Shoup, R. R., Miles, H. T., and Beeker, E. D.: 1966, *Biochem. Biophys. Res. Commun.* **23**, 194.
Simpkins, H. and Richards, E. G.: 1967, *Biopolymers* **5**, 551.
Smith, M. and Khorana, H. G.: 1958, *J. Am. Chem. Soc.* **80**, 1141.

Stephen-Sherwood, E. and Oró, J.: 1973, *Space Life Sci.* **4**, 5.
Stephen-Sherwood, E., Odom, D., and Oró, J.: 1973, in *Proceedings of 4th International Conference on the Origin of Life*, Barcelona, submitted.
Sulston, J., Lohrmann, R., Orgel, L. E., and Miles, H. T.: 1968a, *Proc. Nat. Acad. Sci. (U.S.)* **59**, 726.
Sulston, J., Lohrmann, R., Orgel, L. E., and Miles, H. T.: 1968b, *Proc. Nat. Acad. Sci. (U.S.)* **60**, 409.
Tapiero, C. M. and Nagyvary, J.: 1971, *Nature* **231**, 42.
Tazawa, I., Tazawa, S., Stempel, L. M., and Ts'o, P. O. P.: 1970, *Biochemistry* **9**, 3499.
Tazawa, S., Tazawa, I., Alderfer, J. M., and Ts'o, P. O. P.: 1972, *Biochemistry* **11**, 3544.
Tener, G. M., Khorana, H. G., Markham, R., and Pol, E. H.: 1958, *J. Am. Chem. Soc.* **80**, 6223.
Uhlenbeck, O. C., Borer, P. N., Dengler, B., and Tinoco, I.: 1973, *J. Mol. Biol.* **73**, 483.
Usher, D. A.: 1972, *Nature New Biol.* **235**, 207.
Walker, P. M. B.: 1969, in J. N. Davidson and W. E. Kohn (eds.), *Progress in Nucleic Acid Research and Molecular Biology* **9**, Academic Press, New York and London, p. 301.
Warshaw, M. M. and Tinoco, I., Jr.: 1966, *J. Mol. Biol.* **20**, 29.
Weimann, B. J., Lohrmann, R., Orgel, L. E., Schneider-Bernloehr, H., and Sulston, J. E.: 1968, *Science* **161**, 387.
Weimann, G. and Khorana, H. G.: 1962, *J. Am. Chem. Soc.* **84**, 4329.
Wells, R. D., Büchi, H., Kössel, H., Ohtsuka, E., and Khorana, H. G.: 1967a, *J. Mol. Biol.* **27**, 265.
Wells, R. D., Jacob, T. M., Narang, S. A., and Khorana, H. G.: 1967b, *J. Mol. Biol.* **27**, 237.
Woese, C. R.: 1972, *Brookhaven Symposia in Biology*, Vol. 23, p. 326.
Yang, J. T. and Samejima, T.: 1969, in J. N. Davidson and W. E. Cohn (eds.), *Progress in Nucleic Acid Research and Molecular Biology* **9**, Academic Press, New York and London, p. 223.
Yčas, M.: 1972, *J. Mol. Evol.* **2**, 17.

THE POSSIBLE ROLE OF CLAYS IN PREBIOTIC PEPTIDE SYNTHESIS

MELLA PAECHT-HOROWITZ

Polymer Dept. Weizmann Institute of Science, Rehovot, Israel

Abstract. Hypotheses of macromolecule formations during the prebiotic era are described. The presumed role of minerals and clays in these reactions are: concentration of monomers, proton release by ion exchange whenever the reaction demands it, scattering of the charges of the interacting substances, thus allowing such substances to interact, which in the absence of clays repel each other due to their charges. Because of these reasons the polymerization mechanism in the presence of clays is different from that in their absence. While in the absence of clays only free amino acids or peptides can interact with active amino acid anhydrides, giving thus peptides increased by only one unit, in the presence of clays two molecules of amino acid anhydrides can interact, giving a still active peptide anhydride which can interact with another active peptide. Clays catalyze polymerization only in these cases where the amino acid is small enough to enter between the sheets of the clay. Apparently most of the reactions also occur there and not on the surface of the clay. Copolymerization of different pairs of amino acids proceeds selectively in the presence of clay. The relationship between this selecitvity and prebiotic parent proteins is discussed.

1. Introduction

In 1951, in his lecture on 'The Physical Basis of Life', J. D. Bernal presented the hypothesis that life originated on the sea shores (Bernal, 1951). He introduced the idea that some of the reactants which were the building blocks of biological macromolecules, after having been formed from the inorganic gases which constituted the atmosphere of the Earth in prebiotic times (Miller, 1955; Gulick, 1955; Scott,1956; Oró and Guidry, 1960; Oró and Kimball, 1959; Oró and Kimball, 1961; Oró and Kamat, 1961; Oró, 1960; Palm and Calvin, 1962; Ponnamperuma *et al.*, 1963; Schwartz and Ponnamperuma, 1969; Bar-Nun *et al.*, 1971) fell into oceans or other water reservoirs where, by the movement of waves, they were able subsequently to interact in the aqueous solutions. However, Bernal (1951) pointed out that for any reaction to take place, a minimal concentration of the reactants is needed which is higher than that which occurs in the natural water reservoirs. Therefore he suggested that finely dispersed clay particles, which exist on the sea shores, might have been the means for concentration of the reactants.

This idea, as a matter of fact, is not very different from the Oparin hypothesis that polymerization in the prebiotic era took place inside coacervates (Oparin, 1961), that is, droplets which separate out from solutions in which polymerization takes place in the presence of other polymers. In our view, clay particles which existed in the prebiotic era could have replaced those polymeric substances, the formation of which would in any event require explanation, and could have absorbed the monomers, thus producing very high, local concentrations. Moreover, absorption could have been preferential and if the clays served not only as a means of concentration but also, due to their physical properties, directed the polymerization in one direction or another, a catalytic system would result which would produce polymerizations different from those which are obtained under otherwise identical conditions.

2. The Properties of Clay Minerals as Related to Possible Prebiotic Syntheses

The term clay implies a natural, earthy, fine-grained material which develops plasticity when mixed with an appropriate amount of water. Generally, clay minerals are composed of extremely small particles consisting of a limited number of crystalline minerals. In most clays, the site of the exchange cations are internal and are related to the anisotropy of the crystal in some definite way. The exception is kaolinite clays, in which ion exchanges take place on the external surfaces only. The clay minerals are essentially hydrous aluminium silicates, some with magnesium or iron substituting wholly or in part for the aluminium and some with alkalies or alkaline earths present as essential constituents. A number of clays are composed of a single clay mineral, but in many there is a mixture .

Two structural units are involved in the atomic lattices of most of the clay minerals. One consists of two sheets of closely packed oxygens or hydroxyls in which aluminium, iron or magnesium atoms are embedded in an octahedral conformation so that they are equidistant from six oxygens or hydroxyls (Figures 1a, and b). When aluminium is present, only two-thirds of the possible positions are filled to balance the structure; when magnesium is present, all the positions are filled.

The second unit is composed of silica tetrahedrons, each silica being equidistant from four oxygens or hydroxyls (Figures 1c and d).

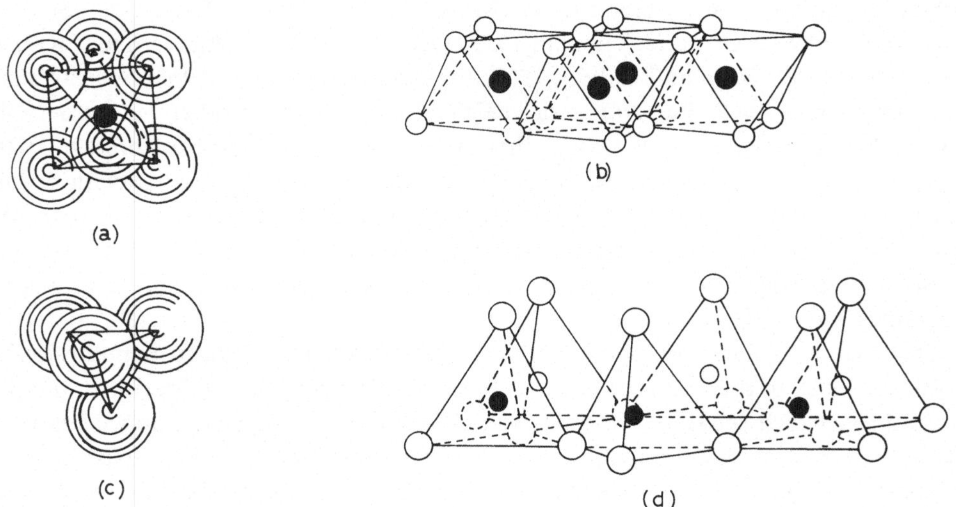

Fig. 1. Diagrammatic representation of single units and sheet structures. (a) Octahedral unit; (b) octahedral sheet structure; (c) tetrahedral unit; (d) tetrahedral sheet structure. ○ and ◌ – oxygens or hydroxyls; ● – aluminium, magnesium etc. ○ and ● – silicons.

Clays consist of either one silica tetrahedral sheet and one alumina octahedral sheet, or of two tetrahedral sheets, with one octahedral alumina sheet between them. The amount of water and its distribution between the layers is very definitive with regard to the properties of clay minerals; it is often the main factor which distinguishes one clay from another. The clay minerals have also ion exchange properties, relating both to inorganic and organic molecules. In the case of the latter, the ion exchange properties depend on the size of the molecule. Relatively large molecules (~80 Å) produce a 'cover up' effect, that is, they mask a large amount of hydrogen in the clay, making it inaccessible to neutralization by other molecules.

Most investigations on ion exchange and absorption effects in clays have been carried out on three-layer clays, mainly montmorillonite. Today it is accepted that organic ions are held in them by van der Waals' in addition to colombic forces (Hendricks, 1941). In general, the larger ions are more strongly adsorbed than the smaller, because the van der Waals' forces are greater; thus larger organic ions are difficult or even impossible to replace with smaller ones, and montmorillonites are used for the extraction of proteins from solutions. If a solution containing proteins is shaken with montmorillonite, the clay absorbs all the protein, and it is not removed by subsequent washings.

3. The Formation of Activated Amino Acids with the Aid of Minerals

The term 'activated amino acid' is used to denote an amino acid in a state where it can undergo polymerization, transpeptidation, esterification, or any other type of reaction. In our work the activation is accomplished by binding the amino acid's carboxyl group to another acid, that is, through the formation of an anhydride bond (Katchalsky and Paecht-Horowitz, 1954). In the laboratory we do it nowadays by organic chemistry methods; Nature does it with the help of enzymes, but how can we envisage the formation of such activated amino acids in the prebiotic period? Kenyon *et al.* (1966) have shown that when dicyanamide is added slowly and at a constant rate to a solution of free glycine at acid pH values, peptides up to tetraglycine are formed, each one reaching in time a constant concentration (Figure 2a). We have obtained the same pattern of peptide formation in a slightly alkaline medium, starting with amino acid adenylates (Paecht-Horowitz and Kachalsky, 1967), which had been added to the aqueous solution at a steady rate (Figure 2b).

The pattern as such is typical of any steady state reaction, but while in Kenyon's system an acid medium was desirable, in our study an alkaline medium gave better results. The difference lies in the fact that in our case the amino acid was already activated; and under these conditions polymerization proceeds much better when the amino group of the amino acid or the peptide to be condensed with it is in its free form, not the ammonium form. In Kenyon's experiments, activation prior to polymerization was necessary. This activation was produced by the action of dicyanamide and an acid pH was required for the reaction to take place.

As the pH of solutions in Nature is usually more or less neutral, we have looked

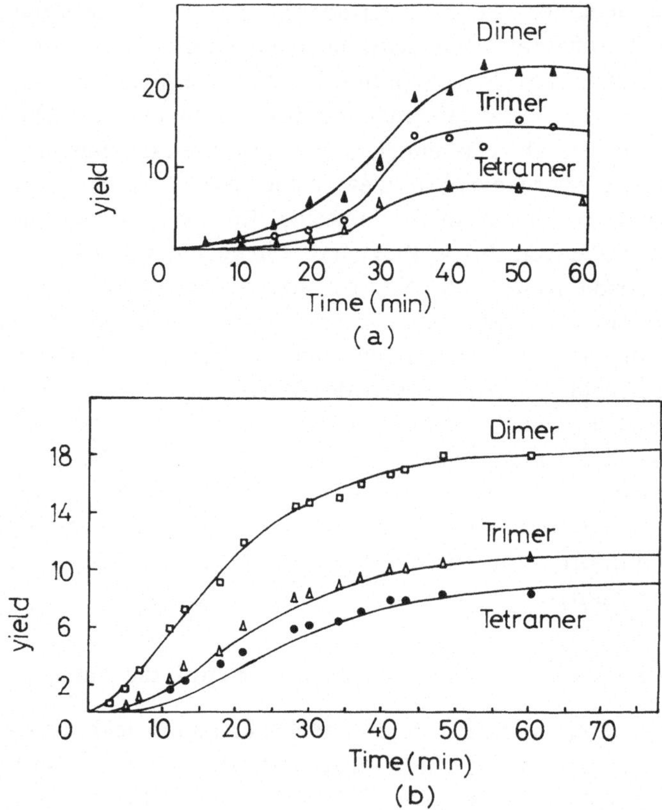

Fig. 2. The formation of peptides as a function of time under steady state conditions. (a) At low pH, starting with glycine and constantly adding dicyanamide (from Kenyon *et al.*, 1966); (b) at slightly alkaline pH, adding amino acid-adenylates to an aqueous solution at such a rate that their concentration stays constant (from Paecht-Horowitz and Katchalsky, 1967).

for a way to produce acid spots on which activation of amino acids might take place in an otherwise neutral medium.

Thus we have started to investigate various minerals for their ion exchange properties.

There is one class of minerals – zeolites – which swell only very slightly in water; they consist of alumina-silicates constructed in such a way that they form canals leading through the whole body of the mineral. These canals are of varying width, according to the original mode of formation of the zeolites and because of them the zeolites serve as molecular sieves. In addition to these properites, which are common to all zeolites, there are some which exhibit strong membrane hydrolysis, so that the salts they form are very readily hydrolyzed, their cations passing quickly into the surrounding aqueous solution. These zeolites will naturally also be very good ion exchangers and might cause the free carboxyl group of the amino acid to react with ATP to give a mixed anhydride. This idea has been tested with an artificial zeolite, whose canals are wide enough for big molecules to pass and it has been seen that indeed mixed anhydrides are formed under these conditions (Paecht-Horowitz and Katchalsky, 1973).

4. Copolymerization of Amino Acids in the Presence of Montmorillonite

In order to study the preferences of the different amino acids for mutual interaction we investigated the polymerization of amino acid-adenlyates (Paecht-Horowitz, 1973a) and the copolymerization of adenylates of several pairs of amino acids in the presence of montmorillonite. For the copolymerization experiments a suspension of montmorillonite of particle size ~150 Å at a concentration of 500 mg l^{-1} was maintained

TABLE I

Relative yields of bonds in the copolymerization reactions of adenylates of pairs of amino acids

Interacting substances	Bonds	Relative yields of bonds (%)
	Al–Al	40
Alanine-Adenylate	Gly–Gly	32
Glycine-Adenylate	Al–Gly	15
	Gly–Al	13
	Al–Al	23
Alanine-Adenylate	Val–Val	52
Valine-Adenylate	Al–Val	12
	Val–Al	13
	Al–Al	47
Alanine-Adenylate	Asp–Asp	49
Aspartyl-Adenylate	Al–Asp	2
	Asp–Al	2
	Al–Al	37
Alanine-Adenylate	Ser–Ser	37
Serine-Adenylate	Al–Ser	12
	Ser–Al	14
	Asp–Asp	55
Aspartyl-Adenylate	Gly–Gly	21
Glycine-Adenylate	Asp–Gly	9
	Gly–Asp	15
	Asp–Asp	59
Aspartyl-Adenylate	Ser–Ser	22
Serine-Adenylate	Asp–Ser	10
	Ser–Asp	9
	Asp–Asp	36
Aspartyl-Adenylate	Hist–Hist	44
Histidyl-Adenylate	Asp–Hist	8
	Hist–Asp	12

by a pH stat at a constant pH of 7.8. The experiment was carried out at room temperature. The mixture of the adenylates of the two amino acids was added in very small portions until the concentration of the amino acids was about 0.05 M. A portion of the montmorilllonite suspension containing montmorillonite of a larger particle size was then added, and subsequently several portions of the amino acid-adenylates. This procedure was continued until about 2 mM of adenylates (1 mM of each amino acid) and fifteen portions of montmorillonite (up to particle sizes of 5000 Å) were added. The mixture was left overnight with stirring, still at controlled pH. The suspension was then centrifuged, the precipitate washed several times and the combined supernatant passed through membranes of various pore sizes.

Each filtrate was concentrated by lyophylization and then passed consecutively through columns of Sephadex of various mesh sizes with water and HCl 0.01 N as solvents. The separations were repeated until the analysis of each fraction was con-

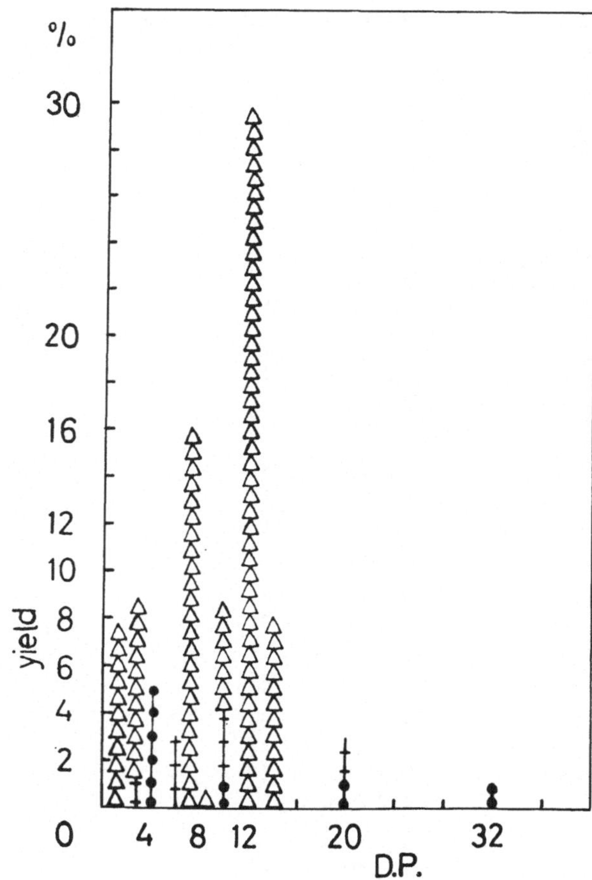

Fig. 3. Products of copolymerization of alanine-adenylate and glycine-adenylate in the presence of montmorillonite. ··· – Alanine peptides; + + + – glycine peptides; △ △ △ – copolymers.

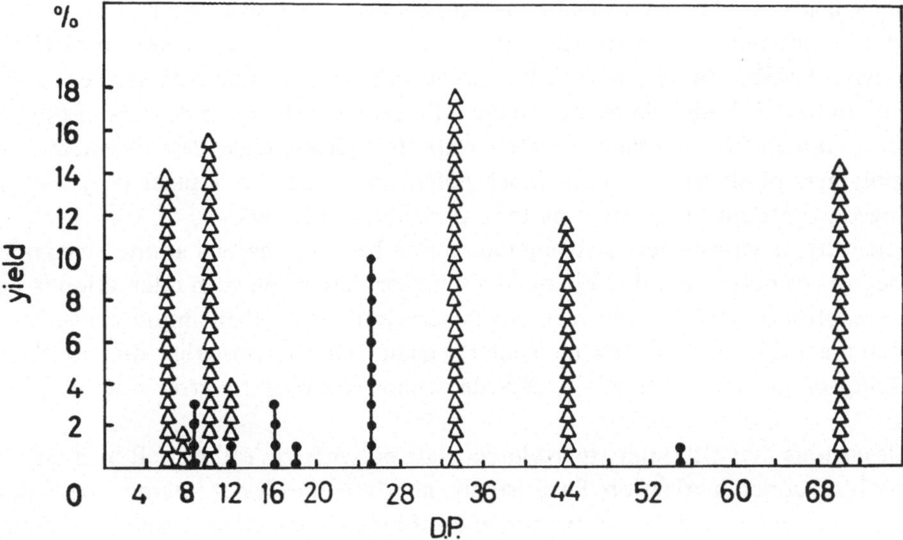

Fig. 4. Products of the copolymerization of alanine-adenylate and valine-adenylate in the presence of montmorillonite. ··· – Alanine peptides; + + + – valine peptides; △ △ △ – copolymers.

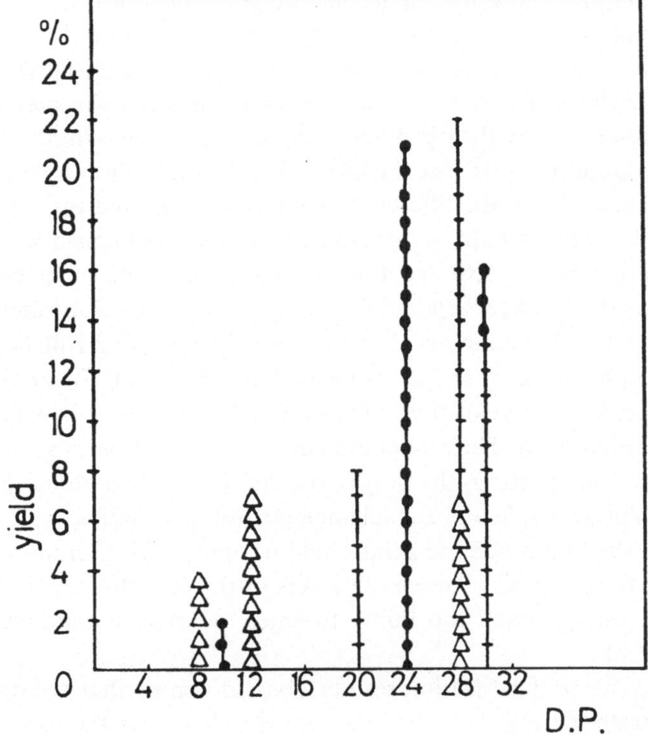

Fig. 5. Products of the copolymerization of alanine-adenylate and aspartyl-adenylate in the presence of montmorillonite. ··· – Alanine peptides; + + + – aspartyl peptides; △ △ △ – copolymers.

stant. When it could be reasonably assumed that a fraction contained only a single substance, the amino acid sequence in every such fraction was determined (Paecht-Horowitz, 1973b). Table I shows the relative yield of bonds of each species for every pair of amino acid-adenylates investigated. Figures 3–9 show the degrees of polymerization obtained for each pair and their respective yields, regardless of whether there are polymers of alternating units, block polymers united in one place only, polymers arranged according to a certain pattern or random polymers.

Generally, in cases where a strong interaction between the two amino acids occurs, the degrees of polymerization obtained are higher than when such interaction is weak. The exception is glycine, which interacts very well with other amino acids, but the resultant peptides are of low molecular weight. This is probably due to the low solubility of glycine peptides – a peptide cannot continue to grow once it precipitates.

It is curious that although aspartyl-adenylate polymerizes rather well by itself, there is very little copolymerization. Surprisingly, not only does it not interact well with the basic histidyl-adenylate, but in the presence of histidyl-adenylate even the homopolymerization of aspartyladenylate is much reduced. When the homopolymerization of histidyl-adenylate in the presence of montmorillonite was investigated, it was found that the mode of polymerization was as when no clay is present; a high degree of hydrolysis occurs and those low molecular weight (not more than 5 mers) peptides which are formed are of consecutive degrees of polymerization.

These data have caused us to revise our former assumption that the absorption of amino acid adenylates on clays is a surface phenomenon – the greater the surface, the more substance will be absorbed and the polymerization will correspondingly increase. If this assumption were true, histidine would behave like any other amino acid, which it does not. As the main difference between histidine and the other amino acids investigated so far was its bulk, we tried to envisage a mechanism where the bulk of the molecule might be decisive. To this end we constructed a molecular model in which two sheets of clay were separated by only the breadth of a water molecule and checked what part of the amino acid-adenylate molecule would fit between the clay sheets. With alanine, glycine, serine, valine and aspartic acid, the amino acid as well as the sugar phosphate part of the molecule fitted in rather well, while the adenine part protruded (Figure 10). Thus we could envisage the molecules as arranging themselves around the clay particle, the polymerization proceeding along the edges of the clay, with the peptide inside and the adenine portion protruding.

In this model the bulkier amino acids could not polymerize on montmorillonite as they are unable to penetrate between the layers of the clay. Indeed, in this molecular model also histidine is much too bulky to enter in between the layers of the clay (Figure 11a and b).

From the negative polymerization results it would appear that not only is histidine unable to penetrate the layers of the clay, but that it blocks them, so that aspartyl-adenylate which, when alone, can penetrate the layers, in the presence of histidine is unable to do so and thus does not polymerize either.

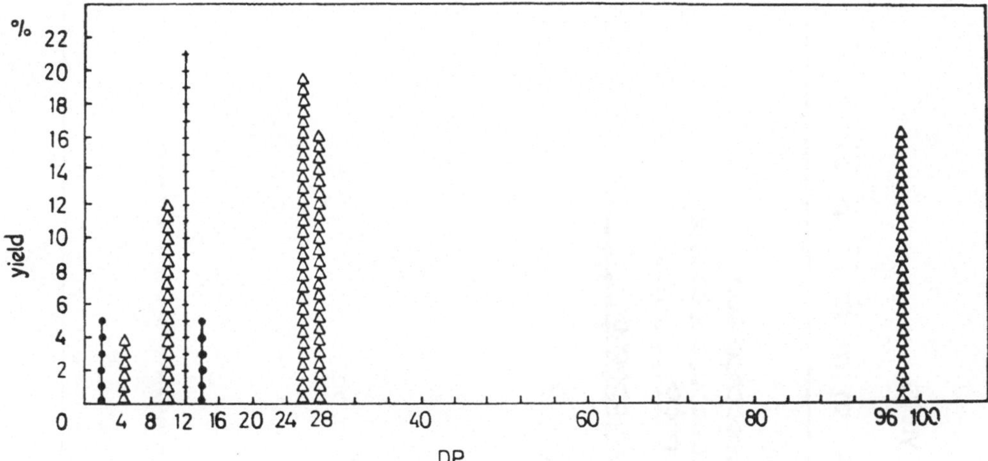

Fig. 6. Products of the copolymerization of alanine-adenylate and seryl-adenylate in the presence of montmorillonite. ··· – Alanine peptides; + + + – serine peptides; △ △ △ – copolymers.

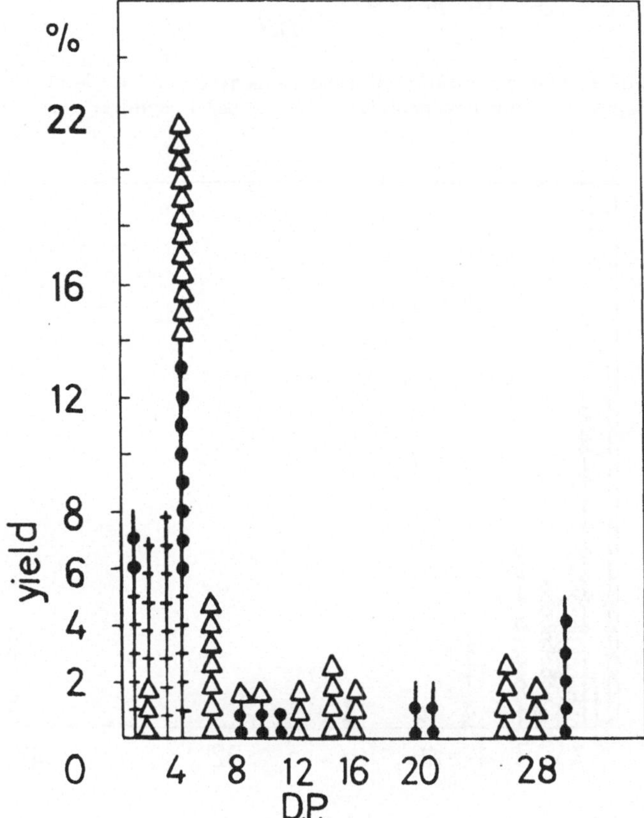

Fig. 7. Products of the copolymerization of aspartyl-adenylate and glycyl-adenylate in the presence of montmorillonite. ··· – Aspartyl peptides; + + + – glycyl peptides; △ △ △ – copolymers.

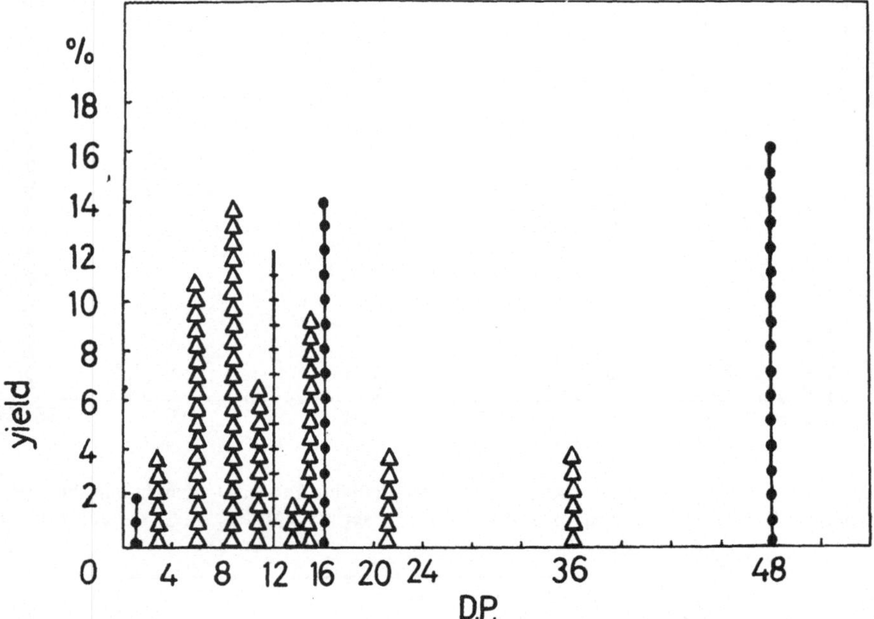

Fig. 8. Products of the copolymerization of aspartyl-adenylate and seryl-adenylate in the presence of montmorillonite. ··· – Aspartyl peptides; + + + – seryl peptides; △ △ △ – copolymers.

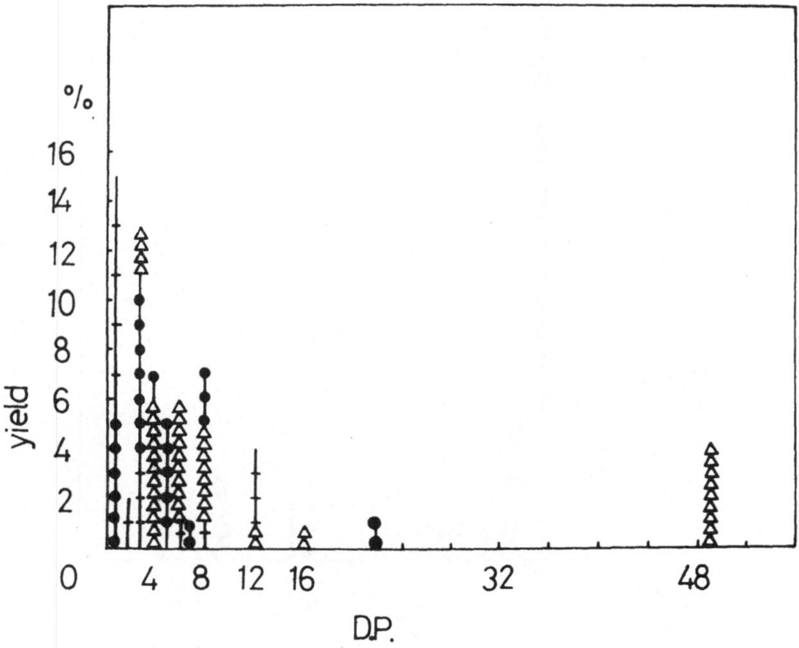

Fig. 9. Products of the copolymerization of aspartyl-adenylate and histidyl-adenylate in the presence of montmorillonite. ··· – Aspartyl peptides; + + + – histidyl peptides; △ △ △ – copolymers.

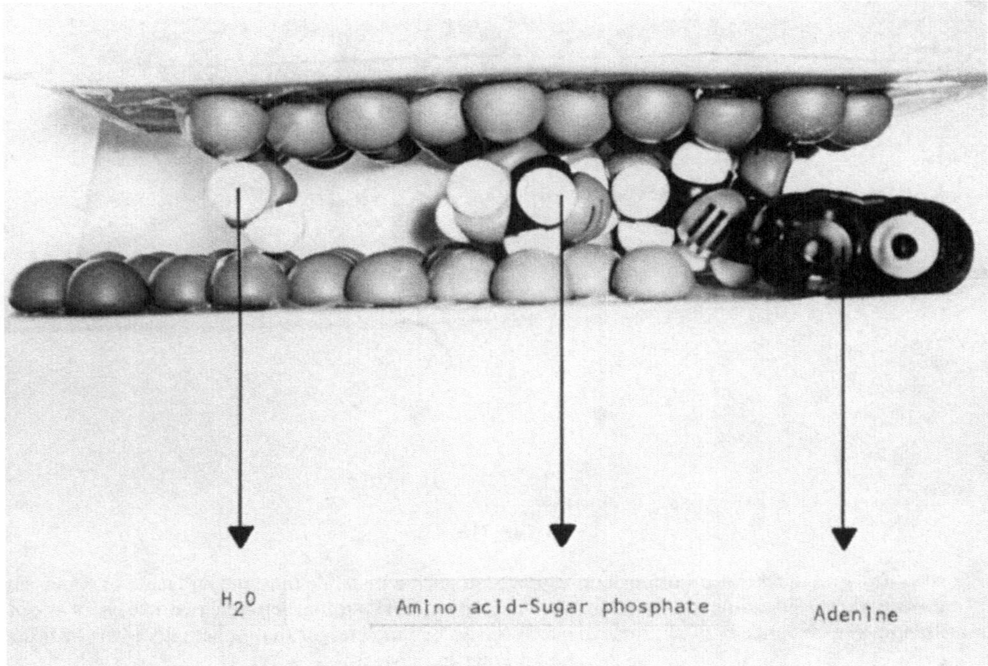

Fig. 10. Alanine-adenylate between two sheets of montmorillonite placed at such a distance that one molecule of water can fit between them easily. Under these conditions, the amino acid and the sugar phosphate part of the molecule can penetrate while the adenine part of the molecule protrudes.

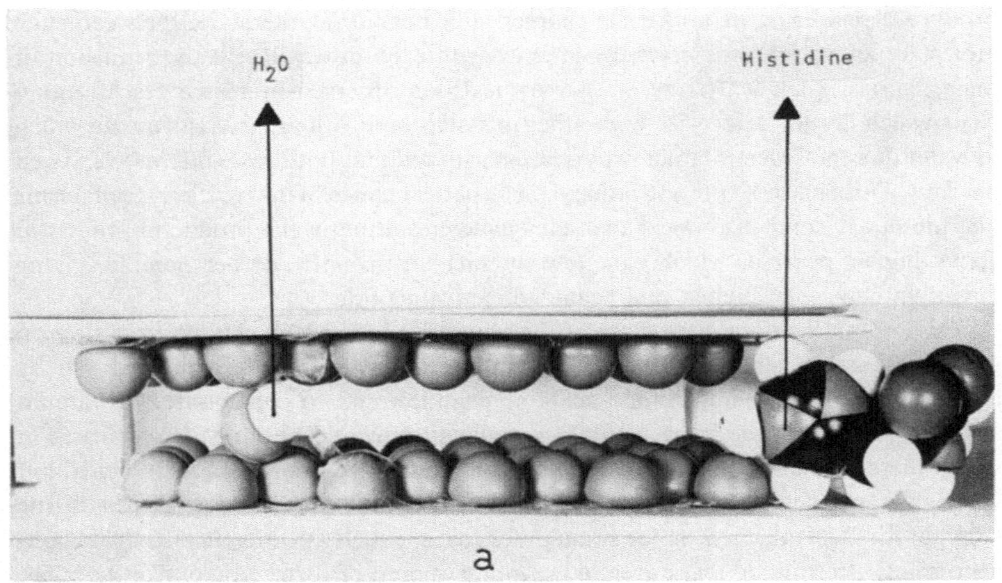

Fig. 11a.

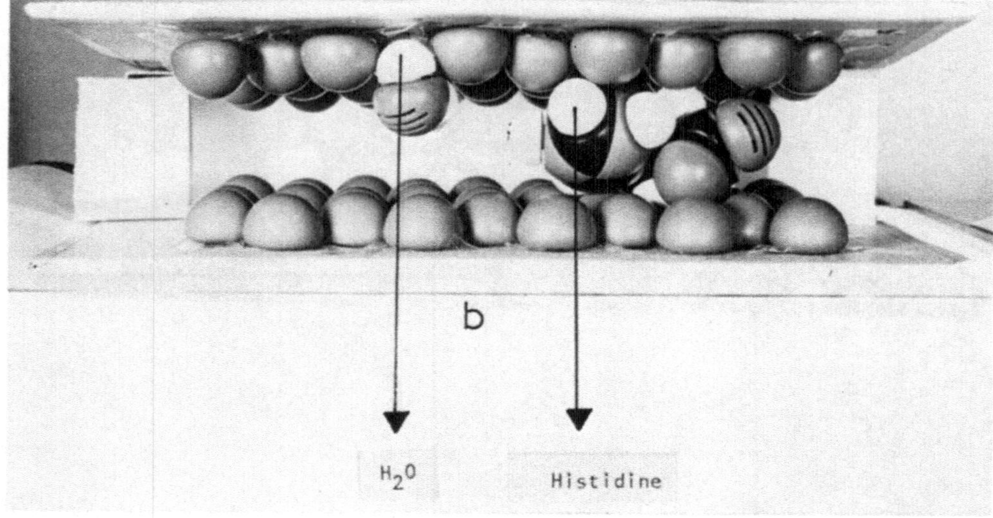

Fig. 11b.

Fig. 11. (a) Two sheets of montmorillonite placed at such a distance that one molecule of water can fit between them. Histidine is too bulky to penetrate. (b) Histidine between two sheets of montmorillonite. The distance between the two sheets has to be much larger than it actually is for histidine to be able to enter in between.

5. Discussion and Conclusions

In addition to their role as a means of concentrating the reactants, what other part could clays have played in prebiotic synthesis? We have seen that they are able to produce a scattering of molecular charges and thus allow interaction between such molecules as whould not normally interact with each other, due to the repulsion of their charges. Such scattering of charges has been observed in adenylates of amino acids which do not react with each other in water, and with aspartyl-adenylate which in water does not even interact with free aspartic acid but both cases interact very well on clays. This charge scattering induces on its part a change in the reaction mechanisms like the direct condensation of two adenylates, resulting in the production of a still active higher peptide, which can now interact again with another peptide, giving directly a new, even higher, and again still active peptide.

There are indications that some of the peptides obtained by copolymerization on clays possess to a certain extent repeating units. So far, this can be explained only by assuming that, in general, amino acids or peptides tend to condensate with amino acids or peptides of the same kind. This explanation would be valid also for cases in which there is relatively very good interaction between two sorts of amino acids, but still better interaction between amino acids of the same kind. Hence, if, due to the configuration of the clay, or for some other reason, such as differences in local concentration, a certain di, tri or even tetrapeptide were to be formed on one of the edges of the clay, there would be a preference for this still active peptide to condense further

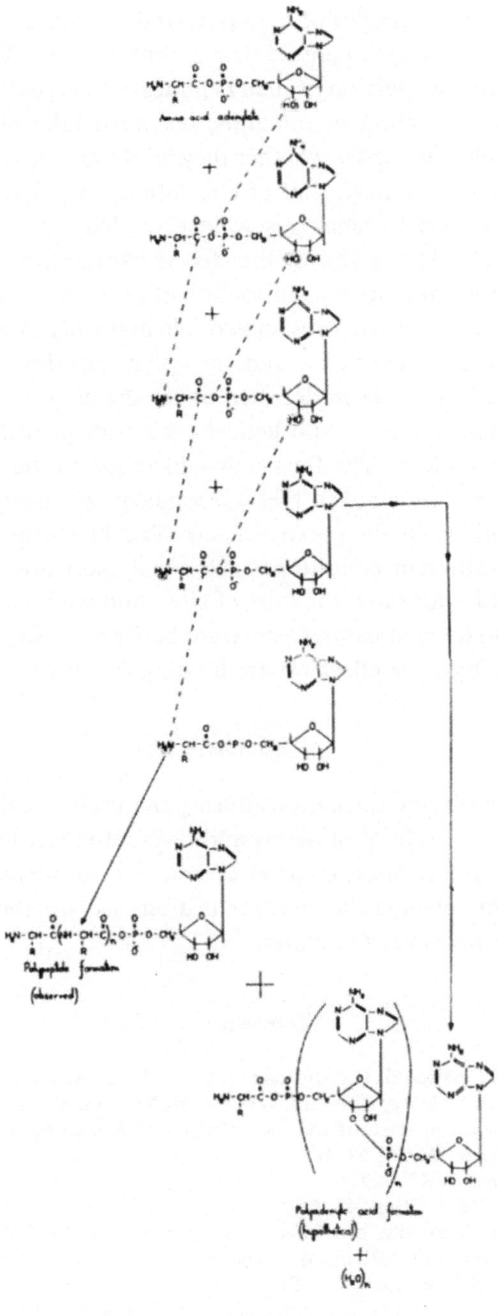

Fig. 12. Schematic representation of simultaneous formation of polypeptides and polynucleic acids from the same precursor.

on the clay or in bulk with peptides of the same kind instead of with peptides of other sequences, thus forming a larger peptide with repeating units. However, it should be stressed that actually the proportion of such peptides with repeating units is rather low.

Since in the presence of clays or minerals, reactions take place which might not otherwise occur, the question arises whether they might not be responsible for another type of polycondensation, namely, that of nucleotides to polynucleotides. For many years researchers have asked themselves what preceded what – the proteins or the nucleic acids? Eigen (1971) has shown that for the formation of life both were necessary. In Figure 12 the simultaneous formation of polyamino acids and polyadenylic acid starting with the same precurors is shown schematically. Yet while the formation of polypeptides has actually been observed, the polymerization of adenylic acid is still only hypothetical and has not been detected under the circumstances described.

It is true that chemically such a hypothetical nucleotide polymerization is much less likely than peptide formation. The former would require ester formation with a relatively far less reactive (compared to NH_2) secondary OH group. But still, the failure to obtain polynucleotides in these experiments (Paecht-Horowitz *et al.*, 1970) does not mean that under different conditions such a polymerization will not take place. So far we have worked with only one type of clay, and we have seen from the molecular model that adenylic acid cannot penetrate between its layers. With other clays, or a combination of clays, the effect we are looking for might be produced.

6. Summary

Hypotheses of macromolecule formations during the prebiotic era are described. The fitting in of the presumed role of minerals and clays into these hypotheses is checked. It is shown that copolymerization of pairs of amino acid adenylates in the presence of clays proceeds selectively and the question is discussed whether clays may not have served as non-specific prebiotic templates.

References

Bar-Nun, A., Bar-Nun, N., Bauer, S. H., and Sagan, C.: 1971, in A. Buvet and C. Ponnamperuma (eds.), *Chem. Evol. and the Origin of Life*, Vol. 1, North-Holland Publishing Co., Amsterdam, p. 114.

Bernal, J. D.: 1951, *The Physical Basis of Life*, Routledge and Kegan Paul, London.

Eigen, M.: 1971, *Naturwissenschaften* **58**, 465.

Gulick, A.: 1955, *Am. Scientist* **43**, 749.

Hendricks, S. B.: 1941, *J. Phys. Chem.* **45**, 65.

Katchalsky, A. and Paecht-Horowitz, M.: 1954, *J. Am. Chem. Soc.* **76**, 6042.

Kenyon, D. H., Steinman, G., and Calvin, M.: 1966, *Biochim. Biophys. Acta* **124**, 339.

Miller, S. L.: 1955, *J. Am. Chem. Soc.* **77**, 2351.

Oparin, A. I.: *Life, Its Nature, Origin and Development*, Oliver and Boyd, Edinburgh and London, p. 47.

Oró, J.: 1960, *Biochim. Biophys. Res. Comm.* **2**, 407

Oró, J. and Guidry, C. L.: 1960, *Nature* **186**, 156.

Oró, J. and Kamat, S.: 1961, *Nature* **190**, 442.

Oró, J. and Kimball, A.: 1959, *Arch. Biochem. Biophys.* **85**, 115.

Oró, J. and Kimball, A.: 1961, *Arch. Biochem. Biophys.* **94**, 217.

Paecht-Horowitz, M.: 1973a, *Angew. Chemi* (int. ed., in English), **12**, 349.
Paecht-Horowitz, M.: 1973b, *Isr. J. Chem.* **11**, 369.
Paecht-Horowitz, M. and Katchalsky, A.: 1967, *Biochim. Biophys. Acta* **140**, 14.
Paecht-Horowitz, M. and Katchalsky, A.: 1973, *J. Mol. Evol.* **2**, 91.
Paecht-Horowitz, M., Berger, J., and Katchalsky, A.: 1970, *Nature* **228**, 636.
Palm, C. and Calvin, M.: 1962, *J. Am. Chem. Soc.* **89**, 2115.
Ponnamperuma, C., Sagan, C., and Mariner, R.: 1963, *Nature* **199**, 222.
Schwartz, A. and Ponnamperuma, C.: 1968, *Nature* **218**, 443.
Scott, J. J.: 1956, *Biochem. J.* **62**, 69.

INTERACTIONS BETWEEN AMINO ACIDS AND NUCLEOTIDES IN THE PREBIOTIC MILIEU

CARL SAXINGER* and CYRIL PONNAMPERUMA

Laboratory of Chemical Evolution, Dept. of Chemistry University of Maryland, College Park, Md. 20742, U.S.A.

Abstract. In the study of chemical evolution we are interested in the path by which nucleic acids and proteins may have arisen under prebiotic conditions giving rise to those interactions which are generally postulated for the threshold of life. Although laboratory experiments have demonstrated that most of the building blocks of life can be generated from the raw material of the primitive atmosphere, in the genesis of polymers, however, the efforts have been less conclusive. It is reasonable to suppose that small peptides and small oligonucleotides may have interacted in the 'primordial soup' giving rise to the earliest association between nucleic acids and proteins. The beginnings of these processes could be related to the properties of individual amino acids and nucleotides. The interaction of oligomers with amino acids has been studied by the use of ion exchange and NMR spectroscopy. The observed affinities appear to depend on the given amino acid and the oligonucleotide chain length. The results so far suggest that direct amino-acid nucleotide interactions could have made a contribution to the early evolution of the Genetic Code.

Perhaps the major problem we face in elucidating the origin of living systems is the origin of nucleic acid directed protein synthesis. At present there exists no compelling hypothesis for how such a system might have arisen.

The nucleic acid double-helical structure and the function of DNA in the cell provide a strong suggestion for how, in principle, polynucleotides might have replicated under primitive abiotic or prebiotic conditions (Howard *et al.*, 1966; Huang and Tso, 1966; Schweizer *et al.*, 1965; van Holde, 1967). The relationship of gene to protein however is far more complex and does not readily suggest the evolutionary steps through which it came into being. In the gene-dependent mechanism for protein synthesis known as translation, information encoded in the primary sequence of nucleotide bases is transferred to the primary sequence of amino acids and consequently to the more or less exact structure of an enzyme capable of specifically interacting with its environment. The coding relationship, or genetic code is manifested by the assignment of one or more specific trinucleotide sequences to each of the 20 amino acids found in protein. In order to achieve this unique and thus accurate highly structured set of amino acid-codon correspondences the cell has evolved in a complex hierarchy of relationships and recognition processes between polypeptides and polynucleotides. It is obvious that such a system could not be the spontaneous product of simple random chemical combinations and so must be the result of a long and complex evolution.

There is no direct experimental evidence at present bearing on the identity of the primitive precursors of the present translation system. It is conceivable that primitive

* Present Address: National Cancer Institute, Bethesda, Md. 20014, U.S.A.

forms of 'translation' could have been rather different from the present one. Speculations as to the nature of protein synthesis in that early evolutionary period before the establishment of the genetic code have taken the following lines (Rich, 1965; Woese, 1967; Crick, 1968). Early translation seems likely to have been quite inaccurate i.e., the gene-protein relationships in a primitive translation system must have been either imprecisely or ambiguously defined. This should follow from the fact that the accuracy and precision of the present translation system depends on the precise function of many if not all of its highly evolved and sophisticated protein components, e.g. activating enzymes and some ribosomal proteins. Since the accuracy and precision of at least some of these proteins are final products of the evolution of the entire system, it is conceivable that the precursors to the present protein synthetic system could have been ones in which precise functional roles of proteins would be of less importance and the role of the nucleic acids corresponding of more importance than seen today. Hypotheses for the origin of the genetic code have, then, to assign importance to the function of nucleic acid in determining codon assignments and the ordered arrangement of the present code.

One of the possibilities that must be explored is that the structure of the set of codon assignments and of the protein synthetic machinery somehow reflects 'recognition' or some other form of catalytic interaction between (some form of) nucleotides and (some form of) amino acids – perhaps only during a primitive stage in the evolutionary history of the cell. To examine this possibility, detailed studies of amino acid – oligonucleotide interactions are required.

Our purpose is thus threefold. First, to establish what types of interactions are possible between these two classes of molecules and how to characterize them. Second, to learn how to use the information from simple interpretable systems to interpret increasingly more complex systems. Third, to determine the degrees of selectivity attainable in systems of increasing chemical complexity.

Previous studies have shown that a variety of interactions can occur between nucleic acids and amino acids depending upon composition, conformation, state of polymerization and environment of the reacting species (see references in Saxinger and Ponnamperuma, 1971). These studies have led to the conclusion that a degree of specificity does exist although its origin has not yet been elucidated.

When the two reacting species are simple (i.e., neither is polymeric) one cannot expect to observe 'specificity' of the sort implied in the biological use of the term. What one can look for on this simplest level is evidence for 'selectivity' of some sort, e.g., cases where the strength of binding of an amino acid to a nucleotide under given conditions, is to some extent a function of the composition of both interaction species. While results of this sort from simple systems would not be spectacular from a biological viewpoint, they are nevertheless a necessary first step to any systematic study of a role for amino acid-nucleic acid interactions in the evolution of living systems. Working with monomeric species in aqueous media readily permits the effects of individual factors to be assessed, and so provides the basic information necessary to interpret more complicated polynucleotide-amino acid interactions. It would be surprising if one did not find some 'selective' interaction of this type in simple (monomeric) systems. The point is to detect these and characterize them sufficiently so that one can begin to ask whether they (or their counterparts that occur when one or both species are polymeric) played a role in determining structure of codon assignments.

If interactions exist, what are they? Are they selective? Can they lead to nucleic-acid directed protein synthesis or help us to understand the system we have? Can they be applied to understanding more complex systems? Can they lead to addition of constraints (experimental) which can achieve genetic coding in a defined prebiotic system? Can they lead to an understanding of the evolutionary history of our present universal genetic code, i.e. through an understanding of the basic interactions, and the constraints required to evolve the genetic coding system to its present form. The experiments to be described should be regarded as undergoing parallel evolution rather than sequential or temporal one.

In our opinion immobilization of the amino acid on a chromatographic support offered the greatest potential in terms of both sensitivity and convenience for an initial characterization of these elusive interactions. Ideally the chromatographic support would allow the attachment of any amino acid in the desired amount by means of a uniform technique. Further, immobilization of an amino acid on this support would cause a minimal perturbation on the potential for specific interaction with (oligo) nucleotides. For these requirements a new, optically-active anion-exchange matrix containing L-amino acids was designed and synthesized (Saxinger, 1969) see Figure 1 (Saxinger *et al.*, 1971). The polyvinylamine matrix contains a large number of primary amine groups (nearly 20 meq g^{-1}) to which acidic groups can be

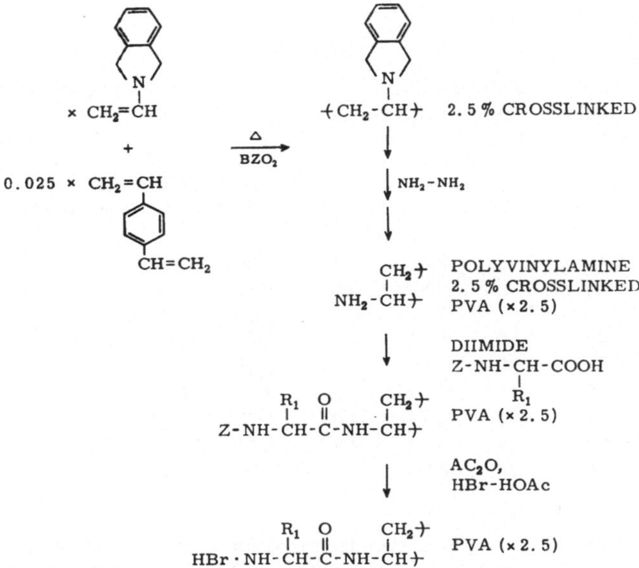

Fig. 1. Scheme for synthesis of amino acid resins.

easily attached by established methods. The form of the amino acid as it appears attached to the gel matrix (amide linkage) is analogous to an N-terminal protein amino acid. It also bears a formal resemblance to biologically activated amino acids. The specificity of the binding of nucleotides by these amino acids was investigated by systematically varying the α-amino acid side chains and the base moieties of nucleoside 5′-monophosphates. Binding of nucleotides to an amino acid resin occurred only when the α-amino group was charged (trp-resin at pH6) and not at pH8, ruling out nonspecific adsorption to the resin matrix. In all cases the observed affinities were dependent on the expected electrostatic interaction but were also dependent on the particular combination of base and side chain (Saxinger *et al.*, 1971).

Apparent association constants can be described by the law of mass action for homogeneous reactions for the binding of nucleotides to the immobilized amino acids by the employment of inorganic orthophosphate binding as an internal reference standard (Saxinger and Ponnamperuma, 1971). Thus nucleotide binding to a particular immobilized amino acid is expressed by a quantity we designated a selectivity coefficient in accordance with the convention of ion-exchange equilibria and is simply the ratio of nucleotide to inorganic phosphate association constants (Table I). Numerically the selectivity coefficient is equal to the nucleotide bound/free ratio divided by the phosphate bound/free ratio which are found after a given amino acid resin is allowed to equilibrate with a solution containing $^{32}P_i$ and ^{3}H-nucleotide.

From a general inspection of Table I it is clear that the observed binding of nucleotides by the amino acids manifests a pronounced system of selectivity, one having several interesting features which appear to be dependent on the nucleotide base and amino acid side chain. A comparison of the basic amino acids is particularly interesting. Lysine and glycine are unique in preferring a pyrimidine, C, to a purine, A, and in

TABLE I

Selectivity coefficients, $K_p{}^N$, for the Binding of ribonucleoside-5′-monophosphates and phosphate to immobilized L-amino acids and solute properties of the amino acids.

Amino acid	Selectivity coefficients[a]					Hydro phobicity (Kcal/residue)	Polar requirement
	5′-Up	5′-CP	5′-Ip	5′-Ap	5′-Gp		
Glycine	0.48	0.58	0.80	0.52	1.00	0	− 7.9
Lysine	0.56	0.66	0.90	0.62	0.92	1.5	−10.1
Proline	1.10	1.07	1.41	1.42	2.10	2.6	− 6.6
Methionine	1.20	1.10	1.37	1.62	2.26	1.3	− 5.3
Arginine	1.38	1.38	2.06	2.19	2.98	0.75	− 9.1
Histidine	1.45	1.51	2.19	2.30	3.20	0	− 8.4
Phenylanaline	1.50	1.39	2.74	3.00	4.30	2.65	− 5.0
Tryptophan	1.99	1.48	5.48	4.17	6.61	3.00	− 5.3
Tyrosine	2.38	2.24	8.96	4.61	6.00	2.85	− 5.7

[a] For glycine through histidine (mean of two determinations; relative average deviations (X%) are less than 4%). For phenylalanine through tyrosine (mean of five determinations; X% are less than 10%).

having the lowest selectivity coefficients. The similarity between lysine and glycine binding is suggestive of a common mechanism, predominately ionic, or interaction. This would indicate that the nucleotide base somehow modulates the ammonium-phosphate interaction, perhaps through water of hydration as suggested by Latt and Sober (1967). In contrast the general pattern or 'spectrum' of nucleotide selectivities manifested by the other basic amino acids, histidine and arginine, resemble that of the aromatic amino acids. A similarity between methionine and proline 'spectra' is also evident. Thus while there is an overall trend in the similarities between the ranking of amino-acids by selectivity coefficient and by their tendency to self associate (hydrophobicity) of their gross solute properties ('polar requirement') these properties alone do not adequately account for the order of binding selectives observed.

It is also clear from the observed nucleotide binding that while the interaction between the two classes of monomers tends to increase with their potential for self-association these properties alone do not adequately account for the order of binding selectives observed. With the exception of the three basic amino acids the binding of nucleotides generally increases with increasing 'hydrophobicity' (Tanford, 1962; Nozaki and Tanford, 1963; Scheraga, 1963) of an amino acid side chain. The binding to lysine is relatively weaker and the binding to arginine, histidine and methionine stronger than would be expected on the basis of hydrophobicity. This correlation is improved in some respects if instead of hydrophobicity, one uses as the criterion of comparison, the amino acid 'polar requirements' (Saxinger, 1969) (Table I). For example, lysine has the largest 'polar requirement' of all the natural amino acids i.e., the least tendency to partition from water into pyridine-like solvents and a correspondingly low tendency to bind nucleotides. It is interesting to note that the ordering of amino acids by polar requirement also bears a striking similarity to the ordering of amino acids by their coding triplets (Woese *et al.*, 1966). Even by this criterion however, arginine and histidine, both of which contain pi electrons in their side chains, remain exceptions to the general trend. In a similar manner, nucleotide binding amino acids follow the tendency of bases to interact with one another by 'base stacking' although once again this tendency is not consistently followed.

Table II shows the apparent association constants measured by Wagner and Arav

TABLE II

Comparison of the binding behavior of immobilized amino acids from Table I and polyamino acids (from Wagner and Arav, 1968, in 0.02 M sodium acetate, pH 7)

Nucleotide[a]	$K_p^N \times 4.00$		$K'_a (\mathrm{m}M^{-1})$	
	L-lysine	L-arginine	Poly-L-lysine	Poly-L-arginine
5′-Up	2.25	5.52	2.25	4.6
5′-Cp	2.64	5.52	2.2	5.3
5′-Ip	3.60	8.24	–	–
5′-Ap	2.48	8.75	2.25	9.4
5′-Gp	3.84	11.9	3.25	14.5

(1968) for nucleotides with poly-L-lysine and poly-L-arginine compared with the scaled nucleotide selectivity coefficients of L-lysine and L-arginine calculated from the data of Table II. Our data are made approximately congruent with theirs by a scaling factor of 4.00. The agreement between the two systems gives strong support for our assumptions and indicates the immobilized L-amino acids are a good model for the amino acids in solution or at least for an amino acid in a polypeptide environment. This comparison further indicates that in these simple cases the relative specificity for the interaction with nucleotides displayed by the polymeric amino acids is to a first order approximation a function of affinities between the monomeric components.

The data presented in Table III shows that amino acid-oligonucleotide binding shows marked dependence on the composition, size, and order of bases within a given

TABLE III

Selectivity coefficients for the binding of oligonucleotides to immobilized amino acids (Saxinger *et al.*, 1973)

	UpGp	GpUp	ApUp	ApApUp	GpApUp	ApGpUp
gly	10.32	14.05	23.6	63.9	16.6	10.9
	13.44	14.39	27.5	60.0	19.4	13.0
trp	95.1	42.1	187.5	2045	60.4	173
	101.3	56.2	177.6	1817	65.4	197

oligonucleotide. These results show binding strengths to increase as a function of chain length of the oligonucleotides in most but not all cases (Saxinger *et al.*, 1971; Saxinger and Ponnamperuma, 1971).

Thus even on a relatively simple level, interactions between amino acids and oligonucleotides are sufficiently strong and of such a nature that a rudimentary 'preferential' scheme can be seen.

A qualitative explanation for our observed results could be in terms of a site-binding model for the nucleotide-amino-acid associations. Acting in concert with the ammonium phosphate electrostatic interaction would be the effect of possible direct interaction between a base and side chain and the resultant effect produced by solvation and organization of the solvent as in the 'stacking' interactions of nucleic acids, and the 'hydrophobic' bonding of proteins. Courtauld molecular models verify that such an association, in which the amino acid side chain and nucleotide base are coplanar and overlapped i.e., stacked, is possible with a nucleotide whose base is in either the syn or anti conformation. In either case a relatively precise co-orientation with maximum overlap is attainable. The extent and stereochemistry of this type of interaction in the case of oligonucleotides would of course depend on the stability of a given oligonucleotide amino acid complex relative to the inherent stability of the oligonucleotide itself. A similar model has been proposed on the basis of fluorescence and P.M.R. studies, for the binding of tryptophan derivatives to nucleic acid components (Helene *et al.*, 1971) in which a charge-transfer interaction takes place between the indole ring and nucleotide base in a stacked structure.

Proton magnetic resonance spectra have been obtained which are consistent with a direct side chain-base interaction. Qualitatively the NMR data show first that mixtures of the monomers yield spectra which are different from the isolated components. Thus the aromatic ring protons of UMP and CMP are shifted upfield increasingly by the phe, tyr, and trp esters (Figure 2, Table IV) The ring protons of the aromatic side chains are not appreciably affected by interaction with the pyrimidines but are shifted upfield by the purine rings. Interaction of phe, tyr, or trp ester with IMP shifts the ring proton positions upfield in that order but only trp causes an appreciable upfield shift of the AMP ring protons (Table V). This result can be explained by the fact that

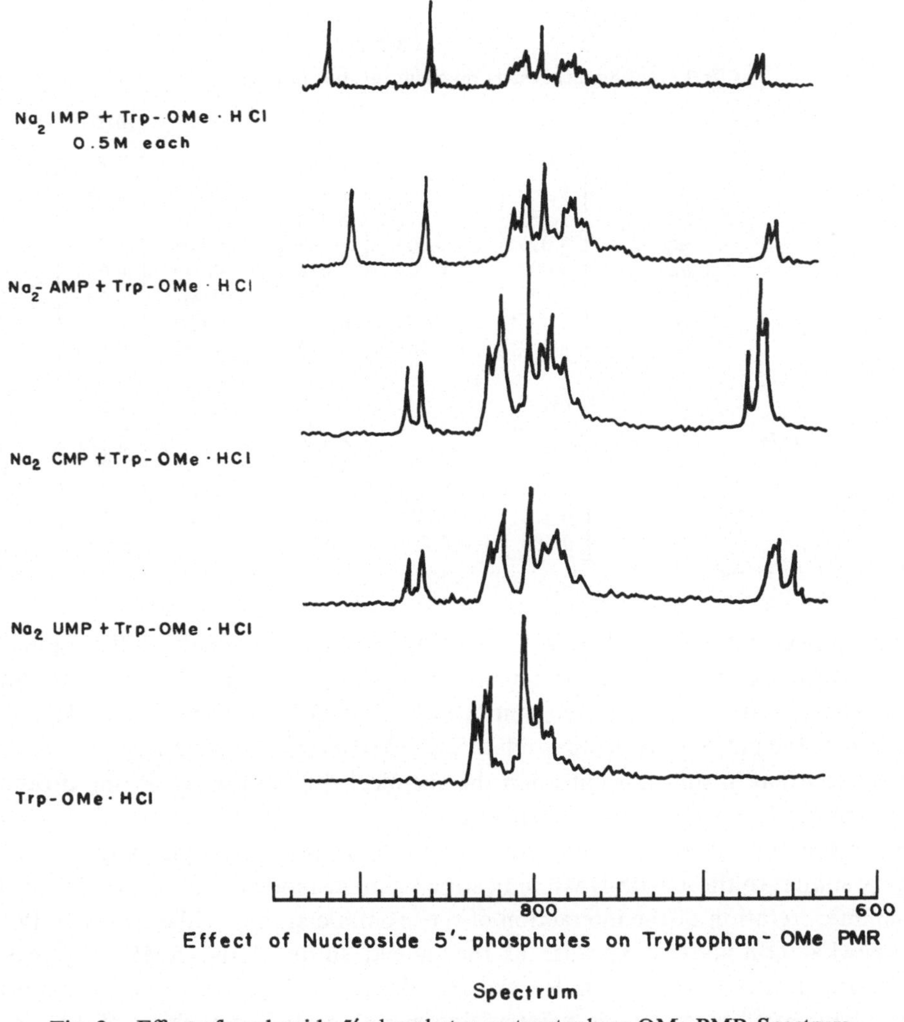

Fig. 2. Effect of nucleoside 5′-phosphates on tryptopham-OMe PMR Spectrum.

TABLE IV

Effect of amino acid esters on chemical shift pyrimidine nucleotides and poly U

	0.05 M *Poly U* δH-6	0.5 M U-5′P δH-6	0.5 M C-5′P δH-6
	8.662	8.892	8.832
0.25 MPhe	8.542	8.777	8.737
Δδ	−0.12	−0.115	−0.095
0.25 MTyr	8.525	8.772	8.722
Δδ	−0.137	−0.100	−0.110
0.05 MTrp	8.405	8.617	8.606
Δδ	−0.257	−0.275	−0.226

TABLE V

Effect of amino acid esters on chemical shift of purine nucleotides (0.5 M ea)

		δH-8	*Δδ*	δH-2	*Δδ*	δH-1′	*Δδ*
AMP		9.099		8.700		6.674	
	met	9.160	0.061	8.792	0.092	6.750	0.076
	lys	9.135	0.036	8.730	0.030	6.575	0.001
	gly	9.095	−0.004	8.695	−0.005	6.570	−0.004
	arg	9.090	−0.009	8.655	−0.045	6.664	−0.010
	phe	9.097	−0.002	8.702	0.002	6.662	−0.012
	tyr	9.090	−0.009	8.710	0.010	6.665	−0.009
	trp	8.970	−0.129	8.540	−0.160	6.550	−0.124
		δH-8		δH-2		δH-1′	
IMP		9.263		8.888		6.773	
	met	9.285	0.022	8.955	0.067	6.805	0.032
	phe	9.230	−0.033	8.840	−0.048	6.735	−0.038
	tyr	9.215	−0.048	8.840	−0.048	6.730	−0.043
	trp	9.080	−0.183	8.490	−0.398	6.595	−0.178

at these concentrations (0.5 M) the purine ring protons are already shifted upfield by self association of the purine monomers (Tso *et al.*, 1969). Furthermore, the interaction of trp with AMP at lower concentrations (0.05 M) causes a much larger upfield shift of the purine ring protons relative to AMP alone at that concentration. It can also be seen that at this concentration the chemical shift of the AMP ring protons is dependent on the concentration of trp (Figure 2).

The interaction of the aromatic amino acid esters with poly U yields spectra which are very similar to those with UMP indicating a similar mode of interaction (Table IV).

The interpretation of the interaction of the aromatic amino acid esters with Poly A is somewhat complicated because of the upfield shifts of the AMP ring protons accompanied by polymerization (Raszka and Mandel, 1971); but the indication is that the mechanism of interaction is by intercalation with partial unstacking of the adenine

TABLE VI

The effect of amino acid esters on chemical shift of poly A and AMP

	Poly A	σ	(0.05 M)		AMP	σ	
	H-8	H-2	(H-8) – (H-2)		H-8	H-2	(H-8)–(H-2)
	8.525	8.430	0.095	0.5 M AMP	9.225	8.950	0.275
).05 MTrp	8.665	8.495	0.265	+ 0.5 M Trp	8.970	8.550	0.420
Δδ	–0.140	–0.030					
				Δδ	–0.255	–0.400	
).25 MPhe	8.715	8.495	0.220				
Δδ	+0.190	+0.065					
				0.05 M AMP	9.099	8.700	0.399
).25 MT yr	8.745	8.520	0.225				
Δδ	+0.220	+0.090		+0.05 M Trp	8.970	8.540	0.430
				Δδ	–0.129	–0.160	

bases. The interaction of typ ester shows (Table VI) a slight upfield shift of the adenine C-2 proton suggesting direct apposition of the adenine and indole rings and a substantial downfield shift of the adenine C-8 proton indicating a relaxation of poly A secondary structure similar to that seen with high temperature melting as previously reported. Phe and tyr esters cause downfield shifts of both adenine ring protons with increased separation of the C-2 and C-8 proton line frequencies as if they too were intercalating but the dominant effect being the lack of mutual shielding by adjacent adenine rings. Thus it appears that amino acid-polynucleotide interaction can compete favorably with the polynucleotide secondary structure. Finally since the amino acid aromatic ring protons are shifted upfield to about the same extent by the interaction with poly A and by AMP (data not shown) it is likely that the mechanisms of interaction are similar.

The purpose of our NMR studies is to learn how amino acids are binding in addition to how strongly. We want to know not only what the binding affinities are, but also whether binding occurs in such a way that catalysis could result or perhaps by knowing how and why they are binding we can decide what sort of perturbations must be made on the interaction to produce the selectivities or reactivities we are seeking. NMR data suggests we also have positional possibility for catalysis in that the amino acid esters bind uniformly and are located close to one another by binding to the polynucleotide. This is suggested by the fact that the adenine ring protons (and uridine) are uniformly shifted upfield as a single peak. We have not attempted to determine from this type of analysis yet whether or not the amino acids are in fact oriented head to tail further facilitating a catalytic effect.

Since we know that amino acids and their derivatives can interact with oligonucleotides by binding with a certain degree of selectivity it would be interesting to know whether this binding can be used to catalyze the reaction of activated amino acid esters and whether the reaction reflects the binding affinities or a new set of selection rules applies.

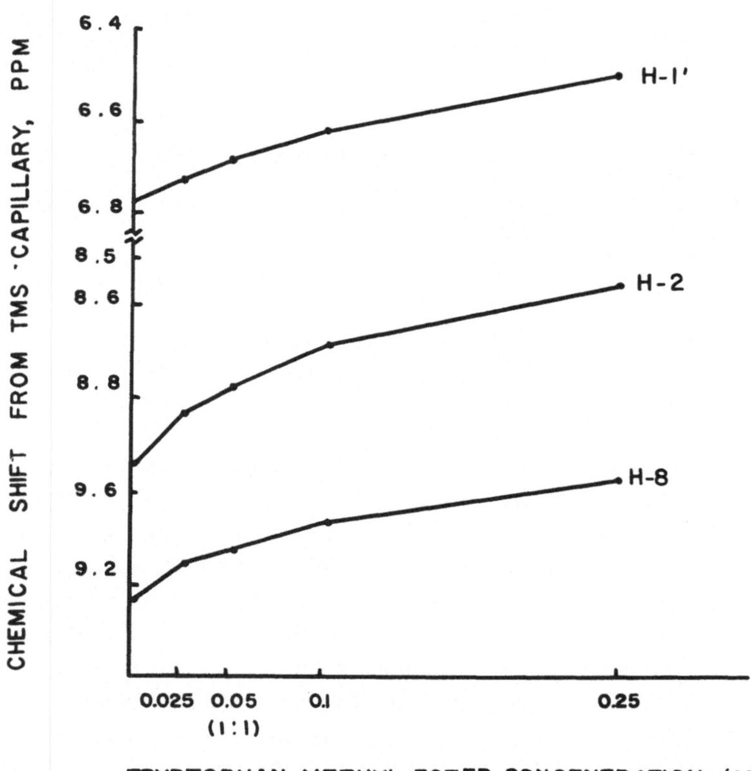

Fig. 3. Chemical shift vs concentration of tryptophan methyl ester for protons of 5′-AMP (35°, 0.05 M). Shifts are relative to RMS by normalization of aceton-in-capillary spectrum to 2.170. The pD was not controlled and varied from 5.7 to 4.9 with increasing tryptophan methyl ester concentration.

To answer this question the thiophenyl esters of a mixture of amino acids (HCL salt) was prepared (Greenstein and Winitz, 1961) and were given DNA, transfer RNA or a synthetic polynucleotide as a possible template. The mixture was neutralized with imidazole and at the end of the 24 hr reaction unreacted amino acids were destroyed by treatment with ninhydrin at low pH, followed by hydrogen peroxide to destroy excess ninhydrin (Markovitz, 1957). The reaction mixture was then hydrolyzed with 6*N* HCL to release free amino acids which were then analyzed by ion-exchange chromatography. Test reactions containing only DNA or RNA yielded essentially no free amino acids at the end of the procedure and conditions were chosen so that reaction of the 'activated' thiophenyl esters without added template was minimal. D,L-Leu was added prior to hydrolysis for internal standardization for the analysis of each reaction mixture. The results shown in Figure 4 were obtained.

It can be seen that the results are not consistent with data from binding affinities

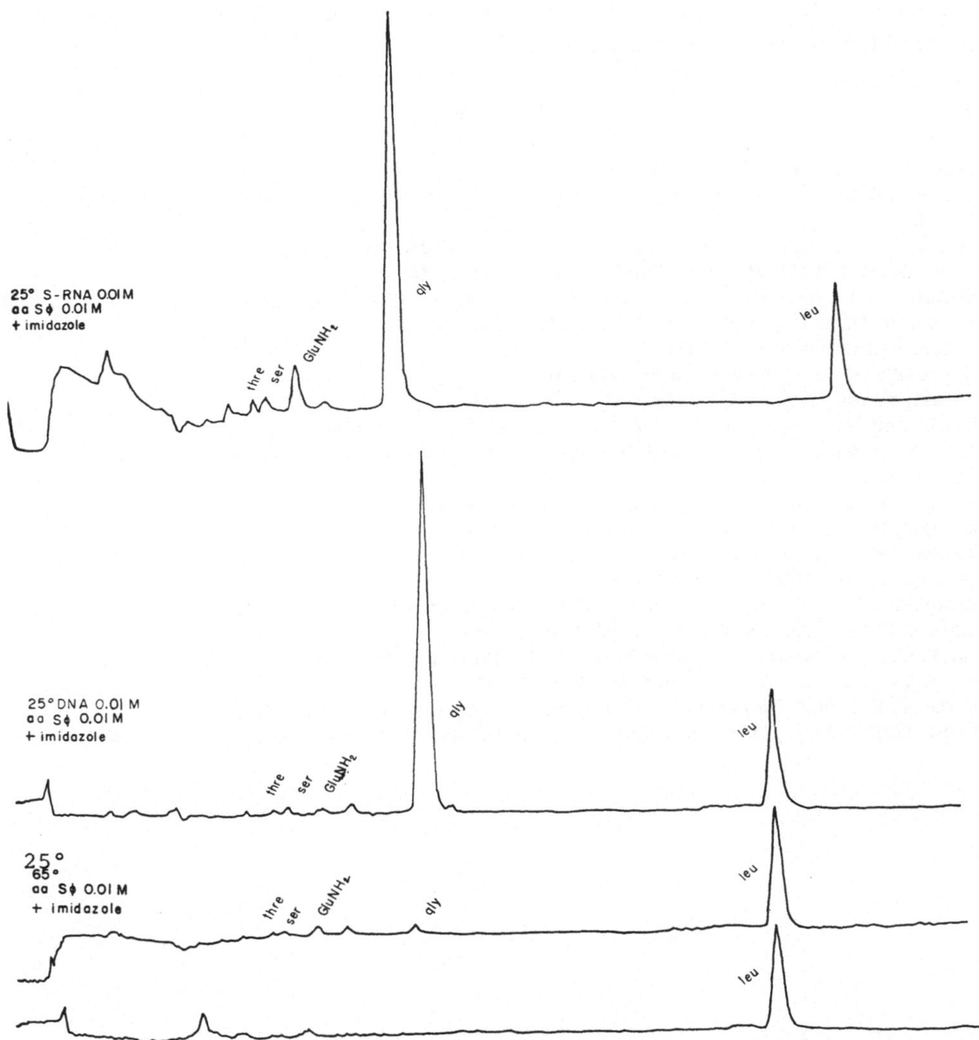

Fig. 4. Catalysis of reaction of amino-acyl thiophenyl esters by nucleic acid 'templates'.

alone. Although both polynucleotides in effect catalyze the reaction of the thiophenyl esters the aromatic amino acids do not appear in the final analysis at all while the glycine appears in relatively great abundance. The effects of both DNA and t-RNA are quite small, but significant compared to the controls. *t*-RNA appears to exert a greater effect on the reaction again by a small but significant margin. These results suggest that the chemical reactivity of the individual amino acids play a dominant role in this rudimentary catalytic system. It is not known what effect varying the composition of the amino mixture would have. Nor is it possible to determine whether the small differences between templates result from differences in secondary structure or from the difference in phosphate-ribose backbone.

Finally it would be interesting to know whether the effect of base composition could influence the outcome of this reaction.

References

Crick, F. H. C.: 1968, *J. Mol. Biol.* **38**, 367.

Greenstein, J. P. and Winitz, M.: 1961, *Chemistry of the Amino Acids*, John Wiley & Sons, Vol. 2, p. 1027.

Helene, C., Dimicoli, J.-L., and Brun, F.: 1971, *Biochem.* **10**, 3802.

Holde, K. van: 1967, *Biochem. Biophys. Res. Commun.* **26**, 717.

Howard, F. B., Frazier, J., Singer, M. F., and Miles, H. T.: 1966, *J. Mol. Biol.* **16**, 415.

Huang, W. M. and Tso, P. O. P.: 1966, *J. Mol. Biol.* **16**, 523.

Latt, S. A. and Sober, H. A.: 1967, *Biochem.* **6**, 3307.

Markovitz, A.: 1957, *Federation Proc.* **16**, 216.

Nozaki, Y. and Tanford, C.: 1963, *J. Biol. Chem.* **238**, 4074.

Raszka and Mandel, M.: 1971, *Proc. Nat. Acad. Sci. (U.S.)* **68**, 1190.

Rich, A.: 1965, in V. Bryson and H. J. Vogel (eds.), *Evolving Genes and Proteins*, Academic Press, New York.

Saxinger, W. C.: 1969, Ph.D. Thesis, University of Illinois.

Saxinger, W. C. and Ponnamperuma, C.: 1971, *J. Mol. Evol.* **1**, 63.

Saxinger, W. C., Ponnamperuma, C., and Woese, C. R.: 1971, *Nature, New Biology* **234**, 172.

Scheraga, H. A.: 1963, *Proteins* **1**, 478.

Schweizer, M. P., Chan, S. J., and Tso, P. O. P.: 1965, *J. Amer. Chem. Soc.* **87**, 5241.

Tanford, C.: 1962, *J. Am. Chem. Soc.* **84**, 4240.

Tso, P. O. P., Schweizer, M. P., and Hollis, D. P.: 1969, *Ann. N.Y. Acad. Sci.* **158**, 256.

Wagner, C. R. and Arav, R.: 1968, *Biochem.* **7**, 1771.

Woese, C. R.: 1967, *The Genetic Code*, Chapter 6, Harper and Row, New York.

Woese, C. R., Dugre, D. H., Saxinger, W. C., and Dugre, S. A.: 1966, *Proc. Nat. Acad. Sci.* **55**, 966.

COACERVATE SYSTEMS AND ORIGIN OF LIFE

T. N. EVREINOVA, T. W. MAMONTOVA, V. N. KARNAUHOV,
S. B. STEPHANOV and U. R. HRUST

Moscow University Lomonosov, Biological Faculty, Dept. of Plant Biochemistry, Dept. of Cytology, and Biophysical Institute, Academy of Sciences of U.S.S.R., Moscow, U.S.S.R.

Abstract. Hydrophilic coacervate systems consist of coacervate drops (0.5–640μ in diameter) and liquid. The most molecules are cooperated into the drops. Defects of such systems and drops are instability. Stable protein-nucleic-acid carbol-hydrate drops are studied. Enzymatical oxidized reactions were fulfield by the peroxidase (1.11.1.7) and the polyphenoloxidase (1.10.3.1) and its substrates (phenol and other ones) in coacervate systems. The drops are getting stable by the oxidized compounds quinones and others. The quinone content of individual drops (more than 5000 drops) was found by means scaning Cytospectrophotometer SIM. The limit of analysis 1×10^{-13}–10^{-14} g, the errors 3 % of the value found. The mathematical equations of the dependence both the size of drops and the content of oxidized compounds were calculated. The structure of stable drops was investigated in electronic microscope. The drops bound one to other by means bridges or hills, and formed colonia. We proposed that different stability of drops, especially stable ones, and colonia are interesting phenomena for the Origin of Life.

1. Introduction

The origin of life and the artificial synthesis of life are among the great problems of science. According to A. I. Oparin's theory of the origin of life (Oparin, 1966) ther were some initial stages of chemical evolution during the prebiological period. The association of molecules, however, could have taken place in microspheres (Fox, 1968), lipoprotein bubbles and coacervate drops (Oparin, 1966). The place of coacervate drops and others of procells models in the evolution of matter is shown in Figure 1 (Evreinova, 1972). Coacervate systems were studied by H. G. Bungenberg de Jong at the beginning of this century. Now, more than 250 hydrophilic coacervate systems are known. They may consist of two or more different molecules. The chemical classification of coacervate systems was published (Evreinova, 1966). The hydrophilic coacervate systems consist of drops (0.5–640 μ in diameter) and equilibrium liquid. There are coacervate drops in protoplasm of living cells. The common property of any coacervate system is the cooperation or association of molecules in the coacervate drops. Only a relatively few polymer molecules exist in the equilibrium liquid; drops may absorb other molecules from equilibrium liquid and the property of drops is used in practice. The most of coacervate drops are unstable ones (Evreinova, 1966).

The purpose of this paper is to show how the enzymatic reactions in coacervate systems affects the stability and the structure of drops.

2. Experimental

Coacervate systems were obtained from aqueous solutions of Histon, DNA, Gum-Arabic (0.5 to 1%) and enzymes (0.03 mg ml^{-1}) Poly phenoloxidase (1.10.3.1)

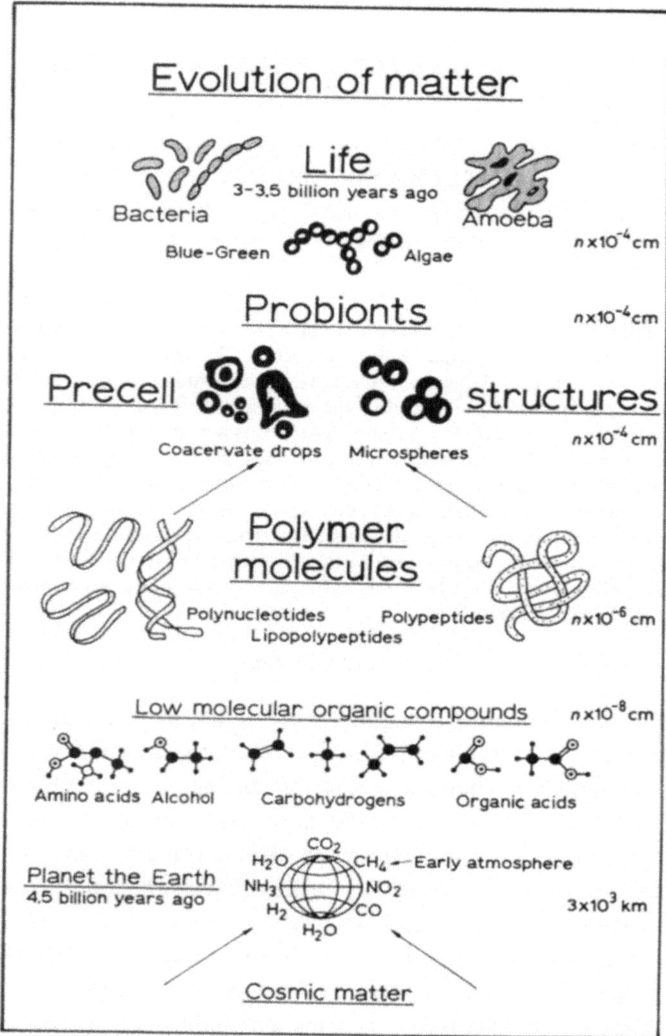

Fig. 1. Evolution of matter and Origin of Life.

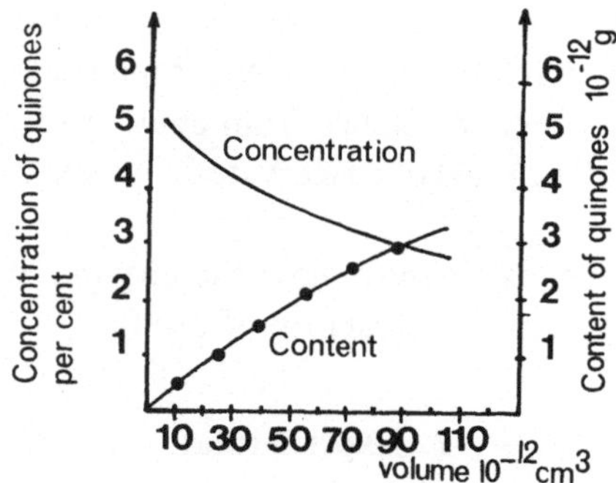

Fig. 2. The content (–·–·–) and the concentration (—) of quinones in individual drops of coacervate system: Peroxidase-Histon-Gum Arabic-Quinones (pH = 6.0).

Peroxidase (1.11.1.7) and their substrates (Phenols, H_2O_2, O-Dianisidine-red). The coacervate drops were formed at the pH = 6.0 and at the temperature 16° to 30°. The size and the oxidized compounds content of individual drops were found by means of a scanning recording cytspectrophotometer SIM. The limit of the analyses was 10^{-13} to 10^{-14} g of oxidized compounds. The error was 2.5% of the value found. Details of composition of coacervate systems and optical methods and calculations for drops have been reported (Evreinova *et al.*, 1972). More than 5000 individual drops were measured. The structure of drops was investigated in electronic microscope (Evreinova *et al.*, 1973). The magnification was equal to 5000 to 50000. Some of the results are illustrated in Figures 2 and 3 and Table I, and in chemical, mathematical Equations (I–V).

The Enzymatic Reactions in Coacervate Systems

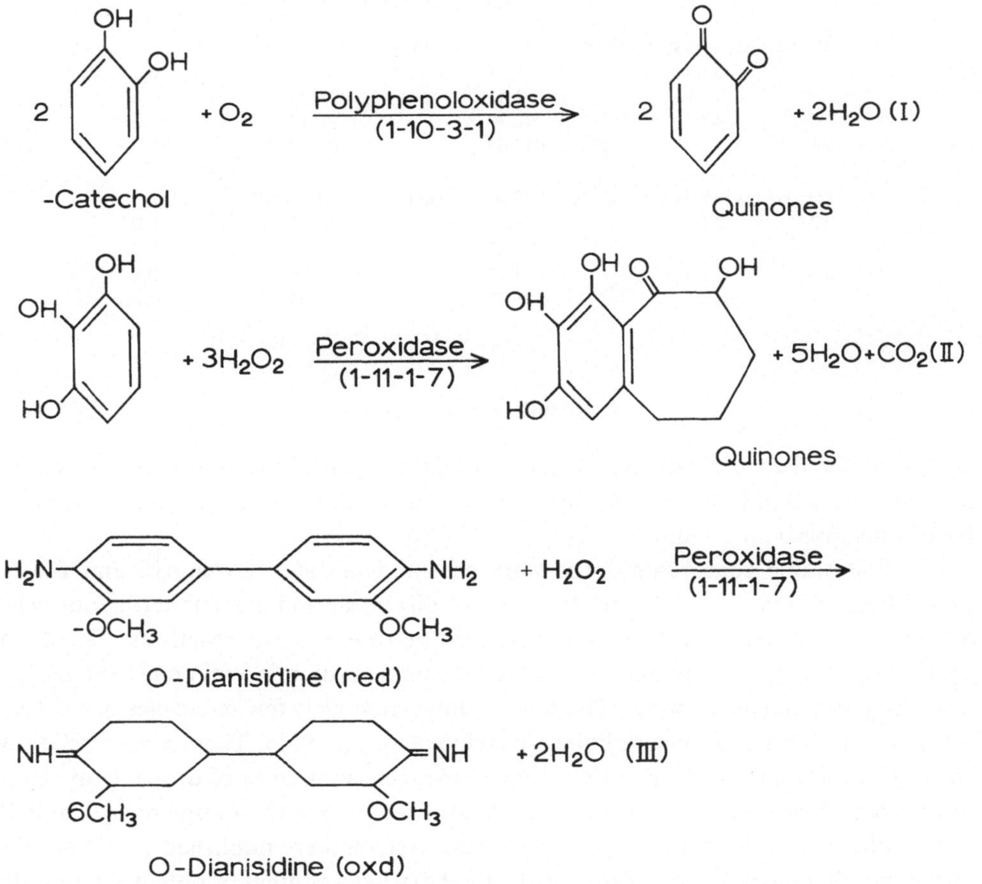

2 -Catechol + O_2 —Polyphenoloxidase (1-10-3-1)→ 2 Quinones + $2H_2O$ (I)

OH, OH, HO + $3H_2O_2$ —Peroxidase (1-11-1-7)→ Quinones + $5H_2O + CO_2$ (II)

H_2N–, –NH_2, -OCH_3, OCH_3 O-Dianisidine (red) + H_2O_2 —Peroxidase (1-11-1-7)→

NH=, =NH, OCH_3, OCH_3 O-Dianisidine (oxd) + $2H_2O$ (III)

Figs. 3A—D. Coacervate drops in electronic microscope: (A) Coacervate drop of Histon-DNA; (B) The part of coacervate drop of Peroxidase-Histon-DNA-Quinones; (C) Coacervate drops-Peroxidase-Histon-DNA-O-Dianisidine (oxd): (D) Coacervate drops-Peroxidase-Histon-Gum Arabic-Quinones.

TABLE I

The content of oxidized compounds in individual coacervate drops

No.	Diameter 10^{-4} cm	Volume 10^{-12} cm^3	Oxidized compounds	
			Weight 10^{-12} g	Concentration %
Peroxidase - Histon - Gum Arabic - Quinones pH = 6.0				
1	4.400	44.604	1.803	4.04
2	4.967	64.126	2.418	3.77
3	5.27	76.81	2.748	3.57
4	6.28	129.56	3.535	2.72
Peroxidase - Histon - DNA - Quinones pH = 6.0				
5	2.76	11.01	0.911	8.27
6	3.20	17.22	1.089	6.32
7	4.62	51.70	2.357	4.55
Peroxidase - Histon - Gum Arabic - O-Dianiside (oxd) pH = 6.0				
8	3.44	21.31	0.160	0.75
9	3.99	33.31	0.220	0.66
10	4.62	51.70	0.314	0.60
11	5.90	107.65	0.597	0.55
Peroxidase - Histon - DNA - O-Dianisidine (oxd) pH = 6.0				
12	2.56	8.81	0.095	1.07
13	3.14	16.21	0.162	0.99
14	3.89	30.77	0.255	0.83
15	4.40	44.69	0.295	0.66

3. Results and Discussion

Drops in trivial coacervate systems settle on the bottom of the vessel and disappear; they convert a thin layer. For example, the drops which consisted of serum Albumin-Histon disappear in 30 min.

The first stable coacervate drops were prepared in 1968 (Evreinova and Bailey, 1968). These drops also settle, but they do not disappear and are stable for four years or more. The stability of drops was reached due to enzymatic reactions. The drops containing oxidized compounds are stable. The content of quinones and O-dianisidine (oxd) in individual drops were 0.003 to 6%. Only a relatively few molecules of oxidized compounds are in equilibrium liquid (Evreinova *et al.*, 1971). There are correlations both in the size and in the concentration of oxidized molecules of drops. Tiny drops contain a high content oxidized compounds per unit volume than large ones (Table I). The results with polyphenoloxidase coacervate systems were published in 1968–1971. The same phenomena takes place for the total dry mass content of polymer molecules in coacervate drops (Evreinova, 1966).

There were many experimental dates (more than 1000 for each coacervate system). That's why it was possible to calculate the content and the concentration of oxidized

PART IV

PRECELLULAR ORGANIZATION

compounds in any drop by the mathematical equations (for example IV–V, and Figure 2).

Peroxidase - Histone - Gum Arabic - Quinones

$$y = 0.089 + 0.050x - 0.00017x^2 \quad \text{(IV)}$$

$$y_1 = 4.805 - 0.019x + \frac{4.6}{x} \quad \text{(V)}$$

y : the content of quinones $n \times 10^{-12}$ g
y_1: the concentration of quinones %
x : volume of a drop -10^{-12} cm^3.

The structure of coacervate drops in electronic microscope is shown in Figure 3A, B, C, D. The stability and the structure of drops are changed by the enzymic reactions. The surfaces of unstable drops (Histon-DNA and others) are homogeneous. There are many hills on the surfaces of the stables drops (Histon-DNA quinones and others). Stable drops are bounded by the hills and the bridges and formed colonia. According to Buvet's and Toupance's theory (Buvet, 1971) the first step was non-enzymic activation of substances; the second step was their oxidation for the prebiological stages. We proposed the stabilisation of molecules in coacervate drops and the formation of colonia are of some interest for solving origin of life; and a number of biological items.

Acknowledgements

We are very grateful to Prof. Buvet. The authors would like to thank Academicians A. I. Oparin and G. M. Frank and Prof. U. S. Chenzov.

References

Buvet, R.: 1971, *Origin of Life and Evolutionary Biochemistry*, Varna, Bulgaria, p. 19.
Evreinova, T. N.: 1966, *Concentration of Substances and the Action of Enzymes in Coacervate Systems*, Nauka, Moscow.
Evreinova, T. N.: 1972, *Dokl. Sci. Biol. U.S.S.R.* **N12**,
Evreinova, T. N. and Bailey, A.: 1968, *Dokl. Akad. Sci. U.S.S.R.* **723**, 179.
Evreinova, T. N., Stephanov, S. B., and Mamontova, T. W.: 1973, *Dokl. Acad. Sci. U.S.S.R.*
Evreinova, T. N., Karnaukhov, W. N., Mamontova, T. W., and Ivanizki, G. R.: 1971, *J. Colloid Interface Sci.* **30**, 18.
Evreinova, T. N., Mamontova, T. W., Littinskaya, K. L., and Hrust, U. R.: 1972, *Dokl. Acad. Sci. U.S.S.R.* **204**, 991.
Fox, S. W.: 1968, *Quart. J. Florida Acad. Sci.* **31**, 1.
Oparin, A. I.: 1966, *The Beginning and the Origin of Life on the Earth*, Medizine, Moscow.

TRANSFER RNA AND THE TRANSLATION APPARATUS IN THE ORIGIN OF LIFE

ALEXANDER RICH

Dept. of Biology, Massachusetts Institute of Technology, Cambridge, Mass. 02139, U.S.A.

Abstract. At present we have some understanding about the mechanisms of prebiotic synthesis and polymerization. Experiments are also being carried out on the prebiotic replication of the nucleic acids. However, the greatest gap in our knowledge is the prebiotic mechanism for the polynucleotide directed synthesis of polypeptides. Some insight into potential mechanisms of this type is afforded by the recent determination of the three-dimensional structure of yeast phenylalanine rRNA. The polynucleotide chain is found to be highly folded and the molecule is found in an L shaped configuration. One end of the L shaped molecule contains the anticodon loop; that arm of the L is made of the double helical anticodon and dihydro U stems. The other end of the L shaped molecule contains the 3′OH adenosine to which the amino acid is added during aminoacylation. That arm of the L is built of the double helical CCA and TΨC stems. The corner of the L is formed by the close approximation of the TΨC loop and dihydro U loop. The distance between the anticodon and the 3′OH adenosine end of the molecule is 76 Å. Comments are made about the manner in which the tRNA molecule may have evolved and in addition, some suggestions are presented about the prebiotic interactions of primitive tRNA polynucleotides and messenger RNA strands.

1. Introduction

One frequently asks at what point in the evolution of living systems did life ‘begin.’ In our present understanding of the emergence of living systems from the nonliving world, the answer to that question is necessarily somewhat arbitrary. We do not know the various stages which occurred in the development of a complex terrestrial chemistry which gave rise to replicating systems of macromolecules. This may have happened several times. It is not known when the nucleic acids were polymerized in the prebiotic world. These molecules contain the capacity for replication in that the purines and pyrimidines contain all of the information necessary to organize the complementary nucleotides from which a complementary nucleic acid can be assembled. Thus, most workers believe that at one stage early in evolution, replicating nucleic acids were present. Was that life? Replicating molecules while essential for life are not necessarily a living system. Life as we know it depends on the expression of genetic information, an expression which is embodied today in the complexities found in the polypeptide chains of protein molecules. These are the molecules responsible for catalyzing large numbers of chemical reactions in cellular metabolism and they are responsible for building many of the stable structures in biological systems. During the past twenty years we have come to understand the nature of the translation mechanism whereby the information encoded in the polynucleotide sequences of the nucleic acids is translated into polypeptide sequences. We would like to know something of the origin of this translation system. Indeed, if one were to assign a time or developmental stage at which life ‘began’, it is perhaps convenient to use as a starting point that stage at which a primitive translation system developed so that polypeptide sequences were being formed under the direction of primitive polynucleotides in an abiogenic world. With the expression of the information encoded in polynucleotides in the form of primitive proteins, the stage is then set for the selection pressures which are known to give rise to the shaping and development of living systems. Thus, in a sense, the development of the translation mechanism is perhaps a reasonable point at which to assign the 'beginning' of life.

It is quite clear that the entire evolutionary process is a continuum, starting from prebiotic molecules of varying degrees of complexity and ending in systems of replicating polynucleotides expressing themselves in terms of polypeptides which in turn substantially influence the manner and degree of genetic expression. It is reasonable to say that the existence of replicating macromolecules such as the nucleic acids are a necessary requirement for a living system of the type that we see on the earth today, nonetheless, by themselves they represent an inadequate manifestation of life. They are inadequate in that it is only through the expression of genetic information in the form of another molecule such as the proteins that makes life possible. The proteins are crucial because of their enormous chemical and physical potentialities which produce the vast variety of chemical environments necessary for the development of a living system.

2. The Replication of the Nucleic Acids

Using a plausible prebiotic reducing atmosphere, a number of investigators have elaborated on experiments first developed by Miller and Urey in which high energy processes are used to convert gases into components of the prebiotic world, including the amino acids, sugars and constituent units of the nucleic acids (Orgel, 1973). A number of experimental conditions have been found in which amino acids are capable of abiogenic polymerization forming a number of polymers. These do not contain information in the sense that they do not have an ordered sequence nor can they be formed reproducibly. In the same way, condensations of purines or pyrimidines with sugars and phosphates have led to the development of nucleotides and in some cases to polynucleotides of varying lengths. It has been clearly demonstrated that the polynucleotides, either ordered or random, can organize their complementary molecules in solution to form double helical molecules using the complementary hydrogen bonding found in the present day nucleic acids. The fact that polynucleotides can organize structures which are biologically relevant today means that these processes are likely to have gone on in the prebiotic world providing one had a steady source of nucleotide monomers plus energy sources. The mechanism for the initial polymerization of nucleotides as well as the conditions favoring the assembly and polymerization of the complementary monomers is largely unknown at the present time. It is of interest, however, to inquire what is the smallest polynucleotide which can form a double helical molecule using the complementary hydrogen bond systems. The answer to this question has been demonstrated recently in our laboratory by the discovery that dimers containing two nucleosides connected by a phosphate linkage are capable of being organized into helical double stranded structures. The structure of two dinucleoside phosphates has recently been solved by X-ray diffraction analysis. These were selected for examination because both of them are self-complementary oligonucleotides. Adenylyl-3′,5′-uridine (ApU) and guanylyl-3′,5′-cytidine (GpC) are both self-complementary in that they have the ability to form pairs of complementary hydrogen bonds with themselves. Thus adenine can hydrogen bond to uracil and guanine to cytosine in a manner analogous to that which is found in the double stranded nucleic acids. The sodium salt of both of these oligonucleotides crystallize in the form of double helical complexes which are virtually identical even though they crystallize in radically different crystal symmetries. ApU crystallizes in a monoclinic space group $P2_1$ with $a = 18.025$ Å, $b = 17.501$ Å, $c = 9.677$ Å and $\beta = 99.45$ Å (Rosenberg *et al.*, 1973). The unit cell contains four molecules of ApU together with 24 water molecules; thus the molecules exist in a very heavily hydrated medium. The form of the molecule can be seen in Figure 1 in which the darker structure is closer to the reader. It can be seen that the adenine bases are hydrogen bonded to uracil using the

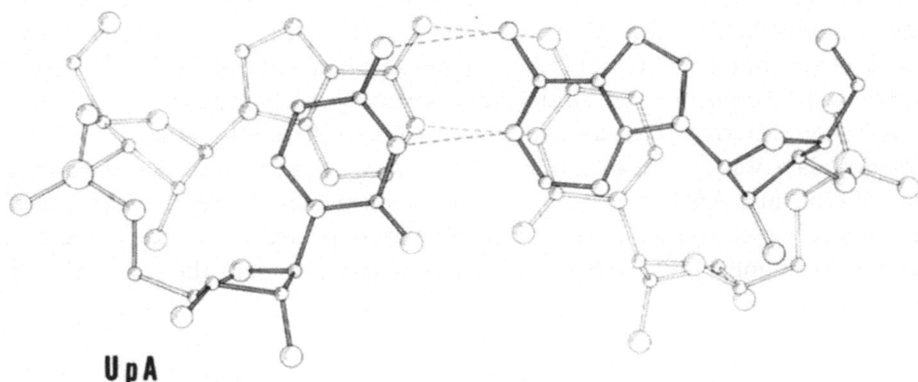

Fig. 1. View of the crystal structure perpendicular to the base planes showing hydrogen bonding (dashed lines) as well as the base stacking interactions. The darkest portions of the figure are those nearest the viewer. The rotational relationship is easily seen by comparing the front (black) and rear (white) glycosideic bonds.

same pair of hydrogen bonds which are believed to hold together the adenine-thymine pairs in double helical DNA or the adenine-uracil pairs in double helical RNA. This is the first visualization of the double helix at atomic resolution. All previous structural studies of helical polynucleotides were carried out on fibers and in these diffraction studies it was impossible to resolve individual atoms and unambiguously fix the mode of hydrogen bonding. It is interesting to note that although the structure of a number of complementary intermolecular complexes have been determined containing both adenine and uracil (or thymine) derivatives (Voet and Rich, 1970), only this crystal which forms a double helical fragment uses the hydrogen bonding on the N_1 nitrogen of adenine which is believed to be found in all double helical nucleic acids. The two molecules of ApU are related to each other by a two-fold rotation axis. This is the same two-fold axis which is believed to exist in DNA or double helical RNA which describes the fact that the strands are anti-parallel. However, in the ApU crystal, even though there are two pairs of ApU in the unit cell, neither of them uses this axis as a symmetry element in building the lattice.

Guanylyl-3′,5′-cytidine (GpC) crystallizes in the space group C2 with $a=2146$ Å, $b=16.93$ Å, $c=9.33$ Å, and $\beta=90.5$ Å (Day *et al.*, 1973). There are 4 GpC molecules in the unit cell together with 36 water molecules. GpC forms a similar double helical complex and a view of its crystal structure is shown in Figure 2 where the darker base pair is closer to the reader. The view is in a direction perpendicular to the base pairs. Here again it can be seen that the guanine and cytosine derivatives form the familiar hydrogen bonded pairs containing three hydrogen bonds and the two base pairs are stacked and helically twisted relative to each other. The two GpC molecules in the helical fragment are related to each other by a twofold rotation axis located between the base pairs. This symmetry element is used in the crystal lattice here. This can be seen more clearly in Figure 3 which shows the view of the two crystalline fragments ApU and GpC, together with the configuration RNA-11 which was derived from a study of double helical viral RNA. It can be seen that all three molecules are similar to each other; however, it should be emphasized that only in GpC and ApU is it possible to fix the positions of all the atoms unambiguously. The atomic positions in RNA-11 were inferred from an analysis of a fiber diffraction pattern in which the diffraction data was at a resolution less than 3–4 Å. It can be seen that the two fragments of GpC and ApU are related to their complementary mates through a twofold rotation axis which results in the anti-parallel arrangement of the backbone chains. Both the ApU and GpC crystals are heavily hydrated so that the sugar phosphate

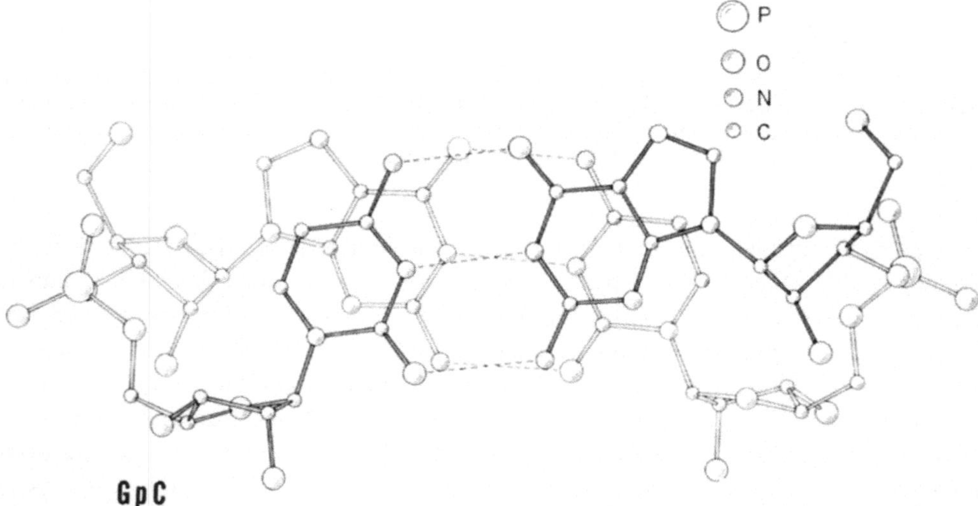

Fig. 2. A view of the GpC double-helical fragment approximately normal to the base planes. The shaded base-pair is nearest the reader.

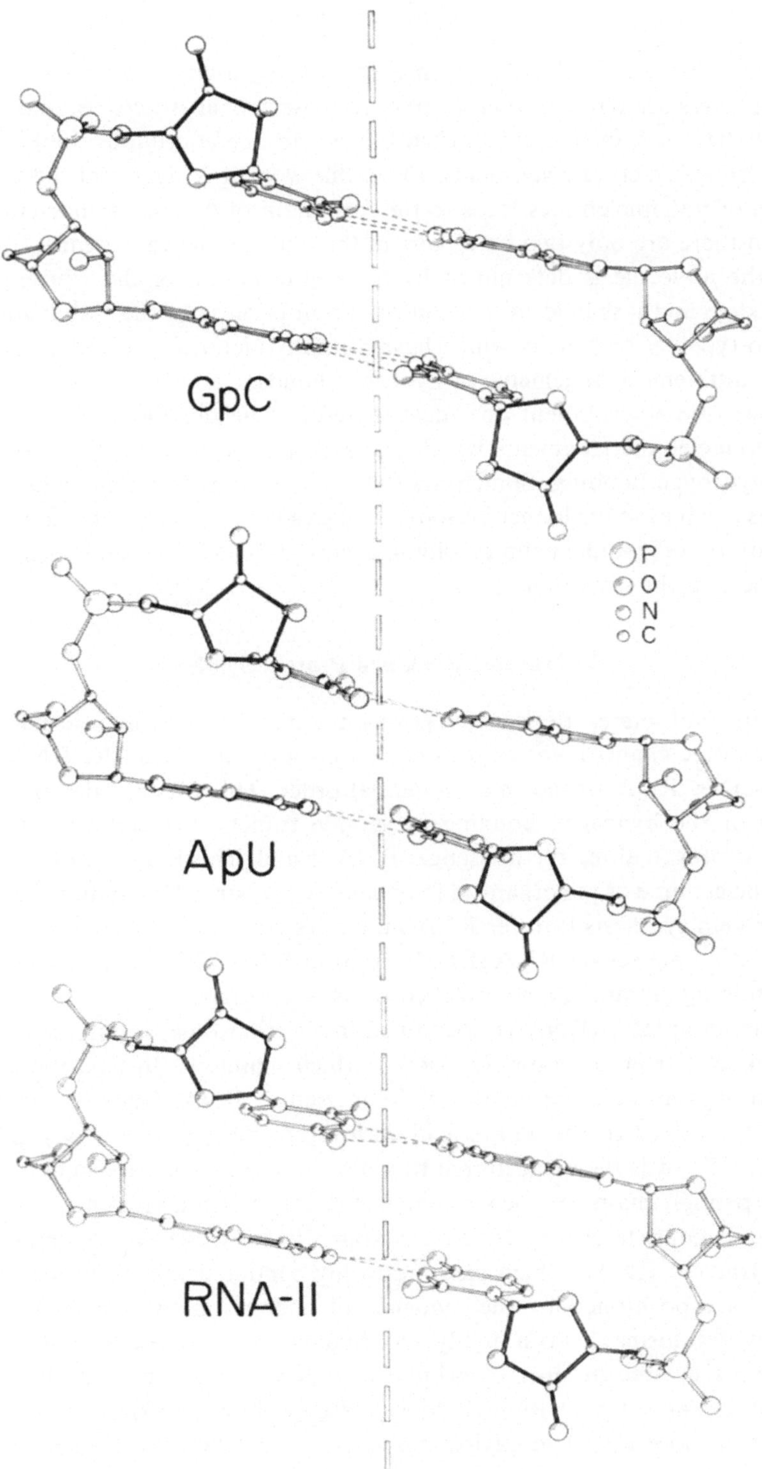

Fig. 3. View of ApU, GpC, and RNA-11 approximately normal to the vertical helix axis. The similarities between the structures are quite visible, as is the overall right-handed helical relationship between the successive residues on each of the chains.

groups are substantially shielded from each other by a large number of water molecules. The bases are parallel to each other in these crystal structures, and in addition the double helical framents are stacked but not in one contunous double helix as is found in the polymeric nucleic acids. From this work it is clear that the natural conformation of polynucleotides leads to the formation of double stranded nucleic acids even when there are only two base pairs in the minimal helical fragment. The helical form of the molecule is determined by the conformation of the ribose-phosphate-ribose linkage and it is held in this conformation because of the stabilizing influence of the two types of base pairs which have virtually identical geometries even though they have different arrangements of hydrogen bonds.

In a prebiotic environment one must therefore consider the possibility of having short oligonucleotide fragments which can associate with each other using complementary hydrogen bonding. Sometimes this association is transient as in solution; in other cases it will exist for longer periods of time espeically where there are appreciable concentrations of complementary oligonucleotides which dry and form complexes such as these in the crystalline state.

3. Transfer RNA and Protein Synthesis

The fundamental intereactions which govern the translation of nucleic acid sequences into polypeptide chains involves the detailed positioning of transfer RNA molecules on a messenger RNA strand in a sequential order. This order is determined by the specificity of the hydrogen bonding nucleotide triplets, the codons, which are sequentially arranged along the messenger RNA strand. This is a copy of the sequences found in the genome of the organism in one of the two strands of double helical DNA. During protein synthesis transfer RNA molecules enter the ribosome where they are positioned on a messenger RNA strand. Transfer RNA (tRNA) comes into the ribosome containing an amino acid attached to its 3′ terminus.

Once the aminoacyl tRNA is positioned in the ribosome, a polypeptide chain is transferred to it from an adjoining tRNA which maintains this position due to its interaction with an adjacent codon of the messenger RNA. Once the peptide chain has been transferred to the amino acid with the formation of a new peptide bond, the peptidyl tRNA is then transferred to a new position in the ribosome where it can donate its peptide chain to a new aminoacyl tRNA which arrives next to it. Protein synthesis thus consists of a cycle of operations which invove the cooperation of two adjoining transfer RNA's which pass a growing peptide chain from one to the other and which are positioned by the sequence of codons on the messenger RNA. In present day organisms this is a highly coordinated and complexely controlled operation involving a large number of protein factors which ensure that the translation process will be carried out with high fidelity. We would like to know how this process occurred in a more primitive environment. Thus, what are the physical constraints which are involved in the interaction of messenger RNA and transfer RNA? And what are the physical constraints which may operate when two transfer RNA's are situated next to each other on adjacent codons of a messenger RNA strand?

In order to understand the physical environment associated with translation, we have to know something about the components, including the ribosome and the transfer RNA. The ribosome is a complicated organelle which has a molecular weight of a few million and contains over 50 proteins as well as macromolecular RNA. Very little is known of its internal organization. The transfer RNA is a molecule containing 75–85 nucleotides and a great deal is known about it. Over 50 transfer RNA's have been sequenced and their sequences can all be written in a typical cloverleaf fashion, such as is shown for yeast phenylalanine tRNA in Figure 4 (Raj Bhandary and Chang, 1968).

In the cloverleaf formulation there are four 'stems' in the molecule in which complementary sequences of bases are found. Three of the stem regions terminate in loops, the dihydrouridine loop (DHU), the TΨC loop and the anticodon loop. The three nucleotides at the bottom of the anticodon loop are the anticondon nucleotides which have been shown to interact with the messenger RNA during translation. Even though a large number of sequences have been found to form a cloverleaf array, the physical basis of this has been revealed only recently as a result of X-ray diffraction studies.

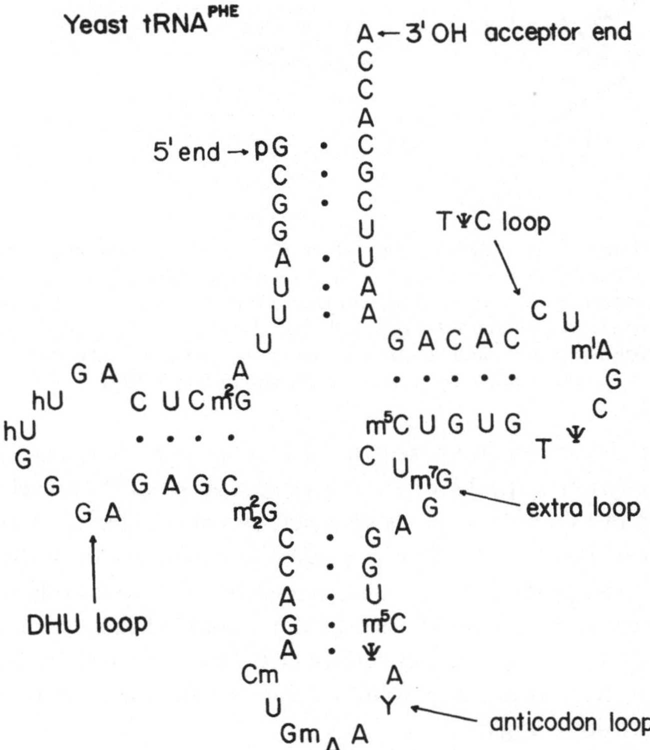

Fig. 4. The sequence of nucleotides in yeast phenylalanine tRNA shown in the conventional cloverleaf diagram.

Yeast phenylalanine tRNA crystallizes in an orthorhombic unit cell with the dimensions $a = 33$ Å, $b = 56$ Å and $c = 161$ Å. Recently the X-ray diffraction pattern has been interpreted to a resolution of 4.0 Å (Kim *et al.*, 1973). This was accomplished through the use of isomorphous heavy atom derivatives of the crystals. At 4 Å resolution it is not possible to differentiate individual bases or sugars in the polynucleotide chain. However, the 48 electrons clustered near each other in the phosphate group comprise an element which scatters X-rays heavily and in the electron density map it appears as a peak. A number of these peaks are found in the map and the three dimensional folding of the chain has been traced. A series of peaks are seen in the map in particular places in a manner shown in Figure 5. Here several sections in the electron density map are shown superimposed on each other. The plus marks indicate the position of the phosphate peaks which average 5.8 Å away from each other. Three chains are shown in Figure 5. It can be seen that two of the chains are related to each other by regions of low electron density passing from one chain to the other. From the geometry of these peaks and the distances between the chains it has been possible to interpret the two joined peaks as arising from a double helical stem in the transfer RNA molecules. The joining electron density arises from the stacked bases. Thus the geometry of these regions at low resolutions agrees with the form of double helical RNA which is shown in Figure 3.

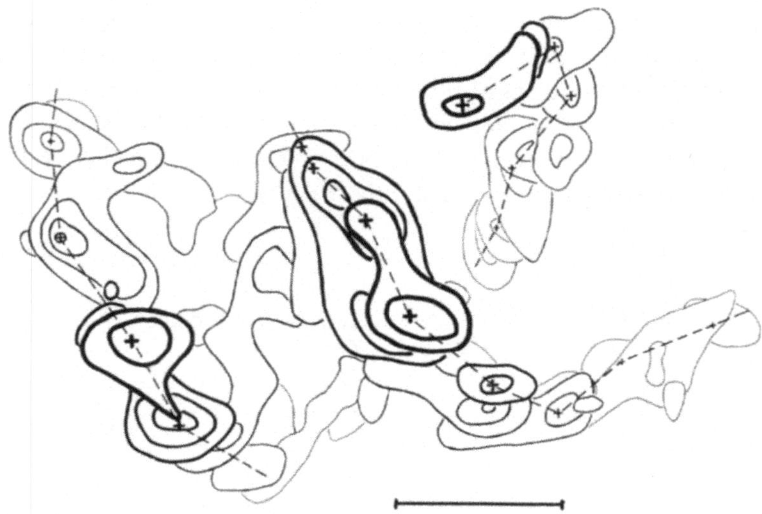

Fig. 5. Sections of the electron density map which illustrates the connections between adjacent segments of polynucleotide chain. A superposition of segments of the electron density map showing portions of three polynucleotide chains. The two chains on the left are joined by regions of lower electron density. The chain on the upper right, although passing nearby, is not joined. X marks the position of the phosphate peaks, and the dotted line shows the continuity of the ribose phosphate polynucleotide chain. The horizontal line is 10 Å.

This has made it possible to trace the entire chain and Figure 6 shows two sides of a solid model of tRNA carved out of the electron density map at 4 Å resolution. tRNA

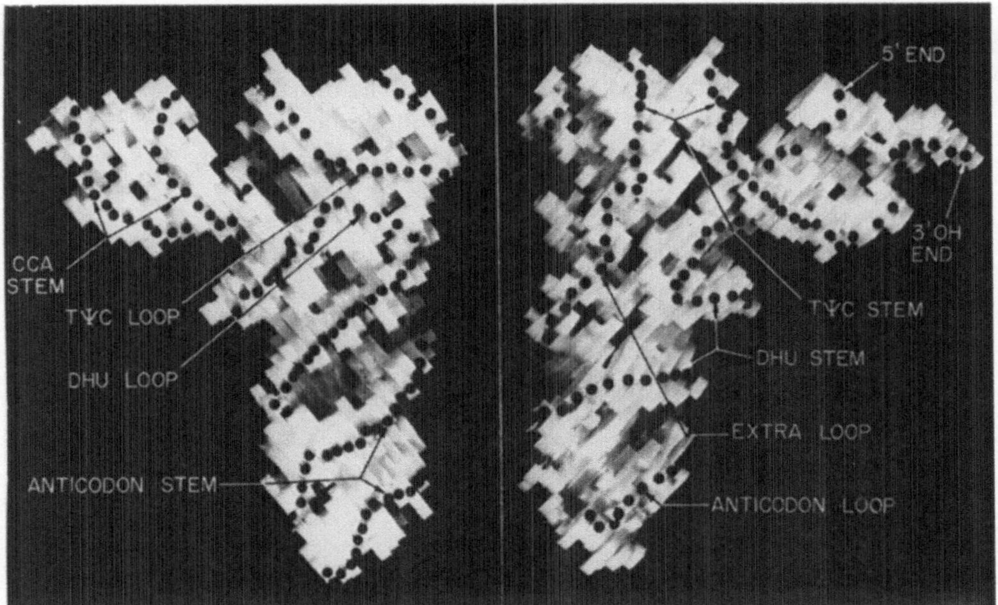

Fig. 6. Two views from opposite sides of a solid molecular model of yeast phenylalanine tRNA as seen at 4.0 Å resolution. The molecule is approximately 20 Å thick in a direction perpendicular to the page. The vertical distance in the molecule is 70 Å. In order to make the tracing of the chains more visible, a series of round headed pins have been inserted into the molecule. These do *not* represent atoms, but are designed to show the folding of the polynucleotide chain. Hydrogen-bonded base paired stem regions can be readily identified because the adjacent polynucleotide chains are connected to each other.

is seen to be a flat molecule, somewhat L-shaped in outline. The thickness of the molecule perpendicular to the page in Figure 6 averages close to 20 Å, approximately the thickness of the double helix of RNA. The chain coils about considerably as can be seen by following the detailed tracing shown in Figure 6 in which the round headed pins follow the course of the ribophosphate backbone. It can be seen that at one end of the L-shaped molecule the CCA end is found, the terminal A stands for the adenosine which is combined with an amino acid during aminoacylation. The upper horizontal part of the molecule consists of two stems organized around a colinear axis, the CCA and the TΨC stems. The vertical part of the molecule in Figure 6 contains the anticodon stem and the dihydro U stem, both of which are organized around the same vertical axis. The anticodon is found at the very bottom of the molecule 76 Å away from the CCA end. At the corner of the L there is a joining of two loops, the dihydro U and the TΨC loops. These come close together to stabilize the form of the molecule, possibly through the mediation of tightly bound cations which hold the chains together.

The horizontal and vertical double helical segments are held together by two covalent chains, one involving residues 8 and 9 in the cloverleaf configuration between the

CCA stem and the dihydro U stem (see Figure 4) while the other involving the 5 nucleotides in the extra loop connecting the TΨC stem at the top of the molecule with the CCA stem near the bottom in Figure 6. The overall dimensions of the L-shaped molecule are 70 Å from the anticodon to the corner of the molecule occupied by the TΨC loop ,and 66 Å from that loop to the CCA end. Figure 7 shows a per-

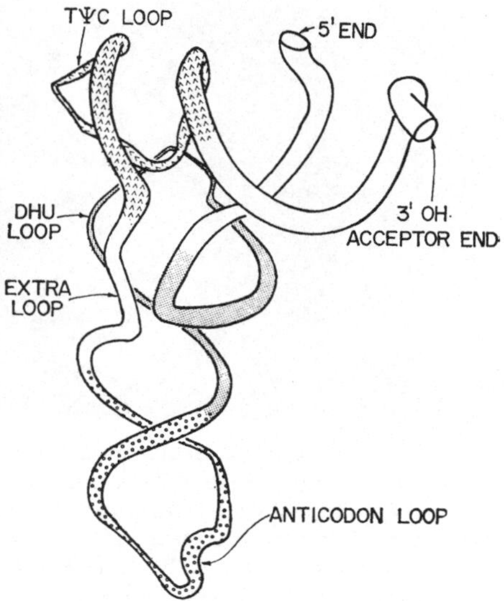

Fig. 7. A perspective diagram in which the polynucleotide chain is represented as a continuous coiled tube. The different shading represents the vaıious loops and stem regions of the tRNA molecule. It can be seen that the TΨC and DHU loops come in very close contact.

spective drawing of the polynucleotide chain traced as a continuous wire which follows the polynucleotide backbone. In this perspective drawing the CCA end is turned toward the reader and the colinearity of the two different stem regions is seen quite clearly as well as the close approximation of the two loop regions.

4. What Happens in the Ribosome?

Once we understand the three dimensional structure of tRNA, we can begin to think rationally about the biological problems concerning its function. Briefly stated, the problem is how can two transfer RNAs act coordinately in the ribosome so that they are both simultaneously coupled to the mRNA strand and at the same time have their CCA ends close enough to transfer the growing polypeptide chain from one tRNA to the amino acid on an adjacent tRNA. The solution of this problem is not at hand. One of the seeming paradoxes is the fact that the molecules themselves are 20 Å across near the anticodon end while the length of a stacked codon or triplet of bases on the mRNA strand is close to 10 Å (3 × 3.4 Å). This suggests that there must be some unusual arrangement of tRNA molecules in the ribosome during protein synthesis. Although the solution of this problem is yet to be uncovered, one can nonetheless begin to think of the tRNA functions in the following terms: a) the aminoacylation of the tRNA, b) the attachment of the anticodon of the tRNA to the message, and c) the transfer of the peptide group from one tRNA to the next.

In contemporary biochemical systems, functions b) and c) both go on in the ribosome. However, tRNA does not only carry on these functions in a concerted manner. For example, tRNA is known to act as an aminoacylating agent without the intervention of either the ribosome or mRNA. tRNA can donate its amino acids either in cell wall peptidoglycan synthesis (Petit *et al.*, 1968) or to cell wall phospholopid synthesis (Nesbitt and Lenarz, 1968). It can also donate its amino acids to the free N-terminus of a completed protein (Leibowitz and Soffer, 1971). All of these functions involve the transfer of an acylating agent to a receptor. However, the receptor is not necessarily a protein in the process of being assembled during ribosomal biosynthesis.

5. tRNA in Primitive Biosynthesis

It is quite likely that tRNA could exist in a primitive form in which it is capable of becoming aminoacylated at its free 3′hydroxy end. One would like to know whether or not it would be possible for tRNAs to associate with each other in such a way that the aminoacyl group could be transferred from one to the other. Thus, a primitive type of synthesis might be carried out if tRNA molecules were able to bind together in pairs, transfer their aminoacyl groups or polypeptide chains to an adjoining tRNA and then dissociate. Continuous repetition of this process could lead to the development of polypeptide chains in the absence of a ribosome. This could occur if there were an inherent self-complementary feature in tRNA molecules which would allow them to associate in a pair-wise fashion. This could be subjected to experimental test to see whether or not it is possible that tRNA molecules have within themsleves a primitive ability to associate and transfer aminoacyl groups or polypeptide chains independent of the large assembly organelle, the ribosome, which is used in contemporary biochemical systems.

This question has not been answered at the present time. Basically we are asking whether or not tRNA by itself is able to carry out a primitive type of aminoacylation and transfer function independent of its ability to associate with an mRNA strand and independent of the existence of a ribosome. One might wonder whether at a later stage in the development of polypeptide synthesis, ribosomes were formed and their interaction with mRNA and tRNA developed.

One way to carry out an investigation along these lines is to look for examples of complementary features on the surface of tRNA. There is an interesting possibility in this regard which should be explored more fully. This has to do with the fact that there is an indentation or depression on the surface of the tRNA molecule near the base of the dihydro U stem where the polynucleotide chain dips toward the center of the molecule where it connects with the CCA stem going toward the 5′ end of the polynucleotide chain. This depression on one side of the molecule is matched by a protuberance on the other side associated with a bending out of the dihydro U loop It is possible to assemble two molecules of tRNA side by side so that this protuberance is complementary to the slight depression of the molecule. In that case one would have a close interaction between that portion of the dihydro U loop of one molecule with the beginning of the extra loop as well as the nucleotide in position number 8 of the other molecule. Once this protuberance and depression are joined together to form a compact tRNA dimer, the CCA ends of the molecule are close enough to pass a polypeptide chain from one tRNA to the other. It would be reasonable to try to carry out such transacylation experiments with tRNA under conditions which might serve to stabilize interactions of these types.

We know very little about the evolutionary history of tRNA. One of the remarkable constants found in molecular biology is the fact that all of the more than 50 tRNAs which have been sequenced can fit into a cloverleaf system similar to that shown in Figure 4. The only differences in these consist of changes in the length of the extra loop and of the dihydro U loop. This invariance in tRNA size is found in all phyla, eukaryotic or prokaryotic. An interesting exception to this may have been reported recently in that size measurements were made on tRNA found in mitochondria. Mitochondria are inclusions within eukaryotic cells which are thought to have an independent system of protein synthesis. According to these recent measurements (Dubin and Friend, 1972), they may have a molecular weight of approximately 19600 as compared to the average of 25000 found for other tRNA molecules. More work has to be done before the significance of these measurements can be assayed. If, however, subsequent experiments bear out these measurements, then we would have to ask what form would the tRNA molecule take if it were deprived of over 20% of its mass. An interesting possibility in this regard is that the molecule may have a form very similar to that seen in Figures 6 and 7 except that the CCA stem containing 14 nucleotides might be missing. If that were so, the molecule would have the same elongated vertical axis going from the anticodon to the TΨC stem and loop, but instead of having an extended CCA stem, the CCA end might then be attached directly to the TΨC stem. A tRNA of this type might possibly function in a ribosomal system in which the peptidyl transferase were moved about 20 Å closer to the tRNA binding site which may be found on the TΨC loop. It is clear that futher work will have to be carried out before one can evaluate a suggestion of this type.

In summary, an understanding of the three dimensional molecular structure of tRNA is extremely valuable in allowing us to assemble a physical picture of this molecule and to describe in concrete terms the interactions which take place between this molecule and either mRNA, the ribosome or its growing polypeptide chain. One of the central unanswered questions of the present time is the origin and development of the translation system. It is not unreasonable to believe that the complete knowledge of the three dimensional form of the tRNA molecule may lead to significant insight into the genesis of this crucial step in the evolution of organic life on this planet.

References

Day, R. O., Seeman, N. C., Rosenberg, J. M., and Rich, A.: 1973, *Proc. Nat. Acad. Sci. (U.S.)* **70**, 849.

Dubin, D. T. and Friend, D. A.: 1972, *J. Mol. Biol.* **71**, 163.

Kim, S. H., Quigley, G. J., Suddath, F. L., McPherson, A., Sneder, D., Kim, J. J., Weinzierl, J., and Rich, A.: 1973, *Science* **179**, 285.

Leibowitz M. J. and Soffer R. L.: 1971, *Proc. Nat. Acad. Sci. (U.S.)* **68**, 1866.

Nesbitt, J. A. and Lennarz, W. J.: 1968, *J. Biol. Chem.* **243**, 3088.

Petit, J. F., Strominger, J. L., and Soll, D.: 1968, *J. Biol. Chem.* **243**, 757.

Raj Bhandary, U. L. and Chang, S. H.: 1968, *J. Biol. Chem.* **243**, 598.

Rosenberg, J. M., Seeman, N. C., Kim, J. J., Suddath, F. L., Nicholas, H. B., and Rich, A.: 1973, *Nature* **243**, 150.

Voet, D. and Rich, A.: 1970, *Prog. Nuc. Acid Res. Mol. Biol.* **10**, 183.

For a general review, see Orgel, L. E.: 1973, *The Origins of Life*, Wiley and Sons, N.Y.

A HYPOTHETIC SCHEME FOR EVOLUTION OF PROBIONTS

A. I. OPARIN

A. N. Bach Institute of Biochemistry, Academy of Sciences of U.S.S.R., Moscow, U.S.S.R.

Abstract. The origin of life cannot be assumed as a single, indissoluble chain of events. The objects we observe on this developmental pathway appeared many times in different biogenetic provinces of the Earth's surface. However, these systems might undergo decomposition and later appeared again. Such process especially might occur with the multimolecular, phase separated systems (probionts), interacting with the surrounding solution in the manner of the open systems. The modelling experiments demonstrate, that basing upon such interaction the systems could well maintain growth and passed through natural selection in the original Darwinian interpretation of this term, i.e. survival of the forms better adapted to the exterior environment. On this basis there proceeded the development of probionts organization, in particular, there arose the 'fitness' of molecular structure of proteins and other polymers. However, the maintenance of steady organization and selfreproduction of probionts were still far from perfection. Their evolution was still connected with the regular and strong decomposition processes. The decomposition products, which had already acquired somewhat better fittness than the original polymers, might either be assimilated by more developed probionts or enter a selfassembly process to form the new systems, like that had occured at the early stage of the development. The appearance of nucleic acids and their coding relations with proteins have provided significantly more high stability of developing probionts and primitive organisms. But on this stage, too, the decomposition and *de novo* formation of the developing systems had to occur. This fact makes us to approach in a new angle the problems of the origin of viruses, symbiosys and parasitism.

I am not going to present here any complete scheme of the evoltuion of organic substances which led ot the origin of life. Rather I hope to avoid the mistakes usually made by authors of such schemes.

The process of biopoesis is often expressed as one indissoluble chain of events where the appearance of any one link obligates the appearance of the following one. Such a concept does not allow for the possibility of a break in the chain or a reverse in the sequence of events; and a concept of this kind is very improbable. It requires either an exceptionally lucky combination of the rarest probabilities, or the existence of some sort of a plan which predetermines the course of events. Such a plan cannot be derived purely from the thermodynamic potentials of the developing systems because these potentials are very large, even in the case of an organic molecule.

Therefore, the probability that system *A* will necessarily change to produce system *B* is very low. But since the probabilities are to be multiplied rather than summed, any multistage process (even the appearance of complex organic substrates) seems to be extremely improbable. So it is usually held that the origin of life was a very uncommon event, which occurred once for the whole period of the existence of our planet.

At the same time, any transition from one stage of biopoesis to the next one is usually connected with the increase of complexity and organization level of the system, and therefore according to the second law of thermodynamics, the synthesis seems to be significantly less probable than the reverse, the decomposition process.

The action of short wave UV light or some other energy source upon the dissolved organic substances in the primitive ocean created a thermodynamic equilibrium. Numerous calculations have shown that under these conditions nothing like the concentrated 'primordial soup' could appear, since under the action of UV the decompositon processes had to proceed much faster than the synthesis. On the Earth's surface such a thermodynamic equilibrium was never and could never be achieved.

Based on the vast geological data in his book, *The Origin of Life by Natural Causes*, the late Prof. M. G. Rutten described the picture of the Earth's surface in the early geosynclinal period of the orogenetic cycle of the planet. In that epoch the balance between water and dry land was changing not only in the littoral zone, but everywhere. Water which impregnated the soil was coming in and out, continually transferring the dissolved substances from the site where they had been formed to the sites where these compounds were concentrated and accumulated. Therefore, although such compounds as amino acids might be formed under very different conditions and energy sources, their quantitative ratios, with the other organic substances and their concentration had to be sharply different in various biogenetic territories of the Earth's surface. In some places the previously formed organic substances, in particular the amino acids and their polymers, had to pass through significant decompositon whereas in the others they could participate in polymerization and complification processes. But even in those sites where for a certain period the decomposition processes took place, the shift in conditions could lead to the synthesis of more complex molecules from the decomposition products to the further complification of organic molecules and multimolecular systems. Thus, at the same time the process of biopoesis had to be at the different developmental stages in different 'subvital areas' of the Earth's surface.

So M. G. Rutten was right when he stated that the origin of life on the Earth could not have occurred only once. Many times the forerunners of living systems – probionts and the primitive organisms which have developed from them – appeared, underwent decomposition, and then appeared again in different places and in different times. Therefore, according to M. G. Rutten's view, the already developed primitive organism might co-exist for a long time (perhaps for many millions of years) with the systems representing more primitive states of biopoesis, which had appeared in the other 'subvital areas'. Certainly such a concept of the multiple formation of living systems completely rules out the hypothesis of the purely accidental character of the origin of life.

The concept of the multiple character of the origin of life is especially important for understanding of the transition from the chemical evolution to the biological one.

The appearance of a multiplicity of phases and the separating out from homogeneous (in particular, water) solution of individual multimolecular systems that are isolated from the surrounding medium by a definite surface of separation is extremely widespread in nature. It can be observed constantly in the laboratory experiments, especially in the work with high molecular organic substances.

The modelling experiments with the phase separated systems, in particular coacervate drops, which are isolated from the water solution of various organic polymers have demonstrated that these systems are capable of the uptake of various energy rich compounds from the surrounding solution, and by this to grow, increasing in volume and mass. The growth rate is determined by chemical and spatial structure of each individual drop; therefore, two different kinds of drops when placed into the same solution behave in a different way. Some of them grow rapidly; the growth of others is retarded, some even to the point of decomposition. Thus, the modelling experiments show the possibility of the primitive natural selection of the drops based on the interaction of this system with the environment. With no doubt such isolation of the phase separated systems from the water solution of abiogenetically formed polymers proceeded on a grandiose scale and have haven place many times on the surface of the lifeless Earth.

The open multimolecular systems formed this way had to vary greatly in their chemical composition as well as in the principles of spatial organization. Many of them were decomposed never to return, but the other ones whose structure could provide their dynamic stability could not just remain under the given environment, but also expanded, growing in their volume and mass. Then under the action of outer mechanical forces they could even divide to produce the daughter systems.

Here at the transition to biological evolution, primitive natural selection of the forerunners of the living organisms, probionts, could appear (in its initial Darwinian interpretation, as 'Survival of the Fittest').

The maintainance of the constant composition and structure of each individual probiont (or daughter system that appeared after division) could be based at that time only upon the dynamic principles. It depended only on a permanent flow of building material and catalysts from the surrounding medium, as well as on the invariable character of chemical reactions occurring in the system and on the combinations of such reactions.

Any change of dynamic or spatial organization of probionts was under permanent environmental control. Such control played a decisive role at that stage of development. It led to the decomposition of the countless number of systems if their structure could not ensure their preservation and growth under given or changing environmental conditions.

Certainly such a pathway of evolutionary development, the pathway of tests and mistakes, was very imperfect. The course of progress based upon it was very slow and was connected with many retreats, failures, and returns to the starting forms of organization. Nevertheless, this process might already serve as a basis for the origin of the characteristic property of living beings, the 'purposeful' organization. By purposeful organization I mean the perfect suitability of the living system as a whole for existence under the given conditions of environment and the adaptability of its individual parts (polymer molecules and molecular complexes) for the fulfillment of the functions they bear in the system.

The opinion is often expressed in the scientific literature that the primitive living systems could arise only after internally organized and purposeful proteins and nucleic acids were formed (still on the molecular level) in the Earth's hydrosphere. It is alleged that the selfassembly of those molecules might lead to formation of the primitive organisms like a machine is assembled from details prepared beforehand. But in the latter example, the details have to be manufactured after a plan or blueprint worked out beforehand. In respect to the origin of life, we cannot speak of such a plan.

Then it becomes absolutely unclear how the primitive nucleic acid and protein molecules, whose molecular structure was not only highly definite and perfect but was well-fitted to the functions these molecules will perform in the integral living systems which could be formed out of them, could arise in the 'primordial soup'.

Today we know that even under primitive abiogenic conditions there could be polypeptides formed with a certain intramolecular structure, i.e. the sequence of amino acid residues. It might cause the enzyme activity of such compounds in some cases. But this activity by itself does not bear any purposefulness because it had no significance for the protein-like molecule. The molecule may acquire its purposefulness only when acting as part of the integral system of the probiont, and only when the reaction catalyzed by this protein plays an important role for the probiont's metabolism which causes dynamic stability and a rapid growth of the system as a whole.

From the evolutionary viewpoint those probionts received exceptionally important privileges. In addition to the 'purposefully' constructed protein-like polymers, the polynucleotides are capable of molecular replication which ensures perfect transfer of properties to the next generations. The perfect fitness of the intramolecular structure of the proteins and nucleic acids to their functions was increasing in the course of evolution on the basis of natural selection. However, there were neither the polynucleotides capable of replication, nor the proteins formed under their influence that were passed through the natural selection, but already rather perfect metabolism which did or did not correspond to the environment.

The appearance of nucleic acids and the development of their coding relations with proteins led to a significant increase of the stability of developing probionts and primitive organisms. However, their decomposition could not be excluded even at this stage. What might happen with the products of such decomposition at various stages of biological evolution? They might be again evolved into the selfassembly process and then give some new systems like that which took place at the beginning of evolution .They might be either spontaneously decomposed under the influence of inorganic factors to form the simplest compounds or be assimilated by more perfect probionts and organisms. But now such self-assembly would proceed on a higher level, since during the evolution of the parent systems the participating components had already acquired some 'purposeful' structure. This circumstance makes us approach in a new way the problems of the origin, in the past, of viruses and microplasma, of parasitism and symbiosis. Moreover, there have appeared some indications on the possibility of similar processes even at the present.

FROM PROTEINOID MICROSPHERE TO CONTEMPORARY CELL: FORMATION OF INTERNUCLEOTIDE AND PEPTIDE BONDS BY PROTEINOID PARTICLES

SIDNEY W. FOX, JOHN R. JUNGCK, and TADAYOSHI NAKASHIMA

Institute for Molecular and Cellular Evolution, University of Miami, Coral Gables, Fla. 33134, U.S.A.

Abstract. Proteinoid microspheres of appropriate sorts promote the conversion of ATP to adenine dinucleotide and adenine trinucleotide. Other microparticles composed of basic proteinoid and enzymically synthesized poly A cause the conversion of ATP and phenylalanine to various peptides of phenylalanine. When viewed in a context with the origin and properties of proteinoid microspheres, these results model the origin from a protocell of a more contemporary type of cell able to synthesize its own polyamino acids and polynucleotides. Related earlier experiments explain in part the origin of the genetic code and mechanism.

Few societies are organized for the study of a single problem, such as that of the origin of life. In the same vein of thought, perhaps no problem has been so thoroughly explored by hypotheses prior to experimentation as has this one. Accordingly, no experimental model can yield results consistent with more than a fraction of the hypotheses that have been so abundant, especially for the emergence of macromolecules and cells. The proteinoid microsphere, as a laboratory model for the replicating protocell (Fox and Dose, 1972; Fox, 1973a) reflects concepts of a primordial sequence (Fox, 1971) as those are implicit in the expressed thought of a number of authors. Among the contributors whose ideas can be interpreted as consonant with that of a cell initiating protein and nucleic acid synthesis and Darwinian selection have been Wald (1954), Van Niel (1956), Oparin (1957), Lederberg (1959), Ehrensvard (1960–1962), and Prosser (1970). The integrated view from their hypotheses and from our experimental results is expressed concisely by the sequence: *protoprotein* → heterotrophic *protocell* → macromolecule-synthesizing *cell*.

The contributions of the above-named authors (and others) have been almost entirely theoretical. An integrated conceptualization benefitted from support by relevant experiments and it was extended by additional concepts that emerged almost solely from experiments. Two of the salient inferences derived from the experiments were (a) the 'self'-ordering properties of reacting amino acids (Dose and Rauchfuss, 1972) and (b) the onset of reproduction as one function associated with others in the proteinoid microsystems assembled from the polymers resulting from the 'self'-ordering reactions (Fox and Dose, 1972). In fact, all that has been learned of this system, including (b), stems from the first experiment based on concept (a) (Fox, 1973a). The experiments of the Oparin school with coacervate droplets have explained the need for an early appearance of boundaried microparticles (Oparin, 1966). The coacervate droplets of Oparin have demonstrated utility in the protection of cellular contents, in the promotion of intracellular reaction sequences, and in favoring of a concentration of growing polymers. Van Niel (1956) has exemplified the combined viewpoints of those who favor the concept of an early minimal cell by stating, "Acceptance of the postulate that chemical evolution preceded biopoesis further suggests that the organized structures representing primitive life were capable of self-reproduction before they acquired mechanisms by means of which they could chemically transform the components of their environment." In 1967, we reported the evidence that proteinoid microspheres are organized structures capable of 'self'-reproduction (Fox *et al.*, 1967; Lehninger, 1970; Szent Gyoergyi, 1972).

In this paper, we report mechanisms by which two kinds of microparticle from proteinoid use *one* component of their environment, ATP, to produce *two* kinds of polymer molecule, polyamino acids and polynucleotides. We have been able to find polymers of amino acids formed by, and oligomers of adenylic acid formed from, the same reactant, ATP. The fact that aminoacyl adenylates (products of amino acids and ATP) contain both amino acid and adenylic acid monomers has provided a basis for understanding how both *cellular* proteins and nucleic acids would have arisen in an almost simultaneous relationship (Nakashima *et al.*, 1970). Ehrensvärd predicted with some accuracy the primacy of enzymically active protein and the emergence of nucleic acids, as modelled (Fox and Dose, 1972) when he stated, "The stabilizers in cells today in the form of genes in chromosome bodies should have originated as *by-products* of early chemical activity in loosely organized units of catalysts" (Ehrensvärd, 1962).

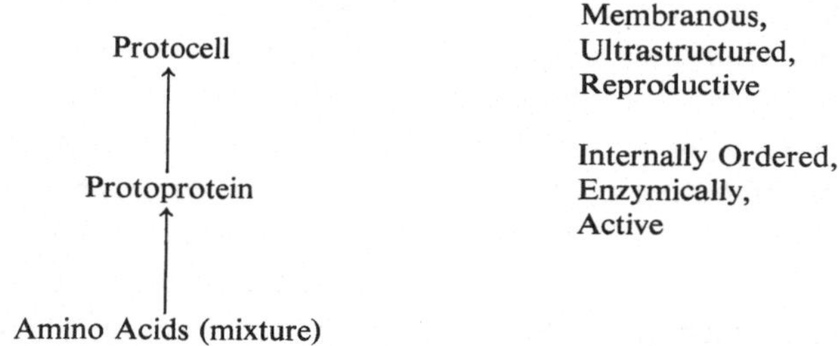

Fig. 1. Properties of protoprotein and protocell models derived from mixed amino acids (Fox and Dose, 1972).

Figure 1 presents some salient aspects of the origin of proteinoid microspheres. As explained elsewhere, evidence has been presented that these microunits are reproductive, evolvable (Fox and Dose, 1972), and heritable (Hsu *et al.*, 1971). All of these properties are derivative of the basic ability to replicate at the systems level, first shown in one way (*cf.* Fox, 1973b) in 1967 (Fox *et al.*, 1967). The systems level, however, is the level at which Darwinian selection operates (Prosser, 1970).

The emergence of the contemporary cell demands two large consecutive steps: (a) the origin of the protocell and (b) an evolution to the contemporary cell. Our understanding of the properties of the protocell permits us to subtract those functions from those of the contemporary cell. When we do, we find that what was needed in step r syntheses of both protein and nucleic acids in a coded mechanism for the energizing of synthesis of anhydro- aper is to describe recent experimental progress toward

the addition of ATP to appropriate basic proteinoid :ide yields adenine dinucleotide. This result is a sign- f proto RNA polymerase. It adds to the now long list teinoids (Dose, 1971) the catalysis of a synthetic reac-

t when ATP is added to microspheres composed of

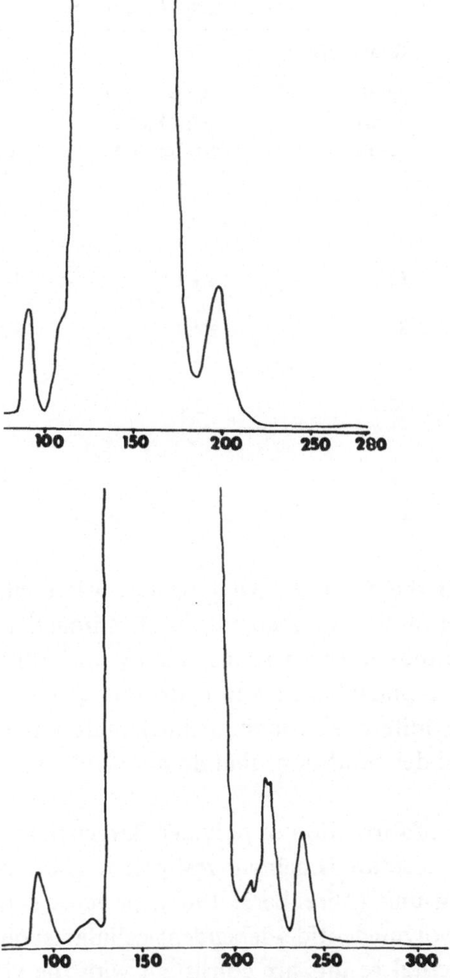

ıine dinucleotide in 20 mM $MgCl_2$ solution. Rightmost peak is ın of adenine dinucleotide and adenine trinucleotide by micro- oids suspended in 20 mM $MgCl_2$ solution. Rightmost peaks are l adenine trinucleotide, from left to right respectively.

acidic and basic proteinoids. In this case a significant fraction of the product is adenine trinucleotide.

The above products were obtained by mere incubation of U-^{14}C-ATP for 24 hr at 37°. These products were fractionated on DEAE-cellulose, and monitored by radioactivity and by UV absorption.

Table I illustrates the results from comparing the reactions of ATP in aqueous solution alone, in the presence of basic proteinoid, in the presence of microspheres of acidic proteinoid, and in the presence of microspheres composed of acidic and basic proteinoids.

TABLE I

Yields of adenine dinucleotide and trinucleotide in various systems with ATP (at 37°, 2 days)

Products	Reactants			
	ATP in aqueous solution	ATP with basic proteinoid	ATP with acidic ptd microspheres	ATP with acidic-basic ptd microspheres
Adenine, adenosine, cAMP, AMP, ADP	12.6%	14.7%	14.8%	13.0%
ATP recovered	86.8	82.9	82.9	84.7
Oligo A eluting beyond ATP	0.7	2.2	2.2	2.3
Trinucleotide / Dinucleotide	0.0	0.0	0.2	0.5

The identity of the di- and tri-nucleotides was confirmed by chromatography on DEAE-cellulose followed by recognition of radioactive product and by UV, by thin-layer chromatography in two systems, and by nucleotide/nucleoside ratios after treatment with alkaline phosphatase and hydrolysis (Junck and Fox, 1973). It is however notable that a definite peak for the trinucleotide has been found only in the microsphere systems, under conditions that do not yield adenine trinucleotide in the absence such particles.

The possibility of formation of polymers larger than trinucleotides both in solution and in a cell-like structure is being investigated. The trinucleotide is already, however, a minimal coding unit (Nirenberg, 1967), perhaps sufficient in size and stability to have initiated a polynucleotide-dependent cellular evolution.

These experimental results are consistent with the view that nucleic acids first appeared in evolution subsequent to cells and their precursor protoprotein, which had also some catalytic activities (Fox and Dose, 1972). This is the sequence of events expressed by Van Niel and by Ehrensvärd, as stated earlier.

The demonstration that adenylic acid is most effectively polymerized in the locale of a microsystem is consistent with thermodynamic reasoning, which indicates that the formation of anhydropolymers is favored under either geological hypohydrous conditions or in hypohydrous zones in cells (Fox and Dose, 1972). The origin of ATP as an energy-rich reactant has been largely modelled (Ryan and Fox, 1973).

Experiments suggesting the origins of cellular protein have employed another kind of microparticle. These particles are models for protoribosomes; they are produced from polynucleotides and basic proteinoids (Waehneldt and Fox, 1968).

When enzymically synthesized poly A is allowed to form particles with lysine-rich proteinoid (37% lysine), microparticles result. When ATP, metal cations, and U-^{14}C-phenylalanine are added to these particles, the phenylalanine is polymerized.

Numerous experiments with cations Na^+, K^+, Ca^{++}, and Mg^{++} and combinations of those cations showed that ATP functions best in combination with K^+ and Mg^{++}, which are significant in contemporary protein biosynthesis. When the microparticles are also added, significant yields of di-phe (III), triphe (IV), and a higher polymer of

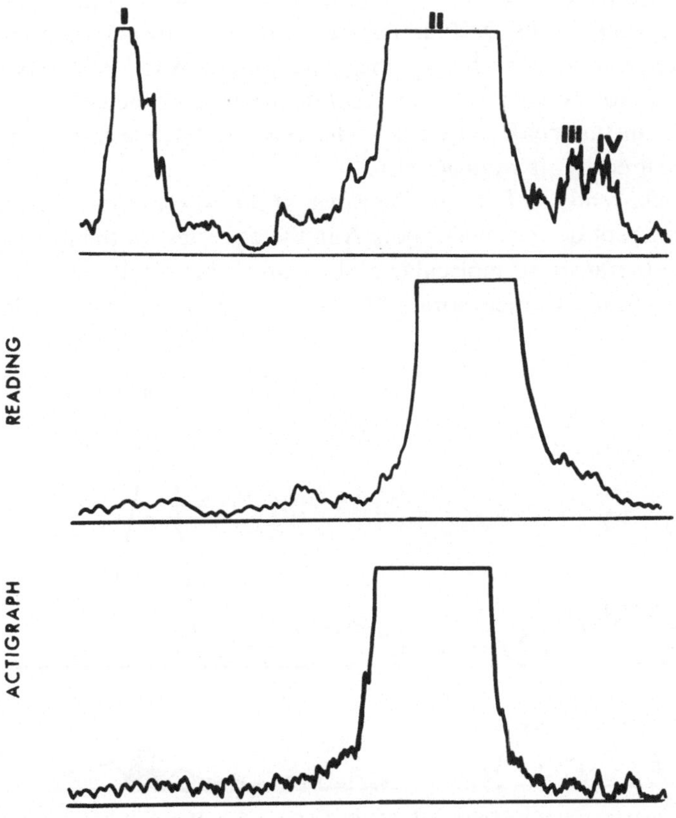

Fig. 3a–b. (a) Production of polyphenylalanines by particles composed of lysinerich proteinoid and poly A. I has *phe/N-*phe = 11 but is associated with nonphe material, II is *phe, III is *phe-*phe, IV is *phe-*phe-*phe. (b) Same as (a), without ATP. (c) Same as (a), without particles.

phe (I) are obtained (Figure 3). Figures 3a and 3b reveal that ATP is necessary to energize the synthesis, since 3b is the same as 3a without ATP. Without the microparticles, but with ATP and K^+ and Mg^{++}, no peptides result from phe, as Figure 3c indicates. Lysine-rich proteinoid alone, instead of microparticles, has almost no activity. Activation by ATP, and hypohydrous or surface conditions provided by an extrasolution particle are thus both necessary to the peptides synthesized.

Analysis of the products of Figure 3a provides information on their identities. Peak III corresponds by R_f to phe-phe-phe. Its identity was confirmed by conversion to its DNP derivative and comparison chromatography with authentic DNP-phe-phe-phe in a solvent of toluene: pyridine: 2-chloroethanol: H_2O : NH_4OH.

Fraction I in Figure 3a proved to have specifications of a higher peptide or peptide mixture. It has been dinitrophenylated, and the DNP derivatives chromatographed on paper with the toluene solvent. When hydrolyzed, the products yield Figure 4. The DNPphe is obtained by extracting the hydrolyzate with diethyl ether and chromatographing the solid obtained on evaporation; the phe in Figure 4 is obtained by chromatography also. When correction is made for the partial hydrolysis of DNPphe during its release from DNP-I (70% recovery), the ratio of phe/N-phe is calculated to be 11/1. Fraction I or its DNP derivative are revealed by gel exclusion studies to be of high molecular weight; perhaps some proteinoid or AMP is closely adherent. The phe/N-phe ratio however provides an accurate measure of the size of the peptide part of the material in this peak since phe is the only radioactive moiety, and each component is measured by its radioactivity.

The total yield tends to fall into the range of 30–40 $\mu\mu$moles of phenylalanine incorporated into peptides per mg of poly A in the particles. In the first case to come to hand from the literature of molecular biology after this result was obtained, fig fruit ribosomes were found to incorporate 31.8 $\mu\mu$moles of phe per mg of RNA (Marii *et al.*, 1972).

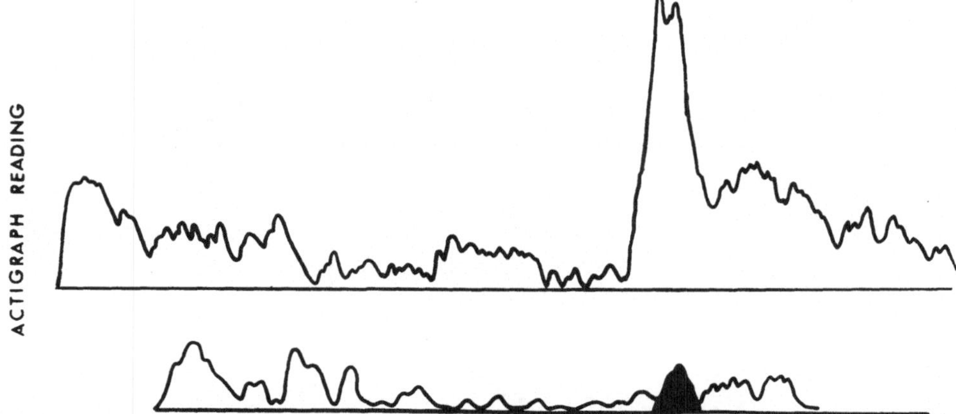

Fig. 4. Radiochromatograms of *phe and DNP*phe from Fraction I of Figure 3a; after hydrolysis. Tall peak in upper chromatogram is *phe. Other peaks may be *phe complexed to fragments of slowly hydrolyzable lysinerich proteinoid. DNP-*Phe in lower chromatogram is dark peak (yellow), after DNPylation of I and subsequent hydrolysis of DNP-I.

The models for protoribosomes, which we call *protosomes*, exhibit a lifetime of a few hours. Figure 5 demonstrates how the peptide-synthesizing activity rises rapidly to a maximum and then falls off with time of incubation at 25°. The need for an early cell, either to protect protoribosomes or to contain their continuous generation, can accordingly be visualized.

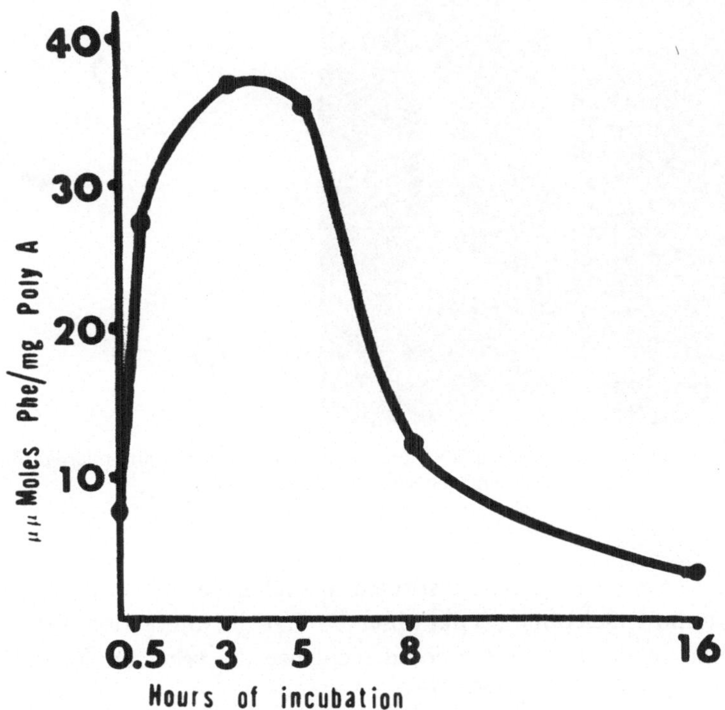

Fig. 5. Rise and decay with time of poly A-lysinerich proteinoid in polymerization of phenylalanine.

The new experiments just described begin to exaplain in part how protein biosynthesis and nucleic acid biosynthesis could have originated in replicating protocells through a contemporary energy-transfer substance, ATP. A number of related questions deserve attention in this context. These include the origin of ATP, the nature of models for protoribosomes, the possibility of selective reactions in the formation and action of such microparticles, the possibility of codonicity in simple systems, and questions of reversibility of direction of recognition from protoproteins to protonucleic acids and protonucleic acids to protoproteins. Data from earlier experimental demonstrations bear on all of these questions.

The common factor in the work on all but one of these questions is the basic lysine-rich proteinoid. Waehneldt and Fox showed in 1968 that proteinoids containing sufficient proportions of lysine would interact with RNA or DNA to yield particulate complexes (Figure 6). These nucleoproteinoid complexes exhibit salt-sensitivity and pH-sensitivity comparable to that of nucleoproteinaceous organelles.

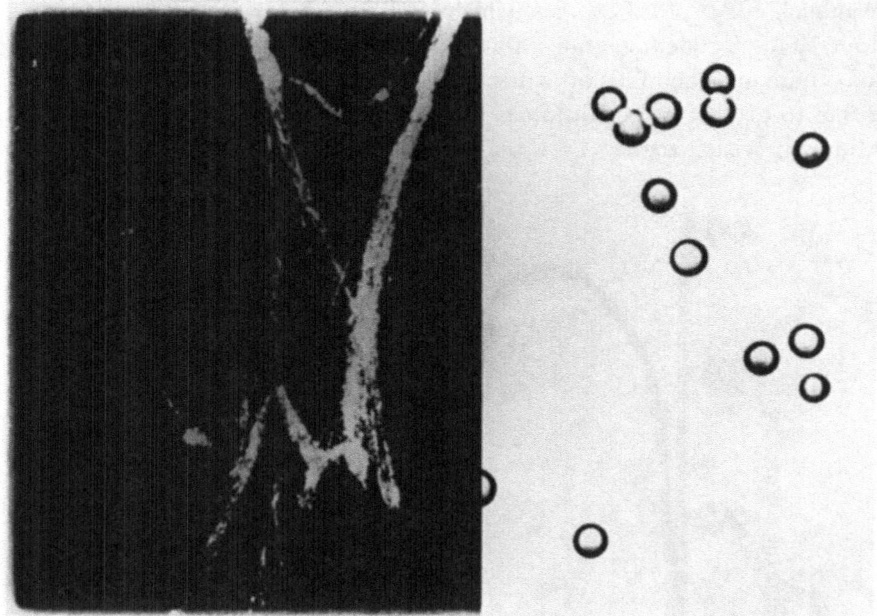

Fig. 6. Microfibers from calf thymus DNA and lysinerich proteinoid, and microglobules from yeast RNA and the same lysinerich proteinoid.

When the nucleic acids were substituted in such experiments by enzymically synthesized homopolynucleotides, Yuki observed that the interaction was selective, and that each kind of macromolecule could recognize the other (Yuki and Fox, 1969). Some data are presented in Table II.

These selective interactions to form microparticles are a kind of result that was not forecast by numerous attempts to observe interactions of polynucleotides with amino acids (Kenyon, 1973; Fox and Dose, 1972). The interaction of polynucleotides with polyamino acids is, however, demonstrable and selective, and this was first shown with

TABLE II

Selective interactions between two basic proteinoids and five homopolyribonucleotides to form microparticles (adapted from Yuki and Fox, 1969)

Polynucleotide	Lysine-rich, arginine-free proteinoid	Arginine-rich, lysine-free proteinoid
Poly C	+++	0
Poly U	+	+
Poly A	0	+
Poly G	0	+++
Poly I	0	+++

○ = no turbidity, + = some turbidity, +++ = much turbidity

synthetic macromolecules in this research. The strong interactions of polyamino acids with polynucleotides, in contrast to the noneffects with free amino acids, are explained as a manifestation of molecular cooperativity.

Polymer-polymer interactions have been demonstrated with DNA and poly-L-lysine (Leng and Felsenfeld, 1966). Also, while significant polynucleotide-amino acid interactions have not been demonstrated, polyamino acid-mononucleotide complexes have been described, both with polyarginine by Woese (1968), and polylysine by Lacey and Pruitt (1970). Indeed, the fact that the polyamino acid-mononucleotide interaction is more easily demonstrated than the polynucleotide-amino acid interaction has been cited as a reason for believing that proteins were the first macromolecules in the origin of the contemporary genetic apparatus (Lacey and Mullins, 1972).

The principal inference, however, is that recognition can occur for either class of macromolecule, polyamino acid or polynucleotide, by the other. We need not accept the hypothesis that the original translation was in the nucleic acid → protein direction; it could have been reverse translation (Fox and Dose, 1972).

An explanation for intermacromolecular recognition leaves at least one fundamental question unanswered. That is the question of how the process first operated dynamically in the cell, which we recognize as itself possessing dynamic biochemical processes (Baldwin, 1957).

This question of dynamic synthesis was first studied by Dr Nakashima in our laboratory through the agency of preformed nucleoproteinoid microparticles, which were suspended in solutions of adenylic acid anhydrides of amino acids. For the adenylates the four homocodonic amino acids: glycine (GGG), lysine (AAA), phenylalanine (UUU), and proline (CCC) were employed. The microparticles were assembled from lysinerich proteinoid of a codonic type (Fox *et al.*, 1971) and each of the enzymically synthesized homopolyribonucleotides. Under an empirically determined set of conditions, in which the ratio of polynucleotide to proteinoid appears to be significant, the polynucleotide which has been most effective in incorporating amino acids from adenylates into the microparticles is the codonically related polymer. This incorporation has been discussed (Nakashima and Fox, 1972).

Instead of continuing those studies as described, we have shifted to using ATP instead of aminoacyl adenylates, for two reasons. One is that aminoacyl adenylates are troublesome to prepare, store, and to monitor in reactions. The other is that contemporary organisms use ATP as their source of energy, and of organic phosphate. We therefore turned our attention to the particle-governed polymerizations described in the beginning of this paper.

In future experiments, we will be interested again in identifying conditions for dynamic codonic and anticodonic interactions, this time with ATP and with examination of the supernatant, or the supernatant and the particles both.

The experiments reported here have described to some degree how a replicating primordial cell would evolve toward a much more fully contemporary cell, i.e. how it would utilize the energy from ATP to polymerize amino acids and mononucleotides. Earlier experiments indicate the intrinsic basis for interactions to constitute a contemporary genetic mechanism and a genetic code. Obviously, some gaps remain in demonstrating the orchestration of processes; other evolutionary innovations, such as the trapping of solar energy, remain largely to be explained by experiments.

In outline, however, experiments demonstrate how there could have occurred an evolution of hydrophobicity from hypohydrous environment to structured cell, from geothermal energy to cellular phosphate energy, from environmental protoinformation (mixtures of amino acids) through protoprotein to RNA-controlled cellular peptide-bond synthesis, and from a largely heterotrophic, replicating protocell to nucleic acid-governed, macromolecule-synthesizing contemporary cell (Figure 7).

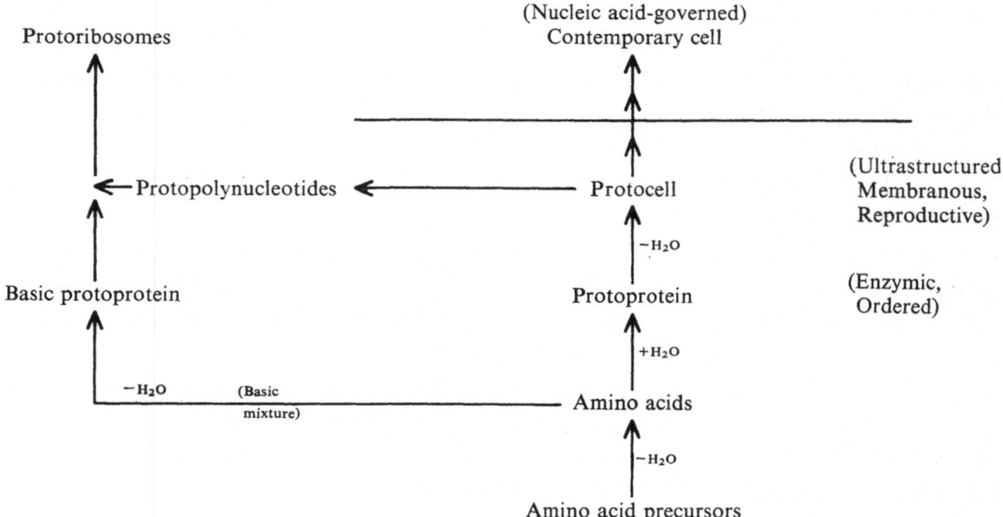

Fig. 7. Flowsheet of experimentally derived concepts, including emergence of protocell and contemporary cell.

The new experiments described are open to interpretation in two contexts. One is for the origin of cellular synthesis of biomacromolecules. What are the minimal requirements for synthesis of phosphodiester bonds or peptide bonds in a minimal cell? The results yield some definition that was previously lacking. The other context is the postulate that the proteinoid microsphere could evolve to a contemporary cell. This possibility has not been fully demonstrated, but evolvability of a model protocell has been demonstrated and the outlines of the total metamorphosis are clearer than they were. We are struck, however, as we were with the other properties of the proteinoid microsphere, by how many associated functions appear simultaneously.

Acknowledgements

The work described was made possible by Grant NGR 10-007-008 from the National Aeronautics and Space Administration. Contribution no. 253 of the Institute for

Molecular and Cellular Evolution. The studies on polymerization of adenylic acid by microspheres are from material in the Ph.D. studies of John R. Jungck; the data on protosomes are from Dr T. Nakashima. We thank Mr C. R. Windsor and Mrs Ania Mejido for technical assistance.

References

Baldwin, E.: 1957, *Dynamic Aspects of Biochemistry* (3rd ed.), Cambridge University Press, London.
Dose, K.: 1971, in A. W. Schwartz (ed.), *Theory and Experiment in Exiobiology*, Vol. 1, Wolters Noordhoff, Groningen, The Netherlands, p. 41.
Dose, K. and Rauchfuss, H.: 1972, in D. L. Rohlfing and A. I. Oparin (eds.), *Molecular Evolution: Prebiological and Biological*, Plenum Press, New York, p. 199.
Ehrensvärd, G.: 1962, *Life: Origin and Development*, University of Chicago Press.
Fox, S. W.: 1971, in R. Buvet and C. Ponnamperuma (eds.), *Chemical Evolution and the Origin of Life*, North-Holland Publishing Co., Amsterdam, p. 252.
Fox, S. W.: 1973a, *Naturwissenschaften* **60**, 359.
Fox, S. W.: 1973b, *Pure Appl. Chem.* **34**, 641.
Fox, S. W. and Dose, K.: 1972, *Molecular Evolution and the Origin of Life*, Freeman and Co., San Francisco.
Fox, S. W., McCauley, R. J., and Wood, A.: 1967, *Comp. Biochem. Physiol.* **20**, 773.
Fox, S. W., Lacey, J. C. Jr., and Nakashima, T.: 1971, in D. Ribbons and F. Woessner (eds.), *Nucleic Acid-Protein Interactions*, North-Holland Publishing Co., Amsterdam, p. 113.
Hsu, L. L., Brooke, S., and Fox, S. W.: 1971, *Currents Mod. Biol.* **4**, 12.
Jungck, J. R. and Fox, S. W.: 1973, *Naturwissenschaften* **60**, 425.
Kenyon, D. H.: 1973, *Science* **180**, 789.
Lacey, J. C., Jr. and Pruitt, K. M.: 1968, *Nature* **228**, 799.
Lederberg, J.: 1959, *Angew. Chem.* **71**, 473.
Lehninger, A. L.: 1970, *Biochemistry*, Worth and Co., New York.
Leng, M. and Felsenfeld, G.: 1966, *Proc. Nat. Acad. Sci. U.S.A.* **56**, 1325.
Marei, N., Gadallah, A. I., and Kilgore, W. W.: 1972, *Phytochemistry* **11**, 529.
Nakashima, T. and Fox, S. W.: 1972, *Proc. Nat. Acad. Sci. U.S.A.* **69**, 106.
Nakashima, T., Lacey, J. C. Jr., Jungck, J., and Fox, S. W.: 1970, *Naturwissenschaften* **57**, 67.
Nirenberg, M.: 1967, in G. C. Quarton, T. Melnechuk and F. O. Schmitt (eds.), *The Neurosciences*, Rockefeller University Press, New York, p. 143.
Oparin, A. I.: 1957, *The Origin of Life on Earth*, Academic Press, New York.
Oparin, A. I.: 1966, *The Origin and Initial Development of Life*, Meditsina Publishing House, Moscow.
Prosser, C. L.: 1970, in J. Moore (ed.), *Ideas in Evolution and Behavior*, Natural History Press, Garden City, N.Y., p. 357.
Ryan, J. and Fox, S. W.: 1973, *Biosystems* **5**, in press.
Szent Gyorgyi, A.: 1972, in D. L. Rohlfing and A. I. Oparin (eds.), *Molecular Evolution: Prebiological and Biological*, Plenum Press, New York, p. 111.
Van Niel, C. B.: 1956, in A. J. Kluyver and C. B. Van Niel (eds.), *The Microbe's Contribution to Biology*, Harvard University Press, Cambridge, Mass., p. 165.
Waehneldt, T. V. and Fox, S. W.: 1968, *Biochim. Biophys. Acta* **160**, 239.
Wald, G.: 1954, *Sci. Am.* **191** (2), 44.
Woese, C. R.: 1968, *Proc. Nat. Acad. Sci. U.S.A.* **59**, 110.
Yuki, A. and Fox, S. W.: 1969, *Biochem. Biophys. Res. Commun.* **36**, 657.

CHEMICAL AND CATALYTICAL PROPERTIES OF THERMAL POLYMERS OF AMINO ACIDS (PROTEINOIDS)

KLAUS DOSE*

Institute for Biochemistry, Joh. Joachim Becher-Weg 28, Johannes Gutenberg-University, Mainz, F.R.G.

Abstract. The significance of thermal polyamino acids (proteinoids) as abiotic predecessors of proteins is reviewed on the basis of new experimental results. Most proteinoids yield only 50 % to 80 % amino acid upon acid hydrolysis. They contain 40 % to 60 % less peptide links than typical proteins, whereas their average nitrogen content is like that of proteins. The arrangement of amino acid residues is nonrandom. The degree of nonrandomness is difficult to determine because unusual crosslinks disturb most of the sequencing methods typically applied in protein chemistry. The products obtained in a polymerization experiment are heterogeneous. They can be separated into a limited number of related fractions by chromatography or electrophoresis and other separation methods applied in protein chemistry. Their molecular weights are typically between 4000 and 10000. The number of free NH_2-groups is usually smaller than in comparable proteins. A significant fraction of NH_2-groups yields imidazole-type bases during the thermal polymerization. Optically active amino acids racemize during the same process. So far no helicity could be detected. Proteinoids are thus clearly distinct from proteins. However, many of them exhibit weak catalytic activities and tend to undergo self-assembly into microstructures. Their properties of which only a few have been mentioned still support their role as possible candidates for ancestors of first proteins.

The 'central dogma' of molecular biology as postulated by Watson and Crick in 1953 (Watson and Crick, 1953) and by Crick in 1958 (Crick, 1958) allows for the transfer of genetic information from nucleic acid to protein, but neither from protein to nucleic acid nor from protein to protein. Lipmann and his associates (Gevers *et al.*, 1969; Lipmann *et al.*, 1971), however, have demonstrated that information can flow from proteins to polypeptides in some contemporary systems.

If we project molecular evolution backwards in order to trace the primary origins of biologically significant macromolecules, we finally face the question of what came first.

The three possible alternatives (proteins came before nucleic acids, nucleic acids came before proteins and both came simultaneously) have been discussed in detail (Lacey and Mullins, 1972; Kaplan, 1971; Calvin, 1969; Eigen, 1971). Experiments designed to simulate the abiotic synthesis of pre-nucleic acids and pre-proteins in the laboratory have been carried out by a large number of investigators (Fox and Dose, 1972a). So far, however, no evidence has been presented indicating a successful a priori synthesis of polynucleotides which exhibit the structural and biological properties of contemporary nucleic acids. The polymerization of amino acids under simulated prebiotic conditions on the other hand has yielded materials which resemble in many aspects polypeptides and proteins such as found in contemporary organisms.

The essential reaction of condensation and its energetics (Borsook, 1953) is depicted

* Dedicated to Boris Rajewsky on the occasion of his 80th birthday

by the equation:

$$^{\oplus}H_3N-CHR-COO^{\ominus} + {}^{\oplus}H_3N-CHR'-COO^{\ominus} \rightleftharpoons$$
$$^{\oplus}H_3N-CHR-CO-NH-CHR'-COO^{\ominus} + H_2O \qquad \Delta G^{\circ} = 2\text{–}4 \text{ Kcal}$$

In aqueous solution the hydrolysis of the peptide is favored. The reaction proceeds from left to right only if the water is removed from the system. Theoretically, this can be achieved by consuming the water in a secondary reaction (use of a condensing agent) or by evaporation. Removal of water by evaporation is operationally very simple inasmuch as the reactant mixture of amino acids has just to be heated above the boiling point of water. The energy of activation for peptide formation is relatively high. If the starting material is an unhydrous mixture of amino acids, the temperature must usually be raised to 180 °C for several hours to initiate the condensation. At such a high temperature amino acids begin to decompose. Thus side reactions compete with the condensation reaction. Some of the difficulties brought about by side reactions were initially overcome by the inclusion of a relatively high proportion of aspartic acid, glutamic acid or lysine (Fox, 1969; Fox, 1956). Other authors have tried to circumvent the barrier between the free amino acids and the peptides by other methods. The use of clays to provide hypohydrous conditions, as suggested by Bernal, has been advanced by the use of montmorillonite with amino acyl adenylates (Paecht-Horowitz and Katchalsky, 1967; Paecht-Horowitz *et al.*, 1970). Another mode is the conduct of the reaction in the presence of anhydrizing agents such as polyphosphoric acid and other polyphosphates (Harada and Fox, 1965), or cyanamide (Ponnamperuma and Peterson, 1965) or dicyanamide (Steinman *et al.*, 1964; Steinman, 1967). Akabori and his colleagues (Akabori and Yamamoto, 1972) have proposed that the 'fore-protein' (pre-protein) could have been formed by the polymerization of amino-acetonitrile followed by hydrolysis to polyglycine and then by the introduction of various side chains onto the methylene group of polyglycine.

Experiments which circumvent the thermodynamic barrier of amino acid condensation have either yielded only extremely small amounts of polymeric materials or – if large amounts could be produced – have lacked widely accepted geological relevance (e.g., experiments with polyphosphoric acid).

Of all methods to produce biologically significant macromolecules under geochemically relevant conditions only the polymerization of amino acids at elevated temperatures has so far yielded products which show a narrow relationship to the corresponding contemporary types, the proteins. I shall now review a number of criteria which support the idea that thermal polymers of amino acids may be regarded as prebiotic foreproteins.

1. Quantitative Composition

Except for serine, threonine, and cystine, the relative composition of almost any contemporary protein has been simulated quite closely by the thermal condensation product in earlier experiments, particularly by Fox and his associates (Fox, 1968;

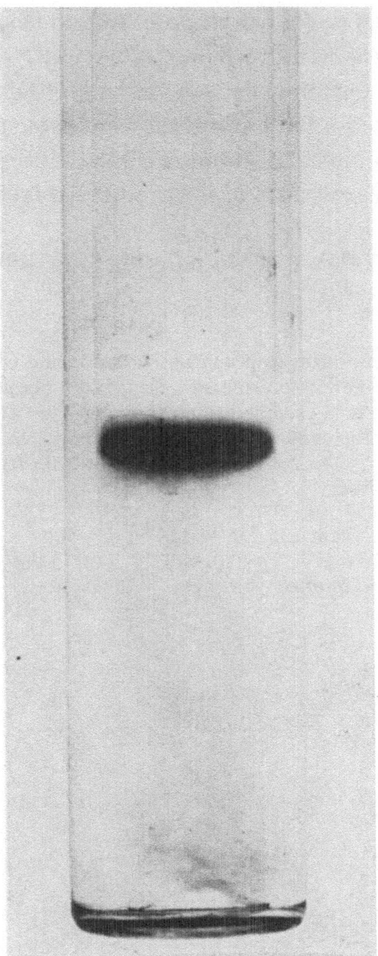

Fig. 1. Disk gel electropherogram of hemoproteinoid. A single bond is seen. Further fractionation yielding a family of closely related polymers is achieved by DEAE-chromatography and similar methods.

Fox and Dose, 1972b). The recovery of amino acids by hydrolysis of a given polymer is not 100% in any case. In some instances only 20% to 50% of the polymeric material yields amino acids upon acid hydrolysis (Dose and Rauchfuß, 1972). Side reactions during the condensation lead to largely unknown products which are linked to the polyamino acids by other than peptidic bonds. The polymers are therefore distinct from pure polypeptides or proteins with respect to the biuret reaction. The color produced by the biuret agent can be taken as a relative measure of the number of peptide bonds in a protein. The biuret tests indicate that thermal polymers of amino acids contain 40% to 60% less peptidic bonds per gram material then bovine serum albumine (taken as a standard protein). The N-content (N-Kjeldahl) on the other hand is within the limits characteristic for proteins.

Some results with respect to the variation of amino acid composition, recovery of amino acids after hydrolysis, relative biuret value and N-Kjeldahl content are summarized in Table I. These experiments were designed to demonstrate that also serine and cystine are incorporated if large proportions of these amino acids are added and a slightly acidic pH is maintained. 12 amino acids have been added only in low proportion, the corresponding hydrolysates of the products therefore contain less than 0.5% of most of them.

Synthesis of polymers A 96a–d: Two parts of four different amino acid mixtures

TABLE I

Amino acid composition of polymers (A96a-d) and corresponding reactant mixtures with increasing amounts of aspartic acid (0.5 gm Na polyphosphate gm^{-1}). All values in gm% (aspartic acid excluded; the percentages for aspartic acid are given relative to the total of all other amino acids); tr corresponds to values below 0.5%. Mean error (reproducibility) within 10%.

a Amino acid	b Reactant mixture (asp omitted)	c A 96a 0.013 gm asp/gm	d A 96b 0.213 gm asp/gm	e A 96c 0.413 gm asp/gm	f A 96d 0.813 gm asp/gm
Asp	1.3	11.0	32.4	46.2	53.5
Thr	1.1	tr	tr	tr	tr
Ser	26.5	8.1	5.6	3.6	2.6
Glu	10.0	22.0	17.4	17.0	16.4
Pro	1.1	tr	tr	tr	tr
Gly	0.7	2.2	2.3	3.0	1.4
Ala	0.7	10.6	12.0	17.8	13.2
Val	1.0	tr	tr	tr	tr
½ Cys	18.1	6.1	4.5	6.9	3.3
Met	1.6	tr	tr	tr	tr
Ile	1.2	tr	tr	tr	tr
Leu	1.2	tr	tr	tr	tr
Tyr	1.8	tr	tr	tr	tr
Phe	1.7	tr	tr	tr	tr
NH_3	–	2.8	3.3	2.8	2.8
Lys	18.1	33.0	34.1	32.2	39.3
His	10.1	7.3	12.3	8.5	11.5
Trp[a]	2.0	tr	tr	tr	tr
Arg	1.7	1.9	2.2	1.7	2.4
other aa	–	<1.0	<1.3	<1.6	<2.0
aa recovered, $\mu gm\ mg^{-1}$ (after hydrolysis)		366	396	461	507
pH of 10 mg reactant ml^{-1} after 30′ at 120°		6.0	5.5	5.0	4.5
relative biuret value, % (bovine serum albumin 100%		55	55	53	40
N-Kjeldahl, %		13.2	12.3	12.8	13.3

[a] Tryptophan is largely destroyed by acid hydrolysis.

(containing increasing amounts of aspartic acid according to Table I) were each heated with one part of sodium polyphosphate at 120 °C for 48 hr under N_2. The raw product was dispersed in distilled water at 40 °C, stirred and cooled to 2 °C. The sediment was homogenized in a Star Mix and filtered. The filtrate was dialyzed for 5 hr against distilled water until the contents in free amino acids (determined by paper electrophoresis and ion exchange chromatography) was less than 3%. Biuret values and nitrogen contents are shown in Table I.

2. Arrangements of Amino Acid Residues

Composition and primary structure of thermal polyamino acids are nonrandom. First evidence has been presented by Fox and his associates. They have demonstrated that the composition of the reactant amino acid mixture differs from the composition of the products. The frequency of the occurrence of a given N- or C-terminal amino acid is not proportional to the molar percentage of the given amino acid in the polymer (Fox and Harada, 1960; Fox and Harada, 1963). Some of our own results on N-terminal amino acids (determined by Sanger's method) are shown in Table II.

Determinations of amino acid sequences of thermal polyamino acids are extremely difficult, because of unusual cross-links and the formation and simultaneous incorpo-

TABLE II

N-terminal amino acids and amino acid composition of thermal polyamino acids (molar percentages).

The amino acid mixture of glycine and lysine, and of aspartic acid (A 119), of alanine (LAG), of valine (LVG) respectively, was heated with 0.5 ml distilled water at 180°C for 5½ hr under N_2. The raw product was dispersed in water and dialyzed for 3 days against distilled water and lyophilized. The dialyzed product was free of amino acids and low molecular weight peptides.

Amino acid	Reactant mixture	Product	N-terminal
Polymer A-119			
Asp	33	13	35
Gly	33	63	20
Lys	33	25	55
Polymer LAG			
Gly	33	35	33
Ala	33	23	33
Lys	33	42	33
Polymer LVG			
Gly	33	47	25
Val	33	16	50
Lys	33	36	25

ration of organic N-bases which are not amino acids. Most thermal polyamino acids are only extremely slowly attacked by the known endo- and exopeptidases. So far, only Fox and Nakashima have been able to demonstrate the sequence of a hexapeptide which they had split from a thermal polyamino acid (Nakashima *et al.*, 1967).

3. Nonrandomness

The degree of nonrandomness found for thermal polyamino acids, particularly thermal proteinoids, has been studied by a number of methods based on different properties. The evidence is summarized in Table III (Fox and Dose, 1972c).

TABLE III

Evidence for nonrandomness of thermal polymers of amino acids

Evidence	Authors and date
Nonrandom sequences by disparity between N-terminal and total analyses in thermal polymers	Fox and Harada (1958)
%s in reaction mixture ≠ %s in polymer	Fox and Harada (1960)
Limited number of fractions on electrophoresis	Vestling (1960)
Nonrandomness on ultracentrifugation	Vegotsky (1961)
Constant composition on repurification from water	Fox *et al.* (1963)
Single band on gel electrophoresis of acidic proteinoid-amide	Fox and Nakashima (1966) (unpublished results)
Nonrandom elution pattern from DEAE-cellulose	Fox and Nakashima (1967)
Symmetrical peaks from DEAE-cellulose	
Almost uniform amino acid compositions	
Stoichiometric amino acid compositions	
Uniform ultracentrifugal patterns of various fractions	
Almost uniform peptide maps in all fractions	
Single spots on high voltage electrophoresis of fractions	
Single species of 'active site' proteinoid	Usdin *et al.* (1967)
Single band for gel electrophoresis of basic hemoproteinoid	Dose and Zaki (1971)

(See also Figure 1.)

4. Molecular Weight

Mean molecular weights of the proteinoids have been determined mainly by end-group assay, by sedimentation analysis in the ultracentrifuge (Fox and Harada, 1960; Fox and Nakashima, 1967) and by the molecular sieve technique (Dose and Zaki, 1971). The average molecular weights of proteinoids made under comparable conditions tend to increase from the acidic types, which are the lowest, through the neutral, which are of intermediate weight, to the basic, which are the highest. The mean molecular weights fall within the lower end of the molecular weight range of contemporary proteins (4000–10000). By the thermal condensation of amino acids a large proportion of oligomers is typically produced, but then disregarded during dialysis. The spectrum of the lower molecular weights of the products, therefore, also depends on the method of dialysis applied for their purification.

5. Solubility and Precipitability

Depending on their amino acid composition and molecular weight, proteinoids correspond to various classes of neutral, acidic and basic proteins. The typical agents used for precipitation and fractionation of proteins have also been applied to proteinoids. Figure 2 and Table IV demonstrate an example for the application of $(NH_4)_2SO_4$ for the fractionation of thermal polyamino acids

Precipitation by various concentration of ammonium sulfate: 100 mg of the pro-

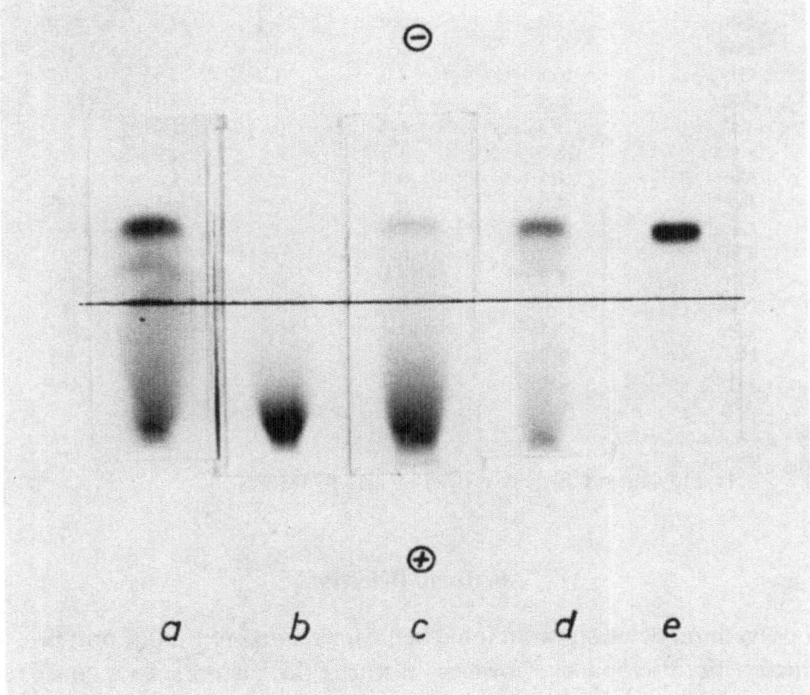

Fig. 2. Electrophoretic control of the fractionation of polymer Aβ with ammonium sulfate. The electropherograms, b, c, d and e represent the analysis of the materials precipitated at 25, 50, 75, and 100% saturation with ammonium sulfate. The mobility of the acidic fraction is about 1.2 times that of bovine serum albumin. The cathodic mobilities of the upper two zones are 0.45 times and 0.2 times that of lysozyme. (pH 8.6; cellulose acetate).

teinoid Aβ were dissolved in 10 ml 0.02 M tris/HCl at pH 7.2. Solid ammonium sulfate was added until the required degree of saturation (25, 50, 75, and 100%) was reached. The precipitations were carried out at 0 °C. The precipitates were collected by sedimentation (1500 × g for 20 min) about 1 hr after the desired amount of ammonium sulfate was added. After sedimentation the precipitates were redissolved, dialyzed for 10–24 hr against distilled water and finally lyophilized.

TABLE IV

Polymer A 61. Amino acid composition of reactant mixture, of Aβ, and fractions obtained by precipitation with $(NH_4)_2SO_4$. F_1 corresponds to 25% saturation with ammonium sulfate, F_2 to 50%, F_3 to 75%, and F_4 to 100% respectively. All values in gm. %; tr corresponds to values below 0.5%. Mean error (reproducibility) within 10%.

a Amino acid	b Reactant mixutre	c Aβ unfract.	d F_1	e F_2	f F_3	g F_4
Asp	1.4	4.1	4.7	4.3	4.7	3.8
Thr	1.2	tr	tr	tr	tr	tr
Ser	27.0	4.7	3.8	3.7	6.0	6.3
Glu	9.8	19.9	23.9	22.2	24.7	15.3
Pro	1.2	tr	tr	tr	tr	tr
Gly	0.8	3.0	3.9	4.1	2.6	2.2
Ala	0.9	14.2	20.4	8.1	18.2	10.5
Val	1.2	1.4	tr	1.9	1.6	0.7
½ Cys	18.5	3.8	tr	4.7	6.4	4.1
Met	1.5	0.9	tr	1.4	tr	tr
Ile	1.3	tr	tr	tr	tr	tr
Leu	1.3	1.0	tr	2.4	tr	tr
Tyr	1.9	tr	tr	tr	tr	tr
Phe	1.7	1.5	tr	2.8	tr	tr
NH_3	–	3.7	tr	tr	tr	tr
Lys	17.5	30.4	31.5	31.9	20.7	42.1
His	9.8	5.9	6.0	6.0	6.5	6.4
Trp[a]	2.1	tr	tr	tr	tr	tr
Arg	1.8	1.9	1.5	1.5	1.8	2.7

[a] Tryptophan is largely destroyed by acid hydrolysis.

6. Ionic Behavior

Thermal polyamino acids show an ionic behavior related to that of proteins. At least one characteristic discrepancy, however, disturbs the picture: As a consequence of the thermal condensation, more than $\frac{1}{3}$ of the ε-amino groups of lysine and N-terminal amino groups disappears and is converted into less basic groups with pK of about 7. The following reaction scheme may interpret this behavior:

```
           R1
           |
 -C-NH-CH-C-NH-
  ||        ||         R1-C=C-NH-
  O         O             |  |
                          N  N-R2
   NH2+           ⇌       \\ /          + 2 H2O
   |                       C
   R2                      |
```

As a consequence of this reaction the isoelectric point of lysine-rich polymers (more than 20% lysine) is not above 11, but only between 8 and 9. As indicated, the reaction may lead to a substituted imidazole. Mild hydrolysis of proteinoids also leads to the splitting of imide-linkages, which also occur in proteinoids. See the following reaction (Rohlfing, 1964):

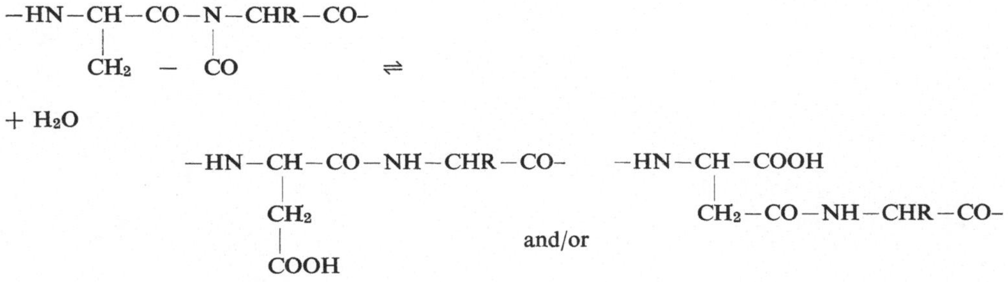

```
–HN–CH–CO–N–CHR–CO–
     |       |
    CH₂  –  CO          ⇌

+ H₂O
          –HN–CH–CO–NH–CHR–CO–       –HN–CH–COOH
              |                          |
              CH₂                        CH₂–CO–NH–CHR–CO–
              |             and/or
              COOH
```

7. Configuration of Residues and Conformation of the Chain

The problem of stereo-enriched poly-α-amino acids and their synthesis under postulated prebiotic conditions has been recently reviewed and re-examined by Rohlfing and Fouche (Rohlfing and Fouche, 1972). These authors have shown that several amino acids are not extensively racemized during thermal co-polymerization at 170–195 °C. Helicity, however, has been sought, but not yet been demonstrated for thermal polyamino acids. Inasmuch as lysyl or aspartyl residues are extensively (many other residues at least partially) racemized, helical structures appear to have a low probability to occur in thermal polyamino acids. All known polyamino acids synthesized so far as models of a pre-protein were either completely racemic or contained substantial proportions of both optical antipodes of amino acids, This fact excludes – in contrast to contemporary proteins – the presence of significant helical regions, but it maintains the possibility of stereospecific conformations in 'catalytically active regions' of the polymers. So far little has been done to seek stereospecific catalysis by proteinoids, however.

8. Linkages and Infrared Spectra

So far the nature of the non-peptidic bonds and cross-links of thermal polyamino acids is poorly established.

Whereas the limited recovery of amino acids from hydrolyzed proteinoids and biuret test support only in part the resemblance of proteinoids with contemporary proteins, the IR-spectra often cannot be distinguished from those of proteins. In Figure 3 the IR-spectrum of a proteinoid (A61) is compared with that of a protein (bovine serum albumin). The IR-spectra indicate that the same types of amide and imide linkages occur in proteinoids which occur in proteins, although the data tell us nothing of the order and stereochemical arrangement of the bonds.

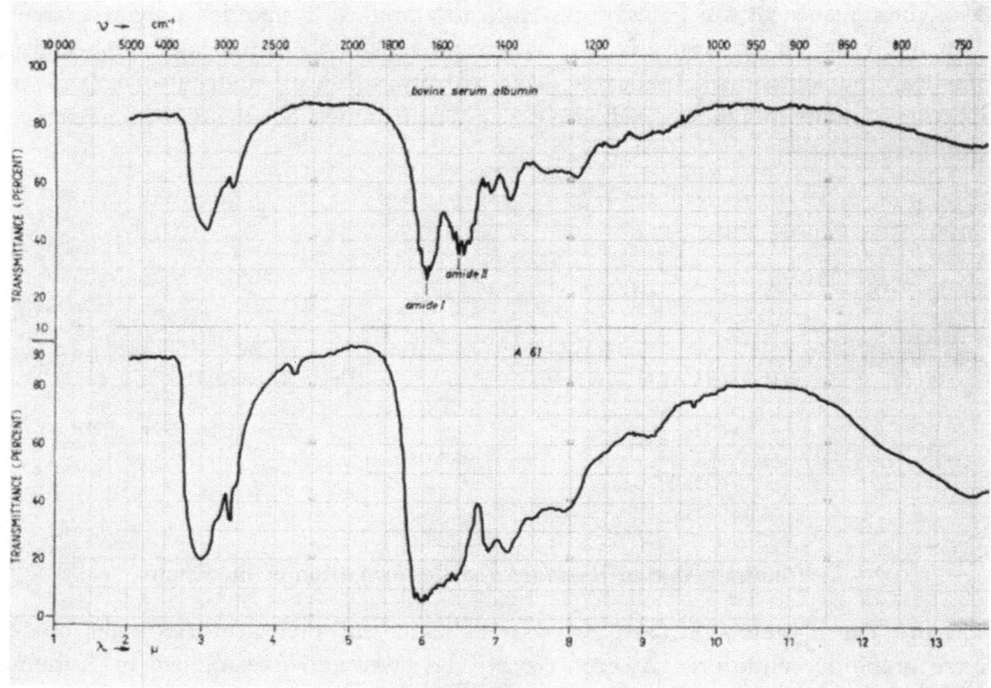

Fig. 3. The IR absorption spectrum of polymer A 61 is compared with the spectrum of bovine serum albumin. In spite of a general agreement of both spectra a close comparison shows several differences, e.g., in the range of the amide I and II absorption. These differences have not yet been explained.

9. Catalytic Activities

The results on various catalytic activities of thermal polymers of amino acids are summarized in Table V.

So far, the iron-porphyrine nucleus is the only known and definable prosthetic group which shows reasonable activity if combined with thermal polymers of amino acids. The iron-porphyrine systems appear to be of unique importance in nature. A related family of compounds, the phthalocyanines, may be visualized as possible candidates as prosthetic groups for redox reactions. But, various phthalocyano-proteinoids which have been synthesized in our laboratory show almost no catalase- or peroxidase-like activity.

Generally, the catalytic activities of thermal polymers of amino acids are relatively low if compared with contemporary enzymes. For a prebiotic system a low activity is of no disadvantage as long as the whole metabolism of such a system is low. Of principal importance, however, is the specificity of catalytic action. With respect to the model systems tested so far, a significant degree of specificity has been demonstrated in many instances. Although the problem of stereospecific catalysis is yet unsolved, the specificity of prebiotic catalysts is generally much broader than the specificity of

TABLE V

Results on various catalytic activities of thermal polymers of amino acids

Substrate	Authors	Active AA in polymer	Activity (max) (μmoles mg^{-1} min^{-1})	Other results and comments
Hydrolytic reactions				
p-Nitrophenyl acetate	Noguchi and Saito (1962)			
	Rohlfing and Fox (1967)	His, Lys, Asp-imide	(min-hr)[a]	1.3 to 10 times more active than free His. Inactivation by heat, Michaelis-Menten kinetics
	Usdin *et al.* (1967)	Glu, Asp, Lys, His, Ser, Thr	1.5×10^{-3} (min-hr)	(Chymotrypsin: 3×10^{-2} μmoles mg min^{-1}). Inhibition by DFB (reversible)
p-Nitrophenyl phosphate	Oshima (1968)	His, other basic and neutral AA cooperative Asp-mide	2×10^{-3} (hr-days)	Free AA inactive. Inhibition by arsenate, Cu^{2+}. 10% degradation by pronase → 30% inhibition. Heat inhibition, Michaelis-Menten kinetics
ATP	Fox and Joseph (1965)	content in Zn^{2+} essent.	(hr-days)	Zn active in ptd microspheres
	Durant and Fox (1966)			Promotion of Zn effects in Lys-rich proteinoids
	Tetas and Löwenstein (1963)	no polymer		Zn^{2+} as active as Zn-ptd microspheres
Decarboxylations				
Glucose or glucoronic acid	Fox and Krampitz (1964)	Lys	Only counts of $^{14}CO_2$ (React. time: hr-days)	Anaerobic conditions. ATP stimulates twofold. Glucose oxidation non-catalytic. Large polymer/substrate ratio. Strictly catalytic? No Michaelis-Menten kinetics
Pyruvic acid	Krampitz and Hardebeck (1967)	Glu, Thr	Only counts of $^{14}CO_2$ (React. time: hr-days)	Activity declines with reaction time. Strictly catalytic?
	Hardebeck *et al.* (1968)	Ile inhibition		Heating reduces activity. Oxalacetic acid not decarboxylated.
$HO-\overset{O}{\overset{\Vert}{C}}-\overset{O}{\overset{\Vert}{C}}-CH_2-\overset{O}{\overset{\Vert}{C}}-OH$	Rohlfing (1967)	Lys	0.3 μmole mg^{-1} min^{-1} (min-hr)[a]	Pseudo-first order kinetics. Thermal polylysine 10 times more active than free Lys. (Enzymes 30–900 μmoles mg^{-1} min^{-1}). No measurable decarboxylation of pyruvic, malic, malonic, α-ketoglutaric, glucuronic, oxalic and aspartic acid.

TABLE V *(Continued)*

Substrate	Authors	Active AA in polymer	Activity (max) (μmoles mg^{-1} min^{-1})	Other results and comments
Pyruvic, Glyoxalic, Glucuronic acid	Weber *et al.* (1968)	Yellow pigment	Photosensitized react. (hr-days)[a]	Up to 80 times the dark reaction.
Amination and deamination				
α-Ketoglutaric acid with urea and similar NH_2-donors	Krampitz *et al.* (1967)	Lys, and other basic AA;	2×10^{-3} (hr-day)[b]	Cu^{2+} required, reducing agent (?) 73% L and 27% D Glu (by Lys-rich ptd)
	Krampitz *et al.* (1968)b	Neutral and acidic ptds not active		Polyanhydrolysine from Leuchs' anhydride inactive. Michaelis-Menten kinetics. Acylation inactivates. No heat inactivation.
Pyruvic, phenylpyruvic and oxalacetic acid + NH_2-donor	Krampitz (1968)		(hr)[b]	
Glutamic acid ↓ α-ketoglutaric acid	Krampitz *et al.* (1968)a)		(hr)[b]	Cu^{+} required, oxidizing agent(?) Strictly catalytic(?) Part of transaminase reaction.
Summary of catalatic and peroxidatic properties of hemoproteinoids				
H_2O_2 (catalatic reaction)	Dose and Zaki (1971)	Phe inhibitory	1 (sec-min)	50% less effective than free hematin. Liver catalase 10^5 times as effective. Pseudo-first order Michaelis-Menten kinetics. Inhibition by complexing agents.
H_2O_2 and H-donor (peroxidatic reaction)	Dose and Zaki (1971)	Lys and other basic AA cooperate with thermal byproduct. Asp-imide inhibitory (?)	0.2 for guaiacol 2.2 for NADH (sec-min)	Up to 50 times more effective than hematin (guaiacol as substrate). Horseradish peroxidase 10^3 times as effective (NADH oxidation). Inhibition by complexing agents. 100% increase of activity after heating. Hematin heated with Leuchs' poly-Lys only as active as free hematin. Pseudo-first order Michaelis-Menten kinetics.

[a] (min-hr) or (hr-days) indicate duration of the particular experiment.
[b] Reaction time.

contemporary enzymes. This property agrees with the concepts of molecular evolution on the premise that enzymes have evolved from lower levels to highly specialized levels. The prebiotic evolution of enzymes could have started with thermal polyamino acids (proteinoids). This thesis is supported by a body of experimental results from the laboratory.

References

Akabori, S. and Yamamoto, M.: 1972, in D. L. Rohlfing and A. I. Oparin (eds.), *Molecular Evolution*, Plenum Press, New York, p. 189.

Borsook, H.: 1953, *Adv. Protein Chem.* **8**, 127.

Calvin, M.: 1969, *Proc. Roy. Soc. Edinburgh*, Sect. B, Biology 70, part 4, 273.

Crick, F. H. C.: 1958, *Symp. Soc. Exp. Biol.* **12**, 138.

Dose, K. and Rauchfuß, H.: 1972, in D. L. Rohlfing and A. I. Oparin (eds.), *Molecular Evolution*, Plenum Press, New York, p. 199ff.

Dose, K. and Zaki, L.: 1971, *Z. Naturforsch.* **26b**, 144.

Durant, D. H. and Fox, S. W.: 1966, *Federation Proc.* **25**, 342.

Eigen, M.: 1971, *Naturwissenschaften* **58**, 519.

Fox, S. W.: 1956, *Am. Sci.* **44**, 347.

Fox, S. W.: 1968, in H. Mark, N. G. Gaylord and Bikales (eds.), *Encyclopedia of Polymer Science and Technology*, Vol. 9, Interscience, New York, p. 284.

Fox, S. W.: 1969, *Naturwissenschaften* **56**, 1.

Fox, S. W. and Dose, K.: 1972a, in *Molecular Evolution and the Origin of Life*, Freeman, San Francisco, p. 135ff.

Fox, S. W. and Dose, K.: 1972b, in *Molecular Evolution and the Origin of Life*, Freeman, San Francisco, p. 152.

Fox, S. W. and Dose, K.: 1972c, in *Molecular Evolution and the Origin of Life*, Freeman, San Francisco, p. 161.

Fox, S. W. and Harada, K.: 1958, *Science* **128**, 1214.

Fox, S. W. and Harada, K.: 1960, *J. Am. Chem. Soc.* **82**, 3745.

Fox, S. W. and Harada, K.: 1963, *Federation Proc.* **22**, 479.

Fox, S. W., Harada, K., Woods, K. R., and Windsor, C. R.: 1963, *Arch. Biochem. Biophys.* **102**, 439.

Fox, S. W. and Joseph, D.: 1965, in S. W. Fox (ed.), *The Origins of Prebiological Systems*, Academic Press, New York, p. 371.

Fox, S. W. and Krampitz, G.: 1964, *Nature* **203**, 1362.

Fox, S. W. and Nakashima, T.: 1967, *Biochim. Biophys. Acta* **140**, 155.

Gevers, W., Kleinkauf, H., and Lipmann, F.: 1969, *Proc. Nat. Acad. Sci.* **63**, 1335.

Harada, K. and Fox, S. W.: 1965, *Arch. Biochem. Biophys.* **109**, 49.

Hardebeck, H. G. and Fox, S. W.: 1967, *3rd Ann. Rept. Inst. Mol. Evolution*, Coral Gables, Florida, p. 18.

Hardebeck, H. G., Krampitz, G., and Wulf, L.: 1968, *Arch. Biochem. Biophys.* **123**, 72.

Kaplan, R. W.: 1971, in R. Buvet and C. Ponnamperuma (eds.), *Molecular Evolution*, North Holland, Amsterdam, Vol. I, p. 319.

Krampitz, G.: 1968, *156th Nat. Meeting Am. Chem. Soc.*, Atlantic City, Abstr. Papers, Div. Biol. Chem., p. 87.

Krampitz, G. and Hardebeck, H. G.: 1966, *Naturwissenschaften* **53**, 81.

Krampitz, G., Diehl, S., and Nakashima, T.: 1967, *Naturwissenschaften* **54**, 516.

Krampitz, G., Haas, W., and Baars-Diehl, S.: 1968a, *Naturwissenschaften* **55**, 345.

Krampitz, G., Baars-Diehl, S., Haas, W., and Nakashima, T.: 1968b, *Experientia* **24**, 140.

Lacey, J. C., Jr. and Mullins, D. W., Jr.: 1972, in D. L. Rohlfing and A. I. Oparin (eds.), *Molecular Evolution*, Plenum Press, New York, p. 189.

Lipmann, F., Gevers, W., Kleinkauf, H., and Roskoski, R., Jr.: 1971, *Adv. Enzymology* **35**, 1.

Nakashima, T., Fox, S. W., and Wang, C.: 1967, *Arbeitstagung Extraterrestr. Biophysik und Biologie und Raumfahrtmedizin*, Tagungsbericht (Chem. Abstr. **72**, 50967x)

Noguchi, J., and Saito, T.: 1962, in M. A. Stahmann (ed.), *Polyamino Acids, Polypeptides, and Proteins*, Univ. of Wisconsin Press, Madison, Wisconsin, p. 313.

Oshima, T.: 1968, *Arch. Biochem. Biophys.* **126**, 478.
Paecht-Horowitz, M., Berger, J., and Katchalsky, A.: 1970, *Nature* **228**, 636.
Paecht-Horowitz, M. and Katchalsky, A.: 1967, *Biochim. Biophys. Acta* **140**, 14.
Ponnamperuma, C. and Peterson, E.: 1965, *Science* **147**, 1572.
Rohlfing, D. L.: 1964, Ph. D. dissertation, Florida State University, Tallahassee.
Rohlfing, D. L.: 1967, *Arch. Biochem. Biophys.* **118**, 127.
Rohlfing, D. L. and Fouche, C. E.: 1972, in D. L. Rohlfing and A. I. Oparin (eds.), *Molecular Evolution*, Plenum Press, New York, p. 219ff.
Rohlfing, D. L. and Fox, S. W.: 1967, *Arch. Biochem. Biophys.* **118**, 127.
Steinman, G.: 1967, *Arch. Biochem. Biophys.* **121**, 553.
Steinman, G., Lemmon, R. M., and Calvin, M.: 1964, *Proc. Nat. Acad. Sci.* **52**, 27.
Tetas, M. and Löwenstein, J. M.: 1963, *Biochemistry* **2**, 350.
Usdin, V. R., Mitz, M. A., and Killos, P. J.: 1967, *Arch. Biochem. Biophys.* **122**, 258.
Vegotsky, A.: 1961, Ph. D. dissertation, Florida State University, Tallahassee.
Vestling, C.: 1960, in S. W. Fox, and K. Harada (eds.), *J. Am. Chem. Soc.* **82**, 3745.
Watson, J. D. and Crick, F. H. C.: 1953, *Nature* **171**, 737.
Weber, A. L., Wood, A., Hardebeck, H. G., and Fox, S. W.: 1968, *Federation Proc.* **27**, 830.

PRE-ENZYMIC ORIGIN OF METABOLIC REDOX PROCESSES AND OF THE ENERGY STORAGE PROCESSES

R. BUVET, L. LE PORT, and F. STOETZEL

Lab. Énergétique Biochimique, Université Paris Val de Marne, 94000 – Créteil, France

Abstract. A treatment of the energetic conditions which regulate the non-equilibrium occurrence of metabolic redox processes associated with degradations of carbon chains or formations of condensed bonds shows that such processes are conditioned by the nature of the substrates and not by the pre-existence of enzymic catalysts or cellular ultrastructures. An experimental program has been initiated to spectrophotometrically, electrochemically and chemically detect which redox components appear during model studies of chemical evolution, operating from simple mixtures of CH_4, NH_3, H_2 and water and from the same mixtures with added phosphorus or sulfur derivatives and metal cations.

All results that have been obtained for the last twenty years from model experiments of chemical evolution clearly demonstrate that the major components of living systems can be obtained from a large range of initially reducing carbonaceous mixtures via many chemical pathways operating in aqueous solutions under the effects of sufficiently hard sources of energy. Therefore, the most overt problem is no longer simply to form new biochemicals under circumstances roughly similar to those which must have prevailed in the primordial terrestrial periphery, but to elucidate how the resultant products must have begun to 'live'; that is to say, to react in a way more or less similar to present-day metabolism. From this viewpoint, non-enzymic archetypes of several categories of biochemical processes have been extensively studied. Thus far, attention has been focused on transacylations and transphosphorylations (Buvet *et al.*, 1971; Le Port *et al.*, 1971; Le Port and Buvet, 1973; Etaix and Buvet, 1973) with emphasis on their importance in the synthesis of biopolymers (Paecht-Horowitz, 1971).

Conversely, despite their essential role in biology, less attention has been paid to the archetypes of present-day biochemical redox processes. Some products, more or less similar to components of the electron transfer chain, were obtained during global model experiments and from simple precursors. Some experimental attempts were developed to find possible archetypes of energy storage processes and of photochemical energy production. There was, however, no systematic attempt to define the role of redox processes in the chemical transformations of the primordial environment.

Let us first consider from the theoretical viewpoint the typically biochemical process of storage in condensed bonds of redox energy. Here, the problem is to decide whether such a process is dependent on the existence of sophisticated compounds or structures, as it is considered in the chemiosmotic theory, or on the contrary presents such a general character that it could have occurred before the formation of such sophisticated biological characteristics. We will give elsewhere (Buvet and Le Port, 1973) a more general discussion from energetic data of the criteria which regulate such processes occurring in biology. Let us here only give the general viewpoint underlying this theory and represent it with a simple example.

Our explanation of the energy storage of redox reactions is based upon the idea that these processes do not depend on the availability of sophisticated biological states, such as implying complex enzymes or ultrastructures, but can in fact be clearly accounted for by considering that the energy which is stored is simply a part of an excess of energy which must in any case be engaged in redox reactions when developing out of equilibrium through multistep mechanisms.

In order to provide an illustration, let us consider, as an example, the case when β-hydroxyacyl thioesters are oxidized. It appears that the normal, i.e., the lowest potential occurring two electron oxidation should be:

$$\begin{aligned} &\mathrm{CHOH - CH_2 - COSR) - 2}e^- - 3\mathrm{H}^+ + \mathrm{H_2O} \\ &\quad \rightarrow (\qquad \mathrm{CO_2^-}) + \mathrm{CH_3\ \ COSR} \end{aligned}$$

which occurs at -0.54 v/S.H.E. at pH7 (Figure 1).

But, in fact, this process is kinetically impossible, as a one-step process, because of the complexity of the atomic rearrangements it implies. Therefore, a primary oxidation process with more favorable kinetic conditions, mainly because it implies a lower number of reagents, is only possible. Nonetheless, this always occurs at higher redox

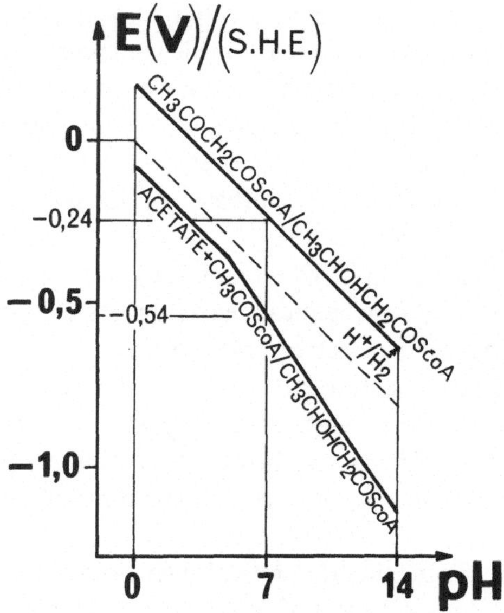

Fig. 1. Energetics of the over-all oxidation of hydroxybutyryl CoA into acetate and acetylcoA as:
$\mathrm{CH_3{-}CHOH{-}CH_2{-}COS\ coA + H_2O \rightarrow CH_3{-}CO_2H + CH_3{-}COS\ coA + 2e^- + (2\ or\ 3)\ H^+}$
and of its oxidation into acetoacetyl coA as:
$\mathrm{CH_3{-}CHOH{-}CH_2{-}COS\ coA \rightarrow CH_3{-}CO{-}CH_2{-}COS\ coA + 2e^- + 2H^+}$.

potentials. In this case, the oxidation of the substrate to a β-ketothioester, as:

$$(\sim\sim\sim \text{CHOH} - \text{CH}_2 - \text{COSR}) - 2e^- - 2\text{H}^+ \rightarrow$$
$$\rightarrow (\sim\sim\sim \text{CO} - \text{CH}_2 - \text{COSR})$$

is possible at -0.24 v/S.H.E. at pH7. It means an excess of energy, by comparison to what should be sufficient for the over-all oxidation:

$$|\Delta G'^{\circ}_{\text{pH7}}| = 2\,\text{F}(0.54 - 0.24) = 13.8 \text{ kcal. mole}^{-1}$$

is necessary to begin the oxidation.

When this is done, this excess, which represents the balance of the relaxation through hydrolysis from β-ketothioester to the degraded substrates as:

$$(\sim\sim\sim \text{CO} - \text{CH}_2 - \text{COSR}) + \text{H}_2\text{O} \rightarrow$$
$$\rightarrow (\cdots\ \text{CO}_2^-) + \text{H}^+ + \text{CH}_3\ \text{COSR}$$

can be recuperated by reagents which are able to attack the keto bond, as:

$$(\quad \text{CO} - \text{CH}_2 - \text{COSR}) + \text{HB} \rightarrow$$
$$\rightarrow \left(\cdots \text{C}(=\text{O})\text{B} \right) + \text{CH}_3\ \text{CO S coA}$$

if the standard free enthalpy of condensation from the shortened liberated acid and HB is lower than $\Delta G'^{\circ}$, which is the case when HB is a thiol (Figure 2). This is in fact

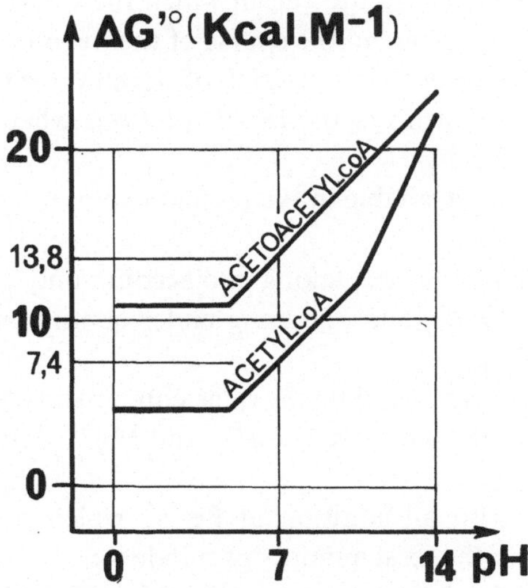

Fig. 2. Comparison, taking into account pH, of the energies liberated in the hydrolytic degradation of acetoacetyl coA, as:

acetoacetyl coA + H_2O → acetate + acetyl coA

and necessary for the condensation leading to acetyl coA from acetate, through:

acetate + HS coA → acetyl coA + H_2O.

the simple reason why the oxidative degradation of fatty acid chains produces a thioester group.

This explanation of energy storage processes, which renews, precises and generalises the principle of the socalled chemical theory of oxidative phosphorylations can be as well applied when other storage processes are considered, such as occurring in the electron transfer chain or when aldehydes are oxidized in the presence of thiols or phosphates (Buvet and Le Port, 1973).

Such developments show that theoretical explanations of such complex biochemical processes can be developed without considering particularities of systems too sharply defined for being conceivable under the primeval conditions, and, then, they pose the problem of the Origins of the Metabolism on a logically solvable ground.

But we must acknowledge that they should remain useless if redox processes were not playing, in the primeval soup, a role comparable to the one they play in the biochemical metabolism. This is why we began a new set of experiments for the purpose of examining if different kinds of manifestations of redox activity can be detected in global model experiments of primary evolution. This program is named CERES, which means Chemical Evolution of REactions into primeval Solutions, by opposition to Chemical Evolution as only manifested by the presence of products. It implies the use of an apparatus partly derived from Miller's design using excitation with sparks in gaseous medium, but including possibilities of frequent determinations of ultraviolet and visible spectra of solutions and of their electrochemical properties. These redox electrochemical properties are determined by polarography on platinum rotating disk electrodes in a separate cell in which the solutions can be temporarily transfered under inert atmosphere in the course of their evolution (Figure 3). In addition, we focused, in the present initial stage of development of the program, on a comparative evaluation of such reactivities of solutions when more or less complex initial conditions are used.

Three groups of data are available in the present exploratory stage of the program. They correspond to:

(1) the simplest initial conditions, taking into account only methane, ammonia and a solution of sodium chloride, and operating under continuous sparkling of the gas and heating of the solution;

(2) a more complex system including the same components, plus small quantities of orthophosphate and sulfide, and of K^+, Ca^{++} and Mg^{++} ions, which was treated in comparable conditions;

(3) the same system with an addition of Fe^{++}, which precipitates as FeS, but corresponds only to 3% of the total quantity of sulfide ions.

In addition the treatment which was imposed in this case to the system was discontinuous, the excitation being maintained only during the working hours. In the first and second cases, the walls of the reactor were covered after about 100 hr with a brown deposit. Figures 4, 5 and 6, respectively, describe the evolution of ultraviolet and visible spectra of the solutions.

On the first one which corresponds to the simplest mixture, we note:

(1) that the optical density regularly increases at all wavelengths to the visible range,

(2) that the spectrum mainly presents an aspect of continuity without major aberration which probably reveals the presence of several unsaturated products whose spectra are overlapping with the exception of a peak at 238 nm whose optical density also regularly increases.

On Figure 5, which corresponds to the mixture including phosphate and sulfide, the strong band of SH^- appears at 230 nm and the general evolution of the spectrum is comparable but not identical to the previous one.

The smooth ondulations are not at the same place, and, the optical density seems to remain lower at a given time which could be due to additions of phosphates or sulfide on double bonds which are present in the atmospheric precursors or in some of their products of transformations in aqueous solutions. Notably, the band of hydrogen

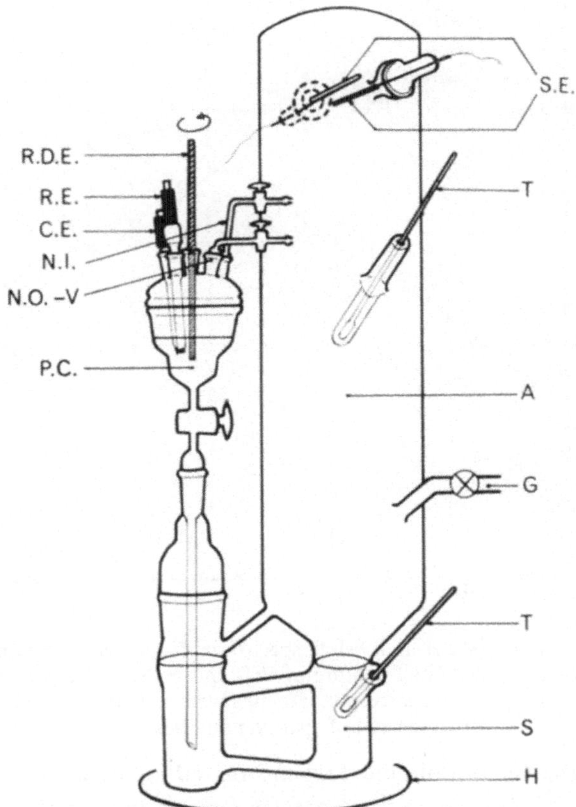

Fig. 3. Apparatus used for CERES experiments: A – Model of primary atmosphere; SE – Sparking electrodes; S – Aqueous solution; *T* – Thermometer; G– Gas inlet and outlet, measurement of pressure; H – Heater; P. C. – Polarographic cell; R.D.E. – Rotating disc electrode (platinum, gold, carbon) R.E. – Reference electrode (calomel); C.E. – Counter–electrode (platinum); N.I.– Nitrogen inlet; N.O.–V – Nitrogen outlet or vacuum.

sulfide ion rapidly disappears thus showing that the sulfide which is not volatile at the pH of the solution must react in aqueous solution on the atmospheric precursors or on products of their transformation. Lastly, the intensity of the 238 nm peak does not seem to be altered by the presence of sulfide since the peak appears as soon as its optical density is higher than the one due to SH^- at about the same level as in the first kind of assay for the considered time of treatment.

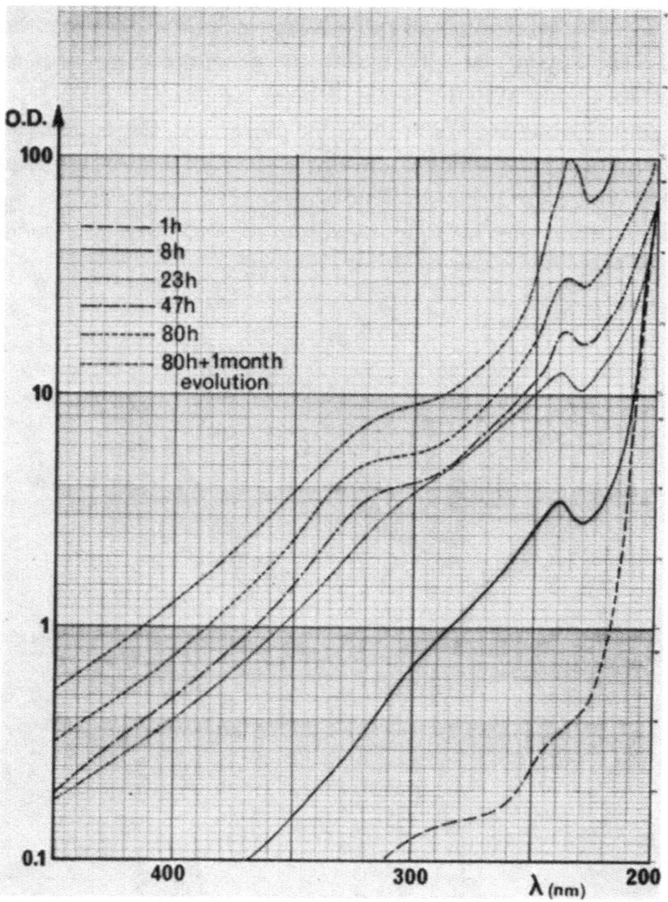

Fig. 4. Evolution of the ultraviolet and visible spectrum of aqueous solutions containing NaCl, heated at 80°C in presence of an initial mixture of CH_4 and NH_3, under continuous sparking. Last curve, noted 80 h + 1 month evolution corresponds to the spectrum of the solution, as recorded 1 month after the end of sparking and heating.

Figure 6 shows the spectra obtained when iron sulfide was present. The most interesting feature which appears on these spectra corresponds to the fact that both the increasing of the optical density due to formed unsaturated products and the decreasing of SH^- band at 230 nm are here much more rapid for a given time of sparking than was the case in the second set of experiments. At the present stage of development of the CERES program we cannot determine if this is connected to the presence of iron sulfide or to the fact that the sparking was discontinuous, thereby allowing more time for products of atmospheric reactions to dissolve in the aqueous solution.

We will end with a short view of the voltamperometric polarograms which were made to search for the appearance of redox activity in the solution. Figure 7 shows the evolution of polarograms in the case when phosphate and sulfide but no iron were

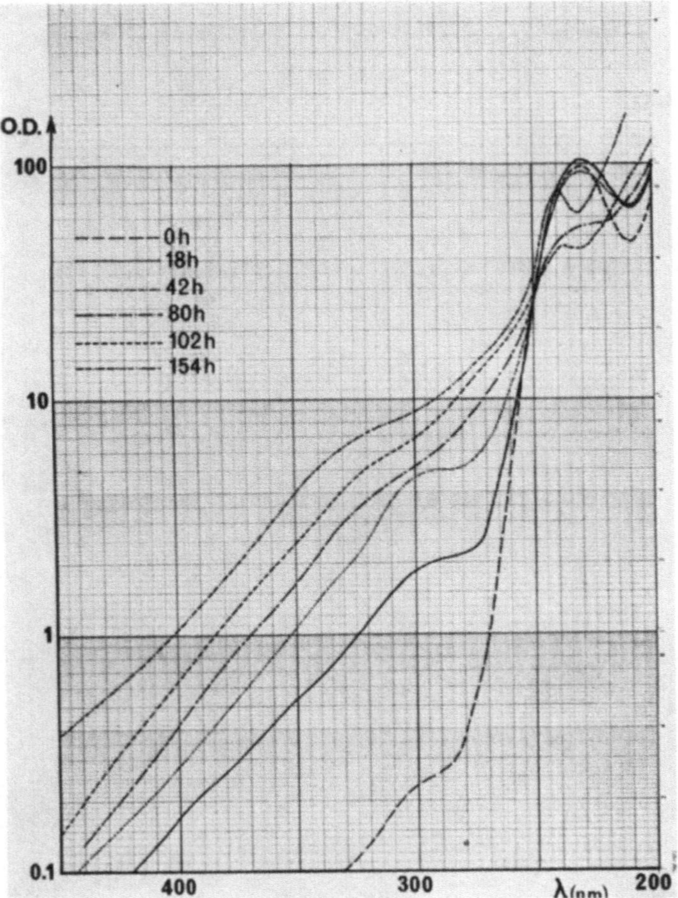

Fig. 5. Evolution of the ultraviolet and visible spectrum of the aqueous solutions, containing Na^+, K^+, Ca^{++}, and Mg^{++} and Cl^-, phosphates and SH^-, heated at 80°C in presence of an initial mixture of CH_4 and NH_3, under continuous sparking.

present. It shows that different kinds of active reductors appear in the solution and are alternatively predominant as treatment time goes on. On the other hand, no oxidizing power, at less detectable from polarographic methods, seems to appear in the solutions.

In Figure 8, the polarograms of the solutions obtained with the three kinds of mixtures at the end of their treatments are compared. Here, the most typical characteristic is the presence of the double peak of oxidation which appears in the case when iron is present at a potential very much lower than the one which corresponds to ferrocyanide ions.

The conclusion that we can draw at the present stage of development of the CERES program is that the most interesting features that are revealed by such determinations of the metabolic properties of primeval solutions are in our opinion the joint presence

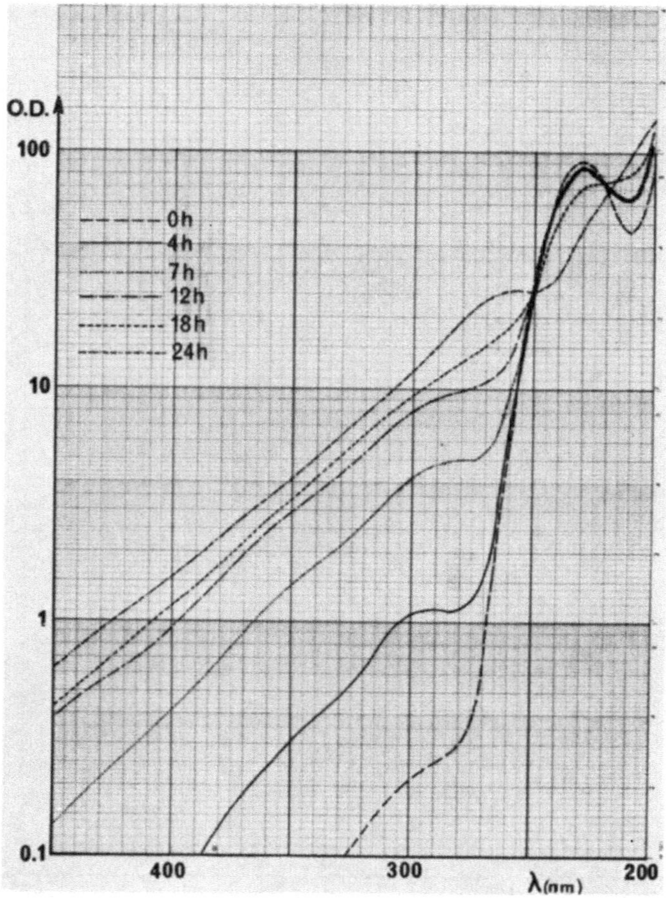

Fig. 6. Evolution of the ultraviolet and visible spectrum of the aqueous solutions, containing Na^+, K^+, Ca^{++} and Mg^{++}, and Cl^-, phosphates and SH^-, and FeS precipitated, heated at 80°C in presence of an initial mixture of CH_4 and NH_3. The sparking of the gaseous mixture and the heating of the solution were occurring only during working hours.

of strong long-wave ultraviolet absorbances and of reducing power which are probably due, at least in some cases, to the same components, thus revealing the possibility of photochemically introducing energy into the solutions in a way more or less similar, as far as its principle, to the present-day photochemical processes of absorption of solar energy which implies a primary act of endergonic departure of an electron from a photorecepting center.

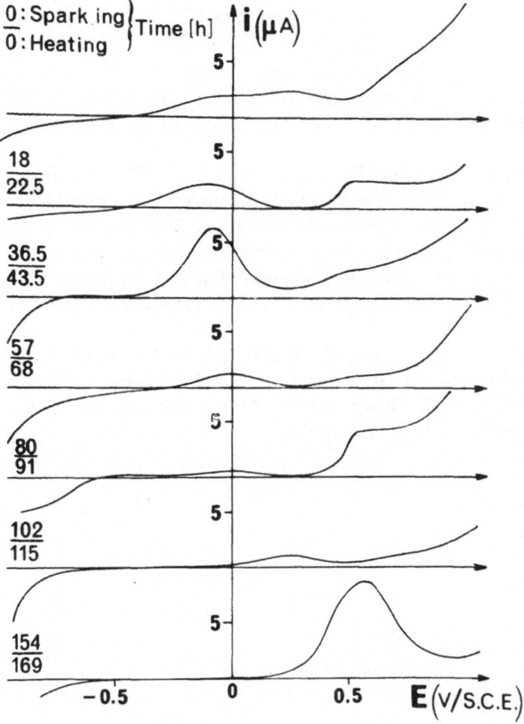

Fig. 7. Evolution of polarograms on a platinum rotating disc electrode of the aqueous solutions, containing Na^+, K^+, Ca^{++}, and Mg^{++} and Cl^-, phosphates and SH^-, heated at 80°C in presence of an initial mixture of CH_4 and NH_3, under continuous sparking.

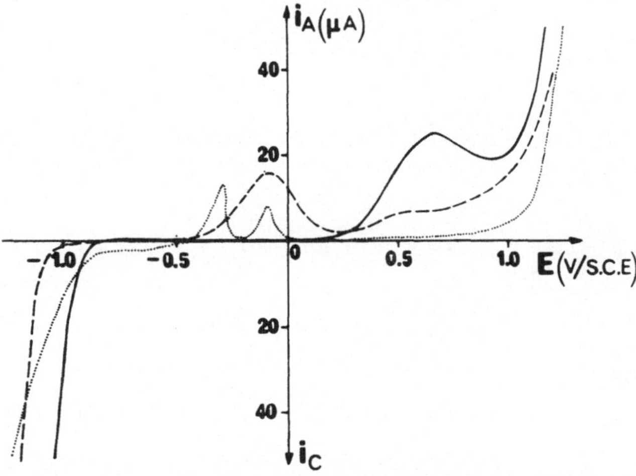

Fig. 8. Comparison of the polarograms on a platinum rotating disc electrode of solutions obtained at the end of the CERES experiments corresponding to assays described in Figures 4, 5 and 6.

— — Conditions as in Figure 4

--- Conditions as in Figure 5

..... Conditions as in Figure 6

References

Buvet, R. and Le Port, L.: 1973, 'Electrochemical Energetics of Biochemical Processes implying Storage of Redox Energy into Hydrolysable Compounds', *Second Int. Symp. on Bioelectrochemistry*, Pont-à-Mousson (F.) 1–5 Oct, 1973. To be published in *Bioelectrochemistry and Bioenergetics*, Birkhauser, 1974.

Buvet, R., Etaix, E., Godin, F., Leduc, P., and Le Port L.: 1971, in R. Buvet and C. Ponnamperuma (eds.), *Chemical Evolution and the Origin of Life*, North-Holland Publ. Co., p. 51.

Etaix, E. and Buvet, R.: 1973, 'Conditions of Occurrence for Primeval Processes of Transphophorylation', *Fourth International Conference on the Origin of Life*, Abstracts, Vol. 93.

Le Port, L. and Buvet, R.: 1973, 'Primitive Transacylation Leading to some Condensed Functions, and to Synthesis of Lipid Carbon Chains', *Fourth International Conference on the Origin of Life*, Abstracts, Vol. 93.

Le Port, L., Etaix, E., Godin, F., Leduc, P., and Buvet, R.: 1971, in R. Buvet and Ponnamperuma eds.), *Chemical Evolution and the Origin of Life*, North-Holland Publ. Co., p. 197.

Paecht-Horowitz, M.: 1971, in R. Buvet and C. Ponnamperuma (eds.), *Chemical Evolution and the Origin of Life*, North-Holland Publ. Co., p. 245.

EXPERIMENTAL ATTEMPTS FOR THE STUDY OF THE ORIGIN OF OPTICAL ACTIVITY ON EARTH

W. THIEMANN and W. DARGE*

Institut für Physik. Chemie, Kernforschungsanlage Jülich, F.R.G.

Abstract. Optical activity of natural compounds is a characteristic of our living world which is based on the asymmetry of the molecular set-up. It is hard to realize a biological cell which would be constructed from racemic compounds alone. Yet it seems attractive to ask why nature preferred only *one* of two possible enantiomers, e.g. the L-amino-acids and D-sugars. Was there or is there a chance for an antipodic biosphere constructed on the basis of the 'unnatural' enantiomers like D-amino-acids and L-sugars on Earth or elsewhere? – The paper presents in its first part a review about hypotheses that would be able to explain the apparent discrepancy between the expectation from laboratory experience and the observation that biological matter consists of extremely asymmetric molecules. The speculations found in literature are divided mainly into two categories: The first one interprets the appearance of optical activity by a chance process and its amplification by suitable means, the second one postulates a cogency leading to the chirality of the biosphere observed today. The discovery of the non-conservation of parity in nuclear physics stimulated a search for related 'asymmetry effects' in chemistry. Experiments were undertaken by some workers to construct possible laboratory models for the evolution of optical activity, but many of them failed due to different causes. On the other hand a number of papers has been published that were not directed specifically to the problem discussed, but could be interpreted on the basis of the various hypotheses. It is particularly interesting in this context to look into papers describing the crystallisation of racemates from solutions, that were published as early as 70 yr ago. – In its second part the paper deals with the study of the polymerization of racemic amino-acids as a model that would possibly allow a decision between the hypotheses for the origin of optical activity, – mere chance or a physical driving force determining the chirality of evolution. Since great care was taken to eliminate all sources of systematical errors, one expected – from the classical standpoint – racemic poly-peptides of absolute zero optical activity. – The monomer amino-acids (α-alanine, α-anino-butyric acid, and lysine) were racemized before the polymerization in order to guarantee 'ideally racemic' substrates. Polymerization was achieved via the N-carboxyanhydrides of the amino-acids. Reaction vessels and measuring cells were thoroughly cleaned with boiling chromic sulfuric acid and kept sealed from the laboratory atmosphere to prevent any contamination. The optical activity was determined in a Cary 60 spectropolarimeter calibrated to detect angles of rotations in the range of 0.5 mdeg with a maximum error of $\pm 50\,\%$. All the poly-amino-acids investigated showed negative angles of rotation at 310 nm between o.25 and 0.84 mdeg that would correspond to an hypothetical asymmetry effect – i.e. the relative difference of the polymerization constants of L- and D-amino-acids – in the order of 8×10^{-6}. We believe that this result emphasises the existence of a physical force that enables a slight accumulation of the L-amino-acids within the high molecular weight polymers in excess to the D-amino-acids and could be of significance for the evolution of the biomass. At this point the experiments do not allow any conclusion about the nature of the observed 'asymmetry effect'.

1. Statement of the Problem

The optical activity of natural compounds is a characteristic of our living world, which is based on the asymmetry of the molecular set-up. It appears that the asymmetry is a necessary attribute of the phenomenon 'life' – although it is not clear, if

* Partially extracted from the thesis of the latter author, RWTH Aachen, 1972.

life is the only system able to create optical activity! Pasteur, who discovered the effect on the crystal level, pointed already to the surprising puzzle, why nature preferred almost exclusively only the one of two possible mirror images (e.g. D-tartaric acid), while in the laboratory the two enantiomers are always synthetised in the same quantity (e.g. racemic acid). The physical reason for this extraordinary property of biological material is not at all understood even today.

The problem may be shortly stated as follows: Abundant optical activity wherever it occurs is rather safe evidence for the action of biological matter. Without its direct or indirect interaction it is highly improbable to create optical activity. Asymmetry and life are interdependent on each other; they are the result of a feed-back process. A complex system like a biological cell cannot be thought to be built up from racemic molecules; there were no secondary, tertiary, quaternary structures of proteins and no helix-shaped nucleic acids, which are all based on the dissymmetry of L-amino-acids and D-ribose. Yet, why is the biosphere constructed apparently from *only* the L-amino-acids and D-sugars? Was there or is there a chance for the biosphere equally well constructed on the basis of the 'unnatural' enantiomers like the D-amino-acids and L-sugars?

2. Speculations on the Origin of Optical Activity

From the classical point of view enantiomer molecules differ from each other in nothing but their sign of chirality, i.e. in the arrangement of atoms around an asymmetric center. They resemble each other like image to mirror image or left to right hand. Their scalar properties, such as molecular weight, density, enthalpy of formation etc. are exactly the same: they are distinguishable only by their handedness. According to experience in working with optical active material this description of enantiomerism has been fully accepted, and as long as there is no experimental evidence against it, there is no logical reason to object to this definition.

2.1. Optical activity by chance

Following this argument of mirror-like properties of enantiomer systems, a living world of left- and right-handed molecules would be equally probable. The chirality of the terrestrial biosphere, resulting from one sort of two architectural designs would then be a chance product of evolution (Wald, 1957).

The mechanism of the appearance of the first optically active substances on Earth by chance may be compared with a precipitation of enantiomer crystals from supersaturated racemic solutions, a method which is utilized widely for the technical production of optically active compounds (Ogawa, 1960). Havinga (1954) and lateron Dongorozi (1969) describe this principle in a critical review of older papers dealing with the subject. The preferential precipitation of one enantiomer from racemic solution (which leads often to high values of optical purity in one step) is caused by inhibition of the establishment of the thermodynamic equilibrium due to the lack of seeds for one of the antipodes. Formally the first asymmetric compounds could have been synthetised or precipitated in this way on the early Earth. Rare organic molecules were built up from smaller entities within a limited space – by the action of heat, electric discharges, etc. – that were important for future evolution. Incidentally they may have belonged to the configuration class L or D observed today. As soon as few of these molecules achieved a primitive capability of duplicating themselves evolution started and the chirality of our present biosphere had been fixed. A parallel spontaneous synthesis of the corresponding antipodic compounds at a somewhat later stage had no chance to initiate an evolution of its own, if the duplication process of the first dissymmetric compounds had proceeded relatively fast. The population of prebiological molecules of the first kind would have been already so large that there was no room for a self-sustaining biosphere of the enantiomer type. The whole evolution followed a yes-no decision in the very early stage (Decker, 1972; Seelig, 1971).

2.2. Optical activity by determination

The above discussed chance process could have determined the chirality of the biosphere only as long as the spontaneous formation of the mentioned 'evolutionary important molecules' occurred very rarely within geological time. An analogous crystallisation experiment may illustrate this: Optically active crystals of accidentally one or the other chirality precipitate from racemic solutions, if formation (or introduction) of seeds is rare compared with the rate of crystallisation itself. Let for instance a typical seeding frequency be of the order of 10 seeds per hour, while the rate of crystallisation be about 10^{20} unit cells of crystal lattice per hour, i.e. the ratio of rate of crystallisation to rate of seeding becomes then 10^{19}. It appears that chance becomes important for the selection of chirality of evolution only as long as the ratio of rate of 'growth' – such as polymerization, association, replication etc. – to that of 'formation' of primordial biological elements were of the order of 10^{19} again. It is a well-established fact that spontaneous synthesis of rather complicated products such as amino-acids, peptides, pyrimidines, carbohydrates etc. from simple substrates proceeds fast and with relative high yields under suitable conditions (among many others: Miller, 1955; Ponnamperuma, 1965). Therefore we believe that the rare formation of primordial biological elements can hardly be maintained as the basis for the significance of the mere chance process of evolution.

Even if we postulated that the first evolutionary important molecules must have been of a very specific poly-peptide type of a defined amino-acid sequence capable of reproducing itself, such macromolecules might yet have occurred sufficiently often within geological times. This comes to mind while investigating the structure of the 'proteinoides' and like products prepared by Fox *et al.* (1970). In contrast to the expectation of a random type polymer mixture made up by the combination of 20 essential amino-acids put into the polycondensation reaction, there appear only a few discrete groups of peptides that seem to contain rather specific amino-acids sequences. If these products were structurally similar to the mentioned evolutionary important system, their spontaneous formation would have been rather frequent, too. It follows that chance alone could not have governed the stereoselectivity of the biosphere.

Hence it seems logical to look for another physical cause that has determined the optical activity of the biomass. Several speculations have been worked out, which are classified into two main categories:

(a) An *external asymmetric physical agent or field* acting on the selection of only one antipodic biosphere.

(b) The asymmetry of the biomolecules as a result of some intrinsic property of matter, in other words the postulation of true *energy differences of enantiomer molecules*, a hypothesis which is in disagreement with the classical description of enantiomers.

Authors who wish to support the first hypothesis refer to the work of Kuhn and Braun (1929); Kuhn and Knopf (1930), who demonstrated that one may 'create' optical activity by transmitting left *or* right circular polarized light through a racemic mixture of suited substances. The circularly polarized light decomposes one enantiomer to a greater extent than the other – dependent on the direction of the polarization – and leaves the lesser affected enantiomer in excess over the more photolyzed one. Since then quite a number of principally similar experiments have been performed that confirmed the earlier works (see f.i. Moradpur *et al.*, 1971). Yet the question remained open, if there existed an excess of left or right circularly polarized light somewhere on earth that would be of any significance for the course of terrestrial evolution. Twenty-five years before the work of Kuhn and coworkers Byk (1904) and more recently Mörtberg (1971) developed models that would allow an abundance of circular polarized light under certain circumstances due to interaction of sunlight with the Earth's ocean and atmosphere.

Other authors think of the possibility that there would be an excess of left or right-handed crystals like quartz or silicotungstates or -molybdates that catalyzed the synthesis of organic molecules in a stereoselective way (Harada, 1970; Dongorozi, 1969). Others think that a priori dissymmetric seeds penetrated from the universe to the Earth in early geological times and determined the chirality of the biosphere by stereocatalysis or preferential precipitation (Oparin, 1971). Also the influence of the Earth's electric and geomagnetic fields has been discussed in the literature (Dongorozi, 1969; Morowitz, 1969).

The discovery of the non-conservation of parity by Lee and Yang (1956) and Wu *et al.* (1957) stimulated the discussion of a pre-determined evolution. The momentum and spin vectors of an emitted electron during a β-decay process are correlated with each other in such a way, that a point on the 'equator of the electron' would describe a left-handed screw around the translation coordinate. Looking at this process in a mirror, one would see an unreal image, which is not realized in nature. In this special case nature treats 'left' and 'right' not equally. Vester (1957) was the first who pointed to this analogy between β-emission and molecular structure of the biosphere. He presented a possible mechanism connecting the dissymmetric properties of a weak interaction with those of molecular structure through decomposition of racemic mixtures by bombardment with β's emitted by radionuclides.

The second different hypothesis that postulates energetic differences within enantiomer systems is, by far, less popular than the above discussed mechanisms. (Most probably this is due to the fact that the generally accepted definition of enantiomerism does not allow such an hypthesis. Overwhelming chemical experience seems to exclude this possibility.) Nevertheless the discovery of the dissymmetry of the weak-interaction leads one to question if the energetic identity of enantiomers were an axiom, which had to be upheld in any case. Could it not be conceivable that our experience with this subject is just a matter of how precise one were able to check the identity of these compounds? Very small differences of thermodynamic data like enthalpies, solubilities, melting points, specific rotations etc. of enantiomers – say on the order of 10^{-6} to 10^{-8} relative – would have hardly been detected in a direct measurement. Yamagata (1966) was the first who drew attention to a model that correlated the chirality observed in nuclear physics with that of the biopolymers and claimed a possible difference in the bond strengths of enantiomer molecules, which would manifest itself in the chirality of our biosphere. The inherent asymmetry of the nucleus would lead to an equally inherent asymmetry of chemical bonds, although this 'chemical non-parity' would be of extremely low purity due to a higher order effect, so that it could hardly be detected.

We shall try to review in the next chapter experiments directed to support one or the other of the theories discussed above, as well as some data and results reported in the literature that were not directed specifically to our problem, but seem worth noting in this context.

3. Review of Experimental Evidence of Asymmetry Effects

3.1. Experiments Directed to Prove a Given Theory of Origin of Optical Activity

Until now all experiments which were worked out to give *direct* evidence for the possible existence of asymmetry between enantiomers have failed more or less. These failures are due to the fact that if there existed any energy differences at all, they were so small that they would be measured only with the greatest effort in choosing a suitable system and the most sensitive experimental arrangement. Yamagata (1966), and also Morowitz (1969), estimate independently that such relative differences of any physico-chemical properties – including those induced by external fields, such as solubilities, melting points, extinction coefficients, enthalpies of dissociation, and bond energies – are expected to be in the range of 10^{-5} to 10^{-8} at most. The same authors suggest how one could amplify these small effects by polymerisation or precipitation, a process that had been most probably used by nature, too, to produce the biopolymers of high optical purity during evolution. Morowitz (1969) describes a very simple experiment that allows the enrichment of optical activity by a precipitation of amino-acids starting from a racemic solution that may deviate from the exact 1:1 ratio by statistical fluctuations only.

A few typical examples are given below that show clearly the inherent difficulties in the interpretation of experimental results. Campell and Garrow (1930) seem to be the first who claimed explicitly to have found differences in the physico-chemical properties of enantiomers. They investigated the solubilities of D- and L-mandelic acids in water and concluded from their measurements that the solubilities of the D- and L-compounds differred from each other up to 20% relatively. Their work was almost immediately criticized by Kortüm (1931) suspecting that their 'pure L- and D-mandelic acid' were no pure enantiomers, but were contaminated by their corresponding antipodes to different degrees. The argument was certainly justified and could not be ruled out, because the most frequently used criterion for optical purity, the constant specific or molar angle of rotation at the yellow sodium line, was rather poorly defined, as it still is today! A series of experiments was published by Ulbricht and Vester (1962) that were directed to show an asymmetric radiolysis or stereocatalytic synthesis by bombarding racemic mixtures with β-particles emitted from radionuclides. Only negative results as to the production of optical activity were reported. Garay (1968) took up similar ideas and irradiated L- and D-tyrosine separately with β-emitting radionuclides. It appeared that after 18 months both samples showed different UV-absorption patterns, so that a stereoselective decomposition should have taken place. Nevertheless the results are not fully convincing, because the separate enantiomers (instead of racemates) were treated independently and the argument of optical contamination cannot be ruled out again, besides the fact that within 18 months an absolute sterility of the samples, which is a necessary prerequisite against bacterial contamination, could not be guaranteed. Some experimental evidence supporting the hypothesis of energetic differences of enantiomers through demonstrating a difference in the solubilities of sodium-ammonium-D- and -L-tartrate, was published in a note from our laboratory (Thiemann and Wagener, 1970).

3.2. Data and results from experiments which were not specifically directed to origin of optical activity

Some arguments against the classical view of the exact identity of enantiomers come as incidental by-products of experiments that were not planned to elucidate any mechanism of the origin of optical activity. A close look into the available physico-chemical handbooks is very impressive*. They contribute substantially to the prevailing confusion in the collection of data dealing with enantiomer compounds. Whenever physical data of enantiomer substances are given, be they solubility, melting point, boiling point, viscosity, index of refraction, or specific rotation at a definite wave length, they are rarely exactly the same for the antipodes as one would expect; they differ in some cases in the second digit of the figures given. It would require too much space in this paper to list even a few of them.

Another pecularity related to the stated problem is the fact that most suppliers of chemicals offer all kinds of 'racemic' substances in their catalogues; yet very often these 'racemic' substances turn out to be not racemic in the true sense of an equimolar mixture of both enantiomers, and an excess of one enantiomer over the other reaches in some cases 7%. Among many examples, we found this with DL-lysine, DL-ephe-

* See e.g. *Handbook of Chemistry and Physics*, 1957–58 (ed. by C. Hodgman), Chemical Rubber Co., Cleveland, pp. 1652–55.

drine, DL-camphor, DL-glutamic acid and others. If those compounds were prepared 'fully synthetically' from inorganic precursors, one wonders how this excess of optical activity appears in the final product.

TABLE I

Experiments, in which a spontaneous precipitation of one antipode occurred in excess over the other one

Substance	Result	Comment of authors	Reference
$K_4(SiW_{12}O_{40})\cdot 18H_2O$	Most frequently dextrogyrous crystals		Wyrouboff (1896)
DL-ammonium-hydrogenmalate	L:D 3:1		van 't Hoff and Dawson (1898)
DL-ammonium-sodium-tartrate	(+)-crystals always in excess to (−)-crystals	Laboratory dust	Kipping and Pope (1903)
$K_4(SiW_{12}O_{40})\cdot 18H_2O$ $K_4(SiMo_{12}O_{40})\cdot 18H_2O$ $K_5(BW_{12}O_{40})\cdot 18H_2O$ $K_6(H_2W_{12}O_{40})\cdot 18H_2O$	Confirming the results observed by Wyrouboff		Copaux (1906–1912)
DL-adrenaline	At first (+)-and (−)-fraction, after recrystallisation only (+)-fractions	Laboratory dust	Darmois (1953)
DL-methy-ethyl-allyl-anilinium-iodide	(+) precipitates in excess to (−), after careful filtration excess decreases	Chance, laboratory dust	Havinga (1954)
2,4-dioxo-3,3-diethyl-5-methylpiperidine	(−) or (+) fraction recovered from technical synthesis	Spontaneous separation	Vogler and Kofler (1956)
DL-aspartic-acid-copper-complex	Always L(+)-enrichment of high optical purity	Dust?	Harada (1970)
DL-ammonium-sodium-tatrate	L(−)-enrichment in precipitate	Energy difference?	Thiemann and Wagener (1970)

Contradictory evidence appears often in crystallisation experiments, where crystals are precipitated spontaneously from racemic supersaturated solutions. In Table I a few papers are summarized, in which precipitation of the same one enantiomer always exceeds the other.

Dongorozi (1969) interprets the data from Havinga (1954), Wyrouboff (1896), and Copaux (1906–1912) as a consequence of the crystallisation of asymmetric crystals in the presence of the combined terrestrial electric and geomagnetic field. The authors, on the other hand, commenting on their experiments, explain their results – almost without exception – by an accidental seeding of laboratory dust particles that give rise to the observed preferential crystallisation of only one sort of enantiomers. Harada (1970) however did not succeed in demonstrating that typical dust particles such as wool, cotton, d- and l-quartz would influence the precipitation in a significant way.

Effects that are similar to precipitation experiments and were due to some asymmetric factors not hitherto fully understood are reported in polymerization reactions of racemic monomers. Blout and Idelson (1956) studied the rate of polymerization and degree of polymerization of γ-benzylglutamate-N-carboxy-anhydride as a function of enantiomer content. They found that the reaction rate as well as the mean molecular weight of the pure L-polymer was considerably larger than that of the pure D-polymer. The difference was about 12%, a figure which seems to be beyond the statistical error of determination. The work was aimed at another problem, namely the variation in the polymerization rates as a function of optical purity, but we believe nevertheless it important to draw attention to these results for two reasons: first because the work was done by one of the best schools in the field of polypeptide research, and second because no comment nor explanation was given at all by the authors, which could mean that there was no interpretation at hand in agreement with classical terms. Wada (1961) arrived in a paper dealing with the same system, γ-benzylglutamate, at qualitatively similar results, which had the same irregularities in common with the above cited work of Blout and Idelson (1956). Again a plot of molecular weight fraction of L- and D-isomer was apparently not symmetric about the point of 0.5 mole fraction. The polymer with the larger L-isomer concentration was of much higher molecular weight than the one with the larger D-content. The same was true for a variation of electric dipole moment with enantiomer content: the plot was again not symmetric about the 0.5 point, the effect being far beyond any limits of statistical errors. In consequence of these results appears the fact that the racemic polymer product – composed of equal amounts of L- and D-residues – shows a significant positive optical rotation dispersion that is not expected from the classical point of view. The author again offers no explanation for this surprising feature. He merely states at the end of the discussion referring to the dipole moment: "... the minimum in μ_0 does not occur at $f_D = 0.5$ within the limit of experimental error, although the figure should be symmetrical about this axis. Both the starting materials (the L- and D-isomers) are pure (99.9%), but it is quite possible that this discrepancy is due to the asymmetry in the synthetizing process of L- and D-NCA, or else to impurity...." What exactly is meant by this is left to the reader. There are more examples like these in the literature, but the above cited data are still by far the most reliable ones to our knowledge. Particularly the work of Wada seems to illustrate the discrepancy of expectation and experimental result as well as the uneasiness of the author to speak out a fact explicitly.

4. Own Work Concerning the Polymerization of Racemic Amino-Acids

Encouraged by the theoretical work and experimental results cited above, we concentrated our investigation on the question of the existence of true asymmetric factors

– intrinsic to the system studied or induced from outside – that governed reactions of racemic compounds. Precautions were undertaken in order to eliminate systematically as many sources of principal errors as possible, so that interpretation of any results obtained would be straightforward and not subject to accidentally varying laboratory parameters.

4.1. General Principle

Precipitation and polymerization of racemates, as indicated in the preceding chapter, have in common that very small differences of physical properties, which were not detectable themselves, can be amplified to such an extent, that they become measurable after reaction. The same principle has been widely used in the enrichment of stable isotopes, where small differences in vapour pressures, solubilities, complex formation, and others are multiplied by suitable techniques, so that at the end of the process the isotopes are recovered in high purity.

Let the rate of reaction of two monomer L-species to a L-dimer be

$$k_{\mathrm{L}} \quad \text{for} \quad \mathrm{L} + \mathrm{L} \rightarrow \mathrm{L}_2$$

and correspondingly for the D-species

$$k_{\mathrm{D}} \quad \text{for} \quad \mathrm{D} + \mathrm{D} \rightarrow \mathrm{D}_2.$$

Furthermore, let us suppose that $k_{\mathrm{L}}/k_{\mathrm{D}} = 1 + \varepsilon$, where ε is a very small figure (one is aware of violating the classical definition at this point explicitly, as mentioned in Section 2!). For a polymerization reaction of the type

$$n\mathrm{L} \rightarrow \mathrm{L}_n$$

and

$$n\mathrm{D} \rightarrow \mathrm{D}_n,$$

where n means degree of polymerization, it follows in a rough approximation that the reaction times t_{L} and t_{D} required to arrive at a given degree of polymerization are given by the equation

$$\frac{t_{\mathrm{D}}(n)}{t_{\mathrm{L}}(n)} = (1 + \varepsilon)^n.$$

(We assume a reversible reaction mechanism and an infinite supply of monomer for this approximation.) After some time t_{max} the degree of polymerization arrives more or less at a maximum value and the amount N of the L- and D-species within the polymer of degree n_{max} is given through

$$N_{\mathrm{L}}/N_{\mathrm{D}} = (1 + \varepsilon)^{n_{max}}.$$

It follows then that the ratio of L/D within each polymer fraction as well as the over-all distribution are specific, due to the hypothetical asymmetry effect, as shown schematically in Figure 1 – except at the intersection of both curves.

In principle the experiment to be performed along this line seems very simple: Polymerize a mixture of a racemic monomer – say any DL-amino-acids according to published procedures – let it react for long periods, in order to produce as large molecular weights as possible under given conditions, and separate the high molecular weight fractions from the unreacted monomers and low molecular weights. Thereafter the high polymers are dissolved in a suitable solvent and their optical activity measured directly in a polarimeter. Since the ratio L/D is biggest at the right end of the distribu-

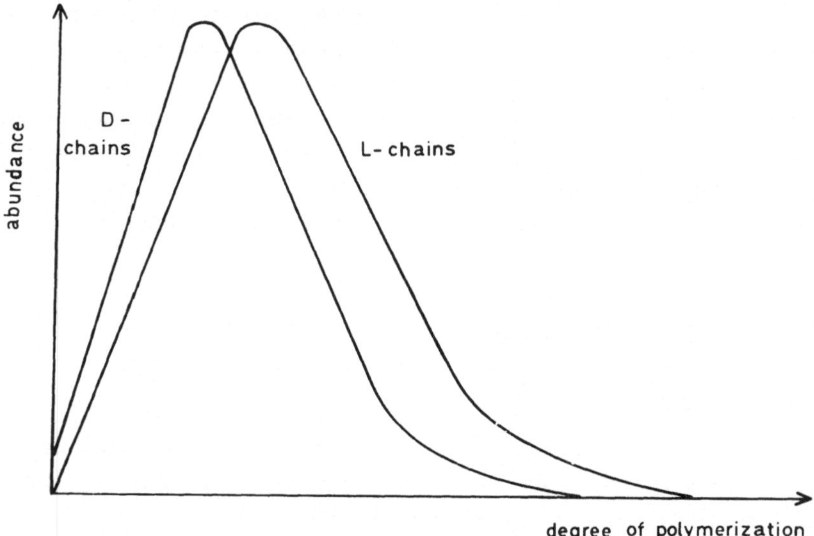

Fig. 1. Schematic plot of the distribution curves of L- and D-residues built into the polymer chains after infinite reaction times, if an asymmetry effect ε is assumed as outlined in the text. x-axis: degree of polymerization; y-axis: abundance of L- and D-residues.

tion curve (Figure 1), according to our assumption, the chance for detection of an excess of one antipode over the other – giving rise to optical activity – is the best for the investigation of the large polymers. Before starting an experiment of this kind, it is worth estimating the magnitudes of the effects to be expected and deciding about the choice of system to be studied. For this purpose we assume the hypothetical asymmetry effect ε to be about 10^{-6} and the polymerization degree n to be about 1000. The ratio N_L/N_D becomes then $(1+10^{-6})^{1000}=1.001$. If the specific rotation of the pure compound under investigation were about 100°, an apparatus capable of detecting a specific rotation of $100°/1000=0.1°$ would suffice to demonstrate the above asymmetry effect of the given magnitude.

4.2. Elimination of Possible Sources of Errors

In planning experiments of this type there arises a serious difficulty. Upon closer inspection there are only a few strictly racemic compounds available per se. We wish to define a term 'ideally racemic' as such an attribute that corresponds to a mixture of enantiomers in an exact molar ratio of 1 : 1. Because of statistical fluctuations and the hypothetical asymmetry effect that may disturb the exact 1 : 1 ratio it makes even more sense to define the term for a mixture which is in thermodynamic equilibrium, i.e. whose entropy is at maximum. The latter definition is more general. Successive addition of one antipode to the other increases the entropy up to the racemic point, at which the racemic state is the most probable one. The rate of spontaneous racemization at a given temperature is mainly determined by its characteristic activation energy. The mechanism of racemization is in most cases sufficiently well described by a first order reaction.

The simplest way to produce an ideally racemic mixture to start with would then be to accelerate the rate of racemization of an enantiomer of undefined purity by increasing the temperature of the system and to let it racemize for a given reaction time which is calculated for an allowed deviation from equilibrium from the half-life of racemization. We allowed a maximum deviation of 10^{-12} relative to the equilibrium condition, a figure that corresponds to the magnitude of statistical fluctuations of 1/100 mole of a given substance, and extrapolated the reaction times necessary to arrive at the ideal racemic state (Wagener, 1971).

Studying the polymerization of amino-acids we dissolved the commercially available so called 'racemic' α-alanine, α-aminobutyric acid, and lysine – containing an excess of up to 4% of one antipode ofer the other – in 6 N HCl to a concentration of about 0.3 mole l^{-1}. – Since reliable and systematic data on the racemization of these amino-acids were not found in the literature, we determined the half-life as a function of temperature in 6 N HCl by following the decrease of the molar rotation of the pure L-α-alanine, L-α-aminobutyric acid, and L-lysine with time. The Arrhenius plot

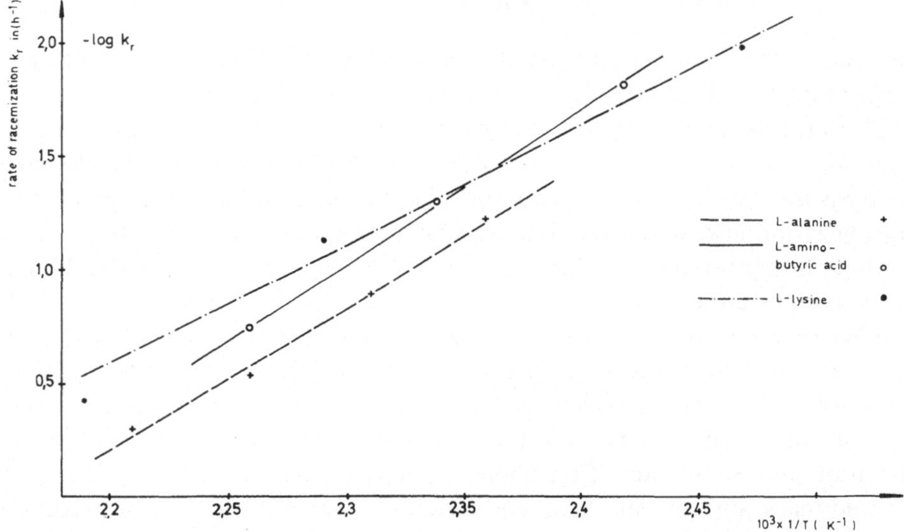

Fig. 2. Arrhenius plot for racemization reaction of L-alanine, L-aminobutyric acid, and L-lysine, x-axis: reciprocal absolute temperature; y-axis: decadic lograithm of rate constant given in reciprocal hours.

TABLE II

Rates, half-lives, and activation energies of racemization for three amino-acids in 6 N HCl

	$k(\mathrm{h}^{-1})$	$t_{1/2}(\mathrm{h})$	E_A (kcal mole^{-1})
α-alanine	0.13	5.3	28.6
α-aminobutyric acid	0.084	8.3	30.6
Lysine	0.074	9.4	24.9

for the three amino-acids is shown in Figure 2. Rate of racemization, half-life of racemization, and activation energy is calculated from Figure 2 and summarized in Table II for a temperature of 160 °C.

We sealed the solutions for racemization in thick-walled Pyrex glass round flasks and put then into a drying oven, where they were left for 16 days at a temperature of 160° ± 5 °C. Substituting the given data into the usual first order law of kinetics

$$c_t = c_0\, e^{-kt},$$

where

$c_{0,t}$ = excess concentration of the L-enantiomer over the D-enantiomer at time 0 and t,

t = reaction time,

one gets for

$$t = 16 \text{ days a ratio of}$$

$$c_t/c_0 = 2.1\ 10^{-22} \quad \text{for α-alanine}$$

$$= 1.4\ 10^{-14} \quad \text{for α-aminobutyric acid}$$

$$= 4.5\ 10^{-13} \quad \text{for lysine.}$$

Recovering the amino-acids from the hydrochloric solutions by evaporization of the HCl to dryness, extraction with benzene and recrystallisation, one obtains monomer substrates for polymerization that meet the standard of the above defined 'ideal racemic state'. (It has to be emphasized that the above result cannot be checked by direct measurement of optical rotation or other parameters, but is extrapolated from the racemization law, which was followed for only shorter time periods. We do not think there is any reason why this extrapolation of the first order law should not be valid for long reaction times.)

Another severe source of error is the requirement of absolute sterility in experiments of the kind described here. A contamination of the system with any bacteria or yeasts and subsequent growth in media that offer optimal nutrition conditions such as aqueous amino-acid solutions would of course lead at once to stereoselective decomposition and production of metabolic chemicals. These would be optically active and would mask any smaller effects completely. To avoid this source of misinterpretation we took great care to clean all reaction vessels and cells with hot chromic sulfuric acid and to wash them with 3 × quartz distilled water. For all reactions and measurements we used exclusively solvents that were most unfavourable to bacterial growth, such as 6 N HCl, dioxane, benzene, trifluoroacetic acid. As to our knowledge there would be not the smallest chance for any biological organism to survive and multiplicate under these conditions.

Again there is the argument of possible accidental contamination with organic or inorganic laboratory dust, which could catalyze the polymerization stereospecifically. Although we know of no paper showing a definite positive influence of any seeds towards the stereoselection of polymers, we took precautions against this phenomenon by oxidizing any carbon containing seeds to CO_2 by the mentioned hot chromic sulfuric acid treatment, by filtering all solutions through the finest glass sinter plates, and by working in closed systems.

Although the idea of *absolutely* dust-free laboratory conditions seems very attractive for future work in this context, one might question the necessity of the tremendous effort of establishing this after studying the paper of Harada (1970), who did not succeed in showing any significant influence of dust on precipitations at all.

4.3. Calibration of the Apparatus for Detecting Small Angles of Rotation

In investigating optically active samples of extremely low optical purity, it is of course of utmost importance to which limit of sensitivity one can trace small angles of rotation. For this purpose we used a Cary 60 spectropolarimeter connected to an A/D-converter, a digital voltmeter, and a tape punch. The data were fed into an IBM-360 computer that allowed us to average the individual data, taken at constant wavelength in the UV-region of the spectrum, and test each set of data for statistical purity. In case they did not obey a normal distribution function, the whole series of measurement was discarded and repeated until any systematic errors due to electric shift, migration of the Xenon arc in the light source, or chemical or optical impurities in the cell were eliminated. Great care was taken to filter the sample solutions, to keep the cylindrical quartz cells free of mechanical stresses, and to allow a few hours' period for warming up the whole measuring device. The effect of warming up as a function of time after starting the high pressure Xe-lamp is shown in Table III.

Each sample was measured for about 15 min, the voltage output recorded every second, and its mean value compared to the value of the inactive solvent, taken by

TABLE III

'Warming-up' of the apparatus

Time after starting (h)	'Angle of rotation' of an inactive solvent at 310 nm $\alpha(m^\circ)$
0	-4.8
0.25	-5.0
0.5	-2.8
1.0	-1.0
2.0	-0.2
3.0	-0.05
4.0	-0.00

definition as standard of exactly zero degree and measured again for 15 min after each unknown sample. Each sample was repeatedly measured according to this procedure at least 10 times.

Three parameters determine the wavelength at which the sensitivity for detection of optical activity is the highest: 1) The molar rotation should be as large as possible, which is the case in the extrema of the Cotton effect of the ORD occurring in the UV-region near the absorption band of aliphatic compounds. 2) Absorption of light should be as small as possible, so that electric amplification is kept low in order to reduce the noise level. 3) Absorption of sample and inactive solvent should be the same, because high absorption may cause some electrically recorded 'pseudorotation' (Hayatsu, 1965).

This was found the best compromise to keep down any systematic errors which were caused by the mentioned electric drift of photomultiplier characteristics, variation of the input voltage, mechanical and thermal stresses of cell due to the rinsing and refilling process, appearance of dust particles or schlieren in the pathway of the light beam, and instability of the Xe-arc due to burning-out of the electrodes after a few hundred running hours. Table IVa gives a typical test for reproducibility of measurements of small angle of rotations.

The sample was discarded and the cell refilled with a second aliquot of the same stock solution, values are given in Table IVb.

A third aliquot gave Table IVc.

In the following test we calibrated the device in the most sensitive range by follow-

TABLE IVA

Test for reproducibility of measuring small rotations

Time (h)	Angle of rotation (m°)	Number of digital data
0	8.27 ± 0.02	417
0.5	8.23 ± 0.09	432
1.0	8.13 ± 0.02	276

Mean 8.21 ± 0.07 m°

TABLE IVB

Repetition of test for reproducibility

Time (h)	Angle of rotation (m°)	Number of digital data
0	8.68 ± 0.09	382
0.5	8.73 ± 0.02	335
1.0	8.63 ± 0.01	514

Mean 8.68 ± 0.05 m°

TABLE IVC

Repetition of test for reproducibility

Time (h)	Angle of rotation (m°)	Number of digital data
0	8.28 ± 0.04	435
0.5	8.15 ± 0.07	278
1.0	8.35 ± 0.06	358
1.5	8.13 ± 0.03	329
2.0	8.24 ± 0.11	329

Mean 8.23 ± 0.09 m°

ing the rotation versus concentration of an optically active sample. For this purpose we dissolved l-ephedrine-hydrochloride in water to 0.05 g l^{-1}, prepared samples in steps of 0.01 g l^{-1} from this stock down to 0 and measured the angle of rotation at 290 nm. In Figure 3 the circles represent the experimental values, the dashed line gives the calculated best fit through them, and the full line gives the theoretical curve as expected from the specific angle of rotation of l-ephedrine·HCl in water. We calculate a statistical error over the whole range from 0 to 1 m° of less than ±15%, an absolute error from the theoretical curve of about +20%.

Thus we are able to determine angles of rotations in the UV-region of the ORD around 0.5 m° with an accuracy bettter than ±50%, around 1 m° better than ±25%, and above 5 m° with an accuracy better than ±10%.

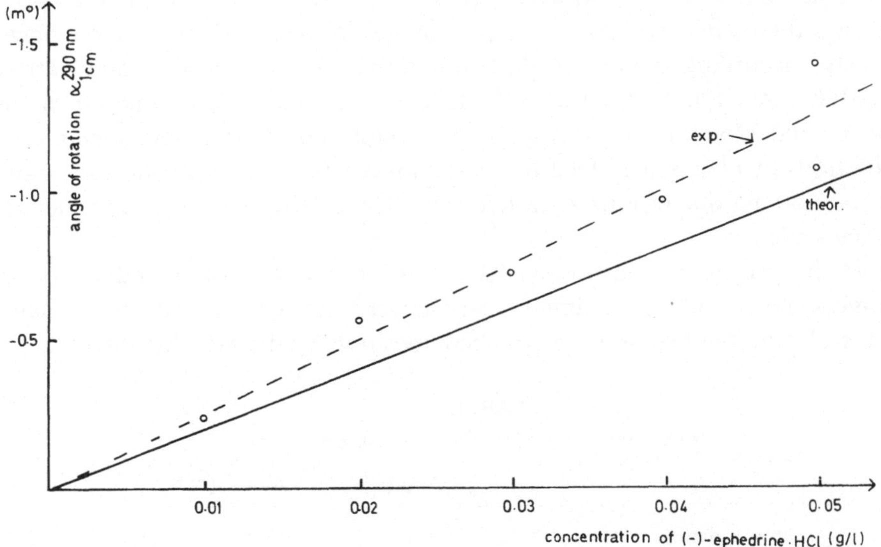

Fig. 3. Calibration of the method for detecting small angles of rotations. x-axis: concentration of (−)-ephedrine. HCl in g/l water; y-axis: angle of rotation determined at wavelength 290 nm, temperature 25°C, and path length 1 cm in (m°). The dashed line gives the best fit through experimental points, the full line gives the theoretical curve calculated from the specific angle of the substance.

4.4. Experimental results

The N-carboxy-anhydrides (NCA) of α-alanine, α-aminobutyric acid, and ε-carbobenzoxylysine were prepared by the Fuchs-Farthing-method, in which the amino-acids are treated with dry phosgene

$$\begin{array}{c} R \\ | \\ H_2N-CH-COOH \end{array} + COCl_2 \rightarrow \begin{array}{c} R \\ | \\ H-N-CH-C=O \\ | \qquad\quad | \\ O=C \text{———} O \end{array} + 2\,HCl.$$

ε-carbobenzoxy-lysine had to be prepared from the racemic lysine in order to mask the second ε-amino-group by the reaction of carbobenzoxychloride with the copper complex of lysine (Neuberger and Sanger, 1943). The mechanism of polymerization of the amino-acids via the NCA's, as proposed by Leuchs (1906–1908), is rather complex and can be summarized in the reaction scheme

$$n \begin{array}{c} R \\ | \\ HN-CH-C=O \\ | \qquad\quad | \\ O=C \text{———} O \end{array} \xrightarrow{\text{(Na-methylate)}} \begin{array}{c} R \\ | \\ (HN-CH-CO)_n \end{array} + n\,CO_2$$

In the final step of polymerization 10 g of each of the ideal racemic anhydrides were dissolved in 1 l dioxane and 2 ml 0.1 M sodiummethylate in methanol added under rigorous stirring. The solutions were stirred for 6–8 weeks at room temperature. After some days there appeared gelatinous precipitates in the solutions, which represented obviously – according to literature data (Bamford *et al.*, 1954; Shalitin and Katchalski, 1960; Stahmann, 1962) – the least soluble high molecular weight homopolymerisates of the L- and D-amino-acids, respectively. These precipitates were separated from the solutions by filtration and washed with dioxane and dichloroacetic acid until only about 60–150 mg dry residue were left over. These polymers were characterized by three methods:

(a) Hydrolysis with concentrated HCl at 100°C for 40 h and identification of the monomers on an automatic amino-acid-analyzer showed that not more than 16% were lost during the reaction, due to decomposition by the HCl treatment.

TABLE V

N-determination of synthetic poly-amino-acids

	N_{exp}(%)	N-$_{\text{theor}}$(%)
Poly-α-alanine	19.1	19.7
Poly-α-aminobutyric acid	17.1	16.4
Poly-ε-CbO-lysine	10.0	10.7
Copolymer of α-alanine and α-aminobutyric acid 1:1	17.1	17.9

(b) Nitrogen was determined by reaction with sulfuric acid at 250–300 °C and subsequent analysis of the produced ammonia. Table V gives the experimental and theoretical N-percentages for the polymers.

(c) Their molecular weight was determined by viscosimetry. The intrinsic viscosities of solutions of 0.2 g polymer in 100 ml trifluoroacetic acid (TFA) were compared to these of standards of known molecular weights in the same solvent. We used as standards poly-DL-alanine of $M=1600$ and poly-L-alanine of $M=2675$ supplied from Miles Research Laboratories. The intrinsic viscosity and molecular weight are related to each other by the equation

$$[\eta]=K\cdot M^{a}$$

The constants K and a were taken from the standards dissolved in TFA as $1.1\ 10^{-3}$ and 0.83, resp. With these parameters substituted into the above equation and similar behaviour assumed for poly-α-alanine and poly-α-aminobutyric acid for lack of a suitable standard we found for the molecular weights

TABLE VI

Molecular weights of synthetic poly-amino-acids

	M	
Poly-α-alanine	12000	
Poly-α-aminobutyric acid	a: 18000;	b: 24000
Copolymer alanine and α-aminobutyric acid	15000	

The molecular weight for ε-carbobenzoxy-lysine was estimated from viscosimetric analysis in dichloroacetic acid as about 30000 by comparison with data taken from Applequist and Doty (1962).

Table VII summarizes the final optical rotation measurements at 310 nm performed the polymers dissolved in TFA, together with the statistical error, concentration, and specific rotations.

TABLE VII

Optical rotations of synthetic poly-amino-acids

Compound	Angle of rotation (m°), 25 °C, 310 nm	Concentration (g/100 ml TFA)	Specific rotation
Poly-α-alanine	-0.84 ± 0.24	0.824	$-1.0\pm29\%$
Poly-α-aba I[a]	-0.51 ± 0.27	0.302	$-1.7\pm53\%$
Poly-α-aba II[b]	-0.63 ± 0.30	0.417	$-1.5\pm48\%$
Poly-α-aba I[a]	-0.56 ± 0.26	0.302	$-1.9\pm46\%$
Poly-ε-CbO-lys	-0.25 ± 0.13	0.503	$-0.5\pm52\%$
Copolymer, ala: aba = 1:1	-0.74 ± 0.31	0.870	$-0.9\pm42\%$

[a] poly-α-aba I stands for poly-α-aminobutyric acid, preparation I.

[b] poly-α-aba II stands for poly-α-aminobutyric acid, preparation II.

For the calculation of the hypothetical asymmetry effect a relation between the observed rotation of the racemic polymer and that of a pure enantiomer homopolymer has to be established. Therefore the standard, poly-L-alanine ($M=2675$, $n=37$), was dissolved in TFA and its ORD recorded as given in Figure 4, together with the ORD of monomer L-alanine for comparison. From this a molar rotation $[\phi]$ of poly-L-

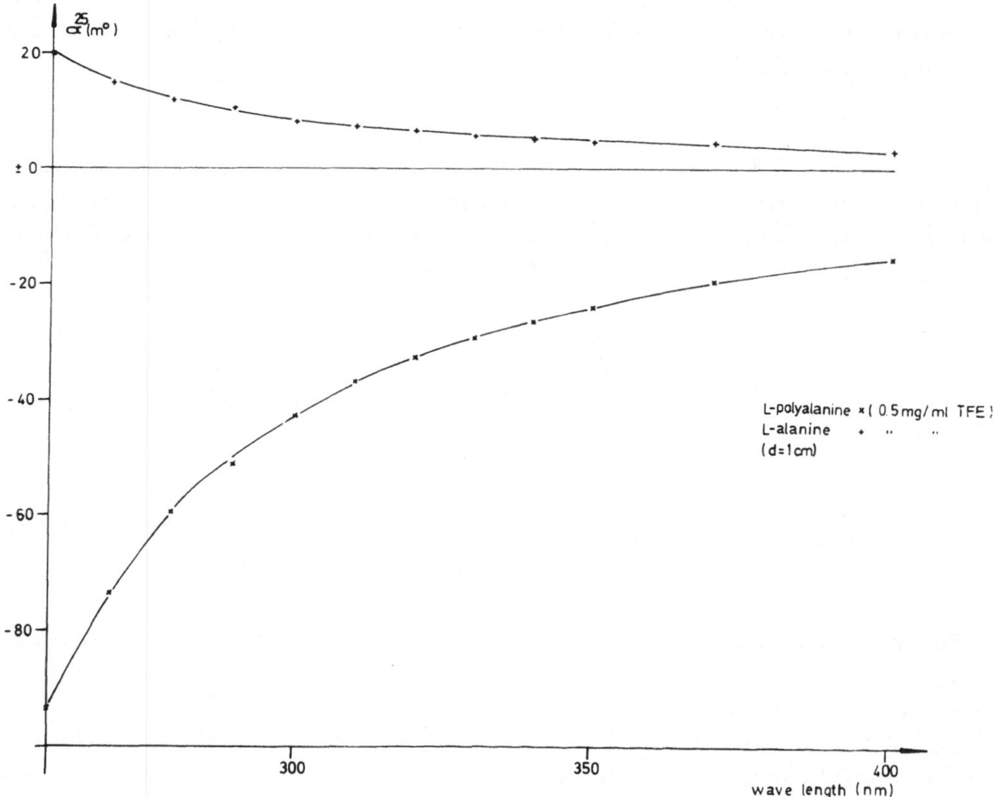

Fig. 4. ORD of poly-L-alanine and L-alanine in trifluoroacetic acid (TFA) between 250 and 400 nm, concentration of both 0.5 mg cm^{-3}. x-axis: wave length in nm; y-axis: angle of rotation in millidegree (m°), path length 1 cm.

alanine at 310 nm and 25 °C of $-20060°$ and a reduced molar rotation $[m']$ for the single monomer within the polymer of $-20060°/37=-542°$ is evaluated. For the racemic poly-DL-alanine we find a molar rotation $[\phi]$ from Table VII of $-120°$ and a reduced molar rotation $[m']$ of $-0.71°$. The asymmetry effect is then calculated through the equation

$$Q = 1 + \frac{[m']_{\text{obs, poly DL ala}}}{[m']_{\text{standard, poly ala}}} \approx (1 + \varepsilon)^n$$

Substituting the above values one gets

$$\varepsilon \approx \frac{[m']_{\text{obs, poly DL alo}}}{[m']_{\text{standard, poly ala}\cdot n}} = \frac{-0.71}{-542 \times 170} = 7.7 \times 10^{-6}.$$

Although exact data are missing on the optical rotations of the other pure poly-L-amino-acids in TFE, they will certainly be of the same order of magnitude as in the case of poly-L-alanine, so that similar relations between observed rotations, standard rotations, and degrees of polymerization will lead to similar asymmetry effects as the one calculated for the case of poly-alanine.

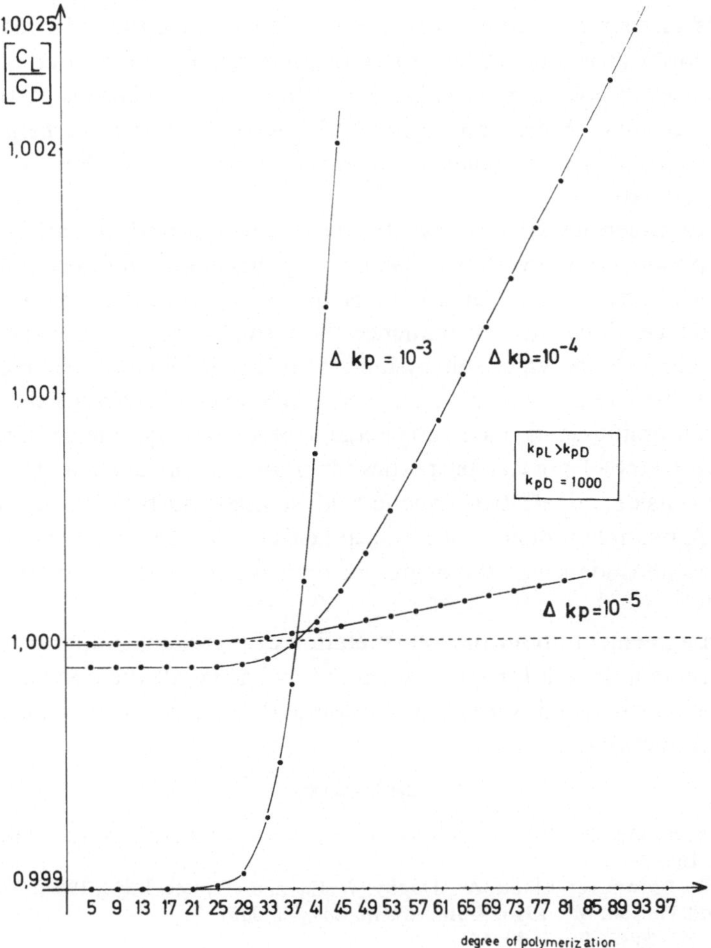

Fig. 5. The theoretical ratio of concentration of L- to D-residues in the polymer chains as a function of the degree of polymerization. The rate of polymerization of D-residues k_{pD} was taken arbitrarily as 1000, while that of L-residues k_{pL} varied between 1000.01 and 1001; the rate of activation of monomer was set to 1. x-axis: degree of polymerization; y-axis: concentration of L-residues to concentration of D-residues.

4.5. Discussion

One might argue, if the very simple relation between an elementary asymmetry effect and its amplification in the polymerization process up to the observed enrichments – as given in 4.1 – is valid at all. A more exact theoretical approach following principally the ideas proposed by Mark and Dostal (1935) leads to the same results for large degrees of polymerization. The reaction is controlled by the ratio of the rate of formation of 'activated monomers' to the rate of polymer growth. Assuming here a constant rate of formation of activated monomers (dependent only on the amount of initiator added) and rate of growth, which depend specifically on the type of monomers – L or D – and differ from each other by the above discussed very small factor ε, the results of such a treatment can be summarized in a plot of concentration of L- to D-polymers as a function of degree of polymerization for three theoretical relative differences in growth rates (Figure 5). The exact calculations according to the modified procedure of Mark and Dostal (1935) were done with the help of an IBM-360 computer. By this more realistic theoretical approach the above made assumptions seem justified.

We want to underline the fact that the enrichments of optical activity observed in the racemic polymers were without exception significant and undirectional. Assuming the exclusion of any basic systematic error in the experimental set-up, we conclude that mere chance alone did not influence the results. There must exist some other factor controlling the polymerizing system. At this point it is still completely open to discussion, if this factor is intrinsic, i.e. a true difference of rates of reaction between the enantiomer amino-acids due to an energetic or kinetic asymmetry effect correlated with the asymmetry of nuclear properties, or if any factors act onto the enantiomer system from outside, i.e. electromagnetic field, sunlight, surface tension, or combinations of all. A possible influence of physical fields could be studied in future work by observing any dependence of the degree of optical enrichment on variations of any applied fields.

Even if an agreement about the interpretation of these and similar results could be achieved, it remains a task for more research to what extent the polymerization reactions of the kind discussed were an important and necessary step in the evolution of the terrestrial biomolecules.

References

Applequist, J. and Doty, P.: 1962, in *Polyaminoacids, Polypeptides and Proteins*, Univ. Wisc. Press, Madison, p. 165.

Bamford, C. H., Brown, L., Elliott, A., Hanby, W. E., and Trotter, I. F.: 1954, *Nature* **173**, 27.

Blout, E. R. and Idelson, M.: 1956, *J. Am. Chem. Soc.* **78**, 3857.

Byk, A.: 1904, *Z. Phys. Chemie* **49**, 641.

Campbell, A. N. and Garrow, F. C.: 1930, *Trans. Farad. Soc.* **26**, 560.

Copaux, H.: 1906, *Ann. Chim. Phys.* **(8) 7**, 129.

Copaux, H.: 1909a, *Ann. Chim. Phys.* **(8) 17**, 234.

Copaux, H.: 1909b, *Compt. Rend. Acad. Sci.* **143**, 633.

Copaux, H.: 1910a, *Compt. Rend. Acad. Sci.* **150**, 475.

Copaux, H.: 1910b, *Bull. Soc. Min.* **33**, 167.

Copaux, H.: 1912, *Z. Kryst.* **50**, 317.
Darmois, E.: 1952, *Compt. Rend. Acad. Sci.* **237**, 124.
Decker, P.: 1972, *Z. Naturf.* **27b**, 257.
Dongorozi, C. S.: 1969, *Rev. Roum. Biochim.* **6**, 297.
Fox, S. W., Harada, K., Krampitz, G., and Mueller, G.: 1970, *Chem. Eng. Ind. News* **48**, 80.
Garay, A. S.: 1968, *Nature* **219**, 338.
Harada, K.: 1970, *Naturwiss.* **57**, 114.
Havinga, E.: 1954, *Biochim. Biophys. Acta* **13**, 171.
Hayatsu, R.: 1965, *Science* **149**, 443.
Kipping, F. S. and Pope, W. J.: 1903, *J. Chem. Soc.* **95**, 103.
Kortüm, G.: 1931, *Ber. dtsch. Chem. Ges.* **64**, 1506.
Kuhn, W. and Braun, E.: 1929, *Naturwiss.* **17**, 227.
Kuhn, W. and Knopf, E.: 1930, *Naturwiss.* **18**, 183.
Lee, T. D. and Yang, C. N.: 1956, *Phys. Rev.* **104**, 254.
Leuchs, H.: 1906, *Ber. dtsch. Chem. Ges.* **39**, 857.
Leuchs, H. and Manasse, W.: 1907, *Ber dtsch. Chem. Ges.* **40**, 3235.
Leuchs, H. and Geiger, W.: 1908, *Ber. dtsch. Chem. Ges.* **41**, 1721.
Mark, H. and Dostal, H.: 1935, *Z. Phys. Chem.* **B29**, 299.
Miller, S. L.: 1955, *J. Am. Chem. Soc.* **77**, 2351.
Moradpur, A., Nicoud, J. F., Balavoine, G., Kagan, H., and Tsoucaris, G.: 1971, *J. Am. Chem. Soc.* **93**, 2353.
Morowitz, H.: 1969, *J. Theor. Biol.* **25**, 491.
Mörtberg, L.: 1971, *Nature* **232**, 105.
Neuberger, A. N. and Sanger, F.: 1943, *Biochem. J.* **37**, 515.
Ogawa, T.: 1960, U.S. Patent **2**, 940, 998.
Oparin, A.: 1971, *Ideen exakt. Wiss.* **1**, 57.
Ponnamperuma, C.: 1965, *Science* **1**, 39.
Seelig, F. F.: 1971, *J. Theor. Biol.* **31**, 355.
Shalitin, Y. and Katchalski, E.: 1960, *J. Am. Chem. Soc.* **82**, 1630.
Stahmann, M. A. (ed.): 1962, *Polyaminoacids, Polypeptides, Proteins*, Univ. Wisc. Press, Madison, p. 8.
Thiemann, W. and Wagener, K.: 1970, *Angew. Chem.* (Int. ed.) **9**, 740.
Ulbricht, T. L. V. and Vester, F.: 1962, *Tetrahedron* **18**, 629.
van't Hoff, J. H. and Dawson, T. M.: 1898, *Ber. dtsch. Chem. Ges.* **31**, 528.
Vester, F.: 1957, Seminar at Yale University, Feb. 7, 1957.
Vogler, K. and Kofler, M.: 1956, *Helv. Chim. Acta* **39**, 1387.
Wada, A.: 1961, *J. Mol. Biol.* **3**, 507.
Wagener, K.: 1971, Private Communication.
Wald, G.: 1957, *Ann. N.Y. Acad. Sci.* **69**, 352.
Wu, C. S., Ambler, E., Hayward, R. W., Hoppes, D. D., and Hudson, R. P.: 1957, *Phys. Rev.* **105**, 1413.
Wyrouboff, G.: 1896, *Bull. Soc. Min.* **19**, 219.
Yamagata, Y.: 1966, *J. Theor. Biol.* **11**, 495.

LIFE'S BEGINNINGS – ORIGIN OR EVOLUTION?

JOHN KEOSIAN

Marine Biological Laboratory, Woods Hole, Mass. 02543, U.S.A.

Abstract. The alleged disproof of the concept of spontaneous generation is examined. The gene theory and the prebiological systems theory of the origin of life are evaluated. The former is considered not feasible. The latter may be feasible but present experimentation tends more to demonstrate the possibility of neobiogenesis (present-day re-origin of life) than of biopoesis (fiirst origin of life). The uniqueness of biopoesis is questioned as is the heterotroph hypothesis. An alternative hypothesis is attempted.

The question is a fundamental one. Did inanimate matter give rise to life more or less suddenly, or did life gradually evolve through stages from inanimate matter? In the first case, a precise definition of life is of paramount importance for it becomes both the experimental goal and the identifying criterion for recognizing life produced experimentally. In the second case neither definitions nor criteria of life are of particular importance for there is no longer any single goal. Rather, the objective is a quest, first, for spontaneously forming multimolecular systems, secondly, a study of their properties, and finally, the determination of the mechanisms involved in driving each successive stage to a higher level of organization. In such a progression it becomes meaningless to draw a line between two levels of organization and to designate all systems below as inanimate and all systems above as living. The meaninglessness of the terms life and living on other grounds was long ago pointed out by Pirie (1937).

Today, the search for origins has narrowed to two main lines of approach. In one, the gene is held to be the basis of life and its origin. In the other, microsystems variously termed microdroplets, coacervates, microspheres or prebiological systems are considered to be the precursors of the first primitive cells. Both approaches, however, assume that the first living things, genes or cells, arose more or less suddenly from their environment. The background and shortcomings of each are discussed below.

The gene theory was inspired by Troland's 'living enzyme' theory (1914, 1916, 1917). His ideas were remarkable for those times for Troland, a Harvard physicist, took a strong stand against his contemporary biologists who had embraced a vitalistic-supernatural-outlook on the subject of life. Troland wrote in part:

> It is the purpose of the present paper to combat the thesis of the new vitalism by showing how a single physico-chemical conception may be employed in the rational explanation of the very life phenomena which the neo-vitalists regard as inexplicable on any but mystical grounds.

Troland put the 'single physico-chemical conception' in these words:

> Let us suppose that at a certain moment in earth history, when the oceans are yet warm, there suddenly appears at a definite point in the oceanic body a small amount of a certain catalyzer or enzyme.... The original enzyme was the outcome of a chemical reaction, that is to say, it must have depended on

> the collision and combination of separate atoms or molecules, and it is a fact well known among physicists and chemists that the occurrence and specific nature of such collisions can be predicted only by the use of the so-called *laws of chance*.... Consequently we are forced to say that the production of the original life enzyme was a chance event.... The striking fact that the enzymic theory of life's origin, as we have outlined it, necessitates the production of only a single molecule of the original catalyst, renders the objection of improbability almost absurd... and when one of these enzymes first appeared, bare of all body, in the aboriginal seas it followed as a consequence of its characteristic regulative nature that the phenomenon of life came too.

Troland considered the normal function of enzymes to be heterocatalysis (metabolic regulation). For his hypothetical 'life enzyme' he proposed the additional property of autocatalysis (self replication). He further proposed that these two properties together constituted the minimum criteria of life. Anything embodying these two properties could be considered alive. He looked upon the 'life enzyme', a protein, as living macromolecule, eventually known as a 'moleculobiont'.

Troland's theory has serious flaws in common with the derivative gene theory. The first detailed statement of the gene theory of life, obviously based on Troland's enzyme theory, was made by Muller (1926, 1929). Muller substituted the 'naked gene' for Troland's 'enzyme bare of all body' and to Troland's two criteria of life, autocatalys and heterocatalysis, he added mutability. The gene theory like Troland's enzyme theory, is untestable for the reason that a meaningful experiment cannot be devised to test an event that depends on remote accidents. But an accidental combination of all of the necessary atoms or molecules to form a life enzyme, or of all of the necessary nucleotides to form a gene, is not a probable event because the synthesis of only a single molecule is postulated. It is an event of vanishing probability. Moreover, a life enzyme or a gene is alive only because the authors so argue.

But granting for the sake of argument, the all-at-once appearance of a gene on the primitive Earth, what are the possibilities of its serving as the ancestor of all life? Ignoring all limitations, it would serve as a template for the synthesis of more of itself at an exponential rate; it would mutate and mutant genes would similarly increase in number and would themselves mutate; the original gene and each mutant variety would be responsible for a particular protein; and, finally, each protein would be specific for some structural or catalytic role. Trying to stay within the confines of these events, it is difficult to see what else would result beside the synthesis, in a short time, of an oceanful of the original gene, lesser quantities of various mutants, and amounts of the corresponding proteins. All of this assumes that the first gene had properties identical with those of modern genes, and overlooks the requirement for specific enzymes in the replication of DNA, as well as an apparatus, a variety of RNAs, and enzymes for the synthesis of proteins. Not insignificantly, the experimental evidence so far for the formation of polynucleotides under prebiological conditions is weak, molecular weights and yields are small, and properties in terms of the goal are lacking.

As long as very little was known about biochemistry, and nothing at all about prebiotic chemical evolution, one could fill the void with broad and logical sounding assertions. It is no longer permissible to dismiss the problem of how a single molecule whether enzyme, hypothetical gene or coded DNA can serve as the ancestor of all life.

The microsystems approach is the one taken by Oparin *et al.* (1936, 1957, 1964), Fox *et al.* (1969, 1972), and others doing experiments on models of prebiological systems. These models are claimed to have many morphological and functional characteristics in common with contemporary cells, including primitive forms of multiplication (Fox, 1969). A serious drawback to considering such models as truly representative of prebiological systems is that they are formed and made to operate under present-day conditions and sometimes with biochemicals of contemporary origin. Also, once microsystems are constructed, they are subjected to various biochemicals and conditions calculated to induce activities known to be essential in some contemporary 'primitive' cell. In such endeavors modern cellular physiology and biochemistry strongly influence the investigators' experimentation. It is an open question whether identical models will form spontaneously and function similarly in solutions exposed to Methane, Ammonia, Hydrogen Cyanide, Hydrogen Sulfide and other gases of the primordial atmosphere with its high flux of short wave ultraviolet rays and ionizing radiation. One can indeed argue that the present experimentation on microsystems is as much an attempt to demonstrate the possibility of neobiogenesis (continuing reorigin of life) as it is of biopoesis (the first origin of life).

The belief in an early and sudden appearance of life under prebiological conditions seems to be inspired more by an overriding desire to rush a living thing onto the scene than by the dictates of the conditions of the primordial Earth. Nevertheless, it is assumed that living things did arise under those conditions. It is further commonly assumed that the first living things (presumed to be similar to present-day 'primitive' heterotrophs) would be destroyed by those conditions especially the high flux of short wave ultraviolet rays. This difficulty is belittled by proposing that the first living things survived in caves or under depths of water not reached by the lethal rays, there to remain until more suitable conditions came about (millions or tens of millions of years?).

The need for such assumptions is strong evidence, it would seem, that what did originate under the harsh conditions were not living things, but a succession of microscopic physico-chemical systems of increasing complexity. The early members in that long line of systems did not need the full complement of biochemicals postulated for the 'hot thin soup'. The same reactants, reactions and forces which together produced organic compounds in the atmosphere and waters, did so, also, in the microsystems. Their chemistry was, in that sense, autotrophic, not heterotrophic. Further, each stage in that progression must be considered to have been stable to the conditions prevailing at the time of its appearance. Changes in the chemical and physical conditions of the environment impressed corresponding changes in the microsystems due to the relatively simple structure and chemistry of those early systems. Homeostasis is a property of complex systems possessing a multiplicity of alternate pathways, a condition not yet characteristic of the early microsystems. Harmony between those systems and environmental conditions (adaptation) was the result of direct, non-Darwinian, interaction.

The chemistry of a microsystem confined by a semipermeable membrane would soon take on a character different from that in the vast environment which tends toward homogeneity. For one, products of reactions can accumulate beyond their solubility products and can precipitate out as solid phases, whereas this may never occur in the outside medium. Likewise, reactants confined in small spaces can soon reach their reaction levels, whereas their dissipation in the outside medium would prevent it. The formation of solid phases within the microsystems would favor the coupling of reactions especially important in the case of exergonic and endergonic reactions. The latter possibility is significant in the eventual change-over from physical energy outside the microsystem, to chemical energy inside it, for the synthesis of compounds. Under such circumstances, the further coupling of reactions into pathways might be favored. *The synthesis of biochemical compounds, and the development of coupled reactions, and of biochemical pathways, must be considered to have been a property of microsystems and not of the structureless medium.*

Considering the primitive method of multiplication of microsystems, the retention of an increasingly complex structure and chemistry by successive generations must have been precarious. This would, in turn, require a long period of nongenetic evolution of microsystems. Microfossils said to be two and a half to three billion years old, may represent various stages in the evolution of microsystems. Structurally, the oldest microfossils have a closer resemblance to clusters of some of Fox's microspheres than their claimed resemblance to primitive algae. The nature of the strata in which those microfossils are found are said to be a characteristic of once-living algae. However, it may be maintained, as well, that such effects were produced by the chemical activities of the more complex microsystems of that era.

Mono-, di-, and oligonucleotides would have a ready role in the chemistry of increasingly complex microsystems. How, when and in what role polynucleotides and nucleic acids became incorporated into that chemistry is hazardous to guess. But it would seem more logical to assume that nucleic acids were fashioned in the chemistry of microspheres as part of their evolving complexity, than to assume that nucleic acids arose independently and spontaneously, 'destined' to take over the activities of microsystems. Further, being the result of the chemical activities of microsystems, nucleic acids would be expected to reflect that chemistry. As for coded nucleic acids, the sudden great proliferation of species, beginning about a billion years ago, is better explained by the appearance, at that time, of coded nucleic acids in already chemically complex microsystems, than by the hypothesis that it was due to the acquisition of oxidative metabolism by already well established heterotrophic cells.

The foregoing observations include reinterpretations of some prevalent views concerning life and its beginnings on Earth. In the following, some other views are examined and reinterpreted.

1. Vitalism and the Spontaneous Generation Question

It was once almost universally believed that there could never be a naturalistic explanation of the origin of life. That belief was based on two firmly held convictions;

first, that only living things could synthesize organic compounds, and secondly, that the theory of spontaneous generation had been disproven beginning with the experiments of Redi (1668) and ending with those of Pasteur (1860, 1862).

The first conviction was shattered by the experiments of Miller (1953, 1955) inspired by Oparin's hypothesis (1936). The second conviction is still held by many although experimenters on spontaneous generation had been few and their experiments were either very limited or inconclusive. The belief in spontaneous generation was nevertheless rendered untenable by advances in the biological sciences. The credit lies not with the few experimenters on spontaneous generation with their limited, inconclusive and sometimes controversial results. Rather, it belongs to the numerous naturalists, anatomists, and physiologists of the seventeenth, eighteenth and nineteenth centuries, whose cumulative work on the anatomy, taxonomy, habitat, reproduction, life cycles, embryology and physiology of species after species of animals and plants shed light on the nature of life and its continuity and dispelled the more mystical beliefs. Thus was the concept of spontaneous generation abandoned, not disproven.

And, finally, Pasteur's conclusion regarding the meaning of his experiments: 'Never will the doctrine of spontaneous generation recover from the mortal blow of this simple experiment' (Vallery-Radot, 1960) was utterly unjustified given the limited scope and number of his experiments. Bastian (1872) and Schäfer (1912) pointed this out (evidently to closed minds) and Moore (1912) succinctly observed:

> Life probably arose as a result of the operation of causes which may still be at work today causing life to arise afresh. Although Pasteur has conclusively proven that life did not originate in certain ways, that does not exclude the view that it arose in other ways. The problem is one that demands thought and experimental work, and is not an exploded chimera. Therein lies the value of Schäfer's contribution to the question, and it is a most refreshing and valuable one.

In more recent times (Keosian 1960, 1964, 1968, 1972) the problem was reviewed in the light of the present era of experimentation on the origin of life. The neobiogenesis controversy was examined in greatest detail in the 1964 reference (pp. 98–111). The point to be raised concerning this episode is not so much the question of the merits or demerits of the spontaneous generation and neobiogenesis hypotheses, but the possible fallibility of even universally held scientific views. This oft-mentioned but rarely observed caution is particularly pertinent today when, with the rapid accumulation of data, views all too soon attain the stagnating status of dogmas.

2. The Questions of Uniqueness

The conditions for the abiotic formation of organic compounds *were* unique and occurred on Earth early in its history and never again. The conditions for the appearance of life are *not* as unique. Primarily necessary is the presence of organic compounds including macromolecules, a condition which has existed ever since those early times. But one must not make the mistake of thinking that each reorigin of life must be a replica of the events of the first instance. On the contrary, each reorigin would reflect the kind of chemicals present and the nature of the physical conditions existing at the time of its origin.

3. Concepts of Life

The terms life and living were at first lay terms and in that context had uncertain boundaries. For centuries science has been attempting to sharpen the lines; to this day attempts continue unsuccessfully. The reason is that any 'universal' definition of life conceptually strips all living things of structures – and therefore the functions of those structures – not essential for the definition. Thus all living things are reduced from what they recognizably are, to the same idealized entity from which the definition is derived. This is indeed meaningless.

Matter driven by energy in an open system can go on to higher and higher levels of organization. The thing to bear in mind is that each level of organization has its own properties by which alone it can best be recognized. Also, each higher level, although incorporating structures and processes evolved at the lower levels, has new properties not predictable from the properties of the lower level. This is true of the whole progression from elementary particles through atoms and molecules to man. Each stage in that progression incorporates structures and processes of the lower level but emerges as a new stage with new properties and the propensity for arriving at a higher level of organization. Where does 'life' fit in this progression? Nowhere. It makes little sense to attempt to squeeze into this hierarchy of stages a nebulous undefinable something called 'life'.

It would appear to be more realistic to approach the problem of life's beginnings not as an attempt to discover the precise point at which lifeless matter gave rise to the 'first living thing', but rather as an examination of the mechanisms operating in the transitions of matter into higher and higher levels of organization. Then the first level of organization that can be considered 'alive' will still be a matter of personal preference, but at least we will all be talking about the same things.

4. The Hot Dilute Soup

Haldane (1928) believed that conditions existed on the primitive earth that converted the primordial oceans into a 'hot dilute soup'. Oparin (1936) proposed a similar hypothesis which would account, he believed, for the formation of an abundance of organic compounds in prebiological times. Laboratory experiments testing that hypothesis do indeed yield a large variety of organic compounds and biochemicals. Two objections may be cited against accepting the hot dilute soup as representative of prebiological waters. For one, experiments are conducted in the confined space of laboratory vessels. The course of chemical reactions under such conditions, as within microsystems discussed previously, may differ greatly from that in the relatively limitless expanse of oceans. The term 'simulated prebiological conditions' is not truly descriptive of such laboratory conditions. Pattee (1971) suggested a setting measured in hundreds of cubic meters with sand and simulated tides as more in conformity with prebiological conditions.

Secondly, there seems to be no general agreement on what the term 'prebiological conditions' signifies. The great variety of compounds claimed to have been synthesized under primitive Earth conditions can be accepted only if we ignore the fact that the list is an accumulation of data from experiments employing a variety of reactants under a variety of conditions, some being mutually exclusive. For example, an experiment designed to produce nucleic acids produces little else. On the other hand, experiments of the Miller type (1953, 1955), while producing many organic compounds, produce only traces of bases and no nucleic acids.

5. The Heterotroph Hypothesis

The thinking and experimentation on the origin of life has been, and still is, strongly influenced consciously or unconsciously, by the heterotroph hypothesis which requires the preexistence of all of the organic and inorganic compounds necessary for the structure, metabolism and nourishment of the first living things. This hypothesis, developed in detail by Horowitz (1945), claims that the first living things had to be completely heterotrophic, i.e. they depended entirely on the enivronment for all of the organic raw materials required by their metabolism. But the simplest heterotrophic cell is an intricate structural and metabolic unit of harmoniously coordinated parts and chemical pathways. Its spontaneous assembly out of the environment, granting the unlikely simultaneous presence together of all of the parts, is not a believable possibility. On the other hand, within prebiological microsystems, syntheses could conceivably be driven by UV rays acting upon the same reactants present in the outside medium. These systems, were, in that sense, autotrophic though not alive. They were not only responsible for the synthesis of more and more complex organic compounds and 'biochemicals', but on disintegration, they enriched the medium with these compounds. As mentioned earlier, the formation of complex organic compounds, of specific biochemicals, and of biochemical pathways was brought about within these autotrophic systems rather than in the structureless outside medium.

6. The Probability of Life Arising From Inanimate Matter

In this conncetion the question that seems to make the least sense is this: 'What is the probability *of the appearance of life* in a context in which its origin seems only remotely likely? That is the approach taken by Troland, Muller and most geneticists. But the question that makes more sense is this: 'What is the probability *of the occurrence of a context* in which the appearance of life is highly to be expected?' It is proposed that an answer to the first question is academic because, from the accumulating evidence, the primitive earth represented an answer to the second question. It was an environment in which the events leading to the formation of living things had a high probability of occurring.

7. The Evolution of Biochemistry

There is only one evolution – the evolution of matter from elementary particles through atoms and molecules to systems of higher and higher levels of organization. Each new level ushers in new properties which could not be predicted from the properties of the lower level. Bernal (1959) put this in the following terms:

> In general, the pattern I propose is one of stages of increasing inner complexity, following one another in order of time, each one including in itself structures and processes evolved at the lower levels. The division into stages is not in my opinion an arbitrary one. Although the evolution of life was continuous, for no stage could have been completely static, it cannot have been uniform. Discontinuities which occurred at later stages of evolution, such as the emergence of airbreathing forms, are likely to have been paralleled at the earlier biochemical stages at such jumps as the genesis of sugars, nucleic acids and fats. One of our major problems is to establish the correct order of the steps inferred from existing metabolism *as well as the postulating and checking of other steps which have been subsequently effaced by the success of more efficient biochemical mechanisms.* (Italics added).

The portion in italics cannot be emphasized too strongly. An emerging concept on the origin of life is that it is a part of the evolution of matter which takes place throughout the universe. The assumption that an almost complete biochemistry evolved, even to the level of biochemical pathways in the absence of living things, is inconsistent with this view. For if that were so, we would be saying that while evolution brought about enormous morphological, functional and psychological developments, biochemistry – the chemistry of living things – changed but little because its fundamental patterns were set before life appeared.

References

Bastian, H. C.: 1872, *The Beginnings of Life*, D. Appleton and Co., New York.

Bernal, J. D.: 1949, *Proc. Phys. Soc. (London)* **62A**, 537–558; **62B**, 597.

Bernal, J. D.: 1959, in F. Clark and R. L. M. Synge (eds.), *Proc. First International Symp. on the Origin of Life on Earth*, Pergamon Press, New York, p. 38.

Evreinova, T. N., Mamontova, T. W., Karnauchov, W. N., and Dudaev, A. N.: 1971, in R. Buvet and C. Ponnamperuma (eds.), *Molecular Evolution*, Vol. 1, North-Holland Publishing Co., Amsterdam, p. 337.

Fox, S. W.: 1969, *Naturwissenschaften* **56**, 1.

Fox, S. W.: 1972, in S. W. Fox and K. Dose (eds.), *Molecular Evolution and the Origin of Life*, W. H. Freeman, San Francisco, pp. 196–236.

Haldane, J. B. S.: 1928, 'The Origin of Life', *Rationalist Annual*, reprinted in *Science and Human Life*, Harper Brothers, New York, 1933.

Horowitz, N. H.: 1945, *Proc. Nat. Acad. Sci. U.S.* **31**, 153.

Kaplan, R. W.: 1971, in R. Buvet and C. Ponnamperuma (eds.), *Molecular Evolution*, Vol. 1, North-Holland Publ. Co., Amsterdam.

Keosian, J.: 1960, *Science* **131**, 479.

Keosian, J.: 1964, *The Origin of Life*, Reinhold Publ. Corp., New York, pp. 98–111.

Keosian, J.: 1968, *The Origin of Life* (2nd ed.), Van Nostrand Reinhold Company, New York.

Keosian, J.: 1972, in D. W. Rohlfing and A. I. Oparin (eds.), *Molecular Evolution*, Plenum Press, New York, pp. 9–21.

Miller, S. L.: 1953, *Science* **117**, 528.

Miller, S. L.: 1955, *J. Am. Chem. Soc.* **77**, 2351.

Moore, B.: 1912, *The Origin and Nature of Life*, Henry Holt and Co., New York.

Muller, H. J.: 1926, 'Address Before the International Congress Plant Science', published in the proceedings of that congress in 1929 (see below).
Muller, H. J.: 1929, *Proc. Intern. Cong. Plant Physiol.* **1**, 897.
Oparin, A. I.: 1936, *The Origin of Life on Earth* (English translation by S. Margulis), McMillan Co., New York, 1938.
Oparin, A. I.: 1957, *The Origin of Life on Earth*, Academic Press, New York.
Oparin, A. I.: 1964, *The Chemical Origin of Life*, Charles C. Thomas, Springfield, Illinois.
Pasteur, L.: 1860, *Compt. Rend.* **50**, 303.
Pasteur, L.: 1860a, *Compt. Rend.* **50**, 849.
Pasteur, L.: 1860b, *Compt. Rend.* **51**, 348.
Pasteur, L.: 1862, *Ann. Chim. Phys.* **3**, 64.
Pattee, H. H.: 1971, in R. Buvet and C. Ponnamperuma (eds.), *Molecular Evolution*, Vol. 1, North Holland Publishing Co., Amsterdam, pp. 42–50.
Pirie, N. W.: 1937, in J. Needham and D. Green (eds.), *Perspectives in Biochemistry*, The University Press, Cambridge, England.
Redi, F.: 1668, *Esperienze intorno alla generatione degl' insetti*, Firenze, Italy.
Schäfer, E. A.: 1912, *Rep. Brit. Assoc. Adv. Sci.*, pp. 3–36.
Troland, L. T.: 1914, *Monist* **24**, 92.
Troland, L. T.: 1916, *Cleveland Med. J.* **15**, 377.
Troland, L. T.: 1917, *Am. Naturalist* **51**, 321.
Vallery-Radot, R.: 1960, *The Life of Pasteur*, Dover Publications Inc., New York. (Translated from French by D. L. Devonshire.)

PART V

EARLY BIOCHEMICAL EVOLUTION

ON THE CHEMICAL CONSTITUTION OF COMETARY NUCLEI

L. BIERMANN and G. DIERCKSEN
Max-Planck-Institut für Physik und Astrophysik, München, F.R.G.

Abstract. The observed properties of comets and the experiments of Anders *et al.* suggest that during the condensation of the outer parts of the solar system from the solar nebula – assumed to contain most of the C in the form of CO, the remaining O mostly in the form of H_2O, and the excess H as H_2 molecules – besides the most stable hydrocarbon (CH_4) also considerable quantities of heavier hydrocarbon were formed. When the surface of the nucleus is warmed up in re-approaching the sun, part of the solar quanta having sufficient energy for dissociating such molecules, exothermic reactions, for instance such leading, to the observed C_2 and C_3 molecules may be expected; an inhomogeneous and porous structure of the nucleus would explain the observed occurrence of short-lived discrete jets, in which a small fraction of the CO could be ionized. Since also N must be present in some quantity (observed mainly was CN and N_2^+), the formation of more complex organic molecules must also have taken place. To which extent complex interstellar molecules can have survived the formation of the solar system is hard to say on the basis of present knowledge of the origin of the solar system.

The molecules observed in cometary atmospheres may be regarded to have been formed by dissociation or cracking from such molecules as are observed in dense interstellar clouds (cf. e.g. Sagan, 1972). Some of these molecules do indeed occur both in cometary atmospheres and in dense interstellar clouds. The presence of N_2^+ and of CO_2^+, for which it is not so easy to indicate a probable parent molecule among those observed in interstellar clouds, do not contradict this statement, since obviously the present list of interstellar molecules is quite incomplete. Only the formation of the C_3 molecule seems to require a different explanation* (see below).

This raises the question whether only similar processes have occurred in interstellar clouds and in the solar nebula during the formation of cometary nuclei $4\frac{1}{2}$ b.y. ago, or whether the material contained in cometary nuclei is essentially that of dense interstellar clouds, with the exception of the excess molecular hydrogen. The first alternative might seem more probable. But before drawing conclusions some further points should be considered:

(1) There appear to be now fairly strong reasons for assuming that CO is a major constituent of cometary atmospheres, comparable to the observed OH (Biermann, 1973). The CO is by far the most abundant heavy molecule observed so far in dense interstellar clouds.

(2) The 'activity' of comets – a common property of bright objects – is now usually ascribed to the energy released in chemical reactions of molecules stored in the cometary nucleus, as first suggested by Donn and Urey (1957) almost 20 years ago.

Though different chemical reactions have been discussed the most likely is the recombination of neutral free radicals with molecules as proposed by Donn and Urey

* The authors are aware of recent experiments of Payne and Stief (1972).

(1957). Free radicals, say 1%, could be stabilized for 4½ b.y. in a matrix at temperatures of the order of 10 K, particularly if the matrix is largely made up of ordinary ice, as suggested by the observed OH molecules. Such a concentration of free radicals in parts of the material of which active comets are built from could arise from the interstellar UV radiation field at the stage when the density of the cloud, from which the solar system formed subsequently, was sufficiently low.

Since the cometary nuclei should have come into existence at relatively large distances from the Sun (Cameron, 1973), it is well possible that the material never reached higher temperatures than a few 10 K. The growth of the cometary nuclei to their present size during the final collapse of the solar nebula itself should in any case have been a rapid process (see again Cameron, 1973), and it is difficult to understand by what process free radicals in the required concentration could be formed at such a stage.

Comet Kohoutek 1973f will provide an excellent opportunity to prove the validity of the arguments presented here by checking for the presence of CO, H_2CO, H_2, CH_4 (or CH_2) and other so far undetected constituents by measurements in the infrared, the mm-, and the cm-range. In the present paper the points outlined above will be discussed in some more detail. The aim is to show that cometary nuclei may well contain prebiotic organic molecules of interstellar origin in some quantity imbedded in ice matrixes. Whether such molecules, set free in the process of dissolution of the comet during later approaches to the Sun, could be related to the origin of life on planets is of course a different question.

The CO^+ ions (which as such cannot be present in cometary nuclei) are usually the most important constituents of the rays and streamers of bright comets which are emitted from the nucleus roughly in the solar direction and subsequently bent backwards into the antisolar direction by the solar wind. In most comets the place of origin of these ions is difficult to observe, because of the brightness of the coma and the obscuration of the inner parts by dust particles. In all comets where such observations were possible it was, however, observed that they originate relatively close to the nucleus down to 500 or 1000 km only, as compared to a diameter of the solid nucleus of the order of magnitude of 1 km. At distances of about 10000 km from the nucleus the observed density of CO^+ ions is of the order of some 10^2 or perhaps 10^3 ions cm^{-3} (Arpigny, 1965). At such a distance the density of the main constituent observed hitherto, the OH molecule, which is most probably formed from H_2O by dissociation, is of the order of 10^5 or 10^6 cm^{-3} for a reasonable bright comet at about 1 AU distance from the Sun. In the last years the rate constants of the reactions between molecular ions and other molecules have become known for several hundred combinations of ions and molecules, which include the reactions between CO^+ on one side, and ordinary water and hydrocarbons on the other. These rate constants are generally of the order of several 10^{-9} cm^3 s^{-1},* which corresponds to an effective cross section of

* The large rate constants of ion molecule reactions used in this paper are approximately independent of the temperature, as has been pointed out by E. Ferguson to one of the authors (L.B.) in a private communication.

10^{-13} cm^2. As a consequence it appears that most of the ions observed in a cometary atmosphere at distances of the order of 10000 km should react and form other ionized molecules (for instance HCO^+ and H_3O^+, both invisible) before reaching larger distances from the nucleus, such that there should be approximate equilibrium between production and elimination of the observed ions. This means that in such comets 0.1–1 ions must be produced per cm^3 and s at that distance from the nucleus. A closer look at the possible mechanisms of ionization reveals that the ionization probability should be at most of the order of 10^{-6} s^{-1}, since otherwise no picture consistent with other evidence emerges. In particular, the observations of the neutral molecules do not indicate the presence of any additional agent (for instance energetic electrons in large numbers*) which would necessarily be more effective in dissociating the neutral molecules than in ionizing CO (and OH, OH^+, though in smaller intensity than CO^+).

This leads to the conclusion that neutral CO is a constituent comparable in abundance with OH and H_2O, consistent with the observations in dense interstellar clouds, where most of the carbon appears to be tied up in CO-molecules**. Such molecules must also be continuously absorbed on the surface of interstellar grains.

Many comets show, when approaching the Sun, remarkable activity: Jets consisting of gas and dust or nearly spherical halos are emitted at intervals down to some fraction of one hour (Donn, 1960). For the geometrical structure of such jets reference is made to the Atlas of Cometary Forms by Donn (1970); Rahe and Wurm (1970), though the much older work of Bredhikhin and Jägermann (1903) which is based mainly on visual observations made during the last century, gives also an adequate impression of the main features. To which extent cometary activity is correlated with solar activity is still unknown, in spite of many efforts made over the last few decades. The only reasonably clear-cut correlation is that between the activity of the gas (plasma) tails of comets and geomagnetic (i.e. interplanetary) events, which provided the first hint to the existence of the quasistationary component of the Sun's corpuscular radiation, now known as the solar wind.

All attempts to account for the existence of cometary activity have led to the assumption that the energy which appears to be released in such events (roughly 10^{10}– (perhaps) 10^{11} erg g^{-1}, involving probably some t or some 10 t of emitted material s^{-1}) can only be set free by chemical reactions. An inhomogeneous structure of the cometary nucleus should be another necessary requirement for the locally restricted jets. A number of exothermic chemical reactions may account for the energy

* Ionizing electrons arising from nonthermal effects (instabilities) in order to increase significantly the probability of photo ionization would have to have a number density 10 cm^3 (for a cross section 10^{-16} cm^2 at an energy of 100 eV). Such electrons would have a lifetime in the region in question of only $\leqslant$ 10 s. The energy requirement to maintain such a state ($\sim 10^{11}$ eV cm^{-2} s^{-1}) seems already too large (compare Brosowski and Wegmann (1972), Biermann and Lüst (1972)).

** CH_4 could still be abundant in interstellar clouds but cannot be detected in the radio spectrum (D. Buhl 1973). A large abundance of CO in comets had been discussed, though on a different basis, by M. Wallis (1968).

release: Heavier hydrocarbons reacting with water molecules forming CO and molecular hydrogen could indeed produce the necessary energy. Since the dust observed in cometary atmospheres – by mass usually of the same order as the gas – may act as a catalyst, it is not entirely inconceivable that such processes do occur. An empirical check should be possible by the present techniques of the observation applied to future comets. It seems, however, more probable that the necessary energy is released by the recombination of atoms, radicals or (unsaturated) molecules. According to the recent work by Greenberg (1972) free radicals together with complex organic molecules have been produced on the surface of grains by uv radiation in the range 1400–2000 Å and embedded in an ice matrix formed around dust grains. Though it is difficult to make reliable estimates for this process, it looks as though 1% concentration of free radicals could be reached in this way. It can then be shown that this concentration would remain essentially unchanged for billions of years as long as the temperature of the grain remains lower than 10 or 20 K. In pure ice the maximum tolerable temperature could be slightly higher (K. Michel, 1973).

In view of the increasing optical thickness during the collapse of the proto-stellar cloud this means that the most favorable conditions for the production of ice coated dust particles with an adequate proportion of free radicals would be given at the initial stage.

The temperature requirements would be met best for the outermost part of the collapsing cloud, which would receive the least amount of heat from the protosun before the cometary nuclei are actually formed.

Some further remarks which seem relevant in this context are the following: Of all molecules observed in cometary spectra, the C_3 appears very hard to account for by the breakup of any organic molecule to be expected in interstellar space. The C_3 molecule is observed in flames where it most probably is formed by radical reactions. The presence of this molecule in sizable quantity might therefore be regarded as an additional indication that radical reactions are important also in comets.

In some large comets the production of gas and dust became noticeable already at distances beyond 5 AU, in one case even at a distance of more than 10 AU, that is at a surface temperature well below 100 K. This indicates – in addition to the presence of very volatile constituents (CH_4 and H_2CO for instance) in large relative abundance – that the chemical reactions taking place are triggered already at such temperatures in line with the picture developed here.

It seems a fortunate coincidence that new techniques of observation in the infrared and in the mm-wavelength range have become available just prior to the appearance of such an exceptional object as comet Kohoutek, which will be very bright around the turn of the year. It is hoped that such observations will shed further light on the hypothesis on the origin of the material from which the nuclei of comets are built, presented in this paper.

Finally it is a pleasure to express the indebtness of the authors to quite a number of colleagues for important comments in particular to E. Ferguson, Rhea Lüst, K. W. Michel and H. U. Schmidt.

References

Arpigny, C.: 1965, *Mém. Acad. Roy. Belgique* **35**, 3.

Bass, A. M. and Broida, H. P.: 1960, *Formation and Trapping of Free Radicals*, Academic Press.

Biermann, L.: 1973, *Copernican Days*, Asiago Astron. Conference.

Biermann, L. and Lüst, Rh.: 1972, Institutsbericht MPI/PAE-Astro 52.

Bredhikhin, T. and Jägermann, 1903, *Mechanische Untersuchungen über Cometenformen*, St. Petersburg.

Brosowski, B. and Wegmann, R. 1972,: Institutsbericht MPI/PAE-Astro 46.

Buhl, D.: 1973, personal communication.

Cameron, A. G. W.: 1973, *Icarus* **18**, 407.

Donn, B., and Urey, H. C.: 1957, *Mém. Soc. Roy. Sci.* **VVIII**, Inst. d'Astrophys.

Greenberg, J. M.: 1972, *Nice Symposium on the Origin of the Solar System*, p. 135.

Michel, K. W.: 1973, personal communication.

Payne, W. A. and Stief, L. J.: 1972, *J. Chem. Phys.* **56**, 3333.

Rahe, J. and Wurm, K.: 1970, *Atlas of Cometary Forms*, National Aeronautics and Space Administration.

Sagan, C.: 1972, *Nature* **238** ,77.

Wallis, M.: 1968, *Planetary Space Sci.* **16**, 1221.

PHOTOCHEMICAL CONVERSIONS OF LOWER ALDEHYDES IN AQUEOUS SOLUTIONS AND IN FOG

T. E. PAVLOVSKAYA and T. A. TELEGINA

A. N. Bakh Institute of Biochemistry, Academy of Sciences, 117071 Moscow, U.S.S.R.

Abstract. One of the possible paths of abiogenic synthesis of biologically important compounds on primordial Earth, as well as under extraterrestrial conditions, might have been the photochemical conversions of polyatomic organic molecule detected at present in the interstellar medium and in cosmic bodies. The important contribution from lower aldehydes is emphasized. Experimental results show that amino acids, peptide-like compounds, N-heterocyclic compounds, etc., are formed by UV irradiation (254 nm) of aqueous solutions either of formaldehyde, or of acetaldehyde with ammonium nitrate. UV-irradiated fog of the same composition yields aminoacids, mostly glycine.

Almost two decades have passed since the start of research on abiogenic synthesis of amino acids under conditions of the primordial atmosphere composition. The attention of research is now again attracted to this problem. The reason is that new facts have been discovered in different fields of science, thus bringing forth again some problems of chemical evolution.

Aliphatic α-amino acids were found in the recent years in the Murchison and other meteorites (Kvenvolden *et al.*, 1970; Cronin *et al.*, 1971; Lawless *et al.*, 1971). What was their origin under extraterrestrial conditions? It was natural to suggest that various chemical conversions of simple gas molecules were responsible for appearance of the α-amino acids in the primordial atmosphere. Experiments have shown again that many amino acids of different structure, similar to those found in a meteorite, were formed under electric discharge in a mixture of CH_4, N_2, H_2O (Ring *et al.*, 1972).

Our knowledge of organic molecules contained in cosmic bodies and in the interstellar medium is becoming wider. The diatomic and polyatomic organic molecules detected in the interstellar medium are evidence of the ocurrence of significant chemical phenomena. Adsorption of interstellar molecules on interstellar 'dust' particles would confer considerable protection against the UV, as stated in *Nature, News and View* **222**, p. 9 (1969). Some molecules, such as OH, CO, and H_2CO, are rather generally distributed over the galaxy (Snyder *et al.*, 1969).

Moreover, investigation of the photochemistry of interstellar molecules gives valuable information on their stability against photodissociation. They are subjected to appreciable action of interstellar radiation fields in the unobscured interstellar regions. Their lifetimes are consequently shorter than 10^2 yr, whereas in regions of moderate density (in clouds) protecting them from the full radiation field it is of the order 10^6 yr (Stief, 1972). The study of photochemistry in comets (Jackson, 1972) and the information on primary photophysical and photochemical processes involving organic molecules, for instance alkanes, in the short UV region (Ausloos, 1972) is closely connected with these problems.

The above discoveries undoubtedly are of great importance for the investigation of chemical evolution. Not only do they confirm the earlier concepts (Urey, 1952; Oparin, 1957) that the primordial reducing atmosphere represented a system of reacting molecules entering into diverse chemical reactions under the action of various energy sources, but also permit more grounded suggestions on the nature of the initial molecules that could form compounds of biological importance.

We are of the opinion that lower aldehydes could play the extremely essential part of active intermediates in the development of abiogenic synthesis on the primordial Earth. At first our idea that formaldehyde synthesis could occur in upper atmospheric layers was based on sparse experimental results (Groth, 1937; Garrison *et al.*, 1951; Dodonova *et al.*, 1961, 1962). Now there is direct information on formaldehyde abundance in nature (Snyder *et al.*, 1969).

A specific structural feature of these relatively stable and at the same time sufficiently reactive compounds is that their molecules contain a carbonyl group. Owing to the electron affinity of oxygen, ionic structure is markedly present in the C—O bond (due to uneven distribution of the π-electrons). It is responsible for the potential chemical activity of the C=O group. Its reactivity is fully displayed in chemical reactions induced on reactants by the electric fields, for instance in those involving HCN, NH_2OH, etc., where electron displacement in the carbonyl group is strong.

It will be noted that the bond energy in a strongly polar carbonyl group is 144 kcal for CH_2O and about 150 kcal in other aldehydes, making it close to the triple bond energy, for example, to that of C≡N in HCN (146 kcal) or in nitriles (149 kcal). This indicates that these molecules are relatively stable, and at the same time their reactivity is high.

The photochemical conversions of lower aldehydes are accounted for by their ability to absorb UV radiation in the region 220–340 nm. The absorption band in this spectral region is ascribed essentially to the $n \rightarrow \pi^*$ transition in the carbonyl group. Aldehydes also exhibit a strong absorptic band in the short UV, close to 180 or 160 nm. It is ascribed to $\pi \rightarrow \pi^*$ and $n \rightarrow \sigma^*$ transitions, respectively.

Photochemical investigation of gaseous CH_2O in the range 330 to 250 nm has shown several basic types of primary aldehyde dissociation. These involve the formation of free radicals R(H) and HCO and of stable hydrocarbon and CO molecules. The process yielding formyl radicals is important over a wide range of wavelengths, as can be seen from comparison of the quantum yields of these processes (Calvert *et al.*, 1966).

Photodissociation of formaldehyde in glass aqueous solutions also yields formyl radicals, as detected by ESR (Brivati *et al.*, 1962; Niskanen *et al.*, 1971).

Evidence has been obtained for the occurrence of a hitherto unsuspected intermediate in gas photochemistry of acetaldehyde and of higher aldehydes at 290 nm. Dioxetane was suggested to be the most probable structure that could be formed by reaction between acetaldehyde molecules in the excited and the ground states. The decay of this high-energy compound occurred with release of triplet aldehyde readily reacting with small amounts of other compounds in the course of co-photolysis (De Groot *et al.*, 1972)

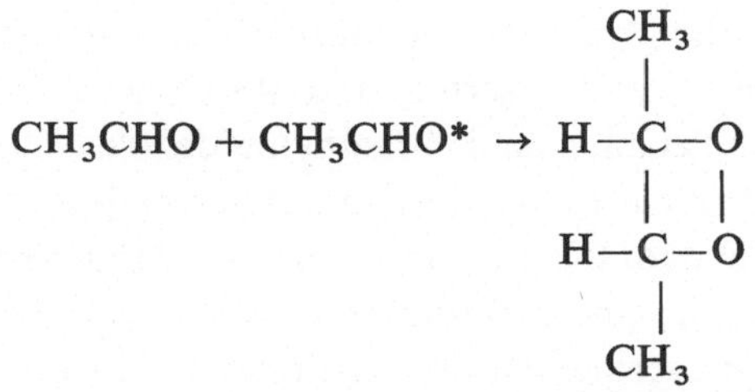

$$CH_3CHO + CH_3CHO^* \rightarrow \begin{array}{c} CH_3 \\ | \\ H-C-O \\ |\quad\ | \\ H-C-O \\ | \\ CH_3 \end{array}$$

3,4-dimethyl-1,2-dioxetane

Thus, the formation of high-energy radical and molecular intermediates responsible for the paths and rates of numerous secondary reactions are the basic steps in the photochemical reaction mechanism.

It was surmised that substances from the cosmos, as well as products of Earth rock surface decay and of inner Earth layers, such as H_2, CH_4, NH_3, CO, H_2S (Urey, 1959), NH_4Cl and other ammonia compounds, and later on $Ca(NO_3)_2$ (Vinogradov, 1959), could accumulate in primordial Earth hydrosphere. The HCN and HCOOH compounds discovered in interstellar medium open up new possibilities for similar investigations of NH_4CN and $HCOONH_4$.

We have studied the photochemical conversions (at 254 nm) of lower aldehydes (2.5%), formaldehyde and acetaldehyde in aqueous solutions, in the presence of ammonium salts (1.5%).

It has been found in earlier research that UV irradiation of aqueous solutions of CH_2O and NH_4Cl or NH_4NO_3 induces synthesis of amino acids resembling in composition (Pavlovskaya *et al.*, 1959).

Synthesis of the amino acids glycine, alanine, serine, threonine, and lysine occurs under the action of UV radiation in systems containing formaldehyde. Neutral amino acids accumulate faster and in larger amounts than do the basic ones (Sidorov, 1965). An important group of compounds was detected along with the amino acids, namely the imidazole compounds that are closely related to formation of nucleic bases, hystidine, and hystamine. These include: imidazole, 4-methylimidazole and its hydroxy and oxo-derivatives (Sidorov *et al.*, 1966). Moreover, irradiation of a solution containing acetaldehyde yields 2-methylimidazole, 4,5-dimethylimidazole, and certain amine derivatives (Pavlovskaya *et al.*, 1968).

We suggest that the initial stages of amino acid and imidazole synthesis are identical and involve the formation of important intermediates, namely the formyl radical and the CH_2OH radical (radical of the formaldehyde hydrate form) and the products of their recombination: glyoxal, methylglyoxal, hydroxy methylglyoxal, etc. The formation of the formyl radical and of glyoxal in UV-irradiated formaldehyde solutions has been observed by us in specific experiments (Niskanen *et al.*, 1971) Most probably, synthesis of the imidazole ring occurs then by the known interaction of dicarbonyl compounds with ammonia and aldehyde, whereas amino acids could be obtained by conversion of glyoxal or of its derivatives to relevant aldehyde or keto-acid with subsequent UV-induced reductive amination.

In case glyoxal is predominant, the glycine and imidazole yields would be highest, hydroxy methylglyoxal would form mostly serine, and 4-oxomethylimidazol and methylglyoxal would form mostly alanine and 4-methylimidazole. The results of qualitative analysis on the reacting micture, and of quantitative analysis of amino acid and imidazole compounds, are in favor of the mechanism proposed (Pavlovskaya *et al.*, 1964; Sidorov, 1965). For instance, glycine and imidazole synthesis in formaldehyde solutions is fastest, and then come serine, alanine, hydroxy methylimidazole, and 4-methylimidazole (Figure 1).

The results obtained seem to show that the highly reactive dicarbonyl compounds

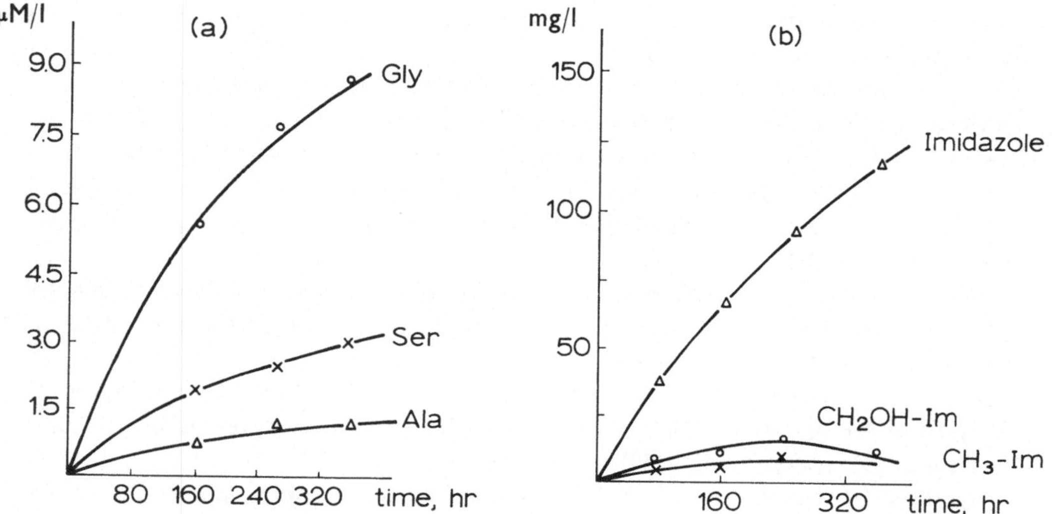

Fig. 1. Rate of formation of amino acids (a) and of diazotizable compounds (b) in UV-irradiated formaldehyde – ammonium nitrate solutions.

can be essential intermediates in UV-induced prebiological synthesis of amino acids and of other biologically important compounds.

Photochemical conversions of acetaldehyde in aqueous solutions in the presence of ammonium nitrate under the action of 254 nm UV light induce the formation also of more complex biologically important compounds (Pavlovskaya *et al.*, 1968). For instance, glutamic and aspartic acid, valine, leucine (or isoleucine) are synthesized, along with glycine, alanine, serine and lysine. It was also found that all amino acids were formed in the bound state as low-molecular peptides (up to tetrapeptides). In these experiments the peptides were isolated by ligand exchange chromatography from the amino acid fraction that was obtained by ion-exchange chromatography of the reaction mixture (Telegina *et al.*, 1974). Under the conditions of acetaldehyde irradiation the compounds of this class might have been synthesized by a mechanism involving photochemical formation of N-acetylamino acids with their subsequent polymerization, as described by Dose *et al.* (1964).

It will be noted that organic compounds of the indole class: indole, triptamine, hydroxy- and oxo-derivatives of indole have been found in the products of photochemical acetaldehyde conversion (Pavlovskaya *et al.*, 1968). In the system involving formaldehyde they were represented only by aliphatic amines and amides (urea). The formation of a condensed heterocyclic indole ring is closely connected with the appearance of an important amino acid, tryptophane, and of pyrrole, a structural unit of porphyrines. The 1,4-dialdehyde, succinic aldehyde, that is known to take part in cyclization reactions to form heterocyclic rings, might be an intermediate in indole ring synthesis.

In the light of recent discoveries, it can be suggested that one of the possible paths of abiogenic appearance of amino acids under primordial terrestrial, and also under extraterrestrial conditions, could be the reactions induced by UV radiation in fog, rain droplets, clouds. The use of intensive ultrasonic waves permits modelling these conditions as follows (Pavlovskaya *et al.*, 1971). An ultrasonic fountain arises at the liquid surface at the site of the ultrasonic wave emergence. The liquid is intensively sprayed around from the upper part of the fountain forming a fog over the surface. Aqueous solutions of 1% formaldehyde and 0.16% ammonium nitrate were subjected to ultrasonic waves of high intensity (vibration frequency 850 kHz, intensity 25 W cm^{-2}). The fog was continuously transported into a vessel by an argon flux, was kept there in suspension, and was irradiated at room temperature by UV waves of 254 nm. The UV intensity at the surface of the quartz container enclosing the lamp was 3.5×10^4 erg cm^{-2} s^{-1}. After irradiation the fog condensed and was collected in the vessel containing the starting solution. In the course of the experiment a fraction of the fog was carried away by argon and was accumulated in a water-filled absorber located at the system outlet. In some experiments this fog fraction was analysed separately, in others it was added to the bulk of the solution. The starting solution blown through with argon for two hours served as control. Glycine was the main reaction product, along with alanine, and possibly threonine (Figure 2). This is in agreement with earlier results on UV irradiation of formaldehyde and ammonium nitrate solutions (Sidorov, 1965; Pavlovskaya *et al.*, 1968). Serine might have been synthesized under these conditions, but its identification was hindered by the low product yields. When the solution ($CH_2O + NH_4NO_3$) was exposed only to ultrasound for two and six hours, this resulted in the formation of indifferent amounts of the amino acids, whereas an increase in the glycine content in the chromatograph concentrate was observed on UV irradiation of fog for two hours.

Irradiation for eight hours increased the amount of synthesized glycine. In this case the latter, as well as other amino acids were partially carried away with fog beyond the zone of the degrading ultrasound effect. We did not observe high amino acid yields. This might have been due to their partial degradation under the action of ultrasound, to the use of a solution with a low ammonium salt content accounted for by fog formation under ultrasonic waves, and also to the short time of UV-irradiation of the fog. Nevertheless, it can be seen that synthesis of amino acids is possible in fog containing aldehyde and an ammonium salt, and that an important factor of their accumulation is the condensation of fog beyond the zone of the UV and the ultrasound action.

Favorable conditions could be created on the primordial Earth by contact of different reaction zones, for instance of atmosphere and of water surface, of cool atmosphere

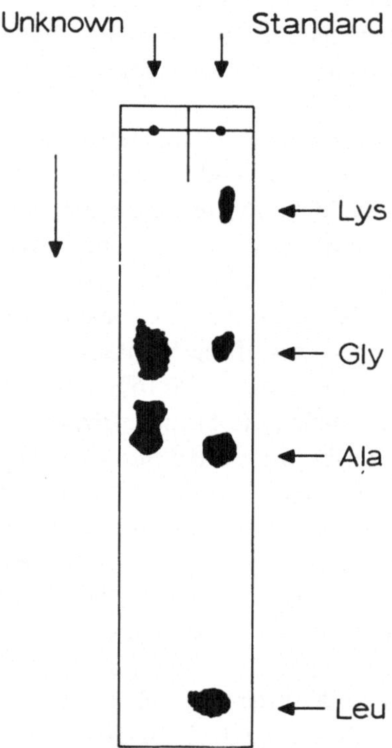

Fig. 2. Chromatogram of amino acids synthesized in fog ($CH_2O + NH_4NO_3$) by UV-irradiation and of a standard solution. Solvent n-butanol – acetic acid – water (4:1:5).

and a zone of volcano eruption. Very essential had to be the absorption of photochemical synthesis products from the atmosphere and their accumulation in oceans (Oparin, 1957), where they were protected by sufficiently thick water layers from degradation by UV rays or ionizing radiation.

In conclusion it can be stated that the part played by lower aldehydes as precursors of biologically important compounds in prebiological chemistry seems to be important. It will also be emphasized that the reducing atmosphere conditions could have favored diverse photochemical processes displaying characteristic homolytic features, owing to the absence of such a strong radical acceptor and inhibitor as is oxygen. Research on photochemical conversions of polyatomic molecules found at present in the interstellar medium and in cosmic bodies casts light upon the possible paths of chemical evolution in the abiogenesis of organic compounds.

References

Ausloos, P.: 1972, *Mol. Photochem.* **4**, 39.

Brivati, J. A., Keen, N., and Symons, M.: 1962, *J. Chem. Soc.*, 00, 237.

Calvert, J. and Pitts, J.: 1966, *Photochemistry*, J. Wiley, N.Y.

Cronin, J. and Moor, C.: 1971, *Science* **172**, 1327.

De Groot, M. S., Emeis, C. A., Hesselman, I. D., Drent, E., and Farenhorst, E.: 1972, *Chem. Phys. Letters* **17**,332.

Dodonova, N.Ya. and Sidorova, A.I.: 1961, *Biofizika* **6**, 149.

Dodonova, N. Ya. and Sidorova, A. I.: 1962, *Biofizika* **7**, 31.

Dose, K., Rajewsky, B. and Risi, S.: 1964. *Proc. Sixth Intern. Congr. Biochem.* **II**, p. 149.

Garrison, W., Morrison, D., Hamilton, J., Benson, A., and Calvin, M.: 1951, *Science* **144**, 416.

Groth, W.: 1937, *Z. phys. Chem.* **37 B**, 307.

Jackson, W.: 1972, *Mol. Photochem.* **4**, 135.

Kvenvolden, K., Lawless, J., Pering, K., Peterson, E., Floress, I., Ponnamperuma, C., Kaplan, I. R., and Moor, C.: 1970, *Nature* **228**, 923.

Lawless, J., Kvenvolden, K., Peterson, E., Ponnamperuma, C., and Moor, C.: 1971, *Science* **173**, 626.

Niskanen, R. A., Pavlovskaya, T. E., Telegina, T. A., Sidorov, V. S., and Sharpatyi, V. A.: 1971, *Izv. Akad. Nauk S.S.S.R., ser. biol.*, No. 2, 238.

Oparin, A. I.: 1957, *Vozniknovenie Zhizni na Zemle*, Izd. Akad. Nauk S.S.S.R., Moscow.

Pavlovskaya, T. E. and Pasynskii, A. G.: 1959, in Oparin, Pasynskii, Braunnshtein, and Pavlovskaya (eds.), *The Origin of Life on the Earth*, Intern. Union of Biochem. Sym. Series, Pergamon Press, New York, N.Y., **1**, p. 151.

Pavlovskaya, T. E. and Pasynskii, A. G.: 1964, *Problemy Evolyutsionnoi i Tekhnicheskoi Biokhimii*, Izd. 'Nauka', Moscow, p. 70.

Pavlovskaya, T. E., Pasynskii, A. G., Sidorov, V. S., and Ladyzhenskaya, A. I.: 1968, *Abiogenesis i Nachalnye Stadii Evolyutsii Zhizni*, Izd. 'Nauka', Moscow, p. 41.

Pavlovskaya, T. E., Telegina, T. A., Sokol'skaya, A. V., and El'piner, I. E.: 1971, *Izv. Akad. Nauk S.S.SR, ser biol.*, No. 6., 922.

Ring, D., Woman, J., Fridman, N., and Miller, S.: 1972, *Proc. Nat. Acad. Sci. U.S.A.* **69**, 765.

Sidorov, V. S.: 1965, *Dokl. Akad. Nauk S.S.S.R.* **164**, 692.

Sidorov, V. S.: 1965, *Thesis*, Institute of Biology, the University, Petrozavodsk.

Sidorov, V. S., Pavlovskaya, T. E., and Pasynskii, A. G.: 1966, *Zhur. Evol. Biokhim. i Fiziolog.* **2**, No. 4, 293.

Snyder, L. E., Buhl, D., Zuckerman, B., and Palmer, P.: 1969, *Phys. Rev. Letters* **22**, 679.

Stief, L. J.: 1972, *Mol. Photochem.* **4**, 153.

Telegina, T. A. and Pavlovskaya, T. A.: 1974, *Zhur. Evol. Biokhim. i. Fisiolog.* **10**, No. 2.

Urey, H. C.: 1952, *The Planets: Their Origin and Development*, Yale University Press, N. Y.

Urey, H. C.: 1959, in Oparin, Pasynskii, Braunnshtein, and Pavlovskaya (eds), *The Origin of Life on the Earth*, Intern. Union of Biochem. Sym. Series, Pergamon Press, New York, N.Y., **1**, 16.

Vinogradov, A. P.: 1959, in Oparin, Pasynskii, Braunnshtein, and Pavlovskaya (eds.), *The Origin of Life on the Earth*, Intern. Union of Biochem. Sym. Series, Pergamon Press, New York, N.Y., **1**, 23.

INFERENCES FROM PROTEIN AND NUCLEIC ACID SEQUENCES: EARLY MOLECULAR EVOLUTION, DIVERGENCE OF KINGDOMS AND RATES OF CHANGE

M. O. DAYHOFF, W. C. BARKER, and P. J. McLAUGHLIN

National Biomedical Research Foundation, Georgetown University Medical Center, 3900 Reservoir Road, N.W., Washington, D.C. 20007, U.S.A.

Abstract. Presently the sequences of more than 150 different kinds of proteins and nucleic acids are known from the many thousands thought to exist in all living creatures. Some few of these have occupied much the same functional niche within the living cell from near the beginning of life. In three of these latter, sequence evidence pointing to duplications of genetic material in a primitive ancestor is available and in the fourth other evidence suggests it. Such a duplication, shared by the many descendant species, permits us to locate the point of earliest time on an evolutionary tree and to infer the actual order of subsequent evolutionary events. The amounts of change which have occurred in each descendant line can be estimated with good confidence. Some inferences can be made of the structure of the ancestral duplicated sequence, the evolutionary mechanisms which have been operative on it, and the functional capacity of the organism in which it originated. We will describe new, sensitive, objective methods for establishing the probable common ancestry of very distantly related sequences and the quantitative evolutionary change which has taken place. These methods will be applied to the four families, and evolutionary trees will be derived where possible. Of the three families containing duplications of genetic material, two are nucleic acids: transfer RNA and 5S ribosomal RNA. Both of these structures are functional in the synthesis of coded proteins, and prototypes must have been present in the cell at the inception of the fundamental coding process that all living things share. There are many types of tRNA which recognise the various nucleotide triplets and the 20 amino acids. These types are thought to have arisen as a result of many gene duplications. Relationships among these types will be discussed. The 5S ribosomal RNA, presently functional in both eukaryotes and prokaryotes, is very likely descended from an early form incorporating almost a complete duplication of genetic material. The amount of evolution in the various lines can again be compared. The other two families containing duplications are proteins: ferredoxin and cytochrome c. Ferredoxin from photosynthetic and nonphotosynthetic bacteria shows clear evidence of a duplication of genetic material. This duplication is very possibly shared by the ferredoxin from plant plastids and the related adrenodoxin from mammalian mitochondria. If so, a chronology of the details of evolution of these groups can be inferred. From these examples of protein and nucleic acid sequence, we conclude that the amount of change in the bacterial lines is less than that in the eukaryote lines. Even though mutant bacteria are easily produced in the laboratory, though their evolutionary adaptation to new drugs is very rapid, and though new virulent strains often appear spontaneously, nevertheless the sequences of ancient structures in the wild types have changed less than those in the eukaryote lines. Cytochrome c sequences from many eukaryotes and the closely related cytochrome c_2 from *Rhodospirillum rubrum* are known. Other types of cytochrome, such as c_{551} and c_{553}, are probably related to these through gene duplication. Knowledge of enough of these structures to establish an early duplication will provide a time orientation for the cytochrome c evolutionary tree. This quantitative tree now contains sequences from animals, fungi, green plants, protozoa, and bacteria, examples from all five biological kingdoms. (Supported by NIH Grant GM-08710, NASA Contract NASW-2288 and HEW GRS Grant RR-05681.)

At present the sequences of more than 150 different kinds of proteins and nucleic acids are known from the many thousands thought to exist in all living creatures (Dayhoff, 1972, 1973). A few of these have occupied much the same functional niche within the living cell almost from the first development of life based on nucleic acid replication. In three of these latter molecules, 5S ribosomal RNA, transfer RNA (tRNA) and ferredoxin, sequence evidence indicating duplications of genetic material in a primitive ancestor is available and in a fourth, the cytochrome c group, additional evidence suggests it. Such duplications, shared by the many descendant species, permit us to locate the point of earliest time on an evolutionary tree derived from sequences of molecules in living organisms and to infer the actual order of the evolutionary divergences. The amounts of change which have occurred in each descendant line can be estimated with good confidence. Some inferences can be made about the structure of the ancestral duplicated sequence, the evolutionary mechanisms which have been operative on it, and the functional capacity of the organism in which it originated.

A useful quantity of sequence data from bacterial groups is just now accumulating. It is already clear that functionally similar sequences in the different orders are so different from each other that sensitive, objective methods are needed to assure ourselves that two sequences show evidence of a common evolutionary origin. Once this relationship is established for a group of sequences, we can refine our understanding of the evolutionary process affecting the early prokaryotic sequences and infer the phylogeny of bacteria and the development of metabolic capacities. In this paper we first describe new objective methods for comparing sequences and then discuss evolutionary inferences from the four groups. Unless otherwise referenced, all protein and nucleic acid sequences mentioned appear in the Data Section of the *Atlas of Protein Sequence and Structure*, Vol. 5 (Dayhoff *et al.*, 1972a) or Supplement I to Vol. 5 (Dayhoff, 1973).

A pair of sequences being compared will fall into one of three general classes: closely related, distantly related or unrelated. In the first group are those sequences which are so similar that sophisticated statistical tools need not be employed to confirm the similarity. It is generally believed that these have been derived from a single ancestral gene either by species divergence or gene duplication and subsequently have accumulated mutations independently. Long regions of such sequences can be aligned without the introduction of any 'gaps' in the sequences. In the third group are sequences whose relationship is either nonexistent or so remote that even the most delicate statistical tools now available cannot distinguish it from chance. If no gaps are permitted in an alignment of typical unrelated protein sequences of equal length, 7% of the residues will match. On the other hand, 23% identities can be obtained by allowing unlimited gaps (Barker and Dayhoff, 1972). For unrelated nucleic acid sequences, 25% match when no gaps are allowed, whereas 48% match if unlimited gaps are permitted. Unrelated sequences may thus appear to be related because most observers focus attention on the number of similarities rather than upon the number of gaps that were required to produce them (see also Cantor, 1968). This illusion has resulted in a number of statistically indefensible reports of relationships in the literature.

The second group of sequences lies between the two extremes. These we call the distantly related sequences. Using statistical methods, the similarity of the sequences can be demonstrated to be greater than that produced by a chance ordering of the residues. These sequences are generally believed to represent genes derived from a common ancestor which have then undergone divergent evolution. Although similar structures could evolve by chance more than once to perform a particular function, such convergent evolution is not thought to play a major role in sequence similarities. So many sequences could be constructed to perform a particular function that independent evolution of the same sequence based on function would be very improbable.

A number of groups including our own have worked out computer methods for the detection of distant relationships between sequences. Papers describing work by Needleman and Wunsch (1970), Fitch (1966, 1970), McLachlan (1971), Haber and Koshland (1970), Sankoff (1972), Gibbs and McIntyre (1970), and Sackin (1971), as well as our own (Barker and Dayhoff, 1970, 1972, 1973) have appeared recently.

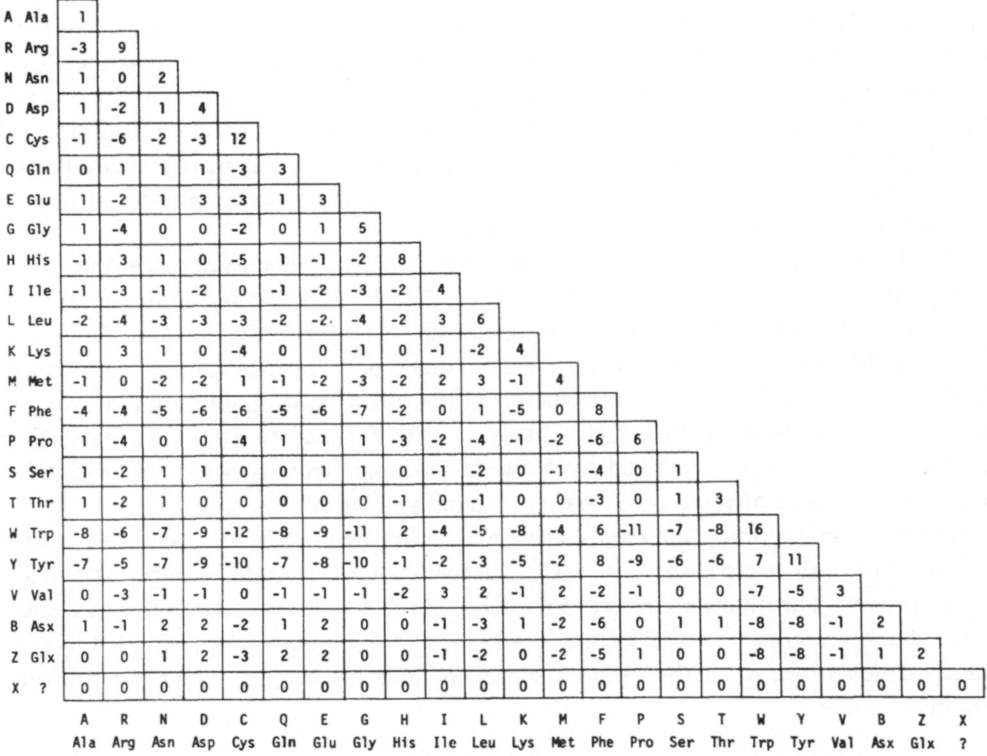

	A Ala	R Arg	N Asn	D Asp	C Cys	Q Gln	E Glu	G Gly	H His	I Ile	L Leu	K Lys	M Met	F Phe	P Pro	S Ser	T Thr	W Trp	Y Tyr	V Val	B Asx	Z Glx	X ?
A Ala	1																						
R Arg	-3	9																					
N Asn	1	0	2																				
D Asp	1	-2	1	4																			
C Cys	-1	-6	-2	-3	12																		
Q Gln	0	1	1	1	-3	3																	
E Glu	1	-2	1	3	-3	1	3																
G Gly	1	-4	0	0	-2	0	1	5															
H His	-1	3	1	0	-5	1	-1	-2	8														
I Ile	-1	-3	-1	-2	0	-1	-2	-3	-2	4													
L Leu	-2	-4	-3	-3	-3	-2	-2	-4	-2	3	6												
K Lys	0	3	1	0	-4	0	0	-1	0	-1	-2	4											
M Met	-1	0	-2	-2	1	-1	-2	-3	-2	2	3	-1	4										
F Phe	-4	-4	-5	-6	-6	-5	-6	-7	-2	0	1	-5	0	8									
P Pro	1	-4	0	0	-4	1	1	1	-3	-2	-4	-1	-2	-6	6								
S Ser	1	-2	1	1	0	0	1	1	0	-1	-2	0	-1	-4	0	1							
T Thr	1	-2	1	0	0	0	0	0	-1	0	-1	0	0	-3	0	1	3						
W Trp	-8	-6	-7	-9	-12	-8	-9	-11	2	-4	-5	-8	-4	6	-11	-7	-8	16					
Y Tyr	-7	-5	-7	-9	-10	-7	-8	-10	-1	-2	-3	-5	-2	8	-9	-6	-6	7	11				
V Val	0	-3	-1	-1	0	-1	-1	-1	-2	3	2	-1	2	-2	-1	0	0	-7	-5	3			
B Asx	1	-1	2	2	-2	1	2	0	0	-1	-3	1	-2	-6	0	1	1	-8	-8	-1	2		
Z Glx	0	0	1	2	-3	2	2	0	0	-1	-2	0	-2	-5	1	0	0	-8	-8	-1	1	2	
X ?	0	0	0	0	0	0	0	0	0	0	0	0	0	0	0	0	0	0	0	0	0	0	0

Fig. 1. Mutation data matrix. This matrix is derived by taking the logarithm of each element of a relatedness odds matrix for an evolutionary distance of 256 accepted point mutations per 100 amino acid links. Each term of the relatedness odds matrix represents the probability that two amino acids would occur at the same position in two proteins because of relationship, divided by the probability that the combination would occur by chance. The matrices that we are currently using are based on 814 mutations observed in deriving the evolutionary histories of eleven groups of closely related proteins. In addition, the relative mutabilities of each type of amino acid and the relative frequencies of occurrence of each were calculated and used in the derivation of the odds matrix. When a protein is compared to another, position by position, one should multiply the odds for each position to calculate an odds for the whole protein. However, it is more convenient to add the logarithms of the odds terms. Therefore, we use the matrix formed by taking the logarithm of each of the odds terms as our mutation data matrix. The matrix is symmetrical. The value of zero for an unknown amino acid and any known residue is arbitary; such a comparison seldom occurs.

We have found the algorithm of Needleman and Wunsch to be sensitive, rapid, objective, and easy to interpret. The work reported here is based on this algorithm. It determines the highest possible score for any alignment (including gaps) of two sequences. This score is then compared with the highest possible scores obtained by aligning pairs of randomized sequences having the same amino acid or nucleotide composition as the two real sequences. We have made 300 such random comparisons for each pair. The scores form a normal distribution for which the mean and standard deviation are calculated.

The sensitivity of the method depends ultimately upon scores accumulated from comparisons of a residue in one sequence with one in the other sequence. The contribution for each pair is specified in a matrix of pair scores which must be supplied to the program. The simplest such matrix, the 'unitary matrix', counts one for identities and zero for non-identities. We use this simple matrix for nucleic acids because there are not sufficient data for more precise estimates. For proteins, we have derived a matrix from mutation data which has proven to be very sensitive for detecting distant relationships (Barker and Dayhoff, 1970). The mutation data result from the combined effects of several factors, including the nature of the genetic code, the rates of mutation at the nucleotide level and natural selection.

The derivation of this matrix, shown in Figure 1, is described in detail in the *Atlas of Protein Sequence and Structure* (Dayhoff *et al.*, 1969; Barker and Dayhoff, 1972). There are many levels of distinction in this matrix. Amino acids which are often conserved in distantly related proteins, such as cysteine or tryptophan, have high scores on the diagonal and low scores elsewhere. On the other hand, highly mutable amino acids such as serine or alanine are as likely to have changed to any of several other amino acids as to have remained unchanged. The alignment score derived with this matrix is very little affected by exchanges among similar amino acids.

We have used the mutation data scoring matrix for the protein comparisons reported here. For each comparison a statistic is derived whose reliability can be estimated and improved. The difference between the score obtained with the two real sequences, s, and the mean score from the 300 pairs of randomized sequences, m, is divided by the standard deviation, σ, of the random scores, to give a score, A, which we call the alignment score.

$$A = \frac{s - m}{\sigma}.$$

The alignment score is thus expressed in units of standard deviations from the mean of random scores. The probability of obtaining an alignment score more than 2 standard deviation units above that from randomized sequences is less than 3%, and the probability of obtaining a score above 3.1 S.D. units is less than 0.1%.

Implicit in the method is a penalty for increasing the overall length of an alignment by inserting gaps in the sequences. We have set this parameter in such a way that the overall length is increased by only a few percent and the resolution of relationships is near optimal. This penalty is applied by adding a constant to every term of the scoring matrix. For amino acids we have used +12 and for nucleic acids, +2.

We have been investigating the possibility that this test for the probability of a distant relationship could be made into a quantitative method for ascertaining the evolutionary distance between two sequences. We have constructed a model sequence of 100 links having an average amino acid composition. From the initial sequence a whole family of other sequences with a known number of point mutations was generated by random processes, with the assumptions that each amino acid has a different probability of mutating in a given interval and each has a distinctive probability spectrum for the replacement amino acids. These probabilities were derived from the changes inferred to have taken place in the many evolutionary trees which we presented in the *Atlas of Protein Sequence and Structure 1969,* as described in Chapter 9 (Dayhoff *et al.*, 1969). This model does not fully express the unique mutability pattern of each position in a protein chain and differs from reality in giving fewer parallel mutations in a group of related sequences and in having fewer positions where no mutations are observed at all. In spite of the simplicity of the assumptions, considerable insight can be obtained with the model.

The dependence of alignment scores on total number of mutations in 100 links is shown by the lower curve of Figure 2. Even after 550 changes have occurred, the alignment score is 3. S.D. units above the score for infinite distance; after 1000 changes the score is 1 S.D. unit above and therefore positive scores are obtained 5/6 of the time. The percent difference between aligned sequences shown in the upper curve of Figure 2, although a good measure at short distance, deteriorates rapidly. At 300 changes the score is 3 standard deviations above the score for infinite distance and

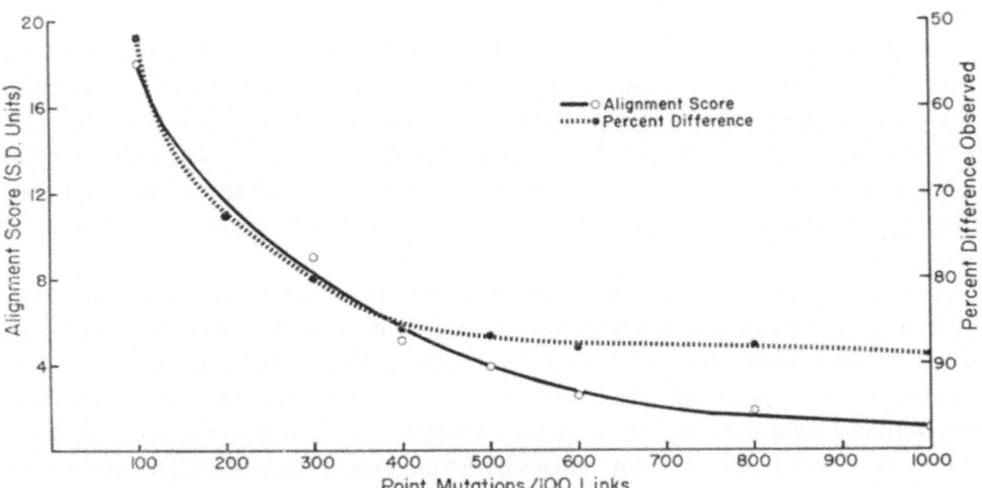

Fig. 2. Dependence of alignment scores on total number of mutations in model protein sequences of 100 links with average amino acid composition. The percent difference between sequences is a good measure of short evolutionary distances; but for sequences that are more than 80% different, the alignment score obtained with our mutation data matrix is a much better measure of relatedness.

after 390 changes it is within 1 standard deviation. For the region with more than 80% difference, the alignment score is a much better measure of relatedness than a count of differences.

It is possible to detect very old relationships using alignment scores. For example, if the related cytochrome sequences behaved like the model sequences, and changed at the same rate that is currently observed in the vertebrate c's, sequences which diverged

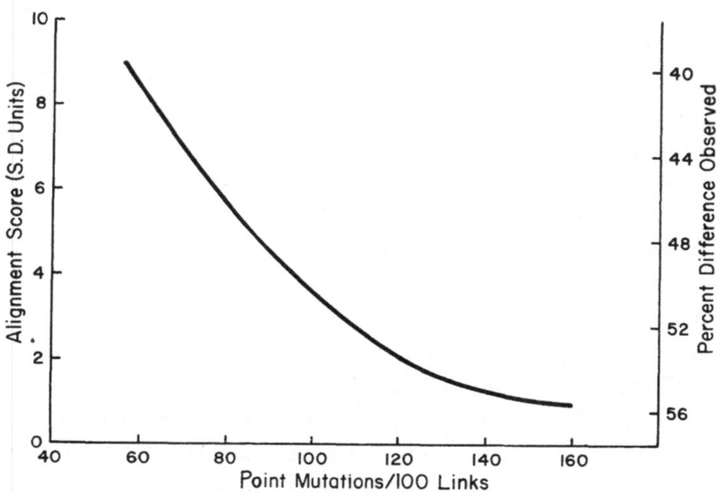

Fig. 3. Dependence of alignment scores and average percent difference on total number of mutations in model nucleic acid sequences of 100 links with equal frequency of occurrence of the four bases. A unitary matrix was used in determining the alignment scores.

3 b.y. ago would give a score of over 12 S.D. units. We do not see such high scores for most of the comparisons. This is in part due to changes in rate and in part because real sequences also suffer changes in length due to the insertion or deletion of genetic material. Such changes degrade the evidence of relationship in the sequence information considerably. In the case of cytochrome c and bacterial c_2 the equivalent number of changes inferred from alignment scores increased by 50% because of the several changes in length.

For nucleic acid sequences, the information degrades much more rapidly than with proteins, as can be seen in Figure 3. In this case we have used a model sequence consisting of 100 residues with equal numbers of each of the four nucleotides and we have assumed an equal probability of change from each nucleotide to each other one. This may not have been the case, but there is as yet insufficient information to estimate the comparison matrix elements more precisely. An alignment score of 3 S.D. units occurs after only 106 changes in a sequence of 100 links and a score of 1 S.D. unit occurs after 160 changes. This poor performance may result from a combination of two factors: that there are only three alternatives for a replacement and that these do not have the chemical distinctiveness of the amino acids.

From Figure 3 it is clear that nucleic acid sequences which are not related at all can still be aligned so that 44% of the nucleotides are identical and there are very few gaps. To protein buffs this may appear to be a good relationship, but in fact it is not meaningful. In spite of the high conservation necessary, we will see that many nucleic acid sequences still preserve recognizable relationships even though they diverged prior to or early in the development of coded proteins.

In the following discussion of the four groups of sequence data, we will use the alignment scores to establish the probability of relationship and the model to calculate approximate amounts of change for the distant sequences.

1. 5S Ribosomal RNA

One of the most interesting molecules, whose structure may antedate the origin of the genetic code, is 5S ribosomal RNA. This molecule, approximately 120 nucleotides in length in the known sequences, has been found in all species which have been examined (Maden, 1971). The genetic code for constructing proteins from nucleic acid templates involves two recognition processes, both involving transfer RNA (tRNA). In the first, each amino acid is recognized by a protein enzyme and bound to an appropriate tRNA. In the second, the anticodon of the charged tRNA binds to the messenger RNA. These recognition processes occur either in solution or on the smaller of the two ribosomal subunits. The 5S RNA, although its exact function is not known, is associated with the larger ribosomal unit which is the site of amino acid polymerization. This part of the process of protein synthesis is independent of the kind of amino acid which reacts and could date back to a more ancient synthetic process preceding that using the code.

TABLE I

Alignment scores of 5S RNA sequences (in S.D. units)

	Hu	Ye	*Ps.*	*E.c.*
Human				
Yeast	8.5			
Pseudomonas	2.8	2.2		
E. coli	4.8	4.6	12.2	

Five 5S RNA sequences have been completely determined, two from bacteria, one from yeast (Hindley and Page, 1972) and two very similar ones from the vertebrates. For simplicity in the following discussion of distant relationships, we will use only the human sequence from the vertebrates. The four sequences are recognizably related to each other, as shown by the alignment scores of the complete sequences in Table I. All pairs have alignment scores greater than 2 S.D. units and all scores involving the *E. coli* sequence are above 4.6. The corresponding probability that the *E. coli* sequence and any of the others could be this similar by chance is less than 10^{-5}. It is immediately evident that the two bacterial sequences are most similar to each other and the human and yeast sequences are most similar to each other. An alignment incorporating all of the sequences can be made, requiring only a few gaps which reflect changes in length.

The *E. coli* sequence was the first to be published, by Brownlee *et al.* in 1968. These authors noted that there were some repeating patterns in the sequence. The nine or ten residues at either end were very similar to each other, being identical at 7 positions. The rest of the molecule seemed to show a doubling. They aligned residues 10 to 60 and 61 to 110 and showed regions where 10 and 8 consecutive residues, respectively, were identical. In addition there were some shorter regions of matching. This amount of similarity is very unlikely to have occurred by chance.

We cut the alignment of the four sequences at the places suggested by the *E. coli* doubled regions, discarded the short terminal sequences, and investigated the relationships of these 'half chains' using alignment scores. The resulting scores are shown in Table II. The first halves show a definite relationship from chain to chain and the relationship among the second halves is also clear. In comparing the set of first halves

TABLE II

Alignment scores of 5S RNA half chains (in S.D. units)

	Hu 1	Ye 1	*Ps.* 1	*E.c.* 1	Hu 2	Ye 2	*Ps.* 2	*E.c.* 2	Anc.
Human 1									
Yeast 1	7.4								
Pseudo. 1	3.2	2.5							
E. coli 1	5.0	3.5	8.3						
Human 2	−0.8	−0.7	0.5	0.0					
Yeast 2	1.8	0.3	−0.3	−0.9	4.1				
Pseudo. 2	0.7	0.4	0.7	1.1	1.2	2.6			
E. coli 2	2.4	2.0	2.5	3.1	1.5	3.2	8.6		
Ancestor	3.3	3.1	4.3	7.4	1.9	3.4	7.0	11.6	

Note: The regions of the sequences used are shown in Figure 4. Where two alternatives appear for the ancestral sequence, we have used the upper character.

with the set of second halves, the *E. coli* halves are definitely related, and many of the other comparisons give positive scores. Some of the scores are low or even negative, indicating that there has been so much change that the common origin is no longer detectable. The degradation of information has involved either point mutations or insertions and deletions of nucleotides. Although the overall length is closely maintained, the alignment of the halves shows evidence of accumulated shifts of portions of the chain.

In order to establish the order of events independently of the somewhat subjective alignment of eight chains, we have used Figure 3 to determine for all pairs the evolutionary distances in terms of point mutations equivalent to the alignment scores. For every pair of organisms, the average evolutionary distance of the corresponding half

```
                                      1                   2                   3                   4                   5                   6
                    1 2 3 4 5 6 7 8 9 0 1 2 3 4 5 6 7 8 9 0 1 2 3 4 5 6 7 8 9 0 1 2 3 4 5 6 7 8 9 0 1 2 3 4 5 6 7 8 9 0 1 2 3 4 5 6 7 8 9 0 1 2

       Human        G C C - A U A C C A C C - - - - C U G A A C G C G C C - - - - - C G A U C U C - G U C U G A U C U C G - G A A G C U A A G C
FIRST  Yeast        G C C - A U(A C,C,A U,C,- - - - U)A G A A A G C A C C - - - - - - G U U C U C C G U C C G A U A A C C U G U A G U U A A G C
       Pseudomonas  A C G A G U A G U G G C - - - - A U U G G A A C A C C - - - - - U G A U C C C A U C C G A A C U C A - G A G G U G A A A C
HALF   E. coli      G C C - G U A G C G C G - - - - G U G G U C C C A C C - - - - - U G A C C C C A U G C C G A A C U C A - G A A G U G A A A C
-----------------------------------------------------------------------------------------------------------------------------------------------------
       Human        A G G - G U C G G G C C U G - G U U A G U A C U U G G A U G G G A G A C C G C C U G - G G A A - - - U - A C C G G G U - G C
SECOND Yeast        U G G - U A A G A G C C U G A C C G A G U A G U G U A G U G G G U G A C C A U A C G - C G A A - - - A - C C U A G G U - G C
       Pseudomonas  G A U - G C A U C G C C - - - G A U G G U A G U G U G - - G G G U U U C C C C A U G - U C A A - - - G - A U C U C G A - C C
HALF   E. coli      G C C - G U A G C G C C - - - G A U G G U A G U G U G - - G G G U C U C C C C A U G - C G A G - - - A - G U A G G G A - A C
-----------------------------------------------------------------------------------------------------------------------------------------------------
       Ancestor     G C C - G U A G C G C C - - - G A U G G U A G U G C G - - G G G U G A C C C C A U G - C G A A C U C A - G U A G N G A - N C
                        G                         -               C   U C     - - -                     C         - - -                     A
-----------------------------------------------------------------------------------------------------------------------------------------------------
       Common                   A         C                                                 C   C           G A                                 C
        to most
-----------------------------------------------------------------------------------------------------------------------------------------------------
```

Fig. 4. Alignment of 5S ribosomal RNA half chains. There are two regions in the *E. coli* sequences where remarkable similarity is preserved; the first 10 residues above are identical and residues 36 to 46 are identical except for the insertion of one residue in the first half chains. It is very probable that these similar regions were produced by a duplication of genetic material. Many mutations are evident in other regions of the alignment. The half-chain sequences from the other organisms, which have changed more than those from *E. coli*, have been aligned to conserve the match to the total *E. coli* sequence. The amino-terminal 9 residues of *E. coli* are almost identical to the carboxy-terminal 10 residues and must have been produced by an event separate from those producing the elongation of the main portion. We have omitted these terminal residues and the homologous regions in the other sequences from the alignment.

chains is always less than the average distances between the first halves and the second halves, which indicates early gene elongation preceding the divergence of species. This result is not closely dependent on the model and is also true if alignment scores themselves are used.

An alignment of the half sequences, shown in Figure 4, follows the main features of the original alignment of Brownlee *et al.* derived from the *E. coli* sequence. It is easy to derive an ancestral sequence from this set, as shown below the actual sequences. In each case of a single entry, the base shown occurs at least once in each set of half chains and is the most frequent, usually occurring in more than half of the sequences. In the cases where two alternatives are shown, each is conserved in at least 3 chains of one half, but does not occur in the other half. Where a decision could not be made, we placed an N. In a number of places insertions and deletions must have occurred. We have postulated a minimum number of these.

The alignment scores of the ancestral sequence with the half chains are shown in Table II. It is clear that the scores with all of the four bacterial half chains are higher than those with the four eukaryote half chains. This indicates that there has been more change in the eukaryote lines since the primordial duplication.

Finally, we have calculated the evolutionary tree using the ancestral sequence method (Dayhoff *et al.*, 1972b; McLaughlin *et al.*, 1972). The approximately 100 residues of each of the four sequences shown in the alignment and the corresponding portion of the toad sequence, as well as the duplicated ancestral sequence, were used. The best estimate of the tree is shown in Figure 5. From the relationships among the half chains,

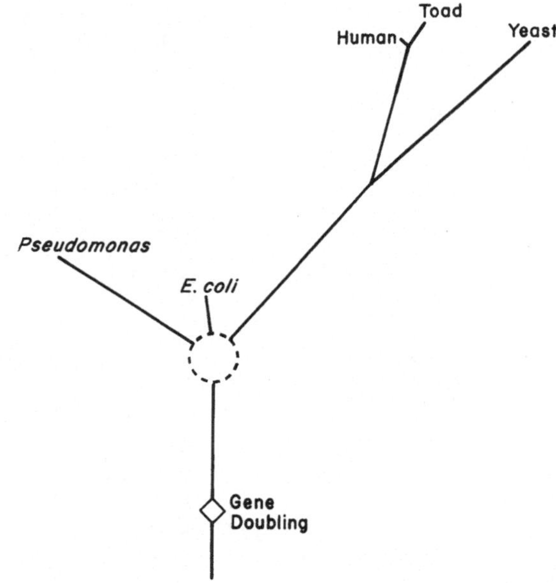

Fig. 5. 5S ribosomal RNA evolutionary tree. The shortened sequences corresponding to the length of the two halves in the alignment of Figure 4, with the addition of a doubled ancestral sequence, were used. The line lengths are suggestive of the number of mutations which are calculated to have occurred.

the time orientation in this tree was established: the chain elongation came first and was succeeded by the divergence of the eukaryote line from the bacterial lines leading to *Pseudomonas* and *E. coli*. Unfortunately, the order of divergence of these three lines cannot be unequivocally placed with this information. Among eukaryotes, first the fungal and vertebrate lines diverged and finally the two vertebrate lines to the mammal and the amphibian separated. The lengths of the lines in the figure are suggestive of the number of point mutations which are calculated to have occurred.

The relative importance of the various mechanisms producing genetic variability may have changed over the time interval represented by this tree. Between the time of the doubling and the divergence of the three lines, the introduction or deletion of genetic material appears to have predominated. During more recent evolution in all the lines, the point mutations far outnumber the changes in amount of genetic material.

This 5S RNA tree shows very different amounts of change in the various lines, the bacteria having changed much less than the eukaryotes. To check possible bias in the method, we also used two alternative lines of reasoning, which omitted all consideration of the doubling, for the upper portions of the tree. A tree made by matrix methods (Margoliash and Fitch, 1967) from the changes equivalent to the alignment scores for complete sequences (Table 1) is in agreement with the corresponding parts of the tree shown. Further, alignments of the five sequences were made under the assumption that all sequences were equally subject to change, and trees were derived from these alignments by the common ancestor method. The *E. coli* arm is still shortest and the *Pseudomonas* arm is shorter than those to the eukaryotes.

TABLE III

Comparisons between tRNA types in eukaryotes and in prokaryotes
(complete sequences without modifications)

tRNA types	Alignment scores		% Identity	
	Yeast–Yeast	*E. coli*–*E. coli*	Yeast–Yeast	*E. coli*–*E. coli*
Asp–Arg	4.2	6.6	50	61
Asp–Gly	3.6	7.1	48	64
Asp–Phe	2.3	5.5	44	53
Asp–Trp	3.1	7.2	47	62
Arg–Gly	2.5	5.3	46	55
Arg–Phe	2.2	4.9	43	55
Arg–Trp	6.6	6.5	61	57
Gly–Phe	4.7	7.7	53	62
Gly–Trp	4.9	8.1	48	64
Phe–Trp	4.1	4.7	53	53
Average	3.82	6.36	49.3	58.6

Note: The sequences which have not been published in the *Atlas of Protein Sequence and Structure* were taken from Harada *et al.*, 1972 (Asp tRNA, *E. coli*), Weissenbach *et al.*, 1972 (Arg tRNA, brewer's yeast) and Yoshida, 1973 (Gly tRNA, yeast).

2. Transfer RNA

The next question which arises is whether great disparity of evolutionary change is usual among bacteria. There is another large body of data for which the time orientation is known – the tRNA sequences. It now seems likely that the divergence of types for the various amino acids occurred very early, long before the divergence of the lines to *E. coli* and the eukaryotes, and that the types have evolved independently since then (Dayhoff and McLaughlin, 1972). Presently 5 types of similar length with homologous codons from both *Saccharomyces* yeast and *E. coli* are known. Alignment scores were derived from all pairs of these within each organism, giving a total of 10 comparisons. We have also derived a measure of percent identity from a single optimized alignment in which the codons and paired regions were matched. Table III presents the results. In both cases, the evidence overwhelmingly supports the view that there is less difference between the pairs of *E. coli* sequences than between the pairs of yeast sequences. Translating the average alignment scores into equivalent mutations, we see that the accumulated change in yeast evolution is 1.4 times larger than that in *E. coli* evolution.

3. The Ferredoxins and Adrenodoxin

The ferredoxins are small iron-containing enzymes which participate in various organisms in such fundamental biochemical processes as photosynthesis, nitrogen fixation, sulfate reduction and other oxidation-reduction reactions. Sequences are known from three main groups: plant plastids, anaerobic photosynthetic bacteria and anaerobic non-photosynthetic bacteria. The sequences of ferredoxins from plant plastids are nearly twice as long as those from non-photosynthetic bacteria, whereas that from *Chromatium*, a photosynthetic bacterium, is intermediate in length. There are also differences between the three types of ferredoxin in content of iron, sulfur, and cysteine and in the functional role of the protein. Nevertheless, portions of the sequences do exhibit similarities when they are aligned.

Adrenodoxin (Tanaka *et al.*, 1973) is an electron transport protein which has been isolated from mammalian adrenal mitochondria, where it acts in the intermediate steps of steroid hydroxylation. It is similar to the plant ferredoxins in length and in

TABLE IV

Alignment scores of mitochondrial adrenodoxin and plastid ferredoxin complete sequences (in S.D. units)

	Bovine adrenodoxin	Algal ferredoxin
Ferredoxin-algal (*Scenedesmus*)	2.8	
Ferredoxin-higher plant (alfalfa)	2.5	19.4

the numbers of cysteines and of bound iron and sulfur atoms which it contains. Alignment scores for adrenodoxin compared to the plant ferredoxins are shown in Table IV. The probabilities of obtaining such scores using randomly scrambled sequences of the same composition are less than 0.01, suggesting a remote relationship. Together with the similarities in overall length, function, and numbers and correspondence of functionally important amino acid residues and prosthetic groups, it seems likely that adrenodoxin shares a distant common ancestor with the plant ferredoxins (Barker *et al.*, 1972).

It has long been recognized that the bacterial ferredoxin sequences display evidence of a gene doubling (Eck and Dayhoff, 1966). Matsubara *et al.* (1967) suggested that the plant and bacterial sequences were recognizably related. Matsubara *et al.* (1968) have advanced a detailed hypothesis regarding the many evolutionary steps linking the two types of sequences. They suggested that the evolution included an early doubling in the ancestral form and the appearance of additional material including smaller duplications and deletions in the plant forms. In what follows we have examined for relationships the portions of the sequences which align with the duplicated bacterial sequences (Dayhoff *et al.*, 1972a, p. D–39); we have omitted the additional terminal material from the plant, *Chromatium*, and adrenodoxin sequences.

Table V gives the alignment scores of the half chains of sequences from *Micrococcus aerogenes* (a non-photosynthetic anaerobic bacterium), *Chromatium* (a photosynthetic anaerobic baterium), plastids of alfalfa (a higher green plant), and from bovine mitochondrial adrenodoxin. The bacterial sequences are definitely related and they share the duplication. Comparison of the sequences from bacteria with those from eukaryote

TABLE V

Alignment scores of ferredoxin half-chains from non-photosynthetic and photosynthetic bacteria, green plant plastids and bovine mitochondria

		First half				Second half			
		Micrococcus	*Chromatium*	Alfalfa	Adrenodoxin	*Micrococcus*	*Chromatium*	Alfalfa	Adrenodoxin
irst alf	*Micrococcus aerogenes* (1–26)								
	Chromatium (1–28)	6.4							
	Alfalfa (32–61)	2.3	4.4						
	Adrenodoxin (37–75)	1.5	2.8	3.4					
cond alf	*Micrococcus* (27–54)	5.6	4.5	1.6	1.8				
	Chromatium (29–65)	3.6	1.3	1.0	0.0	3.4			
	Alfalfa (62–88)	0.8	1.0	0.5	1.0	1.5	1.7		
	Adrenodoxin (76–104)	0.3	0.6	1.5	1.9	0.5	1.0	2.0	

organelles indicates that the relatedness of the first halves is very probable, while that of the second halves is quite distant. The cross terms between the two halves for the bacterial sequences are at a greater distance than the corresponding halves, indicating that the duplication preceded the divergence of these species. While the evidence from the plant ferredoxin and adrenodoxin is not inconsistent with a duplication shared by all the types, neither does it show it conclusively.

A number of bacterial ferredoxin sequences are known and from an alignment of the halves an ancestral sequence can be readily constructed (cf. Dayhoff *et al.*, 1972a, p. D-40). We used such a sequence to obtain the alignment scores of Table VI. If the mitochondrial adrenodoxin and the plastid ferredoxin shared the ancient duplication, it is quite clear that the adrenodoxin sequence has changed most, followed by the plastid sequences, then by that of *Chromatium* and finally by those of the other bacteria.

TABLE VI

Alignment scores of the ancestral ferredoxin half chain with other half chains (in S.D. units)

	First half	Second half
Clostridium butyricum	8.48	6.55
Micrococcus aerogenes	6.80	6.80
Chromatium	7.92	3.16
Scenedesmus	3.95	0.80
Alfalfa	4.36	1.43
Bovine adrenodoxin	1.89	0.22

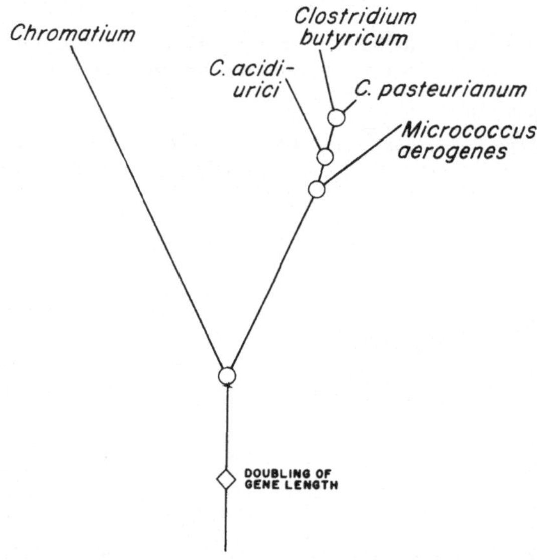

Fig. 6. Evolutionary tree of bacterial ferredoxins. The earliest event was the gene duplication. This was followed by the divergence of the photosynthetic and non-photosynthetic lines. It is impossible to decide from present evidence whether the ancestral organism was photosynthetic or not.

Figure 6 shows the order of events in the bacterial lines. It should be noted that all of the non-photosynthetic forms are on one branch and *Chromatium* is on the other. It is not possible to decide from this whether or not their common ancestor was photosynthetic. The possibility that sequence information could provide an earliest point on the phylogenetic tree of all life and an answer to the question of the primacy of the photoautotroph or the heterotroph is tantalizing.

4. C-Type Cytochromes

Complete sequences of five diverse c-type cytochromes, demonstrably related in sequence to the eukaryote c, are now available. A great deal of information is known about the structure and function of the eukaryote cytochrome c. It is coded in the nucleus and is functional in the respiratory chain of the mitochondrion. Many sequences representing all four eukaryote kingdoms have been educed. Among the bacterial sequences known, by far the most similar to the eukaryote c is cytochrome c_2 from *Rhodospirillum rubrum*. At considerable evolutionary distance from this pair and from each other are three other sequences: c_{551} (Ambler and Wynn, 1973) and c_5 (Ambler and Taylor, 1973) from pseudomonads, functional in respiratory chains; and c_{553} from the plastid of a chrysophycean or golden alga (Laycock and Craigie, 1971; Laycock, 1972), possibly functional in electron transport between photosystems I and II. These cytochromes all have standard reducing potential in the range +0.25 to +0.39 (Mahler and Cordes, 1966; Laycock and Craigie, 1971). Most distant from all of the others is cytochrome c_{553} from *Desulfovibrio vulgaris* (Bruschi and Le Gall, 1972); this form functions in anaerobic metabolism and has a negative standard reducing potential. All six kinds of protein have a heme group covalently attached to cysteine in the amino terminal portion of the chain, and in addition, the c_5 has a second heme similarly attached near the carboxyl end. The alignment scores of these sequences are shown in Table VII. The four remote sequences show >99.9% probability of relationship to either c or c_2.

TABLE VII

Alignment scores of c-type cytochromes

		Horse c	*R.* c_2	Algal plastid c_{553}	*Ps.* c_{551}	*Ps.* c_5	*D.* c_{553}
c	Horse						
c_2	*Rhodospirillum*	11.6					
c_{553}	Algal Plastid	5.2	3.1				
c_{551}	*Pseudomonas*	2.5	3.4	4.3			
c_5	*Pseudomonas*	2.9	4.5	5.4	3.8		
c_{553}	*Desulfovibrio*	3.3	2.5	1.9	2.7	2.0	

Note: Because the molecular architecture of the c_5 sequence is somewhat different from the others, with a longer sequence preceding the first cysteines, we have deleted the first 5 residues for these comparisons. The species from which the sequences were derived include *Rhodospirillum rubrum* (c_2), *Monochrysis lutheri* (c_{553}), *Pseudomonas aeruginosa* (c_{551}), *Pseudomonas mendocina* (c_5), and *Desulfovibrio vulgaris* (c_{553}).

The c and c_2 sequences show marked similarity along their entire lengths, with identical residues at 42 of their 100 aligned residues. The pairs with scores higher than 3 S.D. units show unusual similarity at both ends of the chains, whereas those with scores below 3 S.D. units are similar only in the first third of their sequences where the heme-binding sites are located.

Using the relationship derived from model sequences, we have translated the alignment scores into the equivalent number of point mutations per 100 residues. These range from a total distance of 200 for the c and c_2 sequences to an average of 600 for the *Desulfovibrio* comparisons. From these distances we have estimated a tree, as shown in Figure 7. *Desulfovibrio vulgaris* is an obligatory anaerobe which is generally

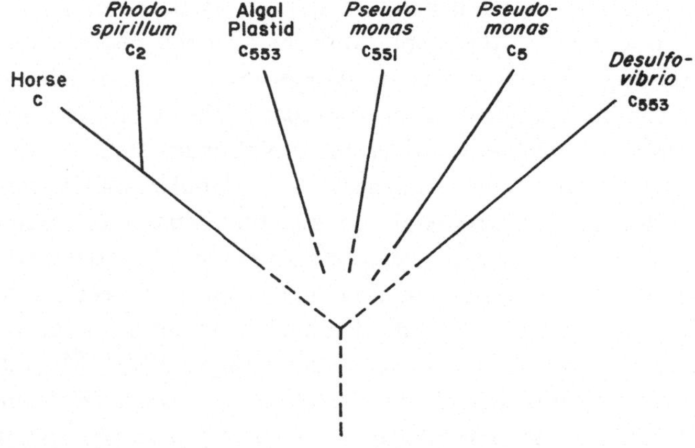

Fig. 7. Estimated evolutionary tree of various distantly related c-type cytochromes. The alignment scores shown in Table VII were translated into the equivalent number of point mutations per 100 residues using the curve in Figure 2. The tree was derived from these estimated distances. This procedure circumvents the necessity of aligning the very distantly related sequences, which is difficult to do. It is evident that c and c_2 are strikingly related whereas all of the other types are very distant from these two and from each other.

conceded to have developed from an early branching during evolution (Margulis, 1970), before the development in other lines of proteins with high positive reducing potentials. We have assumed that an early event on the tree was this species divergence. The great evolutionary distance of the sequences does not conflict with this assumption. The exact order of divergence of the early branches cannot be decided. However, it is clear that c and c_2 are relatively very close together. A great many branches are missing on this tree, so that it is not always possible to distinguish branch points produced by gene duplications within one organism from those produced by divergences of bacterial groups.

Comparison of the evolutionary distances of c and c_2 from each of the lines diverging earlier shows no significant differences in the amount of change in the two lines. We have drawn the tree as though the amounts of change on the two lines were equal.

The complete set of cytochromes which might be recognizably related to this family has not been characterized from any of the groups of organisms mentioned. In eukaryotes there is a c_1 chain which also functions in the mitochondrion, and there is a c_6 from the plastids of higher green plants (Mahler and Cordes, 1966), believed to be an analogue to the algal c_{553} (Laycock, 1972). Ambler has intensively investigated the sequences in pseudomonads. There appears to be at least one additional type, termed c_4 (Ambler and Murray, 1973), for which the amino-terminal sequence was determined and found to be similar to that of c_2 and c.

There are other cytochromes designated broadly as c-type whose sequences do not give significant alignment scores with the c–c_2 group (Dayhoff *et al.*, 1972a; Dayhoff, 1973). These include cytochrome c_3 from *Desulfovibrio*, cytochrome $c_{551.5}$ of undetermined origin and cytochrome cc′ from pseudomonads. Cytochrome b_5 and b_{562} sequences also do not yield significant alignment scores with the c group.

From the c-type cytochromes we can develop an outline of the differentiation of eukaryotes from prokaryotes. This outline is shown in Figure 8, which is an evolutionary tree produced by the ancestral sequence technique. This tree is adapted from a more detailed study of cytochrome c evolution (McLaughlin and Dayhoff, 1973). From this evidence and from that in the previous tree, it is seen that the earliest divergence here is of two lines of prokaryotes. One line leads to *Rhodospirillum* as the present descendant; when other comparable bacterial sequences are known, undoubtedly there will be branches from this line leading to other bacterial groups. The other line evolved into the common ancestor of the eukaryotes. The scheme of evolution

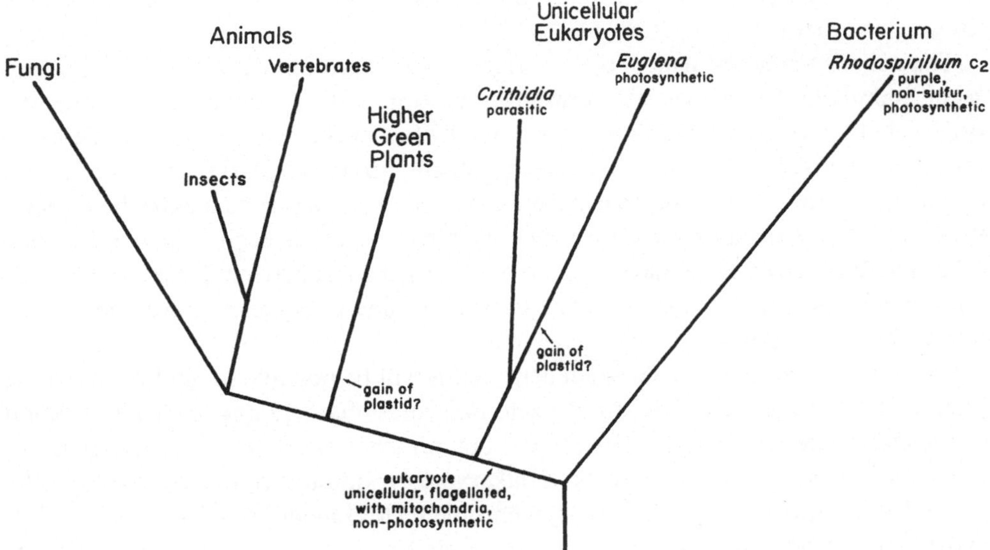

Fig. 8. Outline evolutionary tree of cytochromes c and c_2. The evidence in cytochrome sequences indicates the order of divergence of eukaryote kingdoms shown here. The lengths of the branches are proportional to the number of mutations estimated to have occurred on these branches. The 'stem' of the tree is placed by reference to the preceding tree. The *Euglena* sequence is from Pettigrew, 1973. (Adapted from McLaughlin and Dayhoff, 1973.)

shown by cytochromes slightly favors the symbiotic theory on the origin of plastids in eukaryote cells (Margulis, 1970) and we show the tree for this hypothesis. The common ancestor of the eukaryotes would have been a heterotrophic, flagellated unicellular organism. This ancestor would have had mitochondria which were functioning in electron transport and were already advanced in the process of becoming integrated into the host cell's metabolism. Probably, like *Rhodospirillum* and the eukaryote groups which do not have plastids, this ancestor did not have photosystem II. Separate mechanisms for ingestive nutrition evidently evolved in the diverging protozoa and metazoa. Photosynthetic nutrition would have arisen by the association of blue-green algal symbionts, which became plastids, with heterotrophic hosts. In this tree we can suppose that the acquisition of algal symbionts occurred twice – once on the line leading to the unicellular *Euglena* and another time in an early ancestor of the multicellular plant line.

The mitochondrion, because it is found in all eukaryote species in this tree but not in any prokaryotes, must have appeared in the short evolutionary interval between the divergence of the *Rhodospirillum* line and that of the flagellates. There are two basic theories for this. Either a symbiont was acquired which has since lost or transferred to the nucleus many of its genes and has retained only a few genes and a coding apparatus within itself, or alternatively the eukaryote nucleus and mitochondrion developed from genetic material already in the bacterial ancestor through duplications and reorganization. The correlation of evolutionary trees worked out in the future for various related sequences of prokaryotes and eukaryotes should distinguish between these two theories, or, should some other series of events actually have occurred, provide insight into them.

The answer to the problem of the connection of eukaryote and prokaryote organisms seems imminent. In phylogenetic terms, the question of the symbiotic *vs* nonsymbiotic origin of eukaryote cells becomes a question of whether there are or are not free-living forms on the branches leading to nuclear, mitochondrial and plastid groups of genes. So far, the c-type cytochrome tree has branches leading to a plastid and to the nucleus, with a free-living form on the nuclear branch very close to the beginning of eukaryotic cells. The blue-green algae have long been suggested as free living descendants of the lines which became plastids, but alas, no cytochrome c sequences have been determined from this group.

From these examples, we feel confident that it will be possible to establish the time orientation and the main features of bacterial evolution, including the lines of origin of the eukaryote nucleus, the plastid, and the mitochondrion. We are encouraged in this hope because the main features of eukaryote evolution can, in large measure, be worked out with only one protein, cytochrome c (McLaughlin and Dayhoff, 1973), and because many bacterial sequences show evidence of relationship to each other and the trees derived from them are consistent with other evidence. Further, the bacterial sequences are often more strongly conserved than the eukaryote types, which compensates for the increase in the time span over which evolution has occurred. Thus many 'living fossil sequences' of bacteria await our inquiry.

Acknowledgements

This investigation was supported by Contract NASW-2288 from the National Aeronautics and Space Administration, NIH Grant GM-08710 from the Institute of General Medical Sciences and by NIH Grant RR-05681 from the Division of Research Resources.

We wish to thank Dr Lois T. Hunt for helpful discussions and M. J. Gantt and A. Y. Kulkarni for technical assistance.

References

Ambler, R. P. and Murray, S.: 1973, *Biochem. Soc. Trans.* **1**, 162.
Ambler, R. P. and Taylor, E.: 1973, *Biochem. Soc. Trans.* **1**, 166.
Ambler, R. P. and Wynn, W.: 1973, *Biochem. J.* **131**, 485.
Barker, W. C. and Dayhoff, M. O.: 1970, *Biophys. Soc. Abs.* **10** 152a.
Barker, W. C. and Dayhoff, M. O.: 1972, in M. O. Dayhoff (ed.), *Atlas of Protein Sequence and Structure*, Vol. **5**, National Biomedical Research Foundation, Washington, D.C.
Barker, W. C. and Dayhoff, M. O.: 1973, *Biophys. Soc. Abs.* **13**, 205a.
Barker, W. C., McLaughlin, P. J., and Dayhoff, M. O.: 1972, *Fed. Proc.* **31**, 837Abs.
Brownlee, G. G., Sanger, F., and Barrell, B. G.: 1968, *J. Mol. Biol.* **34**, 379.
Bruschi, M. and Le Gall, J.: 1972, *Biochim. Biophys. Acta* **271**, 48.
Cantor, C. R.: 1968, *Biochem. Biophys. Res. Commun.* **31**, 410.
Dayhoff, M. O., (ed.): 1972, *Atlas of Protein Sequence and Structure*, Vol. 5, National Biomedical Research Foundation, Washington, D. C.
Dayhoff, M. O. (ed.): 1973, *Atlas of Protein Sequence and Structure*, Vol. 5, *Supplement I*, National Biomedical Research Foundation, Washington, D. C.
Dayhoff, M. O. and McLaughlin, P. J.: 1972, in M. O. Dayhoff (ed.), *Atlas of Protein Sequence and Structure*, vol. 5, National Biomedical Research Foundation, Washington, D.C.
Dayhoff, M. O., Eck, R. V., and Park, C. M.: 1969, in M. O. Dayhoff (ed.), *Atlas of Protein Sequence and Structure*, vol. 4, National Biomedical Research Foundation, Silver Spring, Md.
Dayhoff, M. O., Hunt, L. T., McLaughlin, P. J., and Barker, W. C.: 1972a, in M. O. Dayhoff (ed.), *Atlas of Protein Sequence and Structure*, vol. 5, National Biomedical Research Foundation, Washington, D.C.
Dayhoff, M. O., Park, C. M., and McLaughlin, P. J.: 1972b, in M. O. Dayhoff (ed.), *Atlas of Protein Sequence and Structure*, vol. 5, National Biomedical Research Foundation, Washington, D.C.
Eck, R. V. and Dayhoff, M. O.: 1966, *Science* **152**, 363.
Fitch, W. M.: 1966, *J. Mol. Biol.* **16**, 9.
Fitch, W. M.: 1970, *J. Mol. Biol.* **49**, 1.
Gibbs, A. J. and McIntyre, G. A.: 1970, *Eur. J. Biochem.* **16**, 1.
Haber, J. E. and Koshland, D. E. Jr.: 1970, *J. Mol. Biol.* **50**, 617.
Harada, F., Yamaizumi, K., and Nishimura, S.: 1972, *Biochem. Biophys. Res. Commun.* **49**, 1605.
Hindley, J. and Page, S. M.: 1972, *FEBS Letters* **26**, 157.
Laycock, M. V. and Craigie, J. S.: 1971, *Can. J. Biochem.* **49**, 641.
Laycock, M. V.: 1972, *Can. J. Biochem.* **50**, 1311.
Maden, B. E. H.: 1971, *Prog. Biophys. Mol. Biol.* **22**, 127.
Mahler, H. R. and Cordes, E. H.: 1966, *Biological Chemistry*, Harper and Row, New York.
Margoliash, E. and Fitch, W. M.: 1967, *Science* **155**, 279.
Margulis, L.: 1970, *Origin of Eukaryotic Cells*, Yale Univ. Press, New Haven, Conn.
Matsubara, H., Jukes, T. H., and Cantor, C. R.: 1968, *Brookhaven Symp. Biol.* **21**, 201.
Matsubara, H., Sasaki, R. M., and Chain, R. K.: 1967, *Proc. Nat. Acad. Sci. U.S.* **57**, 439.
McLachlan, A. D.: 1971, *J. Mol. Biol.* **61**, 409.
McLaughlin, P. J. and Dayhoff, M. O.: 1973, *J. Mol. Evol.* **2**, 99.
McLaughlin, P. J., Hunt, L. T., and Dayhoff, M. O.: 1972, *J. Human Evol.* **1**, 565.
Needleman, S. B. and Wunsch, C. D.: 1970, *J. Mol. Biol.* **48**, 443.

Pettigrew, G. W.: 1973, *Nature* **241**, 531.
Sackin, M. J.: 1971, *Biochem. Genet.* **5**, 287.
Sankoff, D.: 1972, *Proc. Nat. Acad. Sci. U.S.* **69**, 4.
Tanaka, M., Haniu, M., Yasunobu, K. T., and Kimura, T.: 1973, *J. Biol. Chem.* **248**, 1141.
Weissenbach, J., Martin, R., and Dirheimer, G.: 1972, *FEBS Letters* **28**, 553.
Yoshida, M.: 1973, *Biochem. Biophys. Res. Commun.* **50**, 779.

ON THE POSSIBLE ORIGIN AND EVOLUTION OF THE GENETIC CODE

THOMAS H. JUKES

Space Sciences Laboratory, University of California, Berkeley, Calif. 94720, U.S.A.

Abstract. The genetic code is examined for indications of possible preceding codes that existed during early evolution. Eight of the 20 amino acids are coded by 'quartets' of codons with four-fold degeneracy, and 16 such quartets can exist, so that an earlier code could have provided for 15 or 16 amino acids, rather than 20. If two-fold degeneracy is postulated for the first position of the codon, there could have been 10 amino acids in the code. It is speculated that these may have been phenylalanine, valine, proline, alanine, histidine, glutamine, glutamic acid, aspartic acid, cysteine and glycine. There is a notable deficiency of arginine in proteins, despite the fact that it has six codons. Simultaneously, there is more lysine in proteins than would be expected from its two codons, if the four bases in mRNA are equiprobable and are arranged randomly. It is speculated that arginine is an 'intruder' into the genetic code, and that it may have displaced another amino acid such as ornithine, or may even have displaced lysine from some of its previous codon assignments. As a result, natural selection has favored lysine against the fact that it has only two codons. The introduction of tRNA into protein synthesis may have been a cataclysmic and comparatively sudden event, since duplication of tRNA takes place readily, and point mutations could rapidly differentiate members of the family of duplicates from each. Two tRNAs for different amino acids may have a common ancestor that existed more recently than the separation of the prokaryotes and eukaryotes. This is shown by homology of two *E. coli* tRNAs for glycine and valine, and two yeast tRNAs for arginine and lysine.

1. Introduction

The evolution of the genetic code has been a favorite subject of speculation for some years. Two general approaches are often used. The first consists of attempts to find an affinity between amino acids and nucleic acid bases. The idea is to show how a primitive translation system may have started prior to the origin of life. This approach has the advantage of giving rise to laboratory experiments. The second approach, which I shall discuss first, is to work backward from the present code to something simpler. One attraction of this procedure is that it is linked to actual existence, which makes an author feel more secure, like a rock-climber who uses a rope on a cliff, rather than 'climbs free'.

The present code (Table I) has much to commend it. One of its good features is that no one was able to guess what it was, although various people tried to do so. This may mean that no-one will be able to guess how the code started.

One obvious feature of the code that makes one speculate on its possible evolution is its pattern of degeneracy. Eight amino acids: alanine, arginine, proline, glycine, leucine, serine, threonine, valine are coded by two bases plus 'any base in the third position'. An example is the code GCN, for alanine, where N is any base. It is tempting to suggest (Jukes, 1966) that an earlier code, from which the present code is descended, may have been for only 15 amino acids, each coded by a doublet and any third base to act as a 'spacer'.

TABLE I

The genetic code

UUU Phenylalanine	UCU Serine	UAU Tyrosine	UGU Cysteine
UUC Phenylalanine	UCC Serine	UAC Tyrosine	UGC Cysteine
UUA Leucine	UCA Serine	UAA Chain Termn..	UGA Chain Termn.
UUG Leucine	UCG Serine	UAG Chain Termn.	UGG Tryptophan
CUU Leucine	CCU Proline	CAU Histidine	CGU Arginine
CUC Leucine	CCC Proline	CAC Histidine	CGC Arginine
CUA Leucine	CCA Proline	CAA Glutamine	CGA Arginine
CUG Leucine	CCG Proline	CAG Glutamine	CGG Arginine
AUU Isoleucine	ACU Threonine	AAU Asparagine	AGU Serine
AUC Isoleucine	ACC Threonine	AAC Asparagine	AGC Serine
AUA Isoleucine	ACA Threonine	AAA Lysine	AGA Arginine
AUG Methionine	ACG Threonine	AAG Lysine	AGG Arginine
GUU Valine	GCU Alanine	GAU Aspartic acid	GGU Glycine
GUC Valine	GCC Alanine	GAC Aspartic acid	GGC Glycine
GUA Valine	GCA Alanine	GAA Glutamic acid	GGA Glycine
GUG Valine	GCG Alanine	GAG Glutamic acid	GGG Glycine

An appealing feature of this suggestion is the finding that the same transfer RNA molecule can be the ancestor of two tRNAs, each for a different amino acid. This is shown by comparing the sequences of *E. coli* glycine and valine tRNAs, and of yeast lysine and arginine tRNAs (Table II). Therefore, there is an existing process for switching the translation of a codon from one amino acid to another. However, there is no way of identifying the amino acid that four codons may have specified in a 'earlier code'. For example, were all four codons that start with AG originally assigned to arginine, or to serine, or perhaps to neither one, since arginine and serine each have four other codons?

2. Ambiguous Pairing Between the First Base of Anticodons and the Third Base of Codons

Some anticodons can pair with more than one codon. For example, it was found that the same phenylalanine tRNA would pair with the trinucleotides UUU and UUC; furthermore, no phenylalanine tRNA has been found that will pair with only UUU, or only UUC. When phenylalanine tRNAs were sequenced, the anticodon was found to be GmAA in yeast and wheat (Gm = oxymethyl guanosine) and GAA in *E. coli.* The G or Gm paired with both C and U. This and other findings led Crick (1966) to propose that, while the pairing between the first two bases of the codon and the last two bases of the anticodon followed strictly the regular A·U and G·C rule, there was a certain amount of 'play' or 'wobble' in the pairing between the first base of the anticodon and the third base of the codon. This would be useful biologically, because the utmost fidelity is needed in translating the first two bases of the codon, which are

TABLE II

Comparison of base sequences (unmodified) of *E. coli* tRNAs Valine$_{2B}$ and Glycine$_{3}$, and Yeast tRNAs Lysine$_{H}$, and Arginine$_{3A}$

Val EC2B	G C G U U C A	U A	G C U C	A G U U - G G U - U A	G A G C	A	C C A C C	U U G A C A U	
Gly EC3	G C G G G A A	U A	G C U C	A G U U - G G - - U A	G A G C	A	C G A C C	U U G C C A A	
Val EC2B	G G U G G	G G G - C	G G U G G	U U C G A G U	C C A A U	U G A A C G C	A		% Identical bases = 74
Gly EC3	G G U C G	G G G U C	G C G A G	U U C G A G U	C U C G U	U U C C C G C	U		
Lys YH	G C C U U G U	U G	G C G C	A A U C - G G - - U A	G C G C	G	U A U G A	C U C U U A A	
Arg Y3A	G C G C U C G	U G	G C G U	A A U - - G G - - C A	A C G C	G	U C U G A	C U U C U A A	
Lys YH	U C A	U A A G N U U	A G G G G	U U C G A G C	C C C C U	A C A G G G C	U		% Identical bases = 73
Arg Y3A	U C A	G A A G A U U	A U G G G	U U C G A C C	C C C A U	C G A G U G C	G		

The vertical lines separate helical and non-helical regions.

strictly specific for 8 of the amino acids (see Table I), but the third base of the codon is ambiguous or 'degenerate' in the codes for 18 of the amino acids. Therefore, it might be 'biologically economical' to use the same tRNA for both phenylalanine codons.

By building models, Crick concluded that (i) the alanine anticodon IGC, in which adenine (in adenosine) has been deaminated to hypoxanthine (in inosine), would pair with GCC, GCU and GCA. (ii) either of the two chainterminating codons UAA and UAG would pair with the anticodon UUA. This is thought to be in an *E. coli* serine suppressor tRNA as a result of a mutation from UGA to UUA in the anticodon. (iii) the anticodon CUA, found in another serine suppressor tRNA as a result of a mutation from CGA, would pair only with UAG. Therefore, I would 'wobble' with C, U or A; U would 'wobble' with A or G, but C in the first anticodon position would be specific for G in the third position in the codon.

True to prediction, inosine has been found in the anticodons only of tRNAs for amino acids that have three codons containing U, C and A in the third position. They were alanine, serine, isoleucine, valine and arginine. Most of these tRNAs are from eukaryotic organisms, but one for arginine is from *E. coli,* the tRNAs of which, in bulk, contain very little I.

A new development was reported by Murao *et al.* (1970). They found that uridine-5-oxyacetic acid in the first position of the anticodon oacUAC in *E. coli* valine tRNA would pair with A, G and U, present in valine codons GUA, GUG and GUU, so that the pairing of G in codons was not specific for C in the anticodon. Furthermore, and in contrast, a number of anticodons were found to contain 2-thiouridine in position 1, pairing only with A (Nishimura, 1972), rather than with A and G as in the case of unmodified U (Table III). Evidently, the first base in the anticodon can be modified to produce either greater or less pairing specificity. I conclude that codon-anticodon pairing has changed during evolution to the extent that modification of the bases in anticodons has altered the specificity of pairing with codons in the directions both of more and less ambiguity.

3. The Importance of U·G Pairing in the Evolution of the Code

A striking thing about the code is that eleven amino acids are coded by pairs of codons that end either in a pyrimidine or in a purine, but no amino acid is coded only by a pair of codons one of which ends in a pyrimidine and the other in a purine. To give an example, histidine has the codons CAU and CAC; glutamine, CAA and CAG. Surely this pattern must have some evolutionary significance, because the pattern is a functional one, and is related to codon-anticodon pairing. Codons ending with a pyrimidine will pair with anticodons starting with G. Codons ending with a purine will pair with anticodons starting with U. These two general rules hold except for anticodons that start with modified bases. Such modifications are probably of recent evolutionary origin since they differ in different species.

Crick's proposal shows that this pattern in the code depends on U·G pairing between the first base in the anticodon and the third base in the codon ('1–3 pairing')

TABLE III

First base of anticodons in sequenced tRNAs compared with third base of codons paired

First base in anticodon	Pairing with	Organisms (E = eukaryotes P = prokaryotes)	In tRNAs for
H (I)	U, C, A	E, P	Ile, Val, Ser, (UCN), Ala, Arg (CGN)
oa^5U	U, A, G	P	Val, Ser (UCN)
G	C, U	E, P	Phe, Leu (CUN), Ile, Val, Tyr, Asp, Arg (CGN), Gly
Gm	C, U	E	Phe
G† ('Q')	C, U	P	Tyr, His, Asn, Asp
U	A, G	P	Gly, Term.
s^2U	A	E, P	Gln, Lys, Glu
U†	A, G(?)	E, P	Leu (UUR), Arg (AGR)
C	G	E, P	Leu (CUN, UUR), fMet, Gln, Lys, Trp, Gly
ac^4C	G	P	Met
Cm	G	E	Trp

Abbreviations: H, hypoxanthine; I, inosine; oa^5U, uridine-5-oxyacetic acid; Gm, 2′-0-methylguanosine; G†, 'Q-base' (an unidentified guanosine derivative); s^2U, 2-thiouridine; U†, unidentified uridine derivative; ac^4C, 4-acetylcytidine; Cm, 2′-0′methylcytidine; N, unidentified nucleoside; R, purine.

as follows:

→ Anticodons	GUG	UUG
Codons ←	CAC	AAC
	UAC	GAC
	His	Gln

We shall now examine the possibility for U·G pairing between the *third* base in the anticodon and the *first* base in the codon in an earlier genetic code ('3–1 pairing'). It appears that an elaborate system to *prevent* this ambiguity has evolved in tRNA. This is expressed in special modifications of the nucleoside next following the anticodon (site 40) in tRNA (Table IV). There is, in most cases, an isopentenyl-containing side-chain at site 40 when there is 3–1 pairing between A and U, and a threonyl-containing side-chain at this location when there is 3–1 pairing between U and A, and, in prokaryotes, a methyl group when there is 3–1 pairing between G and C. An obvious role for such side-chains is to prevent 3–1 pairing between G and U, and U and G respectively. The prevention of this pairing is necessary to ensure fidelity of translation of codons starting with U or A, *except* in the case of the initiation of polypeptide chains. In *E. coli* the initiator codon is AUG or GUG, either of which is translated as methionine by

TABLE IV

Modifications of the nucleosides at site 40 in tRNAs as correlated with recognition of codons in *E. coli* and other prokaryotes (Prok.) and yeast and other eukaryotes (Euk.)

	Nucleoside adjoining Anticodon			Nucleoside adjoining Anticodon			Nucleoside adjoining Anticodon			Nucleoside adjoining Anticodon	
Codon	Prok.	Euk.	Codon	Prok.	Euk.	Codon	Prok.	Euk.	Codon	Prok.	Euk.
UUY	ms^{2}i^{6}A	*Y*-base	UCY		i^{6}A	UAY	msi^{6}A	i^{6}A	UGY	ms^{2}i^{6}A	
UUA	ms^{2}i^{6}A		UCA	ms^{2}i^{6}A	i^{6}A	UAA			UGA	-	
UUG	ms^{2}i^{6}A	m^{1}G	UCG	ms^{2}i^{6}A		UAG			UGG	ms^{2}i^{6}A	A
CUY	m^{1}G		CCU			CAY	m^{2}A		CGY	m^{2}A, G	A
			CCC								
CUA			CCA			CAA	m^{2}A		CGA	m^{2}A	A
CUG	m^{1}G		CCG			CAG	m^{2}A		CGG		
AUY	t^{6}A	t^{6}A	ACU			AAY	t^{6}A		AGY	t^{6}A	
			ACC								
AUA		t^{6}A	ACA	mt^{6}A, t^{6}A		AAA	t^{6}A	t^{6}A	AGA		t^{6}A
AUG	t^{6}A		ACG			AAG	t^{6}A	t^{6}A	AGG		
AUG-f	A	t^{6}A									
GUY	A	A†	GCY		meI	GAY	m^{2}A	m^{1}G	GGY	A	
GUA	m^{6}A	A†	GCA	A	meI	GAA	m^{2}A		GGA		
GUG	m^{6}A		GCG			GAG	m^{2}A		GGG	A	

Y-base, a guanosine derivative:
m^{1}G, 1-methylguanosine
A†, unidentified adenosine derivative
meI, 1-methylguanosine
m^{2}A, 2-methyladenosine
m^{6}A, 6-methyladenosine
mt^{6}A, N-methylcarbamoyl in t^{6}A

i^{6}A = 6-(Δ^2-isopentenyl) adenosine
t^{6}A = N-(9-(β-D-ribofuranosyl) purin-6-ylcarbamoyl) -L-threonine[a]
[a] Threonine-containing nucleosides were found in tRNAs accepting arginine and serine
Y = U or C
ms^{2}i^{6}A, 2-methylthio-6-(Δ^2-isopentenyl) adenosine

a special tRNA with the anticodon CAU but *without* modification of the adenosine at position 40.* All such modifications take place by enzymatic mechanisms following transcription. These mechanisms can be presumed to be evolutionary innovations. Prior to their appearance, U·G and G·U 3–1 pairing could have taken place in the primitive prokaryotes. Under these conditions, codons starting with G would pair with anticodons ending with either C or U. Therefore the valine codons, GUN, would pair with the eight anticodons represented by NAC and NAU. Similarly, the leucine anticodons represented by NAG would pair with eight codons consisting of CUN and UUN. According to this relationship, isoleucine and phenylalanine would not have been represented in the code. By applying this model throughout codon-anticodon pairing, a code containing ten amino acids is found to be possible (Table V). Eight more amino acids would become members of the code as a result of the suppression of U·G and G·U 1–3 pairing (Table Vb). Two chain terminator codons are introduced

* Note, however, that in eukaryotes this adenosine is modified, and the initiator tRNA pairs with both AUG and GUG.

TABLE V

(a) The ancestral code for ten amino acids, showing anticodon-codon pairing in the first position of codons that includes ambiguity produced by pairing between G and U in addition to G·C and A·U pairing; (b) code 2, produced as a result of elimination of this ambiguity (c) present code resulting from changes in (b). N = U, C, A or G; R = A or G; Y = U or C.

(a) Ancestral Code:

→		NAG					RUG	YUG				
Anticodons	NAU	NAA	NAG	NAC	RUU	YUU	RUA	YUA	RUG	YUG	RUC	YUC
Codons	NUA	NUU	NUC	NUG	YAA	RAA	YAU	RAU	YAC	RAC	YAG	RAG
←	NUG				YAG	RAG						
Amino acid	Val	Leu	Leu	Val	Asp	Glu	His	Gln	His	Gln	Asp	Glu

→		NGG				NCG		
Anticodons	NGU	NGA	NGG	NGC	NCU	NCA	NCG	NCC
Codons	NCA	NCU	NCC	NCG	NGA	NGU	NGC	NGG
←	NCG				NGG			
Amino acid	Ala	Pro	Pro	Ala	Gly	Arg (Orn)	Arg (Orn)	Gly

(b) Changes in (a) leading to code 2 as a result of enzymatic modification of tRNA base 40 and consequent suppression of U, G pairing.

Anticodons →	NAU	RAA	YAA	NAG	NAC	RUU	YUU	RUA	YUA	RUG	YUG	RUC	YUC
Codons ←	NUA	YUU	RUU	NUC	NUG	NAA	RAA	YAU	RAU	YAC	RAC	YAG	RAG
Amino Acid	Ile	Phe	Leu	Leu	Val	Asn	Lys	Tyr	C.T.	His	Gln	Asp	Glu

Anticodons →	NGU	NGA	NGG	NGC	NCU	NCA	NCG	NCC
Codons →	NCA	NCU	NCC	NCG	NGA	NGU	NGC	NGG
Amino Acid	Thr	Ser	Pro	Ala	Ser	Cys	Arg (Orn?)	Gly

(c) Changes in (b) leading to present code as a result of reassignment of tRNAs.

→		NAU		NCU			NCA	
Anticodons	RAU	UAU	CAU	RCU	YCU	RCA	UCA	CCA
Codons ←	YUA	AUA	GUA	YGA	RGA	YGU	AGU	GGU
Amino Acid	Ile	Ile	Met	Ser	Arg	Cys	C.T.	Trp

at this point. The final step is reassignment of several codons, as shown in (c) Table V, so that methionine and tryptophan are added.

To summarize: codon-anticodon pairing at position 1 of the codon may have been ambiguous at an earlier stage in the evolution of the code. The ambiguity disappeared when tRNA evolved to its present specialized state in which the nucleoside at position 40 is often modified, so as to make the pairing specific between position 1 of the codon and position 3 of the anticodon. I therefore postulate that an earlier form of the code could have been as shown in Table V(a). This could reduce the present list of amino acids used in protein synthesis to ten (Jukes, 1973a).

Further support for the possibility of ambiguity in the first position of an early code may be found in the observation that the second base in the codon confers greater specificity on the code than does the first base or, of course, the third base (Woese, 1965). It is therefore logical to attribute more specificity to the second base than to the first base in earlier codes.

The possibility exists that certain amino acids were formerly in the code but have been discarded from protein synthesis as a result of displacement of them from tRNA combination by members of the present 'group of 20'. A mechanism and an example of this was suggested for ornithine. Other amino acids have been found in mixtures obtained by treatment of simpler molecules. These include α-amino-isobutyric acid, α-amino-*n*-butyric acid, β-hydroxyvaline (Miller and Urey, 1959), and β-alanine (Friedman *et al.*, 1972). Whether the ten amino acids in the 'ancestral code' were actually those listed in Table V(a) is open to question in view of the fact that tRNA molecules, as noted above and elsewhere (Jukes and Holmquist, 1972), can 'switch' from one amino acid to another. The main point of the system outlined in Table V(a) is that it would provide for a code in which only ten amino acids would participate.

4. Arginine as an Intruder into Evolution

About one-tenth of the codons (6 out of 61) specify arginine. Yet most proteins contain less than 10% arginine, and in most cases the content is far below this. We have examined the arginine content of 83 proteins each containing 50 or more amino acid residues (Table VI). Some of the 83 are representatives or averages of a group of homologous proteins, such as cytochromes *c* or hemoglobins. Others have no homologs that have been sequenced, so these were treated as the sole representatives of their groups. The arginine and lysine contents of the list of Table VI are compared in Table VII. Lysine is great favored over arginine in proteins, in direct contrast to the comparative numbers of codons for these two amino acids. The disparity is less marked in eukaryotes than in prokaryotes. Perhaps the presence of the ornithine cycle makes mammals more tolerant to arginine by providing a mechanism for its breakdown.

The calculation in the last line of Table VII assumes equal amounts of A, U, C and G in messenger RNA for these proteins, formed by transcription of DNA. The DNA of vertebrates contains about 56% A+T and 44% G+C. If these values represent messenger RNA containing an average of 28% A, 28% U, 22% G and 22% C, then, for vertebrates, the expected percentage of lysine codons in all amino-acid codons would be 3.9% rather than 3.3%, and of arginine codons, 9.1% rather than 9.8%.

A protein with less than 5% arginine contains about half the arginine expected from a random distribution of codons (assuming A=C=G=U) and a protein with more than 6.6% lysine contains more than twice the lysine expected from such a random distribution of codons. *Clearly lysine is greatly favored over arginine as a basic amino acid.* The code evidently contains more arginine codons than are needed, and arginine codons in DNA are genetically reduced in numbers by Darwinian selection (King and Jukes, 1969). In contrast, lysine, which has only two codons, actually exceeds arginine

TABLE VI

Proteins, Eukaryotic	No. of proteins in class	Average			%	
		No. of residues	ARG	LYS	ARG	LYS
Trypsinogen, bovine	1	229	2	15	0.9	6.6
Lactalbumin	3	123	1.3	11.7	1.1	9.5
Ferredoxin – plants	5	96.6	1.2	4.6	1.2	4.8
Leghemoglobin – plants	1	140	2	14	1.4	10.0
Chymotrypsinogen	2	245	4.5	12.5	1.8	5.1
Caseins	2	204	5	12.5	2.5	6.1
β-Lactoglobulin, Bovine AB	1	162	3	15	1.9	9.3
Myoglobin	8	152.9	2.9	19.6	1.9	12.8
Hemoglobin Alpha Chain	10	141.1	3.0	11.6	2.1	8.2
Hemoglobin Beta Chain	10	145.2	3.3	11.3	2.3	7.8
Globin, Chironomus	1	136	3	10	2.2	7.4
Cytochrome *c*	40	107	2.4	15.6	2.2	14.6
Protease Inhibitor – Lima Bean	1	84	2	4	2.4	4.8
Globin, lamprey	2	146	4.5	13	3.1	8.9
Immunoglobuns, light chains:						
Constant, Kappa, human and mouse	2	106	2.5	7.5	2.4	7.1
Constant, Lambda, human and mouse	2	105	1	7	1.0	6.7
Variable Kappa, human and mouse	18	107.8	4.9	3.7	4.6	3.4
Variable, Lambda, human and mouse	10	109	4.0	3.2	3.6	2.9
mmunoglobuns, heavy chain						
Variable, human	6	109.7	5.8	4.2	5.3	3.8
Constant, human	1	336	7	27	2.1	8.0
Constant, rabbit	1	326	13	19	3.9	5.7
Iaptoglobin Alpha Chain	1	84	2	8	2.6	10.1
Carbonic anhydrase, human	1	260	7	18	2.7	6.9
Iemerythrin, worm	1	113	3	11	2.7	9.7
Glyceraldehyde 3-PO_4 Dehydrogenase	2	332.5	9.5	27	2.9	8.1
Glutamate Dehydrogenase	1	500	30	32	6.0	6.4
'hyrotropin α and luteinizing hormone α	3	96	3	10	3.1	10.4
lcohol Dehydrogenase, horse	1	374	12	30	3.2	8.0
Iuscle triose PO_4 isomerase	1	248	8	20	3.2	8.1
'horionic gonadotropin, human α	1	92	3	6	3.3	6.5
drenodoxin, bovine	1	118	4	5	3.4	4.2
hyrotropin β and luteinizing hormone β	2	116.5	10	6.5	8.6	5.6
'arboxypeptidase, bovine	1	307	11	15	3.6	4.9
rophospholipase A_2, pig	1	130	5	9	3.8	6.9
ibonuclease, pancreatic	3	125	5	9.6	4.0	7.7
eratin, High-sulfur protein	2	153.5	6.5	0	4.2	0.0
rypsin inhibitor, pancreatic	2	56	2.5	3.5	4.5	6.2
rypsin inhibitor, soybean	1	71	2	5	2.8	7.0
rypsin inhibitor, ascaris	1	66	3	7	4.5	10.6
ytochrome B_5	6	87.2	3.6	6.6	4.1	7.8
ee venom phospholipase A	1	128	6	10	4.7	7.8
lastase, porcine	1	240	11	2	5.0	1.2
'oinsulin	3	83.6	4	2	4.8	2.4
ipotropin β and γ	2	90.5	4.5	10	5.0	11.0
'olactin	2	194	10.5	9	5.6	4.5
ıpain	1	212	12	12	5.7	4.7
'ouse nerve growth factor	1	120	7	8	5.8	6.7
rowth Hormone	2	189.5	12	10.5	6.4	5.6

Table VI (*Continued*)

Proteins, Eukaryotic	No. of proteins in class	Average			%	
		No. of residues	ARG	LYS	ARG	LYS
Parathyroid Hormone	1	84	5	9	6.0	10.7
Neurotoxins, snake venom	11	61.2	4.2	5.3	6.2	8.4
Neurotoxin, scorpion	2	63.5	2.5	5.5	3.9	8.7
Avidin, chicken	1	128	8	9	6.3	7.0
Histone II B_2	1	125	8	20	6.4	16.0
Monkey amyloid protein A	1	76	4	4	6.6	5.3
Trypsin inhibitor, basic, bovine	1	58	6	4	10.3	6.9
β chorionic gonadotropin, human	1	139	4	11	7.9	2.9
Lysozyme – Vertebrates	5	129	11.6	6.2	9.0	4.8
Myelin Membrane Encephalitogenic protein	1	170	18.5	12.5	10.9	7.4
Trypsin Inhibitor – Maize	1	65	8	1	12.4	1.5
Histone III, Calf thymus	1	135	18	13	13.3	9.6
Histone IV	2	102	14.5	10.5	13.7	10.8
	Totals	8909	378.5	617.8	Av. 4.25	6.93
Proteins, prokaryotic						
Rubredoxin	2	52.5	0	3	0.0	5.7
Cytochrome C_2	1	112	0	17	0.0	15.2
Cytochrome C_{551}	3	82	0.3	8.3	0.3	10.2
Cytochrome c_3	1	109	0.5	18.5	0.6	18.5
50S Ribosomal protein A_2, *E. coli*	1	120	1	12	0.8	10.0
Neocarzinostatin *Streptomyces*	1	109	1	0	0.9	0.0
Thioredoxin	1	108	1	10	0.9	9.3
Ribonuclease T_1	1	104	1	1	1.0	1.0
Ferredoxin – clostridial type	5	54.8	0.2	0.8	1.0	1.3
Penicillinase *Staph. aureus*	1	257	4	43	1.6	16.7
Subtilisin	2	274.5	3	10	1.1	3.6
Ferredoxin, *Chromatium*	1	1	81	2	2.5	2.5
Azurin	4	125.8	1.2	12.8	1.2	9.3
Acyl carrier protein	1	77	1	4	1.3	5.2
Coat protein – turnip yellow mosaic virus	1	188	3	7	1.6	3.7
Thermolysin, *Bacillus thermoproteolyticus*	1	316	10	11	3.2	3.5
Nuclease, staphylococcal	1	149	5	23	3.4	15.4
Cytochrome B_{562}	1	110	4	16	3.6	14.5
Tryptophan Synthetase A	1	267	11	13	4.1	4.9
Aspartate transcarbamylase R Chain	1	152	8	10	5.3	6.6
Penicillinase, *B. licheniformis*	1	265	15	24	5.7	9.0
Coat Protein – Tobacco Mosaic Virus	5	157.4	9.6	1.6	6.1	1.3
Alpha lytic protease, myxobact.	1	198	12	2	6.1	1.0
Lysozyme, bacteriophage	2	160.5	12.5	12.5	7.9	7.9
	Totals	3628	106.3	262.5	Av. 2.93	7.24

in most proteins. Lysine ($pK_3' = 10.5$) confers important basic properties to proteins through its epsilon amino group, but the guanidine group of arginine is far more basic, perhaps excessively so ($pK_3' = 12.5$). Arginine is formed biochemically from ornithine, ammonia and carbon dioxide via citrulline, or from ornithine and guanidoacetic acid. I suggested (Jukes, 1973b) that ornithine (which is present in peptide linkage in gramicidin) preceded arginine in an earlier genetic code and that the arginine codons were originally assigned to ornithine (Table V(a)). Arginine then appeared as a result of the evolution of the urea cycle, and arginine had a greater affinity than ornithine for the then-existent ornithine tRNAs. Such a phenomenon can occur in enzyme chemistry. Aminopterin (4-amino pteroylglutamic acid) has a much greater affinity than the natural substrate, dihydrofolic acid, for the enzyme dihydrofolic acid reductase.

By replacing ornithine in the aminoacylation of tRNAs, arginine displaced ornithine from protein synthesis. This event increased the sophistication of protein molecules by introducing arginine, a new amino acid with unique properties. It replaced an amino

TABLE VII

Distribution of Arginine and Lysine in 83 proteins containing 8909 codons

	Eukaryotes	Vertebrates only	Prokaryotes
Arginine			
Less than 5 %	39	29	19
5 %–9.9 %	15	15	5
10 % or more	5	4	0
Lysine			
Less than 3.3 %	6	4	6
3.3 %–6.5 %	15	10	6
6.6 % or more	38	34	12
Arginine: Lysine ratio	0.613	0.613	0.405
for random base sequences	3.0	3.0	3.0

acid (ornithine) which was similar to lysine in the same manner that aspartic acid is similar to glutamic acid, for glutamic acid is homoaspartic acid; lysine is homo-ornithine. The deficiency of ornithine in proteins for needed functions was overcome by increasing the lysine content, accomplished by preferentially selecting lysine codons in evolution. The demand for lysine in the biological synthesis of proteins, incidentally, is so great that lysine deficiency is a major global problem in human nutrition.

It is perhaps surprising that the codons AGA and AGG are not the third and fourth codons for lysine, which has similar codons. Yeast tRNA for arginine, pairing with AGA, is similar to the yeast tRNA for lysine, pairing with AAA (Table II). These two tRNAs differ in their sequences by only 27% (Jukes and Holmquist, 1972). The average difference between pairs of tRNAs for different amino acids involved in protein synthesis is 49.4%$\pm$7.0 (Holmquist *et al.*, 1973).

If functional constraints are removed from a protein with low arginine and high lysine content, it should evolve in the direction of higher arginine and lower lysine. I suggested in 1969 that most point mutations in the specificity (S) (variable) regions of immunoglobulins are advantageous, and are rapidly incorporated as evolutionary changes, because 'it is immunologically advantageous to have a large available assortment of different antibodies to cope with various antigenic determinants' while 'those in the constant (C) regions are usually deleterious, thus accounting for the variability of S and the constancy of C sequences,' (Jukes, 1969).

The S and C regions are assumed to have descended from a common ancestor (Hill *et al.*, 1966). The S regions should therefore be higher in arginine and lower in lysine than the C regions as a result of randomly occurring point mutations. Table VI shows that this is indeed the case for both light and heavy human immunoglobulin chains.

5. Aspartic and Glutamic Acids

The similarity of the four codons for the dicarboxylic amino acids is frequently held up as an example of the optimization that exists in the code, on the basis that mutations should be non-injurious because of the similarity of these two amino acids. Aspartic and glutamic acids, by 'internally neutralizing' the basic amino acids arginine, lysine and histidine, are important for formation of zwitterions by proteins, and hence for solubility. However, the similarity may be due to chance, because other pairs of amino acids whose codons are identical except for the third base do not have similar chemical properties. These pairs are asparagine and lysine; histidine and glutamine; cysteine and tryptophan; serine and arginine. I therefore conclude that there is no evidence that the similarity of the Asp and Glu codons is necessarily more than a 'fortunate' accident.

6. The Hydrophobic Amino Acids; Codons with a Middle U

A conspicuous feature of the code is the similarity in chemical properties of the five amino acids having codons with a middle U. These amino acids have the following hydrophobic side chains:

Phe: C_6H_5–CH_2–

Leu: $(CH_3)_2$>CH–CH_2–

Ile: CH_3–CH_2–CH(CH_3)–

Met: CH_3–S–CH_2–CH_2–

Val: $(CH_3)_2$>CH–

The relationship between codons and properties is often considered to indicate optimization in that the codons for all these amino acids are interconnected by single base changes in the first position. The effect of this is alleged to be that mutations involving the first or third positions of these sixteen codons might tend to be neutral or near-neutral rather than deleterious. Here we have a hint of a process in evolution favoring neutral, non-adaptive changes.

Methionine has a single codon and a special function in protein synthesis as the amino acid that initiates the synthesis of polypeptides, following which it is removed enzymatically when or before the protein is completed. Methionine has other special functions in the interior of polypeptide sequences. We suggest that the elaborate mechanism by which methionine initiates polypeptide synthesis is an evolutionary development that appeared after the code reached its present from, and that valine preceded methionine as the 'initiator' amino acid.

Why do some amino acids have six codons? It was pointed out that so many codons seem unnecessary in the case of arginine. Obviously, however, every possible sequence of three bases in a strand of messenger RNA (and hence in the corresponding DNA sequence) must be translated as an amino acid or serve as a termination signal. There should not be a large number of termination codons in the code, or it will be too difficult to maintain the existence of long polypeptide chains against the frequent occurrence of mutations that produce premature termination, with lethal results. The code therefore cannot contain a high proportion of untranslated codons.

Each codon must contain a minimum of 3 nucleotides to provide for distinguishing between 20 different amino acids.

Most mutations in the third codon position are 'silent' or neutral because they do not change the meaning of the mutated codon. It is also of interest that the six codons for serine are such that each of them can be changed to a codon for the similar amino acid threonine by a single-base change. However, it is not clear why there should be six codons for serine and only two for the much-used glutamic acid. Nor is it clear why there are six codons for leucine and only three for isoleucine. Comparisons of the sequences of tRNAs do not resolve these problems, because the sequences of two tRNAs for *different* amino acids usually differ by about 50% and in most cases, the sequences of two tRNAs for the *same* amino acid in two different organisms also differ by about 50% (Holmquist *et al.*, 1973).

We must answer the question: If at one time the number of amino acids used in protein synthesis was 16 or less, why was the code not 'frozen' as a doublet code? Eigen (1971) answered this on the basis of stability constraints, using data compiled from equilibrium and rate studies with oligonucleotides. He comments:

> The most interesting effect is the preference for the triplet, however, not just for the logically obvious reasons, i.e. the prerequisite for the coding of more than 20 symbols, but rather due to mechanistic coincidences. Codons with less than three digits would be very unstable (at least for A and U). Codons with more than three digits, especially for G and C, become too 'sticky'. The life time of a codon-anticodon pair should not exceed milliseconds so that enzymes with corresponding turnover numbers can adapt optimally. The same type of optimization between stability and rate is always found for enzyme-substrate interactions. Any gain in stability means a lowering of complex dissociation rates; these have to match the turnover numbers in order not to become the rate limiting steps for the turnover.

Gatlin (1972) has found the three-digit codon to be informationally optimal for translation on a theoretical basis. Her calculations show that the entropy H_1 for the the protein 'language', $H_1(P)$, reaches its maximum for a 64-variable source when the codons are triplet sequences.

7. Evolution of tRNA

The sequenced transfer RNAs are a group that includes tRNAs for 16 of the 20 amino acids. Their sequences are sufficiently similar that we can conclude that they have descended from a single ancestral molecule by gene duplication followed by mutations. This conclusion is reinforced by the fact that pairs of tRNAs for the same amino acid in the same organism may differ by as little as 3% or as much as 36% (Holmquist *et al.*, 1973), showing that the process of duplication and evolutionary separation of tRNAs is a continuous and ongoing process. The inference that the tRNAs have a single ancestor leads to the additional conclusion that the original ancestral tRNA could not have accepted more than one amino acid, or contained more than one anticodon. The evolution of tRNA therefore leads us back to the beginnings of the present genetic code, which depends on recognition of adaptor molecules by the codons present in the genetic message. We cannot switch from a 3-letter code to a one containing only two letters per codon without destroying all the accumulated hereditary information, and, in effect, originating life over again.

8. Slippage in DNA Replication

Kornberg *et al.* (1964) discovered that enzymatic synthesis of DNA will take place in the absence of a template to produce repetitive sequences of poly $d(AT)_n$, in the presence of dATP and dTTP as substrates. The process was greatly accelerated by adding oligonucleotides with sequences of 6 to 14 alternating deoxyadenylate and deoxythymidylate residues, such as $d(AT)_6$. The authors postulated that successive stages of replication and slippage resulted in the continuous reiteration of the template and the synthesis of a large dAT co-polymer. This phenomenon was utilized by Khorana *et al.* (1966) in experiments to synthesize repetitive sequences in RNA for solving the problem of the genetic code.

DNA 'families' of identical sequences are recognized as being of comparatively recent evolutionary origin (Britten and Kohne, 1965). Some of these families separate from the main bulk of DNA upon centrifugation and are therefore termed 'satellite' DNA. Satellite DNA of guinea pigs contains a repeating sequence of about six nucleotides (Southern, 1970). Such a molecule would be a possible starting point for coding the synthesis of 'new' polypeptides consisting of repetitions of short oligopeptides that if 'biologically useful', could form functional proteins which could eventually become modified by point mutations. Yčas (1972) has reviewed the existing information on proteins that contain periodic repetitions. Some are remarkably simple, such as a lepidopteran silk fibroin consisting mainly of $(\text{Ala-Gly})_n$. Ycas points out that collagen has possibly evolved from $(\text{Gly-Pro-Pro})_n$, changed by a number of replacements of proline residues by other amino acids.

These repetitive proteins are structural rather than enzymatic in function. They are 'new' in the evolutionary sense, and therefore do not cast light on the composition of earlier genetic codes. We do not know whether enzymes were formed by evolutionary changes in long, repetitive peptide chains that were originally without catalytic activity. Another possibility, which seems more feasible, is that enzymes started as short peptides with weak catalytic activity, and that these short peptides evolved by lengthening to proteins with greater activity, and with structures that enhanced their stability and specificity.

9. Evolution in Transfer RNAs

The genetic code is considered to be universal in terrestrial organisms, and to be 'frozen'. This may be true of the assignments of the codons, but the tRNAs, which function in translation of the code, are evidently still in process of evolution as judged by the presence of duplicate forms of tRNA molecules differing only slightly from each other, such as the two serine tRNAs of yeast (Zachau *et al.*, 1966).

The system that synthesizes polypeptides in cells is extremely complex. It involves ribosomes, which contain large numbers of different proteins as well as RNA. Many enzymes and coenzymes participate in the synthesis of proteins upon ribosomes. The entire process is so elaborate that it appears unlikely to resemble in any respect the primitive systems that produced polypeptides during the period when life was first emerging. A discussion of the origin of ribosomes would be pertinent in an essay on the origin of the code, but, in view of the paucity of information on this subject, I shall not undertake it here.

There are two tRNAs in *Staphylococcus epidermidis* that insert glycine into peptidoglycan molecules, which are units of the bacterial cell wall. These tRNAs differ from the tRNAs that participate in the synthesis of polypeptides upon ribosomes, but the essential differences are only in the absence of T and Ψ from loop IV, and U instead of purine at position 40. The remainder of the molecule, including the regular 'cloverleaf' structure, is quite homologous with the other tRNAs (Holmquist *et al.*, 1973). The synthesis of peptidoglycans, however, may well be an evolutionary development that took place subsequently to the appearance of the tRNA-ribosomal system of polypeptide synthesis.

This evolutionary versatility shown by the tRNAs suggests that they could have undergone an epochal event of rapidly-occurring duplication followed by mutations that brought about numerous changes. These changes could have led to the appearance of a large number of different anticodons and different recognition sites for various amino acids.

If this development took place at the same time as the appearance of the ribosomal RNAs, the fundamental units of a polypeptide-synthesizing system, ribosomes and tRNAs, could have made their appearance together.

Holmquist *et al.* (1973) found that pairwise comparisons, of 43 different tRNAs showed that their primary structures had diverged so far that it is impossible to construct a coherent phylogeny for most of the 43. The average divergence, $49.4\% \pm 6.9$, for pairs of tRNAs for different amino acids involved in protein synthesis represents an equilibrium between the restraints of natural selection and the flow of point mutations (Holmquist *et al.*, 1973).

The initial appearance of something resembling the present tRNA on the biological scene must have been a portentous event. It could well have taken place suddenly, because, as Eigen (1971) has shown, a 'clover-leaf' structure emerges from a game played with a random sequence of 80 digits composed of A, U, G and C. Each player throws a tetrahedral die, each face of which represents one of the four letters, and tries to approach a double-stranded structure with as many as possible A·U or G·C pairs. This 'game' in evolution would be played by a system that 'tried to resist' hydrolysis by forming base pairs. Once formed, the tRNA molecule could undergo rapid proliferation by gene duplication, and point mutations would soon produce a series of different anticodons, for the anticodon loop would be exposed in single-stranded form by the very formation of the clover-leaf structure. Such a 'family' of tRNAs could furnish enough anticodons to pair with all the codons, especially if codon-anticodon pairing were only partially specific, as is the case in the third position of codons. If pairing were relaxed in the first as well as in the third position, a pattern could exist as shown in Table V(a). The pattern of amino acid binding by tRNA might develop more slowly than the changes in anticodons; at first, only one or two amino acids might be bound by the primordial tRNAs. Advantages would soon accrue to the system that built up the number of amino acids that were bound by tRNAs and thus were brought into protein synthesis.

The argument that optimization of the code is shown by the nature of amino acid

TABLE VIII

Amino acid interchanges that can occur as the result of single-base changes in the coding triplets

Amino acid	Possible interchanges[a]
Ala	Asp Glu Gly Pro Ser Thr Val
Arg	Cys Gln Gly His Ile Leu Lys Met Pro Ser Thr Trp
Asn	Asp His Ile Lys Ser Thr Tyr
Asp	Glu Gly His Tyr Val
Cys	Gly Phe Ser Tyr Trp
Gln	Glu His Leu Lys Pro
Glu	Gly Lys Val
Gly	Ser Trp Val
His	Leu Pro Tyr
Ile	Leu Lys Met Phe Ser Thr Val
Leu	Met Phe Pro Ser Trp Val
Lys	Thr Met
Met	Thr Val
Phe	Ser Tyr Val
Pro	Ser Thr
Ser	Thr Trp Tyr

[a] The underlined examples have been reported to occur in mutations.

replacements resulting from single-base changes is unconvincing. The 75 possible amino acid interchanges resulting from single-base changes are in Table VIII. It is obvious that many of them represent interchanges between amino acids of widely differing properties. We conclude that the only optimization shown by the code is the fact that many of the changes in the third base of codons do not produce changes in amino acids. This feature may be an incidental result of the spatial nature of codon-anticodon pairing rather than an 'evolutionary optimization'.

10. Origin of a Translation System

The proposal that the genetic code originated by some kind of loose affinity between amino acids and the bases of nucleic acids is an obvious one. Indeed, it was the basis of the first suggestions for a genetic coding system made by Dounce (1952) and Gamow (1954). These suggestions were discarded when they were put to flight by Crick's 'adaptor hypothesis' in 1955. However, although the above proposal is obvious, it runs into very serious practical difficulties. Some of these were discussed by Razska and Mandel (1972). It is fairly easy to show that some amino acid, such as phenylalanine, has a slight preferential affinity for binding weakly to some base, such as uracil, under specialized conditions, but nothing has emerged from these experiments that indicates a special attraction of each individual codon or (just as logically) each anticodon, for the appropriate member of the list of 20 amino acids. This defect is usually countered by saying that even a slight preference in binding, spread over hundreds of millions of years of chemical trial and error, would result in the emergence of a coherent code. Serious objections can be made to this rationalization, because the essence of a successful coding system is *fidelity*, and the errors arising from a pairing mechanism containing a high degree of indefiniteness or ambiguity would be so great as to prevent the emergence of the process of heredity. This objection was raised and discussed by Eigen (1971).

The protagonists of the 'weak binding' theory for the origin of the code include Woese (1967) whose viewpoint was summarized by him in 1967 as follows:

> ...amino acid-nucleic acid 'recognition' interactions (or their equivalent), ... being very weak interactions... could not have given rise directly to the genetic code as it now exists. Instead they must have played the role of constraints operating on the evolutionary process, in this way gradually shaping the form of the code. Since the interactions are weak and therefore cannot manifest an all-or-none sort of specificity with regard to amino acid-codon pairings, it is reasonable to expect that they cannot align amino acids along a nucleic acid template directly, and so this 'recognition' role has been filled by the evolution of an 'intermediary' system, the tRNA's and activating enzymes, that recognize with very high accuracy both an amino acid and its codons.

An objection to this line of reasoning is that it tends to invoke the existence of interactions that are so weak that they cannot be detected except under specialized and artifical experimental conditions. I feel that such an explanation may rely on a desire to postulate non-existent phenomena because we feel intuitively that they ought to have existed since we cannot think of any other way that the genetic coding of proteins might have started. There are infinitely large numbers of ways in which long sequences of variables can be arranged. The task of natural selection working alone to find useful members of an infinitely large set would be impossible. Therefore, there is a tendency to hope that there must have been an orderliness in aligning amino acid with nucleic acids in some primitive system.

A strong objection to the 'stereochemical fit' concept is that the two sets of codons for serine, UCN and AGY, are so dissimilar. Various authors have proposed that there is a relationship between the second base of codons and the chemical properties of the amino acid, and that this relationship has governed the evolution of the code. Such a relationship, in my opinion, is perceptible only for the codons with U in the middle position. The other three 'groups' (Table I), do not share common properties. For example, although serine and threonine are similar, they do not resemble proline; neither are cysteine, tryptophan and arginine similar to each other. Moreover, serine codons are found in two different classes. The 'clustering' of the hydrophobic amino acids could have come about by the phenomenon postulated in Table V, in which all 16 NUN codons are shown as having evolved through three stages of increasingly

TABLE IX

Relative abundance of certain elements in the Earth's crust and in sea-water (After Mason, 1952; Sverdrup *et al.*, 1942; Bowen, 1966)

Element	Presence in Earth's crust ppm	Sea-water pp 10^9
Fe	50000	2 to 20
Mn	1000	1 to 10
Cr	200	0.05
Zn	132	5 to 14
Ni	80	0.1 to 5.4
Cu	70	1 to 9
Co	23	0.1 to 0.27
Mo	15	0.3 to 10
Se	0.09	0.09 to 4

precise codon-anticodon pairing. The fact that five amino acids with hydrophobic side-chains all have codons with a middle U may have been shaped by evolution, or it may be a coincidence.

Mention should be made of the romantic suggestion by Crick and Orgel (1973) that the genetic code could be of pan-spermatic, extra-terrestrial origin; coming from an extra-galactic source via space-ship. In their proposal, Crick and Orgel go to the length of inferring that the beings who sent us the code dwell in an environment that is rich in molybdenum. Their argument is that terrestrial organisms, in contrast to the Earth's crust, are higher in molybdenum than in nickel. However, the proportions of minerals in animals are thought to resemble the composition of these found in sea-water rather than that of the geosphere. The six trace elements that are components of identified enzyme systems are iron, zinc, copper, manganese, molybdenum and selenium. A seventh, cobalt, is not utilized as an inorganic element; animals (but not green plants) use it as a component of vitamin B_{12}, which they obtain from micro-organisms. It is interesting to compare the six in terms of their abundance in the Earth's crust and in sea-water (Table VI). They show a tremendous range in their abundance in the earth's crust, extending between six and seven orders of magnitude. However, their concentrations in sea-water, in which life is thought to have originated, are remarkably uniform. The average values for each of the six 'enzymatic' trace elements extend from 1 to 11 parts per 10^9. Note that chromium occurs below this range, and so does cobalt. Molybdenum occurs in sea-water at higher concentrations than nickel or chromium. It is unnecessary to postulate that molybdenum stars might have served as a jumping-off point for pan-spermatic organisms that 'infected' the Earth. In any case, the mineral content of protoplasm evidently is greatly influenced by natural selection, judging from anomalies such as the iodine content of the thyroid gland, the vanadium content of tunicates, etc.

11. Summary

(1) Suggestions are made for possible pathways of evolution of the genetic code, assuming that the present code was preceded by codes that contained fewer amino acids. It is recognized, however, that it is possible that earlier codes contained amino acids that are not currently used in protein synthesis.

(2) As an example of the latter phenomenon, it is suggested that ornithine preceded arginine in the code. Support for such a suggestion is found in the observation that the average arginine content of proteins is only about half the value expected from the fact that arginine has six codons. It is proposed that the introduction of arginine led to a diminished use of the codons CGN and AGR, and an evolutionary selection of more lysine resulted than would be expected from the fact that it has only two codons.

(3) Pairing between G and U (or U and G), in the first position of codons and the third position of anticodons could reduce the number of amino acids involved in protein synthesis to ten. It is suggested that this formed the basis of an earlier code.

(4) The introduction of tRNA into protein synthesis may have been a cataclysmic and comparatively sudden event, since duplication of tRNA takes place readily, and point mutations could rapidly differentiate members of the family of duplicates from each other.

Acknowledgements

This work was supported by NASA Grant NGR 05-003-460 to the University of California, Berkeley.

I wish to dedicate this essay to the memory of Prof. J. B. S. Haldane, who asked me to write a manuscript for him during the meeting on 'The Origins of Prebiological Systems and of their Molecular Matrices' at Wakulla Springs, Florida, November 1963.

References

Bowen, H. J. M.: 1966, in *Trace Elements in Biochemistry*, Academic Press, London and N.Y.

Britten, R. J. and Kohne, D. E.: 1965–1966, *Carnegie Institute*, Washington, D.C. 20015, Year Book, pp. 78–106.

Crick, F. H. C.: 1966, *J. Mol. Biol.* **19**, 548.

Crick, F. H. C. and Orgel, L.: 1973, *Icarus* **19**, 341.

Dounce, A. L.: 1952, *Enzymologia* **15**, 251.

Eigen, M.: 1971, *Naturwissenschaften* **58**, 465.

Friedman, N., Haverland, W. J., and Miller, S. L.: 1972, in R. Buvet and C. Ponnamperuma (eds.), *Molecular Evolution* **1**, 123, North-Holland, American Elsevier Publishing Co., Inc., New York.

Gamow, G.: 1954, *Nature* **173**, 318.

Gatlin, L. L.: 1972, *Information Theory and the Living System*, Columbia University Press, New York, pp. 1–210.

Hill, R. L., Delaney, R., Fellows, R. E., Jr., and Lebovitz, H. E.: 1966, *Proc. Nat. Acad. Sci. U.S.A.* **56**, 1762.

Holmquist, T., Jukes, T. H., and Pangburn, S.: 1973, *J. Mol. Biol.* **78**, 91.

Jukes, T. H.: 1966, *Molecules and Evolution*, Columbia University Press, New York and London, pp. 1–285.

Jukes, T. H.: 1969, *Biochemical Genetics* **3**, 109.

Jukes, T. H.: 1973a, *Nature* **246**, 22.

Jukes, T. H.: 1973b, *Biochem. Biophys. Res. Commun.* **53**, 709.

Jukes, T. H. and Holmquist, R.: 1972, *Biochem. Biophys. Res. Commun.* **49**, 212.

King, J. L. and Jukes, T. H.: 1969, *Science* **164**, 788.

Khorana, H. G., Buchi, H., Ghosh, H., Gupta, N., Jacob, T. M., Kossel, H., Morgan, R., Narang, S. A., Ohtsuka, E., and Wells, R. D.: 1966, *Cold Spring Harb. Symp. Quant. Biol.* **31**, 39.

Kornberg, A., Bertsch, L. L., Jackson, J. F., and Khorana, H. G.: 1964, *Proc. Natl. Acad. Sci. U.S.A.* **51**, 315.

Mason, B.: 1952, *Principles of Geochemistry*, John Wiley and Sons.

Miller, S. L. and Urey, H. C.: 1959, *Science* **130**, 245.

Murao, K., Saneyoshi, F., Harada, F., and Nishimura, S.: 1970, *Biochem. Biophys. Res. Commun.* **38**, 657.

Nishimura, S.: 1972, *Progr. Nucl. Acid. Res. Mol. Biol.* **12**, 49.

Raszka, M. and Mandel, M.: 1972, *J. Mol. Evol.* **2**, 38.

Southern, E. M.: 1970, *Nature* **227**, 794.

Stewart, T. S., Roberts, R. J., and Strominger, J. L.: 1971, *Nature* **230**, 36.

Sverdrup, L., Johnson, W., and Fleming, J.: 1942, *The Oceans* Prentice Hall.

Woese, C. R.: 1965, *Proc. Nat. Acad. Sci. U.S.A.* **54**, 1546.

Woese, C. R.: 1967, *Progr. Nucl. Acid Res. Mol. Biol.* **7**, 107.

Ycas, M.: 1972, *J. Mol. Evol.* **2**, 17.

Zachau, H. G., Dütting, D., and Feldmann, H.: 1966, *Z. Physiol. Chem.* **347**, 212.

GENETICS AND THE ORIGIN OF THE GENETIC CODE

G. W. R. WALKER

Genetics Department, University of Alberta, Edmonton, Alberta, Canada

Abstract. The genetic code has been analysed by a method similar to that used by Gregor Mendel. The current codon catalogue is shown to be symmetrically subdivisible into two discrete subcatalogues of eight quartets each by classifying the quartets as *monocoding* (for one amino acid only) vs *heterocoding* (for two amino acids or for amino acid plus nonsense). The internal symmetries of the two subcatalogues are identical, and are governed by two common parity rules. These rules, together with one governing the subdivision itself, can be explained by the hypothesis that two primaeval sets of polynucleotide-borne anticodons, corresponding closely but not exactly with the subcatalogues originated independently and separately (were not originally together within any replicating pre- or proto-biont). The discorrespondence between the primaeval sets and the subcatalogues is itself symmetrical, involving quartets sharing identical locations in the two subcatalogues. The primaeval sets correspond exactly with the subdivisions of the catalogue proposed by Skoog and co-workers on the basis of the presence vs the absence of cytokinins or "cytokininlike bases" adjacent to the anticodons. A molecular model for the origin of the primaeval anticodon sets is described, and the relationship of the hypothesis with the origin of life, together with some possibilities for testing it, are discussed.

This is a report of a study of the origin of the genetic code. I would like to introduce it with a partial quotation from Crick's well-known paper on the genetic code at the 1966 Cold Spring Harbor Symposium on Quantitative Biology (Crick, 1966). He says, "I hope... that when people put forward detailed theories about the origin of the genetic code, they will try if possible to produce ones which can be tested in some way or other." And he suggests "experimental evidence, either frozen in the present organisms or from dramatic experimental results." It has always been apparent, I am sure, that such theories, however, need cold, hard support in an overriding and simple logic, mathematical or otherwise.

In this study I have tried to be faithful to this principle, using the logical methods of Boole and Mendel first, and then whatever I might glean of logic from studies of the code. Boolean and Mendelian logic both require three things. First, the simplest possible grouping of 'Yes' and No' alternatives in the experimental data. Second, the derivation of rules governing the grouping. And last, the consistent and logical interpretation (or explanation) of the rules.

There are few who would not agree that the Genetic Calendar is the tabular results of one of the most remarkable experiments of this century. The version shown in Figure 1 is a faithful replica of the data in the 'Crick' calendar, but for a reversal of A and G in the letter margins. The simplest, and the only possible logical grouping comes from asking a simple question about each of the sixteen codon quartets, or boxes. The question is simply 'Is there only one amino acid present, and nothing else?' Enclosed boxes give 'Yes', the others give 'No' answers; I call the first monocoders, (I) and the second heterocoders (N). One can see the monocoder group has two axes of

Letter 1	Letter 2								Letter 3
	U(A)		C(G)		G(C)		A(U)		
U(A)	N	Phe Leu	I	Ser	N	Cys Trp ?	N	Tyr Non	U C G A
C(G)	I	Leu	I	Pro	I	Arg	N	His Gln	U C G A
G(C)	I	Val	I	Ala	I	Gly	N	Asp Glu	U C G A
A(U)	N	Ile Met Ile	I	Thr	N	Ser Arg	N	Asn Lys	U C G A

Fig. 1. The current codon catalogue with marginal letters A, G reversed and with probable anticodon letters 1 and 2 in parentheses.

symmetry. If we remove the heterocoders to a separate but equally valid subcalendar with the inner and outer margin letters reversed, as in Figure 2, this group shows the same symmetry. This evidence of order has never been described before. We want to know then, the rules governing it. The marginal ratios are expansions of the binomial $(1+1)$ which means letter parity across the axes of symmetry. The exponents give the number of these rules, which amount to 10, plus an additional pair (the '2+0' rules)

Letters 1	Letters 2				Totals
	U_2	C_2	G_2	A_2	
U_1		I			1
C_1	I	I	I		3
G_1	I	I	I		3
A_1		I			1
Totals	2	4	2	0	

Letters 1	Letters 2				Totals
	C_2	U_2	A_2	G_2	
C_1			N		1
U_1		N	N	N	3
A_1		N	N	N	3
G_1			N		1
Totals	0	2	4	2	

Fig. 2. Sub-catalogues for monocodons (I) and heterocodons (N) produced by marginal transformation.

caused by separating the subcalendars. These are the 'results' of the analysis, and it is clear that they refer only to letters 1 and 2 of codons.

Seeking to explain them I have again used Mendelian logic. Thus if data divide into equal subgroups a segregating mechanism is implied. Mendel's mechanism proved later to be meiotic disjunction. I am here suggesting that the subcalendars had separate origins, either in space or in time. Whether separation means 'in different protobionts' or 'in different regimes' is beyond consideration at this stage. Each subcalendar would nevertheless appear to require cohesion of some kind, and at first I searched

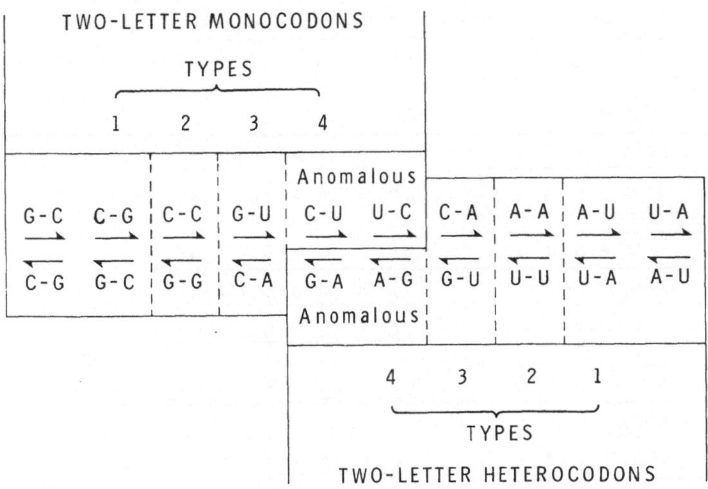

Fig. 3. Dinucleotide-hybrid categories and distribution. Types are: (1) Watson-Crick pair in sequence. (2) Duplicate bases in sequence. (3) Non-pairing pyrimidine-purine set in sequence. (4) Different purines/pyrimidines in sequence. For 'anomalous' doublets see text.

exhaustively for some hypothetical single-stranded type of polynucleotide whose base-composition would sequester the appropriate subgroups as 2-letter sequences. There are none.

Then I happened to assemble what may be called 'an ordered arrangement of 2-letter hybrids' as shown in Figure 3. The important observation from the figure is that both complementary strands of a hybrid usually go to the same subcalendar. Although this rule is disobeyed by four anomalous strands, the observation suggested that I should examine double-stranded rather than single-stranded polynucleotides.

Four models were found that possessed interesting sequestering properties. These are shown in Figure 4 to the left. The first, which I have called 'hybrid poly (G-C) plus U' has stretches of G-C pairing interspersed with U's, but possesses no A's and therefore had to be synthesized in the absence of adenine. Similarly, the second, called 'hybrid poly (A-U) plus G' has no C's, having been synthesised in the absence of cytosine. Thus uracils in the first model system and guanines in the second have pairing partners other than their normal ones, A and C. These 'unusual' or 'non

Watson-Crick' bases are symbolized 'X' in these models, signifying pyrimidine or purine base derivatives. The two further models are referred to later.

These models possess a remarkably good ability to sequester subcalendars, as is shown in Figure 4. Each carries seven two-letter sequences of a subcalendar. The eighth is the anticodon sequence required to complete the subcalendar. There are no other sequences present. It turns out that they can explain the last letter of the triplet

ABSENT BASE	HYBRID POLYNUCLEOTIDE	LETTER DOUBLETS SEQUESTERED*	Proposed Third Letter (X) Additions to Anticodons
A	Poly -(G-C) + U	for Monocoders:	
	X G C U G G U	CC + CG + GC + GG +	
	U C G X C C X	CU + GU + UC + UG.	
C	Poly -(A-U) + G	for Heterocoders:	
	X A U G A A G	UU + UA + AU + AA +	
	G U A X U U X	UG + AG + GU + GA.	
U	Poly -(G-C) + A	for Monocoders:	
	X G C A G G A	CC + CG + GC + GG +	X = I
	A C G X C C X	CA + GA + AC + AG.	
G	Poly -(A-U) + C	for Heterocoders:	
	X A U C A A C	UU + UA + AU + AA +	X = O-methyl G?
	C U A X U U X	UC + AC + CU + CA.	

Fig. 4. Hybrid Polynucleotides and their subcatalogue sequestening capacities. Underlined doublets are anticodon doublets.

code quite well also, if we change our consideration from codons to anticodons. There are, indeed, very excellent reasons for concluding that the origin of anticodons and their 'adaptor' carriers preceded, or coincided with, that of codons (see Woese, 1967; Ycas, 1971).

Here we must, however, consider the next pair of models, 'hybrid poly (G-C) plus A' and 'hybrid poly (A-U) plus C'. We have reversed the deficits from A to U and C to G. These now sequester seven anticodons plus one codon of a subcalendar. Furthermore, referring to these two hybrid polynucleotides themselves, it is to be noted that X bases may lie to either side of any doublet sequence, and consequently may occupy sites corresponding to the last letter of anticodons (viz., that opposite the third codon base). Thus one such X base might be hypoxanthine, which, as the nucleotide inosinic acid, currently occupies the last letter position of anticodons for four monocoders, as seen in Figure 5, and it will be noted that other base derivatives (G*, O.M.G. C*) are also present at this site in some anticodons of three of the four heterocoding groups for which sequencing data are available (*TYR*-AMBER-OCHRE; *PHE*-LEU;

ILE-*MET*-ILE). Such current base derivatives represent modifications during RNA maturation rather than constant elements of the genetic code, hence their presence in anticodons can only suggest that X bases performed critical functions in the origin of the code. Yet it is clear, from an extensive exploration of other models, without success, that their presence in these models is an absolute requirement. The only other alternative is the presence of gaps opposite the 'unpaired' bases, a possibility that has to be rejected since the gaps can in no manner be seen capable of faithful replication.

AMINO ACID	SOURCE	CATALOGUE SEQUENCE	AMINO ACID	SOURCE	CATALOGUE SEQUENCE
ALA	yeast	C G I*	TYR 1	yeast	A Ψ G
VAL	yeast	C A I*	TYR 1 su⁻	E. coli	A U G*
VAL	torula yeast	C A I*	PHE 1	yeast	A A G+
VAL	torula yeast	C A I*	PHE 1	wheat	A A G+
SER	yeast 1	A G I*	PHE 1	E. coli	A A G
SER	rat	A G I*	MET	E. coli	U A C*
			f. MET	E. coli	U A C
ILU 1	torula yeast	U A I*	LYS	yeast	U U C

Fig. 5. Anticodon sequences –, * and +; base derivatives (I* = inosinic acid, G+ = O-methyl G); Ψ = pseudouridylic acid.

An interesting feature of the Figure 4 models is seen when they are considered in pairs. Thus the first and third models combined account for *all* the letter 1, 2 doublets of monocoders, both in codons and in anticodons, and no others; and the same is true of the second and fourth models with respect to heterocoders. In other words, these combinations can account exactly for the partitioning of the genetic calendar, and the possibility that they might ensue from inter-model hybridization is now being investigated.

In conclusion, this model system in my view supersedes a slightly less efficient one presented at Mol, Belgium, in 1970 (Walker, 1971). Neither system, it is to be noted, makes the pretension of predicting specific amino acid assignments. They do, however, provide the type of definitive evidence 'frozen in the present organisms' mentioned earlier as a requirement of theories of code-origin. Furthermore, the simple combining of simple models, as indicated above, creates a nucleic acid basis, not only for primaeval sets of adaptor molecules corresponding exactly, anticodonwise, to the current tRNA's, but also for classes of potential messages, bearing codons specific to each adaptor set. Thus, because of their simplicity and definitiveness, they argue strongly against contentions (see Woese, 1967) that primaeval coding was ill-defined, and required complex and stochastic processes to achieve the orderliness of present-day coding. Finally, they provide simple statements of requirement for base-deficit and double-strandedness at the origin of life that are testable in the laboratory and should be considered in any general theory of life-origin.

References

Crick, F. H. C.: 1966, *Quant. Biol.* **31**, 3.
Walker, G. W. R.: 1971, in *Informative Molecules in Biological Systems*, Proc. Int. Symp. on Uptake of Informative Molecules by Living Cells, Mol., Belgium, 1970 (ed. by L. Ledoux), North Holland, Amsterdam-London, pp. 148–156.
Woese, C. R.: 1967, *The Genetic Code: The Molecular Basis for Genetic Expression*, Harper, New York
Ycas, M.: 1971, *The Biological Code*, North-Holland, Amsterdam-London.

ORIGIN OF THE GENETIC CODE: A PHYSICAL-CHEMICAL MODEL OF PRIMITIVE CODON ASSIGNMENTS

JOSEPH NAGYVARY

Dept. of Biochemistry and Biophysics, Texas A & M University, College Station, Tex. 77843, U.S.A.

and

JANOS H. FENDLER

Dept. of Chemistry, Texas A&M University, College Station, Tex. 77843, U.S.A.

Abstract. Selective compartmentalization of amino acids and nucleotides according to their polarities is proposed as a physical-chemical model for the origin of the genetic code. Assumptions made in this hypothesis are: (1) an oil-slick covered the surface of the primitive ocean, constituents of which formed association colloids or micelles at the water-oil-air interfaces; (2) depending on the polarity of the media, these aggregates possessed hydrophilic and hydrophobic interiors where selective uptake of amino acids and nucleic acid constituents could take place; and (3) condensation and polymerization in the micellar phase were enhanced. According to the chromatographically observed polarities, for example, lysine and uridylate fall into the hydrophilic compartment, and phenylalanine and adenylate are enriched in the hydrophobic environment. These components could eventually be condensed to form a charged adaptor loop with an anticodon which is complementary to the presently valid codon. Only two groups of amino acids, hydrophilic and hydrophobic, were recognized by the primitive translation mechanism. Implications of this hypothesis for the further development of the genetic code is discussed. The catalytic power of micelles have been substantiated by successful synthesis of nucleotides under relatively mild conditions using thiophosphates as high energy phosphates.

No rational interpretation has been advanced to-date for the observed codon assignments (Woese *et al.*, 1966; Orgel, 1968; Crick, 1968). Direct interactions of amino acids to oligonucleotides which would have been most satisfying could not be found (Woese *et al.*, 1966). The reported codonic alignment of AMP on polylysine is probably due to a simple effect of the polycation (Lacey and Pruitt, 1969). One of the meaningful experimental approaches, the polynucleotide-directed incorporation of amino acids into protenoid microspheres is not free of ambiguity and requires itself an interpretation (Nakashima and Fox, 1972). It is generally assumed that the code in its most primitive form could only differentiate between two classes of amino acids, i.e. hydrophilic and hydrophobic (Orgel, 1968; Crick, 1968; Orgel, 1972). The evolution of the precise assignments could have been accidental (Crick, 1968). The grouping of codons and amino acids by similar polarity criteria has been advocated (Woese *et al.*, 1966).

Here we propose a physical-chemical model involving the selective uptake of amino acids and nucleotides in compartments composed of simple, and progressively more complex, association colloids. Subsequent to compartmentalization, polymerization and condensation could be enhanced in the microenvironment of such colloidal aggregates. This hypothesis is based on the suggested composition of the primordial ocean

(Lasaga *et al.*, 1971) and on the recognized properties of relatively simple synthetic and biologically occurring aqueous, or 'normal', and nonaqueous, or reversed micelles and selective substrate solubilization therein (Fendler and Fendler, 1970; Fendler *et al.*, 1973). The proposed oilslick of 1 to 10 m thickness on the surface of the primitive ocean presumably predominantly consisted of saturated and unsaturated hydrocarbons and contained amines, bicarbonates, nitriles, carboxylic acids, and heterocyclic compounds in low concentration which were capable of forming, and/or reacting with hydrocarbons to form, association colloids and hence micelles. Advantages of our proposal are that

(1) it includes some of the essential features of the present code,
(2) rules out accidental allocation of amino acids to codons, and
(3) most significantly, it could be approached experimentally.

In dilute solutions in the absence of enzymes or other suitable templates, the simplest way to ensure the charging of the primitive adaptor with the amino acid is the compartmentalization of these components. Neglecting the actual mechanism of amino acid and nucleotide condensation for the time being, if a compartment contains only nonpolar amino acids such as phenylalanine and leucine and mostly adenylate, this system can lead to the predominant synthesis of a charged adaptor with an A-A-A anticodon (loop). The same holds for some other amino acids and their respective anticodons. Assuming simple partitioning, the amino acids and nucleotides would be distributed on the basis of their solubilities, the order of which is akin to that of the chromatographic values in a similar system. Accordingly, phenylalanine and adenylate would be enriched in the oil phase while uridylate and lysine would remain in the water phase. Rudiments of nucleotide and amino acid alignments according to their polarity behavior have been suggested by Woese previously (Woese *et al.*, 1966). Such idea has not found fruition until now, primarily because the amino acids and codons were assumed to be of the same polarity instead of the more propitious grouping of amino acids with anticodons. Some of the assigned polarities of nucleotides are in fact in error (Woese *et al.*, 1966). On theoretical grounds uridylate was listed, for example, as less polar than adenylate, whereas chromatography establishes the reverse order (Wyatt, 1955; Lohrmann *et al.*, 1966) (Table I). Simple distribution of amino acids and nucleotides in the aqueous and non-aqueous phases according to their polarities is clearly insufficient to bring about the desired interactions since the concentration of nucleic acid constituents would be undoubtedly very low in each phase. Compartmentalization of substrates in relatively high concentration in the micellar pseudophases renders such interactions highly feasible. Formation, structure and physical chemical properties of simple aqueous and non-aqueous micelles as well as substrate solubilization are well established (Fendler and Fendler, 1970; Fendler *et al.*, 1973). Reversed micelles contain a hydrophilic cavity which can solubilize polar compounds such as uridylate and lysine. Less polar amino acids and nucleotides, on the other hand, are likely to be enriched in aqueous micelles. The selectivity for the nucleotide monomers would probably be surpassed by an even stronger discrimination at the level of dimers and trimers and their amino acid anhydrides. Since a

TABLE I

R_F values of nucleotides

	Solvent	
	Isobutyrate-H_2O	Phenol-H_2O
Adenosine-2′-phosphate	0.49	0.70
Guanosine-2′-phosphate	0.24	0.46
Uridylic acid	0.24	0.35
Cytidylic acid	0.37	0.57
U-U-U	0.42	
A-A-A-	2.22	
C-C-C	1.45	
G-G-G	0.25	

All data taken from Wyat (1955) and Lohrmann *et al.* (1966).

variety of different types of micelles is feasible with differing surfactant structure and medium polarity, the uneven distribution of all existing amino acids and nucleotides seems possible. The principle of *similis simili gaudet* would govern the aggregation of micelles and their captive ingredients as well. Although our discussion will be restricted to the A-U system, the enrichment of G, I and C can also be envisioned. G and I could be taken up into particular micelles, especially in the presence of aromatic compounds.

At this point one should remember the significant differences between the individual amino acids and nucleotides with respect to their chromatographic mobilities (Woese *et al.*, 1966). The separation of tRNAs via chromatography and counter-current distribution is even more remarkable if their similar size is considered. The numerous substituents such as acyl, alkyl and thio groups occurring in tRNAs without much apparent functional justification could have had their origins as discriminators with respect to partitioning.

The postulation of micelles as the crucial loci in prebiotic evolution has other merits apart from the opportunity of compartmentalization. Micelles have catalytic powers which at times approximate those of enzymes (Fendler *et al.*, 1973). The relative exclusion of water could have facilitated the condensation reactions which belong to the least understood chapter of prebiotic chemistry. The size of the intra-micellar cavity could have been a decisive factor in determining the size of the first adaptors. Additionally, the protection of labile compounds against hydrolysis by reversed micelles could have functioned as a further selective force.

For our present discussion it is not very critical to elaborate on how and where the condensation of nucleotides with each other and with amino acids could have taken place. For instance, it is also possible that small oligomers were first formed in the aqueous phase and subsequently compartmentalized for further modification. Elsewhere, we shall argue that thiophosphates and thionucleotides could have played an important role in the process. The intermediary formation of aminoacyl nucleotide anhydrides, one of which was first proposed by Paecht-Horowitz *et al.* (1970), is particularly attractive because such compounds could be polymerized to oligonucleotides bearing an activated amino acid.

The evolution of the ideal system could have progressed rapidly from relatively simple micelles through larger ones to the coacervates of Oparin (1965) or to the microspheres of Fox (Nakashima and Foe, 1972). The prebiotic random synthesis of macromolecules must have preceded this step. The crudest form of primitive cell contained, in our opinion, both aqueous and reversed micelles in its 'cytoplasm', where the polynucleotide template eventually associated with the various charged adaptors. Figure 1 illustrates such a primitive system of translation machinery. The adaptors could have been simple loops held together by A-U base pairs. In addition to the Phe-AAA adaptor, which is shown, the nonpolar compartment could have also contained the anticodons AAU, AUA, and UAA in different adaptors, also charged by leucine, isoleucine and tyrosine. Conversely, the adaptors in the reverse micelle could have possessed UUU and AUU in their loops and lysine and asparagine at their end. One may extend this model to other nucleotides and amino acids. Glycine and cyti-

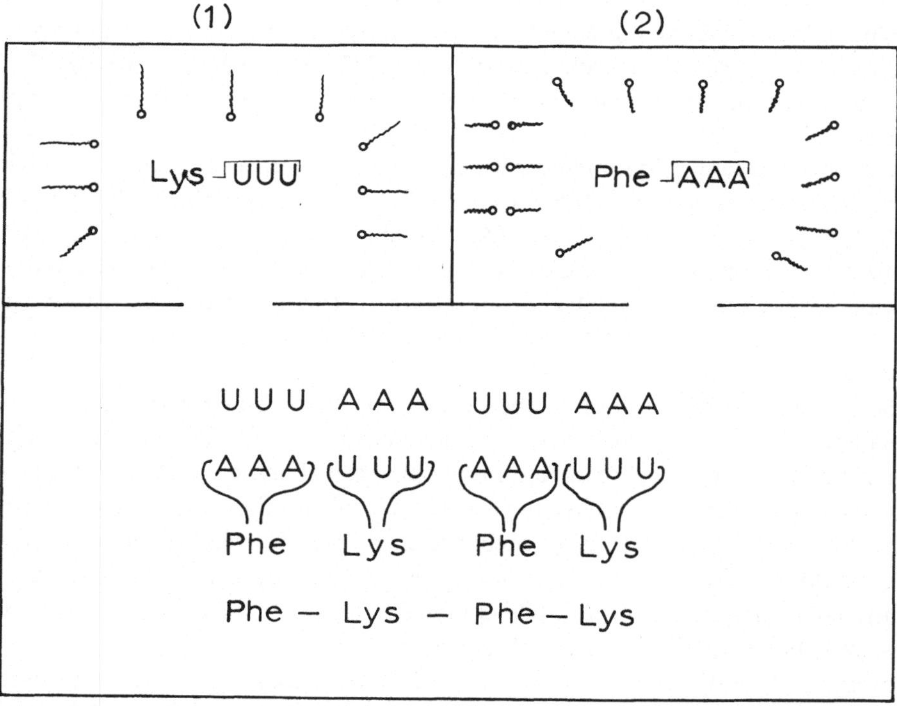

Fig. 1. Scheme of primitive translation machinery. Compartment 1 contains polar cavities which can be provided by reversed micelles and/or membranelike aggregates. Compartment 2 possesses a hydrophobic environment. Compartment 1 is capable of enriching lysine, uridylate and their short oligomers which are eventually condensed with some adenylate to form the adapter complex. The symbol Lys—UUU stands for as mall loop which is held together by A-U pairs. Phe—AAA may be formed in compartment 2 in a similar fashion. The arbitrary template **5** UUU-AAA-UUU - would be translated to Phe-Lys-Phe.

dylate, and proline and inosinate, respectively, could have fallen into the same compartments. The polypeptide which was eventually produced on the template should have reflected the composition of the message, albeit not necessarily with great fidelity. For the evolution of more discrimination adaptors the peptide products must have first enlarged and differentiated the micellar compartments, and then, eventually, developed a specific compartment of their own, i.e. recognition site for aminoacyl-adenylate and tRNA. The capacity of the adaptors to fit in a geometrical sense into compartments has continuously evolved, and so did the ability of the anticodon loop to discriminate, starting from a simple partitioning and culminating in an accurate recognition of a molecular environment. It is amazing that even in the present tRNAs the sum of hydrophobic and hydrophilic interactions, as evidenced in countercurrent distribution mobilities, reflects the composition of the anticodon (Woese *et al.*, 1966).

This hypothesis is intended to be, first of all, an interpretation of some basic codon assignments. It suggests that the amino acid-codon relationship, at least with respect to polarity, is not accidental, but it is the result of seemingly trivial physical-chemical factors which originally did not possess the criteria of direct recognition. We suggest that the accurate and specific ligases evolved from simple association colloids in a less obvious manner in accordance with the Principle of Continuity. The asymmetric peptides proposed by Orgel (1972), or some of the proteinoids prepared by Fox (Nakashima and Fox, 1972) may contain regions of discriminatory tendencies in the sense as suggested above. Because of their simplicity and similarity to biological membranes reversed micelles would offer a first approximation and obvious starting point for the study of primitive adaptors.

Experimental verification of all aspects of this hypothesis may well be time-consuming, but it is quite feasible. It is necessary to demonstrate that all condensation reactions, *viz.* phosphorylation of nucleosides, formation of aminoacyl nucleotides, and oligonucleotides, can be accelerated by colloid systems. We are currently using the energy of thiophosphate and S-cyanoethyl phosphorothioate to effect such condensations. So far we have found that all ribo- and deoxyribonucleosides can be converted to natural nucleotides at 60° and 70° in organic solvents; the 5′ nucleotides are the dominant products. Preliminary experiments also show that nucleoside phosphorothioates can be polymerized to form natural oligonucleotides. Selective uptake of polar nucleotides by reversed micelles has also been demonstrated in our laboratory.

Acknowledgement

This work was supported by the Robert A. Welch Foundation.

References

Crick, F. H. C.: 1968, *J. Mol. Biol.* **38**, 367.

Fendler, E. J. and Fendler, J. H.: 1970, *Adv. Phys. Org. Chem.* **8**, 271.

Fendler, E. J., Chang, S. A., Fendler, J. H., Medary, R. T., El Seoud, O.A., and Woods, V. A.: 1973, in E. H. Cordes (ed.), *Reaction Kinetics in Micelles*, Plenum Press, New York, p. 127.

Lacey, J. C. and Pruitt, K. M.: 1969, *Nature* **233**, 799.
Lasaga, A. C., Holland, H. D., and Dwyer, M. J.: 1971, *Science* **174**, 53.
Lohrmann, R., Soll, D., Hayatsu, H., Ohtsuka, E., and Khorana, H. G.: 1966, *J. Amer. Chem. Soc.* **88**, 819.
Nakashima, T. and Fox, S. W.: 1972, *Proc. Nat. Acad. Sic. (U.S.A.)* **69**, 106.
Oparin, A. I.: 1965, in S. W. Fox, (ed.), *The Origins of Prebiological Systems*, Vol. I, Academic Press Inc., New York, p. 331.
Orgel, L. E.: 1968, *J. Mol. Biol.* **38**, 381.
Orgel, L. E.: 1972, *Israel J. Chem.* **10**, 287.
Paecht-Horowitz, M., Berger, J., and Katchalsky, A.: 1970, *Nature* **228**, 636.
Woese, C. R., Dugre, D. H., Dugre, S. A., Kondo, M., and Saxinger, W. C.: 1966, *Cold Spring Harbor Symp. Quant. Biol.* **31**, 723.
Wyatt, G. R.: 1955, in E. Chargaff and J. N. Davidson (eds.), *The Nucleic Acids*, Academic Press Inc., New York, p. 256.

THE IRON-SULPHUR PROTEINS: EVOLUTION OF A UBIQUITOUS PROTEIN FROM MODEL SYSTEMS TO HIGHER ORGANISMS

D. O. HALL, R. CAMMACK, and K. K. RAO

University of London, King's College, Dept. of Botany, London SE24 9JF, England

Abstract. Ferredoxins are Fe–S proteins with low molecular weight (6–12000) which act as electron carriers at very low redox potentials eg. -300 to -500 mV, in diverse biochemical processes such as bacterial and plant photosynthesis, N_2 fixation, carbon metabolism, oxidative phosphorylation and steroid hydroxylation. They are found in a wide range of organisms from the 'primitive' obligate anaerobic bacteria, through photosynthetic bacteria, blue-green and green algae, to all higher plants and animals. Three types of ferredoxins are known – 8 Fe + 8 S, 4 Fe + 4 S and 2 Fe + 2 S. All three have been found in bacteria while the 2 Fe and some 8 Fe ferredoxins have been found in plants and animals possibly representing an evolutionary sequence. The 8 Fe ferredoxin may all be composed of two 4 Fe units. We have proposed that because of the simplicity of the 8 Fe ferredoxins (only 9 common simple amino acids in clostridia, 6 of which have been detected in the Murchison meteorite) they may have been amongst the earliest proteins formed during the origin of life. A simple peptide of about 27 amino acids could incorporate inorganic Fe + S (or possibly an existing Fe–S complex) into it non-enzymatically under anaerobic conditions to form a protein carrying one or two electrons at the potential of the H_2 electrode. More than ten Fe–S model compounds have been proposed as analogues of the 4 Fe or 2 Fe containing active centres; inorganic, organometallic and peptide complexes have been synthesized. A few have many of the properties of ferredoxins but none as yet fulfills a sufficient number of criteria to substitute for ferredoxins chemically and biologically – a goal which will provide many clues to primitive peptide systems undergoing biological electron transfer reactions.

1. Introduction

The iron-sulphur proteins are a group of proteins which include the ferredoxins, rubredoxins, high potential iron proteins, and iron-sulphur flavoproteins. There are a number of reviews which may be consulted for further details and references [1–7]. See Table I for a summary of the properties of iron-sulphur proteins.

The iron-sulphur proteins all contain non-haem iron in the active centre which is co-ordinated to cysteine sulphurs. In addition to specific numbers of iron atoms they all (except rubredoxins) contain an equivalent number of sulphide (inorganic sulphur) atoms, e.g. 2 Fe + 2 S and 4 Fe + 4 S. In the case of rubredoxins, the molecule does not contain inorganic sulphur but has a single iron co-ordinated to four cysteine sulphurs. Other notable properties are (i) ability to act as redox carriers at specific potentials between -490 and $+350$ mV, (ii) characteristic EPR spectra at liquid N_2 and He temperatures in either the reduced form, e.g. ferredoxins at $g = 1.94$, or in the oxidized form, e.g. HIPIP at $g = 2.03$, (iii) the proteins are acidic and show characteristic behaviour on DEAE-cellulose columns, (iv) the Fe (plus S) can be readily removed from the active centre to form an apoprotein which can be reconsitituted with Fe and sulphide.

TABLE I

Properties of iron-sulphur proteins

Type of Fe–S chromophore	Source	No. of Fe and S	MW	Approx. no. of amino acids	Redox potential mV	No. of electrons transferred	Type of reactions catalyzed
4[Fe–S]							
(a) Clostridial type (8 Fe) ferredoxin	Anaerobic bacteria	8 Fe, 8 S	6000	55	−390	2	Phosphoroclastic reaction, CO_2 reduction
	Green photo-synthetic bacteria.						N_2 fixation, photosynthetic CO_2 fixation
	Chromatium	8 Fe, 8 S	9000	81	−490	2	N_2 fixation, photosynthetic CO_2 fixation
(b) 4 Fe Ferredoxins	*D. gigas*	4 Fe, 4 S	6000	56	–	–	Sulphite reduction
	B. polymyxa	4 Fe, 4 S	9000	79	−390	1	N_2 fixation, NADP photoreduction
(c) HIPIP	*Chromatium*	4 Fe, 4 S	9600	86	+350	1	Photosynthetic electron transport
2[Fe–S]							
(a) Plant ferredoxin	Algae and plants	2 Fe, 2 S	10500	97	−430	1	NADP photoreduction, photo-synthetic-phosphorylation, N_2 fixation (Blue-green algae)
(b) Adrenodoxin	Mammalian mitochondria	2 Fe, 2 S	12500	114	−270	1	Hydroxylation of steroids
(c) Putidaredoxin	*P. putida*	2 Fe, 2 S	12500	114	−240	1	Hydroxylation of camphor
1[Fe]							
Rubredoxin	Anaerobic bacteria	1 Fe	6000	53	−60	1	–
Rubredoxin	*P. oleovorans* (aerobic)	1 or 2 Fe	19000	174	−60	1 per Fe	Oxidation of hydrocarbons
Complex Fe–S proteins							
Xanthine oxidase	Bacteria	8 Fe, 8 S					
	Mammals	2 Mo, 2 FAD	275000	–	–	–	Oxidation of xanthines, aldehydes

Iron-sulphur proteins have only recently been recognised as a ubiquitous group of proteins which occur in obligate anearobic bacteria, photosynthetic bacteria, fungi, algae, plants and animals – in fact in all forms of organisms. They are a class of electron carriers which act in an analogous manner to cytochromes, but are more widespread and often present in larger amounts than the cytochromes. They are involved in such diverse reactions as the phosphoroclastic reaction of pyruvate, nitrogen fixation, bacterial and plant photosynthesis, mammalian steroid hydroxylation, etc. So we know less about their structures than about the cytochromes. Moreover, their active centre is more labile than the haem group and therefore to study them we must use physico chemical rather than organic chemical techniques.

In this article we will discuss the idea that a bacterial ferredoxin may be considered as a simple type of redox protein which may have been amongst the earliest proteins formed during the origin and evolution of life [8, 9]. The ferredoxin type of active centre may have been a model which was derived from existing Fe–S compounds, and subsequently evolved into the Fe–S proteins of higher forms of life.

2. Early Life Conditions and the Origin of Ferredoxins

If one examines the amino acid sequences of the ferredoxins thus far sequenced from the five species of clostridia (obligate anaerobic fermentors) it will be seen that there is a striking homology, Figure 1, refs. [10–12]. These ferredoxins contain 8 Fe, 8 S and 8 cysteines – the cysteines are all invariant, and the molecule, which is only 55 amino acids long, consists of two similar halves, indicative of gene duplication. Of the 55 amino acids, 38 occupy invariant positions in the molecule and only 9 of the 20 different protein amino acids are common in all the 8 Fe ferredoxins. If one examines the composition of *C. butyricum* ferredoxin (Table II) it is seen that these 9 amino acids comprise 91% of the molecule and we thus consider it to be the simplest ferredoxin sequenced thus far. Six of these nine amino acids have been detected in the Type I carbonaceous chondrite meteorites, viz., gly, ala, val, glu, asp and pro (column II of Table II); these six comprise twothirds of the total amino acid content of *C. butyricum ferredoxin.* These six amino acids, and the remaining three amino acids, viz., cys, ser, ile, have all been synthesized under simulated primitive earth conditions (columns IV–VII of Table II) [8, 12]. It is also noteworthy that there is a complete absence or very low content of lysine, histidine, arginine, tryptophan and methionine in these proteins.

Orme-Johnson [13] has cleaved the apoprotein of *C. acidi urici* ferredoxin into two fragments, containing residues 1–29 and 30–55, using trypsin, and reconstituted them with iron and sulphide. He concluded that 'the half chains of at least this contemporary ferredoxin retain some capacity for forming iron-sulphur clusters that transfer electrons'.

Thus we may propose that once the required nine simple amino-acids were available a 27 amino acid ferredoxin prototype may have been formed under the anaerobic conditions which existed. Iron and sulphur existed in abundance on the primitive earth and

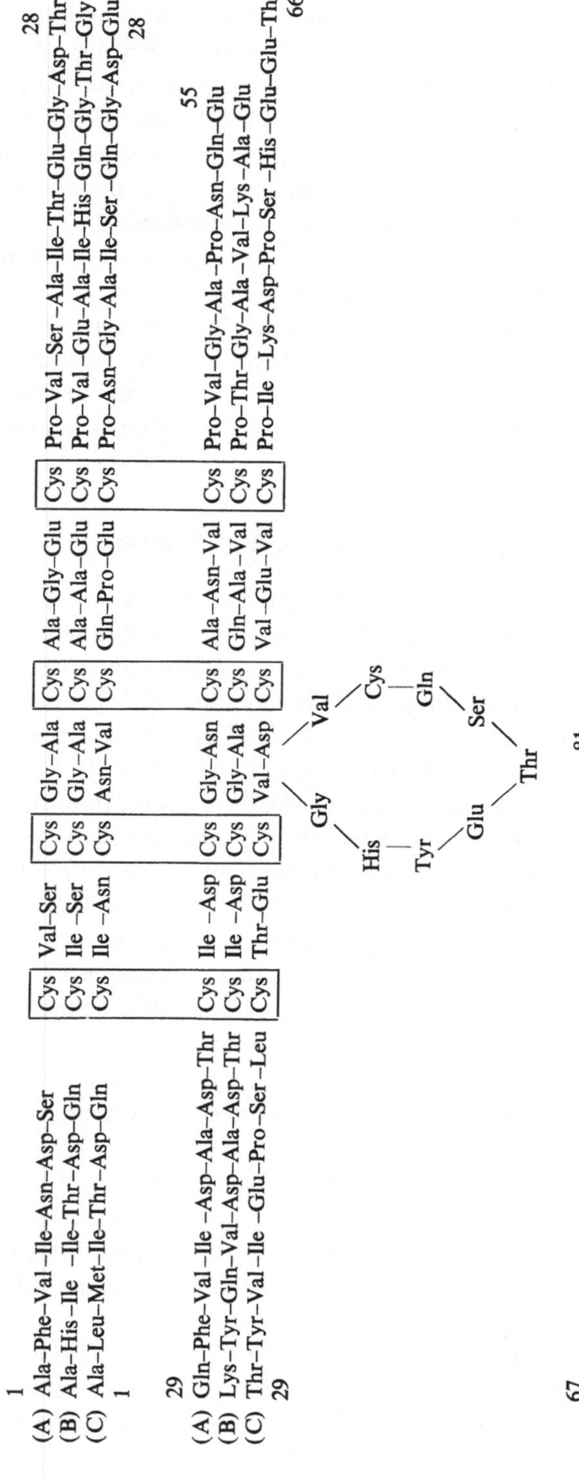

(A) *C. butyricum*: (B) *C. tartarivorum*: (C) *Chromatium*:

Fig. 1. Comparison of sequences of *Clostridium* and *Chromatium* ferredoxins (from Hall *et al.*) [16].

TABLE II

Amino acid composition of a ferredoxin, extraterrestrial material, and prebiotic syntheses

	Clostridium butyricum ferredoxin	Murchison, Murray and Orgueil meteorites	Lunar soil (Apollo 11, 12 and 14)	CH_4, C_2H_6, H_2O, H_2S and NH_4 – UV irradiation	CH_4, N_2, H_2O, and NH_4 – sparked	HCHO, NH_3, H_2O – heat, 185°	CH_3OH, NH_3, HCOOH – UV irradiation
Amino acid	(I)	(II)	(III)	(IV)	(V)	(VI)	(VII)
Lys	0						
His	0						
Arg	0						
Trp	0						
Asp	⑨	⊕	+	+	+	+	+
Thr	3		+		+		+
Ser	③		+	+	+	+	+
Glu	⑤	⊕	+	+	+	+	+
Pro	③	⊕			+	+	+
Gly	⑤	⊕	+	+	+	+	+
Ala	⑦	⊕	+	+	+	+	+
Cys	⑧			+			
Val	⑥	⊕			+	+	+
Met	0						
Ile	④		+		+		+
Leu	0		+		+		+
Tyr	0						+
Phe	2						+

○ = 9 'common' amino acids of clostridial ferredoxins comprising 91% of *C. butyricum* ferredoxin.
⊕ = 6 amino acids comprising 64% of *C. butyricum* ferredoxin.

may have been incorporated as a pre-exisitng Fe_4S_4 molecule or as individual atoms into an orientation predetermined by the amino-acid chain. This formation of an active ferredoxin from an apoprotein plus Fe and sulphide has been shown to occur non-enzymatically under anaerobic conditions [14, 15].

It has been proposed that there is an evolutionary development of ferredoxins from the obligate anaerobic bacteria, through the green and red photosynthetic bacteria and the sulphate-reducing bacteria (all anaerobic), to the blue-green algae and thence to higher plants and animals (Figure 2) – [3], [6], [10], [12] and [16] for literature and discussions. The role of the 8 Fe, 4 Fe and 2 Fe ferredoxins in this scheme will be discussed later.

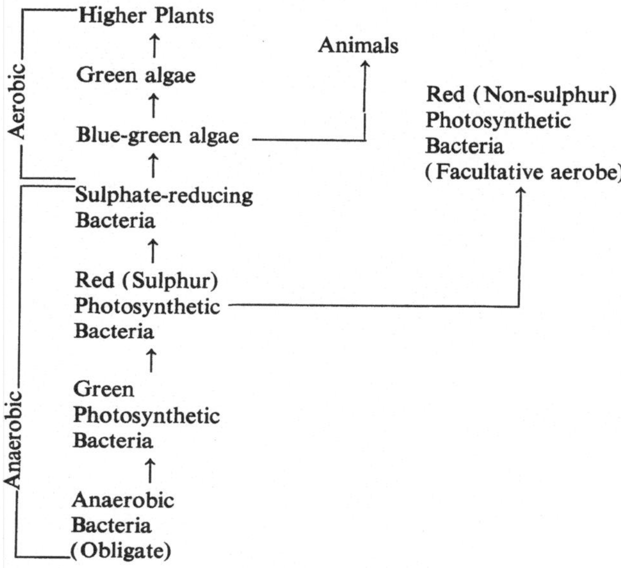

Fig. 2. Evolutionary development of ferredoxins (Hall *et al.*) [12].

3. Geological and Model Fe-S Compounds

The formation of a number of iron sulphides has been studied in detail by Rickard [17]. Their interrelationships in aqueous solution at 25 °C are shown in Figure 3. He has concluded that no differences can be observed between the mechanisms of formation of biogenic and abiogenic sulphides. It is interesting that the formation of greigite (Fe_3S_4) is brought about by heating ferrous ammonium sulphate and sodium sulphide in aqueous solution at 190 °C – these two same reagents are used to reconsitute ferredoxins. Mackinawite (FeS) is formed by the reaction between ferrous and sulphide irons at room temperature in aqueous solution. Pyrite (FeS_2) can be formed by the reaction between mackinawite and polysulphide ions.

Thus it is possible to imagine that a preexisting iron sulphide complex could have been incorporated as such into an apoportein to form a ferredoxin-type protein. The

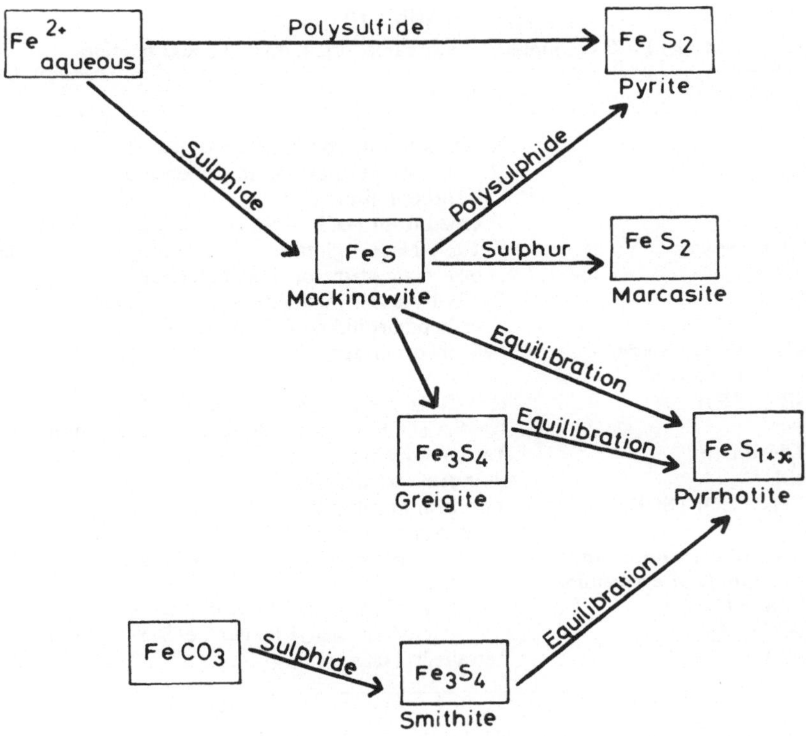

Fig. 3. Iron sulphide interrelationships in aqueous solutions (redrawn from Rickard, 1969) [17].

physico-chemical properties of these iron sulphide compounds may not exactly mimic the Fe–S active centre of ferredoxins but their presence in a given protein environment may be sufficient to produce the known properties of ferredoxins' active centres. We shall see later how the redox state of the 4 Fe + 4 S proteins can be dramatically altered by varying the environment of the protein with, for example, agents such as dimethylsulphoxide.

Since the early recognition of the unique physico-chemical properties of Fe–S proteins, model compounds have been synthesized so as to attempt to mimic the Fe–S active centres of the 4 Fe and 2 Fe proteins. More than a dozen compounds (inorganic, organic and peptide complexes) have been studied (Table III) [18–34]. A few have some of the required properties, but none as yet fulfills a sufficient number of criteria, to substitute for Fe–S proteins chemically and biologically – a goal which undoubtedly will provide many clues to primitive peptide systems undergoing biological electron transfer reactions.

The three most promising compounds seem to be those numbered 8, 9 and 10 in Table III – their structures are shown in Fig. 4. They are all 4 Fe and 4 S analogues. None of the analogues compatible with a 2 Fe + 2 S centre seems very promising as yet; this is a great pity as we have X-ray structures of 4 Fe + 4 S centres but none is

TABLE III

Model Fe–S compounds – inorganic, organometallic and peptides

Complex	Comment	Reference
1) $(KFeS_2)_n$	K Dithioferrate (polymeric with tetrahedral Fe)	18, 19
2) $Fe_3L_4S_6H_2$	$[L=S_2C_2(CF_3)_2]$ Diiron in presence of isopropyl disulphide	20
3) Fe–S	Formed from $FeCl_3+Na_2S+$ mercaptoethanol	21
4) peptide Fe–S chelates	$FeCl_3+Na_2S+$ glutathione or mercaptopropionate or acetylcysteine or a pentapeptide	22
5) Dithiol–Fe–S	$FeCl_3\pm Na_2S+$ butanedithiol or pentanedithiol or heptanedithiol or dithiothreitol	23
6) Pentapeptide and Decapeptide analogs of rubredoxin	Fe^{2+} incorporated	24
7) $[Fe(S_2CSR_2)(SR)]_2$	$R=C_2H_5$, n-C_3H_7, n-C_4H_9, $C_6H_5ClCH_2$	25
8) $[Fe_4S_4L_4]^{n-}$	$L=S_2C_2(CF_3)_2$ Tetrameric dithiolene complex	26, 27
9) $[Fe_4S_4(SCH_2Ph)_4]^{2-}$	Tetrameric anion	28
10) $[C_5H_5\text{–}FeS]_4$	Elongated tetrahedron	29, 30
11) $[Fe(h^5\text{–}C_5H_5)(CO)\text{–}(SCH_3)]_2^+$	Two iron moieties linked by two bridging mercapto ligands	31
12) Fe^{2+} complex of cysteamine and hydropersulphide of cysteamine $(RS_2\text{–}Fe^{2+}\text{–}S\text{–}R)$	Cysteine methyl ester can be used instead of cysteamine. EPR g-values 1.92, 1.94 and 2.00	32
13) $[CpFe\text{–}(SC_2H_5)S]_2$	Cp = cyclopentadienyl Planar Fe–S–S–Fe bridge	33
14) FeS_3Se and FeS_2Se_2	Tetrahedral clusters	34

available of a 2Fe+2S centre because of the difficulty so far in obtaining crystals of sufficient size and stability for X-ray analysis.

4. The 8Fe+8S Ferredoxins

This type of Fe–S protein has thus far only been isolated from bacteria (both non-photosynthetic and photosynthetic) e.g. anaerobic fermentors like the clostridia and peptococci, photosynthetic bacteria like the green *Chlorobium* and the red *Chromatium*, and in aerobic nitrogen fixers like *Azotobacter*. All the 8Fe ferredoxins which have been analyzed, except *Chromatium* ferredoxin, consist of about 55 amino acids with a molecular weight of approximately 6000 daltons. They have a redox potential of about −390 mV (again except for *Chromatium* which is −490 mV) and transfer 1 or 2 electrons in biological reactions. An X-ray investigation of crystalline *P. aerogenes* ferredoxin has shown that the 8Fe and 8S atoms exist as two identical clusters of 4Fe+4S atoms (Figure 5) [35]. It had already been shown [36] by proton magnetic resonance techniques that the iron atoms were probably bound to the eight cysteines which are in invariant positions in the bacterial ferredoxins which have been sequenced (Figure 1).

The 8Fe ferredoxins seem to be rather simple proteins which do however, catalyze a large number of reactions which are vital to the metabolism of anaerobic bacteria [5, 7]: (i) carbon metabolism, e.g. the pyruvate phosphoroclastic reaction, CO_2 reduc-

Synthetic Analogues of 4Fe+4S iron-sulphur proteins

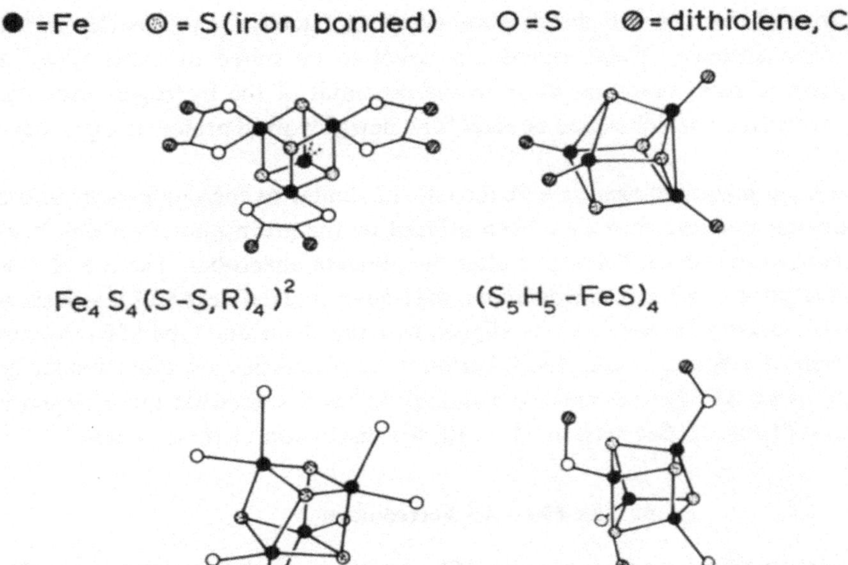

Fig. 4. Model Fe–S compounds (redrawn from Balch [26]; Schunn *et al.* [29]; Bernal *et al.* [27]; Herskovitz *et al.* [28]).

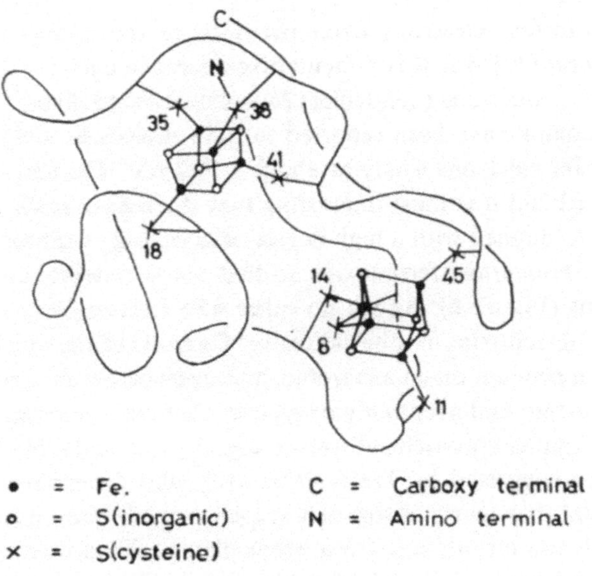

Fig. 5. The structure of *P. aerogenes* 8 Fe ferredoxin from X-ray analysis (redrawn from Jensen *et al.*) [35].

tase, pyruvate synthase, α-ketoglutarate synthase, (ii) hydrogen metabolism, e.g. H_2 as a reductant in N_2 fixation, H_2 evolution in the presence of excess reductant, (iii) light activated electron transfer in photosynthetic bacteria, e.g. possibly as the primary electron acceptor. These ferredoxins seem to be suited to many types of 1-and-2 electron transfer reactions close to the potential of the hydrogen electrode (−420 mV) – attributes which would be ideal for a newly evolved protein in a primitive environment.

It has also been proposed that an 8 Fe ferredoxin similar to the one present in the obligate anaerobic bacteria may have been utilized by the green photosynthetic bacteria (e.g. *Chlorobium*) which developed after the obligate anaerobes. The red photosynthesizers (sulphur type), e.g. *Chromatium*, may have evolved next and it is seen in Figure 1 that its ferredoxin can be easily aligned with the clostridial type of ferredoxin by the insertion of a 9 amino-acid loop; certainly its properties are quite similar in most respects to all 8 Fe ferredoxins even though it has a somewhat more negative redox potential (Table I). See reviews [3, 6, 10] for discussion of these points.

5. The 4Fe+4S Ferredoxins

These iron-sulphur proteins which contain 4 Fe and 4 S per mole have only been discovered recently. They have properties similar to the 8 Fe ferredoxins and their properties are being actively investigated because the structure of their iron-sulphur centre may be far easier to resolve than the 8 Fe ferredoxins. Although they may be considered as half an 8 Fe ferredoxin their protein chain is in fact about the same size. They have been isolated from (i) the obligate anaerobe *Desulphovibrio gigas* [37] which is a sulphate reducer, (ii) *Bacillus polymyxa* [38, 39] which is a facultative N_2 fixer where the ferredoxin transfers electrons from pyruvate to the nitrogenase enzyme, (iii) *Spirochaeta aurantia* [40] which is a facultative anaerobe and (iv) *Bacillus stearothermophilus* facultative anaerobe (denitrifier) (our unpublished data).

The 4 Fe ferredoxins have been reported to have molecular weights of 6000–9000 daltons and transfer electrons singly at about −390 mV. The ferredoxin of *D. gigas* has been sequenced and it is most interesting that the amino acids in the first half of the molecule can be aligned with a high degree of homology with the first half of both *Clostridium* and *Chromatium* ferredoxins so that the 4 cysteine residues are again in invariant positions (Figure 6). As yet no other 4 Fe ferredoxin has been sequenced.

The 4 Fe type of ferredoxins may be indicative of a mode of life which is in the process of becoming less dependent on an anaerobic, hydrogen-based metabolism and tending to use sulphate, nitrate and eventually oxygen as electron acceptors in the process of electron transfer from energy-rich substrates, e.g. organic acids, NADH, etc. [41–43]. Indeed it has been suggested by Travis *et al.* [37], who determined the sequence of *D. gigas* ferredoxin, that 'parts of this new sequence exhibit an unusually high degree of homology with the ferredoxins from green plants'. Thus one may consider 4 Fe ferredoxins as a link between the 8 Fe ferredoxins of obligate anaerobes and the 2 Fe ferredoxins of plants and animals. However, this attractive idea is complicated by the

C. butyricum

1 ... 28

Ala–Phe–Val–Ile –Asn–Asp–Ser [Cys] Val –Ser [Cys] Gly–Ala [Cys] Ala–Gly–Glu [Cys] Pro–Val –Ser –Ala–Ile –Thr–Gln–Gly–Asp–Thr

D. gigas.

Pro–Ile –Gln–Val –Asp–Asn [Cys] Met–Ala [Cys] Gln–Ala [Cys] Ile –Asn–Glu [Cys] Pro–Val–Asp–Val–Phe–Gln–Met–Asp–Glu–Gln–Gly–Asp

Spinach

Glu [Cys] Pro–Asp–Asp–Val–Tyr–Ile – Leu–Asp–Ala–Ala–Glu–Glu

17 ... 30

D. gigas

30 ... 56

Lys–Ala–Val–Asn–Ile – Pro–Asn–Ser–Asn–Leu–Asp–Asp–Glu [Cys] Val–Glu–Ala–Ile–Gln–Ser–Cys–Pro–Ala–Ala–Ile–Arg–Ser

Spinach

Glu–Gly–Ile –Asp–Leu–Pro–Tyr–Ser–Cys–Arg–Ala –Gly–Ser [Cys] Ser–

31 ... 45

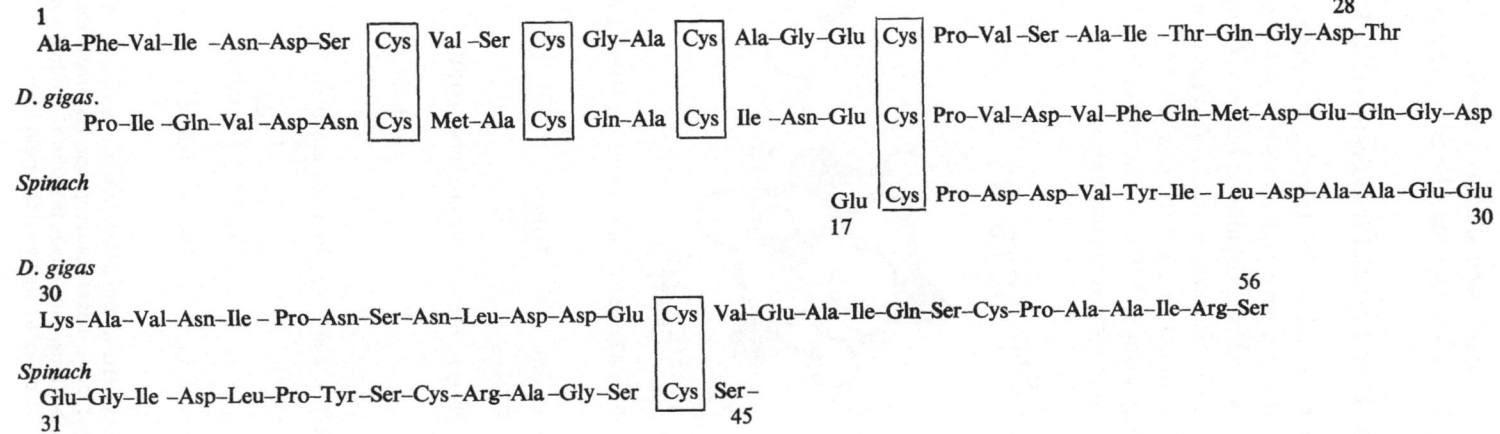

Fig. 6. Sequence of *D. gigas* ferredoxin compared with analogous segments of the ferredoxins of *C. butyricum* and Spinach (from Hall, 1972) [16].

fact that anaerobic bacteria have been shown to possess 2 Fe ferredoxins (sequences unknown) and we don't know whether the 4 Fe and 8 Fe ferredoxins occur in aerobic organisms.

6. High Potential Iron Protein (HIPIP)

This 4 Fe and 4 S protein resembles the ferredoxins quite remarkably but differs in one significant respect – it has a very high (positive) redox potential of +350 mV. It has been isolated from two red photosynthetic bacteria *Chromatium* and *Rhodopseudomonas*. The HIPIP from *Chromatium* contains 86 amino acids including four cysteine residues with a molecular weight of 9650 daltons [44, 45].

As isolated from the organism HIPIP exists in the 'reduced' state; it can be oxidized by a single electron transfer with reagents like potassium ferricyanide. Unlike the

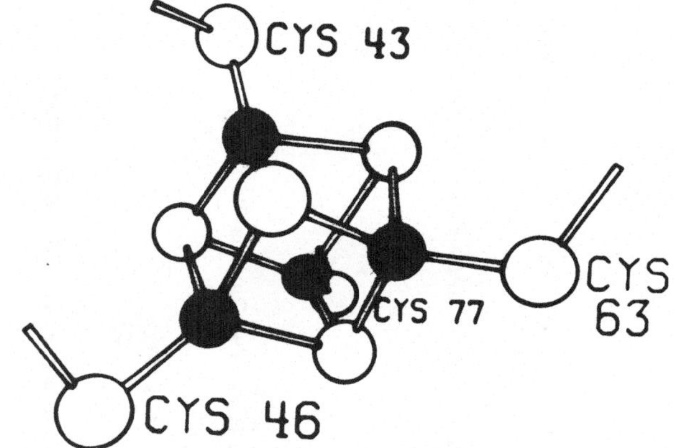

Fig. 7. The structure of the 4 Fe centre of *Chromatium* HIPIP (from Carter *et al.*) [46].

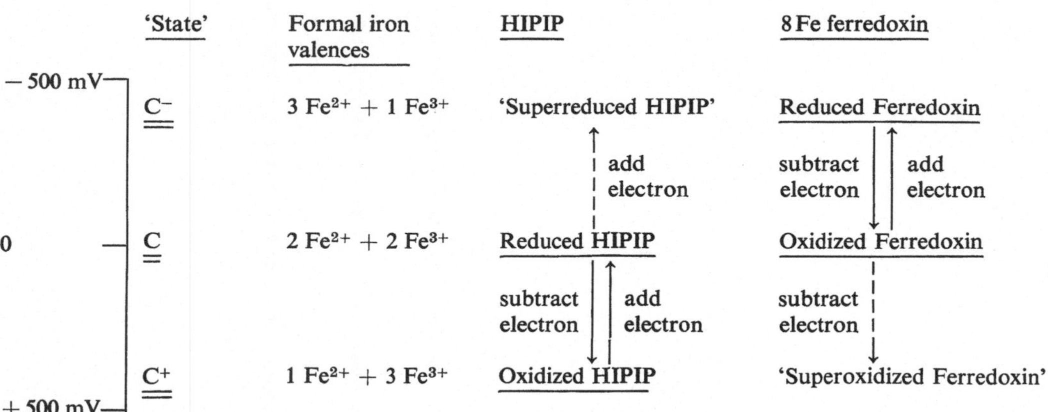

Fig. 8. Proposed redox states of HIPIP and 8 Fe ferredoxin both with 4 Fe centres (after Carter *et al.* [47], and our unpublished data). Note that the states proposed *ferredoxin* atoms are 'formal' valences only. The effects of covalent bonding may mean that the individual iron atoms are not as clearly differentiated as this would imply.

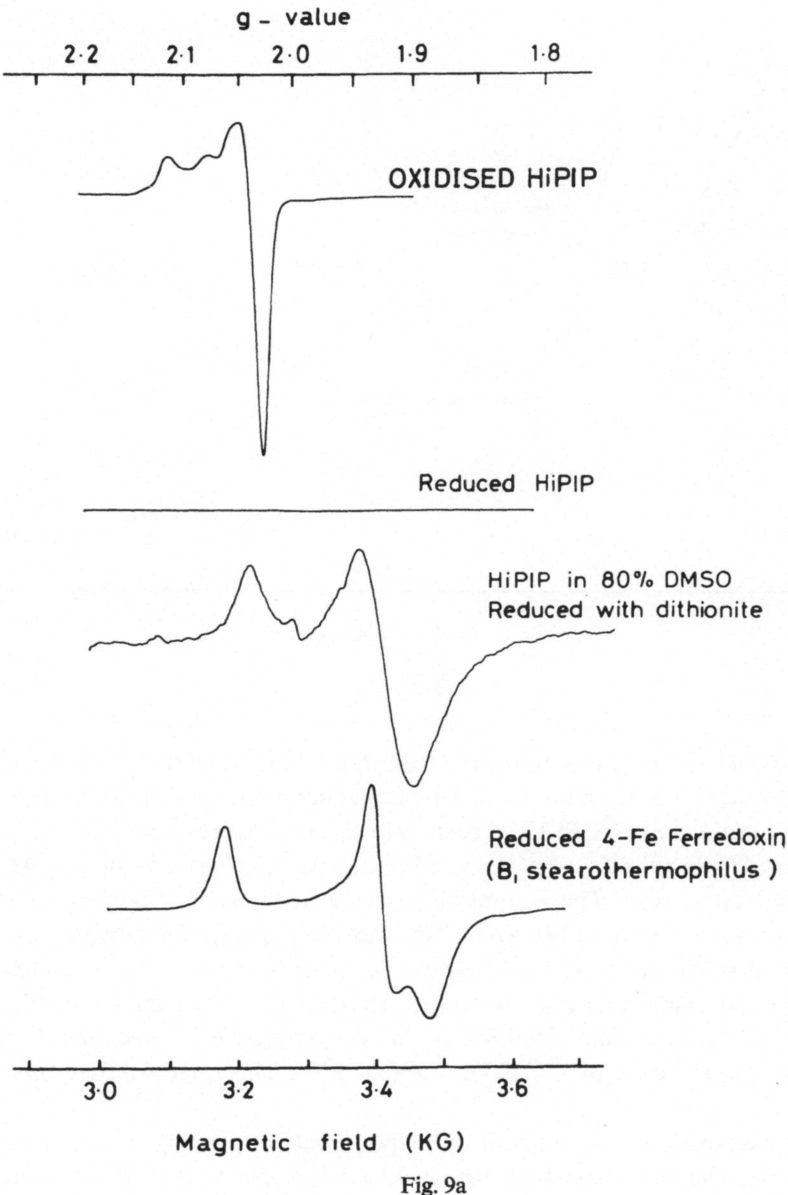

Fig. 9a

Figs. 9a–b. (a) EPR and (b) visible spectra of *Chromatium* HIPIP to show influence of DMSO on Fe–S active centre (after [98]). The spectrum of reduced four-iron ferredoxin from *B. stearothermophilus* is shown for comparison.

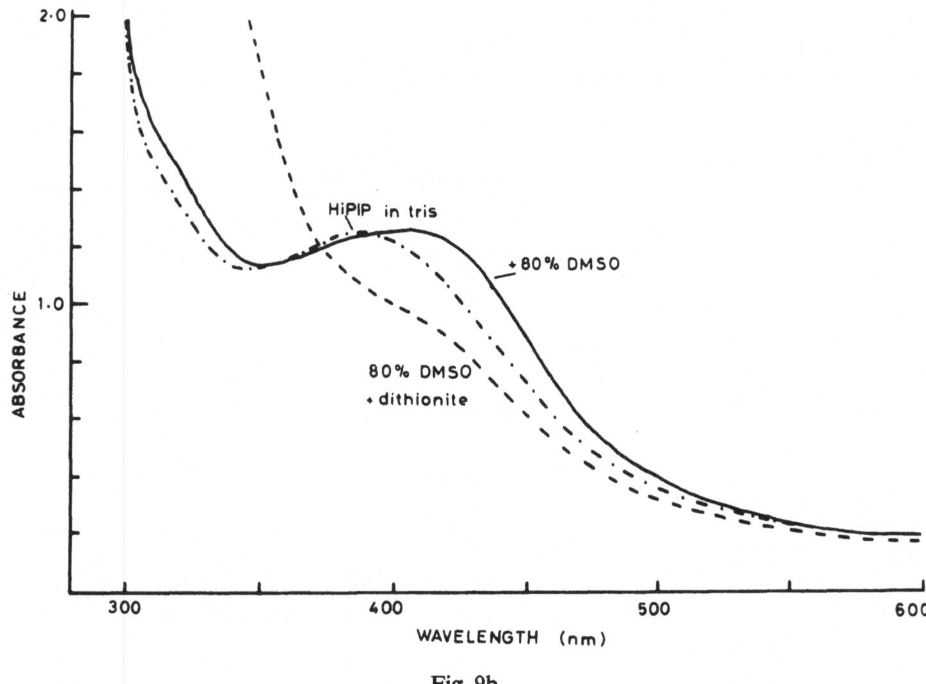

Fig. 9b.

ferredoxins which are magnetic in the reduced state ($g = 1.94$), HIPIP is non-magnetic in the reduced state but the oxidizied protein is magnetic with an EPR signal at $g = 2.03$.

The X-ray structure of HIPIP has been determined (Figure 7) [46] and it has also been shown [47] that the 4Fe+4S cluster is indistinguishable from the two 4Fe+4S cluster of 8Fe ferredoxins. This is somewhat unexpected in view of the large difference in redox potentials (740 mV) between HIPIP and ferredoxin. To explain this, it has been proposed [47] that the 4Fe+4S cluster can exist in three oxidation states. Thus it is assumed that oxidized ferredoxin and reduced HIPIP occupy the same redox state called state 'C' (with formal valencies of the iron atoms $2Fe^{2+}$ and $2Fe^{3+}$). Ferredoxin can be reduced to state 'C^-' ($1Fe^{2+}$ and $3Fe^{3+}$). This is shown diagramatically in Figure 8.

This proposal explains the different redox potentials very nicely. We have now obtained evidence that reduced HIPIP (as isolated) can indeed be further reduced by dithionite to the redox level of reduced ferredoxin. What is required apparently to perform this 'superreduction' is to distort the protein conformation of HIPIP by using a mild denaturing agent such as dimethylsulphoxide (DMSO). As seen in Figure 9 when the reduction is done in the presence of DMSO the EPR spectrum resembles that of a reduced 4Fe ferredoxin. The visible spectrum of the reduced HIPIP is also changed in the presence of DMSO to resemble that of a 4Fe ferredoxin.

Thus we have a 4Fe+4S cluster which can assume redox potentials of +350 or −390 mV by being associated with different protein 'backbones'. The advantage to primitive and evolving organisms in using such a cluster to catalyze different types of electron transport would be great. A further point of considerable importance is that the synthetic model Fe_4S_4 compound of Herskovitz *et al.* [28] (shown in Figure 4) has been reported to be geometrically nearly identical to that of the 4Fe+4S clusters of HIPIP and 8Fe ferredoxin.

Until recently no clear physiological function has been described for HIPIP but Evans *et al* [48] have observed by EPR spectroscopy that HIPIP in *Chromatium* chromatophores undergoes oxidation-reduction changes on illumination, or by treatment with artificial electron donors. They present data showing that HIPIP is a component of the photosynthetic membrane system functioning in the photosynthetic electron transport chain.

7. The 2Fe+2S Ferredoxins

This type of Fe–S protein was originally thought to be confined to O_2-evolving photosynthetic organisms, e.g. blue-green algae, other algae and higher plants, but it is now known that they can occur in aerobic and anaerobic bacteria and also in animals (Table I). They thus appear to be widespread but their role and significance in many organisms is not yet known.

The plant ferredoxins have been extensively investigated but as already noted no X-ray structure of the molecule has yet been reported because of difficulties in obtaining a large enough crystal stable for a sufficient length of time. These proteins consist of about 97 amino acids and have a molecular weight of about 10500 daltons. They have a redox potential of about −430 mV and transfer electrons singly while catalyzing photosynthetic electron transport. Another type of ferredoxin is probably the primary electron acceptor in the chlorophyll-catalyzed light reaction [49–51].

The primary structures of five plant ferredoxins are known (Figure 10) [10, 11, 16]. There is extensive homology in the structure from green algae (*Scenedesmus*), spinach (*Spinacea*), alfalfa (*Medicago*), taro (*Colocasia*), a monoctyledon, and a leguminous tree (*Leucaena*). About 60% of the residues are invariant; significantly four cysteines are invariant (cys 18 is absent from *Equisetum* [52]) and they are thought to bind the 2Fe's of the active centre. Figure 11 shows a model [53] of the 2Fe+2S active centre which was originally proposed by Gibson *et al.* [54]. This model seems to describe the active centre of all 2Fe ferredoxins from diverse organisms [55–59].

It is interesting that the ferredoxin from blue-green algae [60, 61] is very similar to that from higher plants where it is localized in the chloroplast. This supports the idea that chloroplasts are derived from a symbiotic association between a blue-green alga and a nonphotosynthetic eukaryotic cell [62, 63]. We have also shown a very stable ferredoxin is present in the blue-green alga *Spirulina* [64] which has been shown by Schopf and Blacic [65] to occur in very similar form in the Bitter Springs (Australia) micro-fossils believed to be about 900 m.y. old.

The 2Fe ferredoxin which has been isolated from mammalian adrenal glands (called adrenodoxin) [66] contains an active centre almost identical to that of plant ferredoxins [57, 67]. The adrenodoxin, which catalyzes steroid hydroxylation, contains

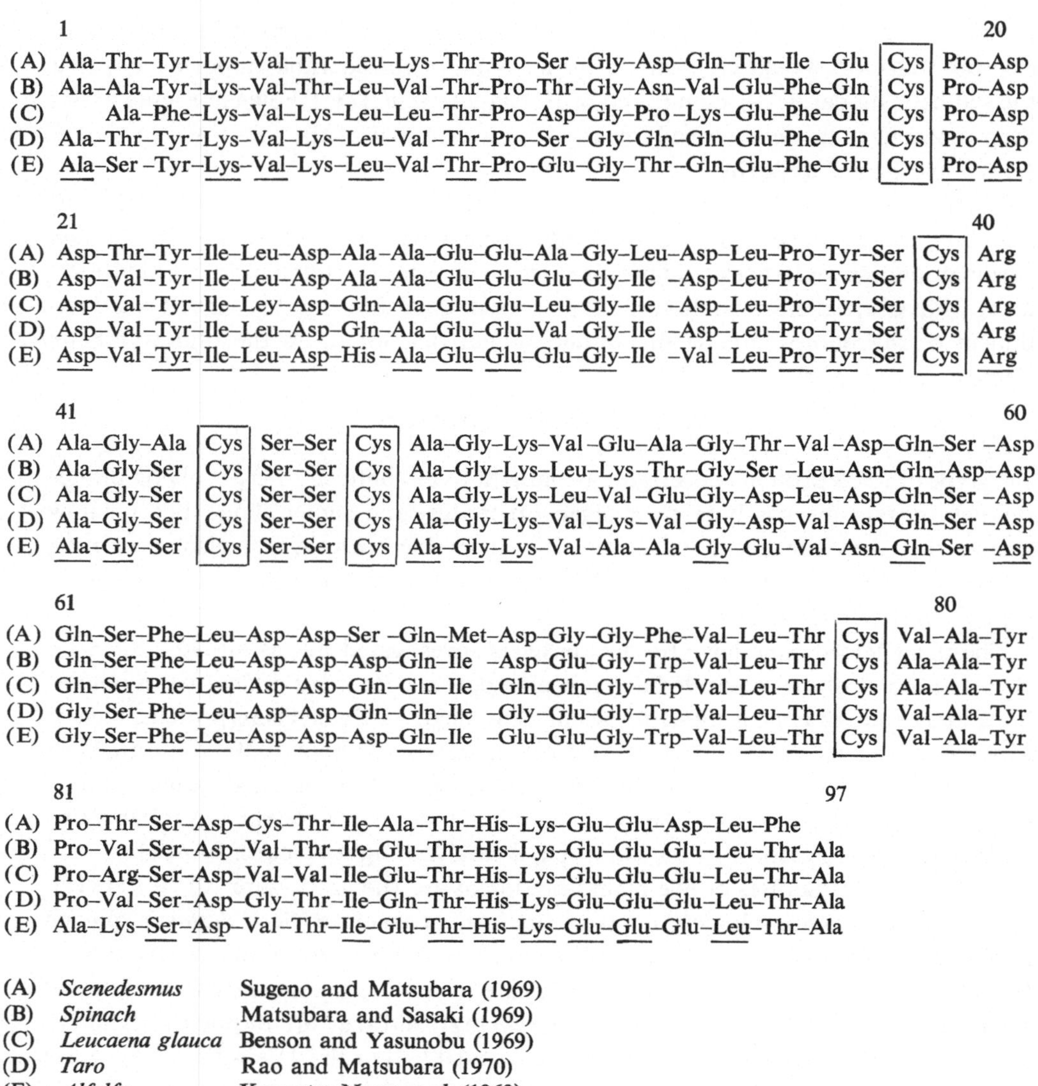

1 20
(A) Ala–Thr–Tyr–Lys–Val–Thr–Leu–Lys–Thr–Pro–Ser –Gly–Asp–Gln–Thr–Ile –Glu Cys Pro–Asp
(B) Ala–Ala–Tyr–Lys–Val–Thr–Leu–Val –Thr–Pro–Thr –Gly–Asn–Val –Glu–Phe–Gln Cys Pro–Asp
(C) Ala–Phe–Lys–Val–Lys–Leu–Leu–Thr–Pro–Asp–Gly–Pro –Lys–Glu–Phe–Glu Cys Pro–Asp
(D) Ala–Thr–Tyr–Lys–Val–Lys–Leu–Val –Thr–Pro–Ser –Gly–Gln–Gln–Glu–Phe–Gln Cys Pro–Asp
(E) Ala–Ser –Tyr–Lys–Val–Lys–Leu–Val –Thr–Pro–Glu–Gly–Thr –Gln–Glu–Phe–Glu Cys Pro–Asp

21 40
(A) Asp–Thr–Tyr–Ile–Leu–Asp–Ala –Ala–Glu–Glu–Ala –Gly–Leu–Asp–Leu–Pro–Tyr–Ser Cys Arg
(B) Asp–Val –Tyr–Ile–Leu–Asp–Ala –Ala–Glu–Glu–Glu–Gly–Ile –Asp–Leu–Pro–Tyr–Ser Cys Arg
(C) Asp–Val –Tyr–Ile–Ley–Asp–Gln–Ala–Glu–Glu–Leu–Gly–Ile –Asp–Leu–Pro–Tyr–Ser Cys Arg
(D) Asp–Val –Tyr–Ile–Leu–Asp–Gln–Ala–Glu–Glu–Val –Gly–Ile –Asp–Leu–Pro–Tyr–Ser Cys Arg
(E) Asp–Val –Tyr–Ile–Leu–Asp–His –Ala–Glu–Glu–Glu–Gly–Ile –Val –Leu–Pro–Tyr–Ser Cys Arg

41 60
(A) Ala–Gly–Ala Cys Ser–Ser Cys Ala–Gly–Lys–Val –Glu–Ala –Gly–Thr –Val –Asp–Gln–Ser –Asp
(B) Ala–Gly–Ser Cys Ser–Ser Cys Ala–Gly–Lys–Leu–Lys–Thr–Gly–Ser –Leu–Asn–Gln–Asp–Asp
(C) Ala–Gly–Ser Cys Ser–Ser Cys Ala–Gly–Lys–Leu–Val –Glu–Gly–Asp–Leu–Asp–Gln–Ser –Asp
(D) Ala–Gly–Ser Cys Ser–Ser Cys Ala–Gly–Lys–Val –Lys–Val –Gly–Asp–Val –Asp–Gln–Ser –Asp
(E) Ala–Gly–Ser Cys Ser–Ser Cys Ala–Gly–Lys–Val –Ala–Ala –Gly–Glu–Val –Asn–Gln–Ser –Asp

61 80
(A) Gln–Ser–Phe–Leu–Asp–Asp–Ser –Gln–Met–Asp–Gly–Gly–Phe–Val–Leu–Thr Cys Val–Ala–Tyr
(B) Gln–Ser–Phe–Leu–Asp–Asp–Asp–Gln–Ile –Asp–Glu–Gly–Trp–Val–Leu–Thr Cys Ala–Ala–Tyr
(C) Gln–Ser–Phe–Leu–Asp–Asp–Gln–Gln–Ile –Gln–Gln–Gly–Trp–Val–Leu–Thr Cys Ala–Ala–Tyr
(D) Gly–Ser–Phe–Leu–Asp–Asp–Gln–Gln–Ile –Gly–Glu–Gly–Trp–Val–Leu–Thr Cys Val–Ala–Tyr
(E) Gly–Ser–Phe–Leu–Asp–Asp–Asp–Gln–Ile –Glu–Glu–Gly–Trp–Val–Leu–Thr Cys Val–Ala–Tyr

81 97
(A) Pro–Thr–Ser–Asp–Cys–Thr–Ile–Ala –Thr–His–Lys–Glu–Glu–Asp–Leu–Phe
(B) Pro–Val –Ser–Asp–Val–Thr–Ile–Glu–Thr–His–Lys–Glu–Glu–Glu–Leu–Thr–Ala
(C) Pro–Arg–Ser–Asp–Val –Val –Ile–Glu–Thr–His–Lys–Glu–Glu–Glu–Leu–Thr–Ala
(D) Pro–Val –Ser–Asp–Gly–Thr–Ile–Gln–Thr–His–Lys–Glu–Glu–Glu–Leu–Thr–Ala
(E) Ala–Lys–Ser–Asp–Val –Thr–Ile–Glu–Thr–His–Lys–Glu–Glu–Glu–Leu–Thr–Ala

(A) *Scenedesmus* Sugeno and Matsubara (1969)
(B) *Spinach* Matsubara and Sasaki (1969)
(C) *Leucaena glauca* Benson and Yasunobu (1969)
(D) *Taro* Rao and Matsubara (1970)
(E) *Alfalfa* Keresztes-Nagy *et al.* (1969)

Fig. 10. Sequences of plant ferredoxins (from Hall *et al.*) [16].

114 amino acids and transfers electrons singly at a potential of −270 mV. By using statistical analyses, Barker *et al.* [68] have pointed out that the adrenodoxin sequence 'can be aligned with ferredoxins so that four cysteines are common to all sequences' suggesting that the 'plant ferredoxin sequences are about equally distant from those of animal adrenodoxin and bacterial ferredoxins'. This has been confirmed by Dayhoff recently – see her contribution in this volume. However, a corrected sequence of adrenodoxin has recently been reported by Tanaka *et al.* [69a] who state that 'there is no evident homology (of adrenodoxin) with the sequence of any of the known iron-

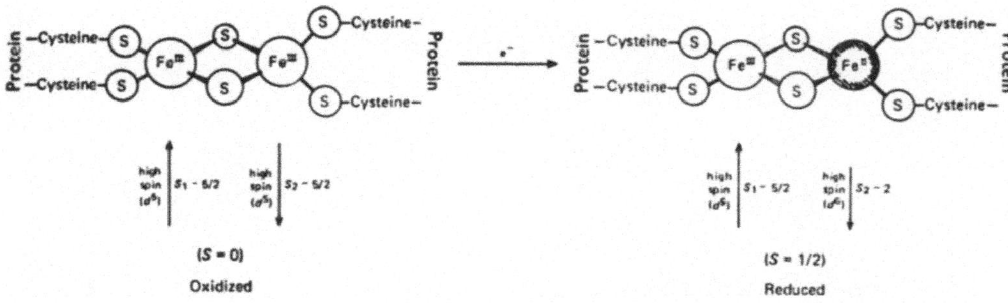

Fig. 11. Model of the active centre of 2 Fe ferredoxins (after Rao *et al.*) [53].

sulphur proteins'. If some homology does exist between adrenodoxins and other Fe–S proteins it must show only a very distant evolutionary relationship.

Two other 2 Fe iron-sulphur proteins have been isolated from animals (from mitochondria) but we know nothing of their sequences. These are the Complex III 'iron-protein' [69] which has a redox potential of +280 mV and the subunit from succinic dehydrogenase [70] (Complex II) which has a redox potential of +30 mV. Although not much detail is known of these proteins they may be a further case of big variations in the redox potential of a given active centre by varying the protein 'backbone'.

A 2 Fe ferredoxin has been isolated and characterised from *Pseudomonas putida* and termed putidaredoxin [71]. It catalyzes camphor (substrate) hydroxylation in an analogous manner to the mechanism of action of adrenodoxin. The ferredoxin has a potential of −240 mV and contains 114 amino acids whose sequence has been partially determined [72]. Tsai *et al.* [72] have identified a number of homologous segments in the primary structure of putidaredoxin and adrenodoxin which also exhibit a fair degree of homology with the sequences of plant ferredoxins.

Lastly, 2 Fe ferredoxins have also been isolated from *Clostridium* (three different types) and aerobically grown *E. coli, Azotobacter, Rhizobium, P. aminovorans* and *Agrobacterium* (Table I of Ref. [7] and Figure 12). However, not much is known about these proteins as yet; it would be highly desirable to sequence ferredoxins from a number of these bacteria.

8. Rubredoxin

These are Fe–S proteins which contain only 1 Fe per molecule and have no labile sulphur. They are classified as Fe–S proteins because the Fe atom is bound by four cysteine sulphurs in an analogous manner to the ferredoxins. Rubredoxin from *Clostridium* and other anaerobic bacteria consists of 55 amino acids, molecular weight 6000 daltons, and transfers electrons singly at a potential of −60 mV (see refs. [73–75], for further details). The X-ray structure [76, 77] of the molecule is shown in Figure 13. This information, together with other physico-chemical data [78, 79, 75] has been used to describe the molecule and its mode of action in some detail and this is proving useful in interpreting data from the ferredoxins where the Fe is also bonded to cysteine sulphurs.

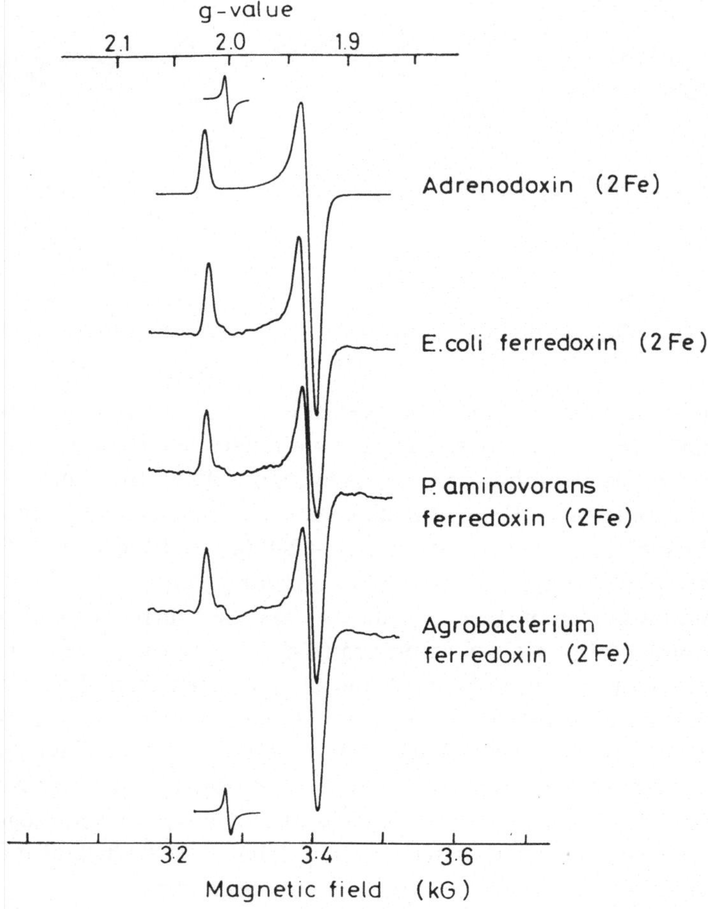

Fig. 12. EPR spectra of reduced two-iron Fe-S proteins (from Hall *et al.*) [7].

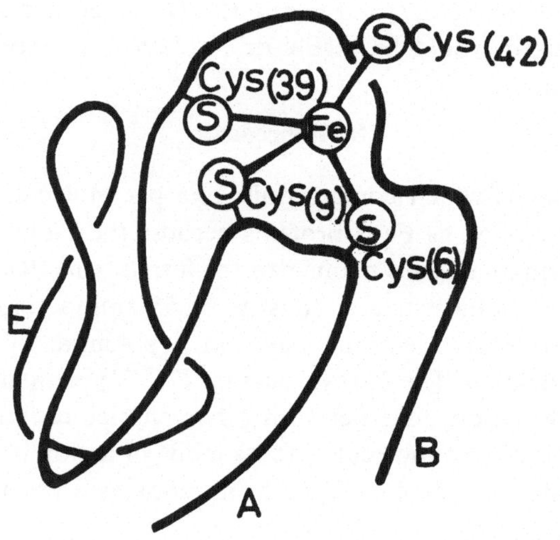

Fig. 13. The structure of the 1 Fe rubredoxin of *C. pasteurianum* (redrawn from Herriott *et al.*) [76].

A rubredoxin has also been isolated from an aerobic bacterium *Ps. oleovorans* [80]. This protein can contain 2 atoms of Fe and is larger (molecular weight 19000; 174 amino acids). Amino acid residues 1–53 and 119–172 can be aligned with the rubredoxins *Peptostreptococcus elsdenii* and *P. aerogenes,* indicating homology and the occurrence of gene duplication during the evolution of *Pseudomonas oleovorans* rubredoxin, i.e. evolution from anaerobic to aerobic bacteria [73].

The *Ps. oleovorans* rubredoxin is an electron carrier in the alkane and fatty acid hydroxylating systems [81] but no role is yet known for the rubredoxins from the anaerobic bacteria.

9. Complex Fe–S Proteins

A number of Fe–S proteins have been characterised which have flavins (FMN or FAD) and/or molybdenum associated with the Fe–S active centre. At least five proteins of this type are known [7]. It seems as if the basic Fe–S centre can be utilised in combination with other electron carriers such as flavin and Mo to form a complex multienzyme system catalyzing electron transfer from a number of substrates to wide range of oxidants. This would have been a great advantage in evolution of more complex electron transfer systems using Fe–S centres as a basic unit.

Probably the most studied Fe–S enzyme has been xanthine oxidase (found in liver, milk, bacteria, etc.) which contains 8Fe+8S+2FAD+2Mo – see ref. [82] and [83] for recent literature. The enzyme molecule consists of a multicomponent electron transfer chain which can catalyze the oxidation of a number of substrates, e.g. xanthine, purines, pyrimidines, pterins, etc., and can reduce a number of electron acceptors, e.g. O_2, ferricyanide, cytochrome c, dyes, etc. The electrons from the substrate xanthine are transferred to Mo, thence to the flavin and then to the Fe–S complex, and finally to O_2 if it is to be the oxidant. The Fe–S component of xanthine oxidase has been shown to have a very similar EPR signal [84] to that of spinach ferredoxin – the Mössbauer [85] and circular dichroism spectra also show great similarities [86].

Another extensively studied complex Fe–S protein is the molybdoferredoxin component of the nitrogenase enzyme catalyzing N_2 fixation in anaerobes, e.g. *Clostridium* [87], facultative anaerobes, e.g. *Klebsiella* [88], and aerobes, e.g. *Azotobacter* [89, 90]. Nitrogenase catalyzed N_2 fixation also occurs in blue-green algae, e.g. *Anabaena* [91] and in the red photosynthetic bacterium *Chromatium* [92, 93]. Molybdoferredoxin contains 20Fe and 20S and 2Mo in a tetramer of molecular weight 220000. The fixation of N_2 requires the co-operation of this molybdoferredoxin with another Fe–S protein called azoferredoxin. A recent reoprt [94] has claimed to show non-enzymatic N_2 fixation using model Fe–S protein (numbers 8 and 10 of Table III, Figure 4); this may be of great practical significance.

The presence of at least 8 or 9 Fe–S centres in mitochrondria (from yeast, plants, and animals) is also an excellent example of the use of the Fe–S unit catalyzing electron transfer at various redox potentials (Figure 14). Davis and Hatefi [95] have isolated a subunit from succinate dehydrogenase of beef heart mitochondria and shown that it is an Fe–S flavoprotein with 1 flavin, 4Fe+4S per molecular weight of 70000.

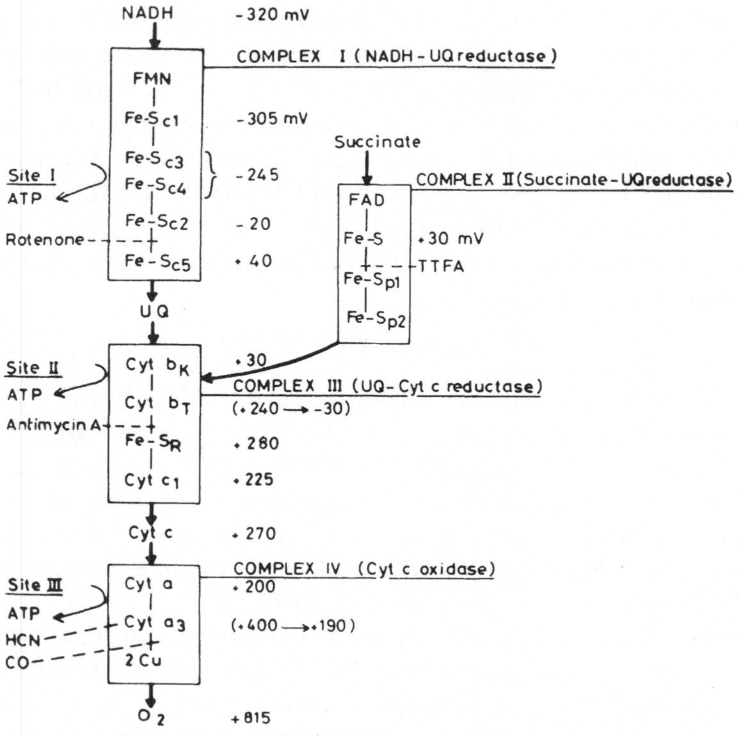

Fig. 14. Scheme of electron transport in oxidative phosphorylation (from Hall *et al.*) [7].

10. The Special Significance of the Position of Cysteine in the Protein

If one examines the sequences of a wide variety of Fe–S proteins it immediately becomes clear that the cysteines occupy invariant positions in the sequences of related proteins (see Figures 1, 6 and 10) with at least one –Cys–X–X–Cys– grouping in each sequence. In Table IV we have compiled the information on the positions of cysteines in all the known sequences of Fe–S proteins.

Proteins containing two Fe–S centres, e.g. *Clostridium* 8 Fe ferredoxin, have two Cys groupings. The 4 Fe ferredoxins only have one such grouping – the *Desulphovibrio* ferredoxin cysteines being closely homolgous to the 8 Fe ferredoxins, but the HIPIP showing no such homology. The 2 Fe ferredoxins of bacteria and animals contain two Cys groupings while the plant ferredoxins only show one such grouping. This reflects a number of detailed differences in the properties of the 2 Fe plant ferredoxins and the 2 Fe ferredoxins of animals and bacteria, e.g. EPR spectra, redox potentials, optical absorption spectra, stability etc. The three varieties of 2 Fe ferredoxins are not functionally interchangeable probably because of their different protein sequences and redox potentials.

The rubredoxins have two cysteines groupings for each sequence of 53 amino-acids – the anaerobic rubredoxins have 1 Fe for each unit sequence while the rubredoxin from

TABLE IV

Positions of cysteines in sequences of Fe–S proteins

Protein	No. of residues	Fe/mole	Cys/mole	Positions of Cys
Clostridium ferredoxin	55	8	8	8, 11, 14, 18, 35, 38, 41, 45
Desulphovibrio ferredoxin	56	4	6	7, 10, 13, 17, 43, 50
Chromatium HIPIP	86	4	4	43, 46, 63, 77
Pseudomonas putidaredoxin	114	2	6	28, 32, 35, 79, 82
Bovine adrenodoxin	114	2	5	46, 52, 55, 92, 95
Plant ferredoxin	97	2	4	39, 44, 47, 77
Micrococcus rubredoxin	53	1	4	6, 9, 39, 42
Pseudomonas rubredoxin	174	1 or 2	10	5, 8 38, 40, 41 123, 126 156, 157, 159

the aerobe *P. oleovorans* has two such unit sequences and can contain 2 iron atoms per molecule.

Thus it would seem that in all Fe-S proteins there is a minimal requirement for one Cys–X–X–Cys grouping per unit active centre whether it contains 4, 2 or 1 Fe. In designing model Fe–S compounds the structural requirements predetermined by these Cys arrangements should be – and have been in a few instances – kept firmly in mind.

11. Conclusions

It is evident from the above discussion that all the chemical ingredients and energy requirements necessary for the formation of a ferredoxin-type molecule were available during the period of the origin of life, and that possibly a ferredoxin-type of iron-chelated peptide would have functioned as electron carrier in the most primitive organisms. When these organisms began to develop the mechanism of nucleic acid-directed protein synthesis the information necessary to transcribe the synthesis of ferredoxin-type peptides was incorporated into their genetic machinery. By the processes of gene duplication, mutation, deletion, etc., organisms evolved biologically, and simultaneously, the ferredoxin molecule also evolved from a simple peptide of say about 27 amino acids to long chain proteins consisting of 100 and more amino acid residues.

Until a a few years ago there were only two major classes of ferredoxins known, the 8 Fe proteins found in bacteria and the 2 Fe proteins found in algae and plants. But recently ferredoxins with a 4 Fe + 4 S cluster per molecule have been isolated from some obligate and facultative aerobes and two-iron ferredoxins have been characterized from many species of aerobic and anaerobic bacteria. Still we consider that those ferredoxins with two 4Fe+4S clusters are the earliest of the extant iron-sulphur proteins to evolve. These are mainly found in the anaerobic bacteria and also in the nitrogen fixing aerobe, *Azotobacter*. It is interesting that the red non-sulphur type photosynthetic bacterium *R. rubrum* (which is a facultative anaerobe) produces an eight iron ferredoxin only when grown anaerobically [96]. Later on, one of these (4Fe+4S) clusters was lost, and ferredoxins with a single 4Fe+4S centre, capable of transferring one electron were developed. These are found in present day obligate and facultative aerobes like *D. gigas* and *S. aurantia*. These proteins may be the evolutionary precursors to the two Fe ferredoxins, with longer amino acid chains, found in algae and higher plants, which transfer electrons singly. The development of the ferredoxin molecule seems to have culminated with the formation of the plant ferredoxins. However, parallel to the development of the algae and plants from the photosynthetic bacteria, evolution of the mammalian system took place. Proteins resembling the ferredoxins in structure and function occur in the mammalian system as such (adrenodoxin) and also as larger molecules complexed not only to iron and sulphur, but also to other cofactors like Mo and flavins. Hatefi *et al.* [97] have found that succinic like Mo and flavins. Hatefi *et al.* [97] have found that succinic dehydrogenase, a complex membrane bound iron-sulphur protein, isolated from *R. rubrum* can substitute for the succinic dehydrogenase of the bovine heart electron transport chain, supporting the idea of an evolutionary conservation in these electron carriers. Thus the iron-sulphur proteins occur in the most primitive up to the most evolved organisms.

References

[1] Hall, D. O. and Evans, M. C. W.: 1969, *Nature* **223**, 1342.
[2] Tsibris, J. C. M. and Woody, R.: 1970, *Coordin. Chem. Rev.* **5**, 417.
[3] Buchanan, B. B. and Arnon, D. I.: 1970, *Adv. Enzym.* **33**, 119.
[4] Buchanan, B. B. and Arnon, D. I.: 1971, *Meth. Enzymol.* **23**, 413.
[5] Yoch, D. C. and Valentine, R. C. 1972, *Ann. Rev. Microbiol.* **26**, 139.
[6] Hall, D. O., Cammack, R., and Rao, K. K.: 1973, *Pure Appl. Chem.* **34**, 553.
[7] Hall, D. O., Cammack, R., and Rao, K. K.: 1973, 'Non-haem Iron Proteins', Chapter 8 in A. Jacobs (ed.), *Iron in Biochemistry and Medicine*, Academic Press, London (in press).
[8] Hall, D. O., Cammack, R., and Rao, K. K.: 1971, *Nature* **233**, 136.
[9] Lipmann, F.: 1972, in A. Buvet and C. Ponnamperuma (eds.), *Chemical Evolution and the Origin of Life*, North-Holland, Amsterdam.
[10] Matsubara, H.: 1972, *Aspects Cell Mol. Physiol.* **19**, 31.
[11] Dayhoff, M. O.: 1972, *Atlas of Protein Sequence and Structure* **5**, Natl. Biomed. Res. Fndn., Georgetown University, Medical Center, Washington, D.C., 20007 U.S.A.
[12] Hall, D. O., Cammack, R., and Rao, K. K.: 1973, *Space Life Sc.*, (in press).
[13] Orme-Johnson, W. H.: 1972, *Biochem. Soc. Trans.* **1**, 30.
[14] Malkin, R. and Rabinowitz, J. C.: 1966, *Biochem. Biophys. Res. Comm.* **23**, 822.
[15] Hong, J. S. and Rabinowitz, J. C.: 1967, *Biochem. Biophys. Res. Comm.* **29**, 246.
[16] Hall, D. O., Cammack, R., and Rao, K. K.: 1972a, in A. W. Schwartz (ed.), *Theory and Experiment in Exobiology* **2**, p. 67.
[17] Rickard, D. T.: 1969, *Stockholm Contrib. Geol.* **20**, 50.

[18] Blomstrom, D. C., Knight, E., Phillips, W. D., and Weiher, J. F.: 1964, *Proc. Nat. Acad. Sci. U.S.A.* **51**, 1085.
[19] Cowan, D. O., Pasternak, G., and Kaufman, F.: 1970, *Proc. Nat. Acad. Sci. U.S.A.* **66**, 837.
[20] Rubinson, K. A. and Palmer, G.: 1972, *J. Am. Chem. Doc.* **94**, 8375.
[21] Yang, C. S. and Huennekens, F. M.: 1970, *Biochem.* **9**, 2127.
[22] Sugiura, Y. and Tanaka, H.: 1972, *Biochem. Biophys. Res. Comm.* **46**, 335.
[23] Sugiura, Y., Kurishima, M., and Tanaka, H.: 1972, *Biochem. Biophys. Res. Comm.* **48**, 1400.
[24] Ali, A., Fahrenbols, F., Goring, J. C., and Weinstein, B.: 1972, *J. Am. Chem. Soc.* **94**, 2556.
[25] Coucouvanis, D., Lippard, S. J., and Zubieta, J. A.: 1970, *J. Am. Chem. Soc.* **92**, 3342.
[26] Balch, A. L.: 1969, *J. Am. Chem. Soc.* **91**, 6962.
[27] Bernal, I., Davis, B. R., Good, M. L., and Chandra, S.: 1972, *J. Coord. Chem.* **2**, 61.
[28] Herskovitz, T., Averill, B. A., Holm, R. H., Ibers, J. A., Phillips, W. D., and Weiher, J. F.: 1972, *Proc. Nat. Acad. Sci. U.S.A.* **69**, 2437.
[29] Schunn, R. A., Fritchie, C. J., and Prewitt, C. T.: 1966, *Inorg. Chem.* **5**, 892.
[30] Wei, C. H., Wilkes, G. R., Treichel, P. M., and Dahl, L. F.: 1966, *Inorg. Chem.* **5**, 900.
[31] Connelly, N. G. and Dahl, L. F.: 1970, *J. Am. Chem. Soc.* **92**, 7472.
[32] Bayer, E., Eckstein, H., Hagenmaier, H., Josef, D., Koch, J., Krauss, P., Röder, A., and Schretzmann, P.: 1969, *Eur. J. Biochem.* **8**, 33.
[33] Kubas, G. T., Spiro, T. G., and Terzis, A.: 1973, *J. Am. Chem. Soc.* **95**, 273.
[34] Schneider, J., Dischler, B., and Räuber, A.: 1968, *J. Phys. Chem. Solids* **29**, 451.
[35] Jensen, L. H., Sieker, L. C., Watenpaugh, K. D., Adman, E. T., and Herriott, J. R.: 1973, *Biochem. Soc. Trans.* **1**, 27.
[36] Poe, M., Phillips, W. D., McDonald, C. C., and Lovenberg, W.: 1970, *Proc. Nat. Acad. Sci. U.S.A.* **65**, 797.
[37] Travis, J., Newman, D. J., LeGall, J., and Peck, H. D.: 1971, *Biochem. Biophys. Res. Comm.* **45**, 452.
[38] Shethna, Y. I., Stombaugh, N. A., and Burris, R. H.: 1971, *Biochem. Biophys. Res. Comm.* **42**, 1108.
[39] Yoch, D. C. and Valentine, R. C.: 1972, *J. Bacteriol.* **110**, 1211.
[40] Johnson, P. W. and Canale-Parole, E.: 1973, *Arch. Mikrobiol.* **89**, 341.
[41] Broda, E.: 1970, *Progr. Biophys. Mol. Biol.* **21**, 143.
[42] Schlegel, H. G.: 1971, 'Origin of Life and Evolutionary Biochemistry', Int. Symp., Varna, Bulgaria.
[43] Yamanaka, T.: 1972, *Adv. Biophys.* **3**, 229.
[44] Dus, K., Tedro, S., Bartsch, R. G., and Kamen, M. D.: 1971, *Biochem. Biophys. Res. Comm.* **43**, 1239.
[45] Bartsch, R. G.: 1971, *Methods Enzymol.* **23**, 644.
[46] Carter, C. W., Freer, S. T., Xuong, Ng. H., Alden, R. A., and Kraut, J.: 1971, *Cold Spring Harbour Symp. Quant. Biol.* **36**, 381.
[47] Carter, C. W., Kraut, J., Freer, S. T., Alden, R. A., Sieker, L. C., Adman, E., and Jensen, L. H.: 1972, *Proc. Nat. Acad. Sci. U.S.A.* **69**, 3526.
[48] Evans, M. C. W., Lord, A. V., and Reeves, S. G.: 1973, *Biochem. J.* (in press).
[49] Malkin, R. and Bearden, A. J.: 1971, *Proc. Nat. Acad. Sci. U.S.A.* **68**, 16.
[50] Bearden, A. J. and Malkin, R.: 1972, *Biochem. Biophys. Res. Comm.* **46**, 1299.
[51] Evans, M. C. W., Telfer, A., and Lord, A. V.: 1972, *Biochem. Biophys. Acta* **267**, 530.
[52] Aggarwal, S. J., Rao, K. K., and Matsubara, H.: 1971, *J. Biochem. Tokyo* **69**, 601.
[53] Rao, K. K., Cammack, R., Hall, D. O., and Johnson, C. E.: 1971, *Biochem. J.* **122**, 259.
[54] Gibson, J. F., Hall, D. C., Thornley, J. H. M., and Whatley, F. R.: 1966, *Proc. Nat. Acad. Sci. U.S.A.* **56**, 987.
[55] Fritz, J., Anderson, R., Fee, J., Palmer, G., Sands, R. H., Tsibris, J. C. M., Gunsalus, I. C., Orme-Johnson, W. H., and Beinert, H.: 1971, *Biochim. Biophys. Acta* **253**, 110.
[56] Dunham, W. R., Bearden, A. J., Salmeen, I. T., Palmer, G., Sands, R. H., Orme-Johnson, W. H., and Beinert, H.: 1971, *Biochim. Biophys. Acta* **253**, 134.
[57] Cammack, R., Rao, K. K., Hall, D. O., and Johnson, C. E.: 1971, *Biochem. J.* **125**, 849.
[58] Johnson, C. E., Cammack, R., Rao, K. K., and Hall, D. O.: 1971, *Biochem. Biophys. Res. Comm.* **43**, 564.
[59] Munck, E., Debrunner, P. G., Tsibris, J. C. M., and Gunsalus, I. C.: 1972, *Biochemistry* **11**, 855.

[60] Mitsui, A. and Arnon, D. I.: 1971, *Physiol. Plant.* **25**, 135.
[61] Rao, K. K., Smith, R. V., Cammack, R., Evans, M. C. W., Hall, D. O., and Johnson, C. E.: 1972, *Biochem. J.* **129**, 1159.
[62] Sagan, L.: 1967, *J. Theor. Biol.* **14**, 225.
[63] Taylor, D. L.: 1970, *Int. Rev. Cyt.* **27**, 29.
[64] Hall, D. O., Rao, K. K., and Cammack, R.: 1972, *Biochem. Biophys. Res. Comm.* **47**, 798.
[65] Schopf, W. and Blacic, J. M.: 1971, *J. Paleont.* **45**, 925.
[66] Kimura, T.: 1968, *Structure Bonding* **5**, 1.
[67] Eaton, W. A., Palmer, G., Fee, J. A., Kimura, T., and Lovenberg, W.: 1971, *Proc. Nat. Acad. Sci. U.S.A.* **68**, 3015.
[68] Barker, W. C., McLaughlin, P. J., and Dayhoff, M. O.: 1972, *Fed. Proc., Abs. 837*, No. 3514.
[69] Rieske, J. S., Zaugg, W. S., and Hansen, R. E.: 1964, *J. Biol. Chem.* **239**, 3023.
[69a] Tanaka, M., Haniu, M., and Yasunobu, K. T.: 1973, *J. Biol. Chem.* **248**, 1141.
[70] Davis, K. A. and Hatefi, Y.: 1971, *Biochemistry* **10**, 2509.
[71] Gunsalus, I. C., Lipscomb, J. D., Marshall, V., Frauenfelder, H., Munck, E., and Greenbaum, E.: 1972, in G. S. Boyd and R. M. S. Smellie (eds.), *Biological Hydroxylation Mechanisms*, Academic Press.
[72] Tsai, R. L., Gunsalus, I. C., and Dus, K.: 1971, *Biochem. Biophys. Res. Comm.* **45**, 1300.
[73] Benson, A. M., Tomoda, K., Chang, J., Matsueda, G., Lode, E. T., Coon, M. J., and Yasunobu, K. T.: 1971, *Biochem. Biophys. Res. Comm.* **42**, 640.
[74] Meyer, T. E., Sharp, J. J., and Bartsch, R. G.: 1971, *Biochim. Biophys. Acta* **234**, 266.
[75] Rao, K. K., Evans, M. C. W., Cammack, R., Hall, D. O., Thompson, C. L., Jackson, P. V., and Johnson, C. E.: 1972, *Biochem. J.* **129**, 1063.
[76] Herriott, J. R., Sieker, L. C., Jensen, L. H., and Lovenberg, W.: 1970, *J. Mol. Biol.* **50**, 391.
[77] Watenpaugh, K. D., Sieker, L. C., Herriott, J. R., and Jensen, L. H.: 1971, *Cold Spring Harbor Symp. Quant. Biol.* **36**, 359.
[78] Blumberg, W. E.: 1967, in A. Ehrenberg, B. G. Malström, and T. Vanngard (eds.), *Magnetic Resonance in Biological Systems*, p. 119, Pergamon Press, Oxford.
[79] Poe, M., Phillips, W. D., Glickson, J. D., McDonald, C. C., and San Pietro, A.: 1971, *Proc. Nat. Acad. Sci. U.S.A.* **68**, 68.
[80] Lode, E. T. and Coon, M.: 1971, *J. Biol. Chem.* **246**, 791.
[81] Boyer, R. F., Lode, E. T., and Coon, M. J.: 1971, *Biochem. Biophys. Res. Comm.* **44**, 925.
[82] Edmondson, D., Massey, V., Palmer, G., Beecham, L. M., and Elion, G. B.: 1972, *J. Biol. Chem.* **247**, 1597.
[83] Lowe, D. V., Lynden-Bell, R. M., and Bray, R. C.: 1972, *Biochem. J.* **130**, 239.
[84] Gibson, J. F. and Bray, R. C.: 1968, *Biochim. Biophys. Acta* **153**, 721.
[85] Johnson, C. E., Bray, R. C., Cammack, R., and Hall, D. O.: 1969, *Proc. Nat. Acad. Sci. U.S.A.* **63**, 1234.
[86] Garbett, K., Gillard, R. D., Knowles, P. F., and Stangroom, J. E.: 1967, *Nature*, **215**, 824.
[87] Mortenson, L. E., Zumft, E. G., Huang, T. C., and Palmer, G.: 1973, *Biochem. Soc. Trans.* **1**, 35.
[88] Eady, R. R., Smith, B. E., Thorneley, R. N. F., Ware, D. A., and Postgate, J. R.: 1973, *Biochem. Soc. Trans.* **1**, 37.
[89] Burns, R. C., Holsten, R. D., and Hardy, R. W. F.: 1970, *Biochem. Biophys. Res. Comm.* **39**, 90.
[90] Benemann, J. R., Yoch, D. C., Valentine, R. C., and Arnon, D. I.: 1971, *Biochim. Biophys. Acta* **226**, 205.
[91] Smith, R. V., Noy, R. J., and Evans, M. C. W.: 1971, *Biochem. Biophys. Res. Comm.* **253**, 104.
[92] Winter, H. C. and Arnon, D. I.: 1970, *Biochim. Biophys. Acta* **197**, 170.
[93] Evans, M. C. W., Telfer, A., and Smith, R. V.: 1973, *Biochem. Biophys. Acta* **310**, 344.
[94] Van Tamelen, E. E., Gladysz, J. A., and Miller, J. S.: 1973, *J. Am. Chem. Soc.* **95**, 1347.
[95] Davis, K. A. and Hatefi, Y.: 1971, *Biochemistry* **10**, 2509.
[96] Shanmugam, K. T. and Arnon, D. I.: 1972, *Biochim. Biophys. Acta* **256**, 489.
[97] Hatefi, Y., Davis, K. A., Baltscheffsky, H., Baltscheffsky, M., and Johansson, B. C.: 1972, *Arch. Biochem. Biophys.* **152**, 613.
[98] Cammack, R.: 1973, *Biochem. Biophys Res. Comm.* **54**, 548.

A NEW HYPOTHESIS FOR THE EVOLUTION OF BIOLOGICAL ELECTRON TRANSPORT

HERRICK BALTSCHEFFSKY

Botaniska och Biokemiska institutionerna, Stockholms universitet, Stockholm 50, Sweden

Abstract. A new hypothesis for the evolution of biological electron transport is presented. According to this hypothesis biological electron transport originated close to the potential of the hydrogen electrode and evolved in various advantageous directions including, when molecular oxygen became available on the Earth, that of the oxygen electrode. This implies stepwise evolution along and across the potential scale. The hypothesis is based mainly on existing information obtained from studies of primary and tertiary structural relationships of proteins. It is hoped to provide a framework for closer understanding of both evolution and mechanisms of cellular oxidation-reduction as well as energy coupling reactions.

1. Introduction

Biological electron transport chains are known to consist of several kinds of electron carriers which are usually classified on the basis of the active or prosthetic group or the coenzyme involved. From the evolutionary point of view it is, of course, important to consider also, and even especially, the protein part of various electron carriers.

Evolutionary relationships for certain selected electron carriers from different organisms have been well documented by analysis and comparison of total amino acid sequences. It has been possible to construct rather detailed evolutionary trees for the electron carriers cytochrome c (Fitch and Margoliash, 1967) and, more recently, ferredoxin (Dayhoff, 1971).

Studies of the three-dimensional structure of proteins have become a most important complement to amino acid sequence determination for detection of evolutionary relationships. The more so, as it has been possible to show that the tertiary structure of a protein may evolve more slowly than the amino acid sequence (McLachlan, 1972; Buehner *et al.*, 1973). Thus, in cases where no conservation of primary structure has been detected even with sensitive statistical tests, retention of essentially identical three-dimensional structural features has allowed the identification of, for example, probable gene duplication and maybe even triplication (Kretsinger, 1972) as well as evolutionary connection between proteins with different, although related, functions (Buehner *et al.*, 1973).

I would like to propose the following new hypothesis: biological electron transport originated close to the potential of the hydrogen electrode and evolved in various advantageous directions including, when molecular oxygen became available, that of the oxygen electrode. This implies evolution both along, from one and often one type of electron carrier to another, and across the potential scale. It will be shown in this presentation that the hypothesis is based mainly on information obtained from studies of primary and tertiary protein structural relationships.

2. The Biological Electron Carriers

The discussion of the new hypothesis will be facilitated by consideration of how the biological electron carriers can be grouped together. From a classical structural point of view one may distinguish between iron binding proteins and nucleotide binding proteins, which in fact constitute the two major classes .Each of these may be divided into two groups, all four of them being most important in biological electron transport. To the iron binding proteins belong the nonheme iron proteins and the heme proteins. To the nucleotide binding proteins belong the flavin nucleotide binding proteins and the nicotinamide adenine nucleotide binding proteins. These groups of electron carriers are shown in Table I. In addition to the proteins belonging to these groups one has, in photosynthetic systems, the chlorophylls and the proteins to which they are bound in the photosynthetic membranes as well as, special for the green plant type

TABLE I

Biological electron carriers. Groups and individual carriers

Iron binding proteins	Nucleotide binding proteins	Other carriers
(a) Non heme	(a) Flavin	Chlorophylls
(b) Heme	(b) Nicotinamide adenine	Plastocyanin
		Quinones

electron transport, the copper containing protein plastocyanin. Furthermore, there are the quinones (menaquinone, plastoquinone, ubiquinone) which do not seem to be linked to any particular protein. They usually occur in relatively high concentrations, which may be related to their apparently quite special pool function. Also these electron carriers are listed in Table I. For reasons which I believe will become obvious, the discussion will focus on the proteins from the two main classes, the iron and the nucleotide binding proteins.

To the nonheme iron binding electron carriers belong the ferredoxins, which contain low molecular weight protein, iron and 'labile sulfur' (Fe:S = 1:1). Among the bacterial ferredoxins especially those from *Clostridia* are characteiized by their in many ways quite unique 'primitiveness' when compared to other known proteins. They lack several of the 20 common amino acids, and their amino acid composition is strikingly similar to that found in some analysed meteorites (Hall *et al.*, 1971, 1973). Furthermore, with a total of only about 55 amino acids, they show strong evidence for gene duplication and other repetitivity of amino acid sequence, the latter concerning in particular the 8 cysteines which are involved in the binding of the 8 Fe atoms in the two (4Fe + 4 labile S) clusters (Sieker *et al.*, 1972; Jensen *et al.*, 1973). It should be noted, that the oxidation-reduction potential at neutral pH value of these ferredoxins is exactly at, or veiy close to, −420 mV, which is the same as that of the hydrogen electrode. It may suffice here to mention in addition to the ferredoxins, among the great number of nonheme iron proteins which have become known in the last 10 years the putidaredoxin and the adrenodoxin, which both contain the (Fe:S = 1:1) group and are components of mixed function oxidase systems and, last but not least, the rubredoxins, which contain Fe but no labile S and which in most known cases have a molecular weight very similar to that of the *Clostridia-type* ferredoxins.

The vast amount of diffeıent cytochromes are the heme protein electron carriers, among which it will be useful for the forthcoming discussion to distinguish between the low potential multiheme c-type cytochromes and all the other cytochromes (cytochromes a, b, c, etc. etc.)

The wide variety of the nucleotide binding proteins, with FMN or FAD as the prosthetic group of the flavoproteins, and with NAD or NADP as the coenzyme of the NAD (P) binding proteins, is well known. It may suffice to mention in this connection that

(1) the flavodoxins, with FMN as prosthetic group, are acidic proteins which can replace bacterial ferredoxin in various reactions (including evolution of molecular hydrogen, nitrogen fixation, and pyruvate oxidation); and

(2) both flavodoxin and some NAD binding dehydrogenases have recently been analysed in great detail with respect to both primary and tertiary protein structure.

3. Evolutionary Considerations

According to the new hypothesis for the evolution of biological electron transport the 'primitive' ferredoxins (or their ancestral counterparts), with oxidation-reduction potentials at or very close to that of the hydrogen electrode, served as 'growing points' foı emerging electron transport chains in living cells. (It may be recalled that reducing conditions prevailed under the early periods of the history of the Earth.) What evidence can be found to support or contradict this hypothesis? The whole problem has two parts, one concerning evolution of other nonheme iron proteins from the low molecular weight ferredoxins, and the other concerning the possibility that more or less distant evolutionary relationships exist and may still be traced between ferredoxins and other types of electron carriers such as, for example, cytochromes, or even nucleotide binding proteins.

The first part of the problem, concerning other nonheme iron proteins, would appear to be *a priori* an easier requirement to satisfy. Indeed, electron transport chains involved in, for example, hydrogen evolution, nitrogen fixation, photosynthesis, and respiration have been found to contain two or more nonheme iron proteins. However, the structures of the actual proteins involved are not yet sufficiently known for answers to be given to the question of whether evolutionary relationships exist between such neighbor links in biological electron transport chains. In the respiratory chain of yeast mitochondria as many as 11 or 12 nonheme iron sulfur centers have been identified (Ohnishi, 1973) but none of the actual proteins has so far been purified and sequenced. On the other hand, evidence for evolutionary relationship between ferredoxins and adrenodoxin has been obtained (Barker *et al.*, 1972). As evolutionary links within this

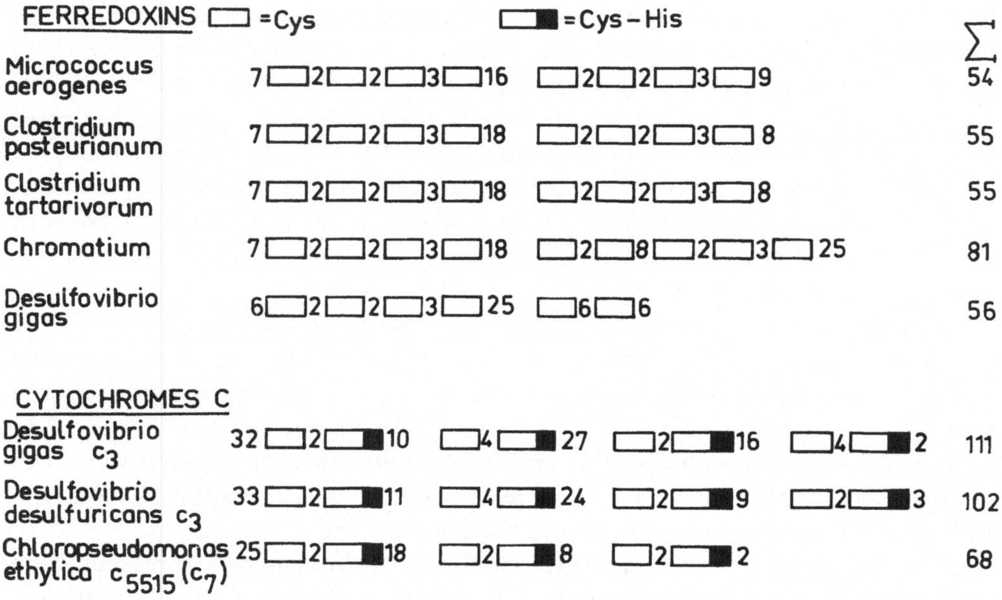

FERREDOXINS □ = Cys	□■ = Cys – His				Σ
Micrococcus aerogenes	7□2□2□3□16	□2□2□3□9			54
Clostridium pasteurianum	7□2□2□3□18	□2□2□3□ 8			55
Clostridium tartarivorum	7□2□2□3□18	□2□2□3□8			55
Chromatium	7□2□2□3□18	□2□8□2□3□25			81
Desulfovibrio gigas	6□2□2□3□25	□6□6			56
CYTOCHROMES C					
Desulfovibrio gigas c_3	32□2□■10	□4□■27	□2□■16	□4□■2	111
Desulfovibrio desulfuricans c_3	33□2□■11	□4□■24	□2□■9	□2□■3	102
Chloropseudomonas ethylica c_{5515} (c_7)	25□2□■18	□2□■8	□2□■2		68

Fig. 1. Locations of cysteine residues in some bacterial ferredoxins and low potential multiheme c-type cytochromes. The numbers in the polypeptide 'sequences' indicate numbers of amino residues. The numbers under the sigma-sign are the total numbers of amino acid residues in each sequence.

area were discussed earlier today by Hall *et al.* (1974), and as a rather detailed treatment of it is given elsewhere (Baltscheffsky, 1974), only the second, apparently more far-fetched part of the problem, concerning possible evolution of different types of electron carriers from ferredoxin, will be treated here at some length.

As is shown in Figure 1, certain similarities exist with respect to especially the positions of cysteine residues in the polypeptide chains of some bacterial ferredoxins and some bacterial low potential multiheme c-type cytochromes. The total amino acid sequence is known for each of the electron carriers listed in Figure 1. In nearly all known cases both the nonheme iron atoms of ferredoxins and the heme groups of c-type cytochromes are bound to their respective proteins by sulfur atoms of cyteines. Closer inspection of the polypeptide chains with the well-known Accepted Point Mutations (PAM) method of Dayhoff (1972) provides quite good although not entirely convincing evidence in support of our earlier assumption that evolutionary relationship between ferredoxin and c-type cytochrome exists (Baltscheffsky, 1972). As is shown in Table II (the data were obtained in collaboration with Dr M. O. Dayhoff), an alignment score of greater than 2.0 is obtained when a selected amino acid sequence from ferredoxin is compared with one from cytochrome c_3, both isolated from *Desulfovibrio gigas*. Using 30 random scores, the probability of obtaining an alignment score greater than 2.0 is only about 2.5% (Dayhoff, 1972). This would appear to be at least a clear indication of evolutionary relationship. However, final evaluation of the concept, advanced also by Yamanaka (1972) and by Egami (1973) at this meeting, that ferredoxins and low potential multiheme c-type cytochromes show

TABLE II

Alignment scores for ferredoxin and cytochrome c_3 from *Desulfovibrio gigas*

Ferredoxin	Cytochrome c_3	Alignment score
Whole protein	Whole protein	1.6
Amino acid residues 4–56	Amino acid residues 30–111	2.2

evolutionary relationship will be possible only when more structural data become available.

Have ferredoxins evolved to proteins not belonging to the iron binding protein groups? A number of reasons exist for putting flavoproteins into focus in connection with this question. It may be recalled that: (a) flavoproteins are often immediate acceptors of electrons from ferredoxins and other nonheme iron proteins; (b) certain flavoproteins (flavodoxins) may substitute for ferredoxins, in thiosulfate and sulfite reduction; (c) certain flavoproteins (flavodoxins) show a number of similarities to ferredoxins, *i.e.* low oxidation reduction potential close to that of the hydrogen electrode, high number of acidic groups, and low molecular weight; and (d) several flavoproteins exist where the ratio flavin:nonheme Fe:labile S is 1:4:4, indicating the occurrence of polypeptide structures which are capable of binding both flavin and the (4Fe+4S) cluster. As was noted by Fox and Brown (1971), *Clostridia* ferredoxin and flavodoxin showed no obvious relationship with respect to amino acid sequences. With support from the recent flavodoxin sequence data of Dubourdieu *et al.* (1973), I have found what may perhaps be considered a hint that possibly rubredoxin may be or may indicate a link on an evolutionary chain connecting ferredoxin and flavodoxin. This is shown in Table III, which gives the amino acid sequences at the amino terminal end of ferredoxin, rubredoxin, and flavodoxin from *Clostridium pasteurianum* as well as flavodoxin and cytochrome c_3 from *Desulfovibrio vulgaris*. In addition to indicating the possible place of rubredoxin on an evolutionary pathway from ferredoxin to flavodoxin, Table III also shows a common –Pro–Lys–Ala–sequence in flavodoxin and cytochrome c_3 from *Desulfovibrio vulgaries*, which could be more than a coincidence and thus tend to add support to the hypothesis which is expressed in Figure 2. Here is shown my tentative model, a minimum scheme, for the possible evolution of c-type cytochrome and flavoprotein from ferredoxin (or, more generally,

TABLE III

Some ferredoxin, rubredoxin, flavodoxin, and cytochrome c_3 amino acid sequences at the amino terminal ends

Ferredoxin	(*Clostridium pasteurianum*):	Ala –Tyr–Lys–Ile –Ala –
Rubredoxin	(*Clostridium pasteurianum*):	f–Met–Lys–Lys–Tyr–Thr –
Flavodoxin	(*Clostridium pasteurianum*):	Met– –Lys–Val–AsN–
Flavodoxin	(*Desulfovibrio vulgaris*):	Met–Pro–Lys–Ala–Leu –
Cytochrome c_3	(*Desulfovibrio vulgaris*):	Ala –Pro–Lys–Ala–Pro –

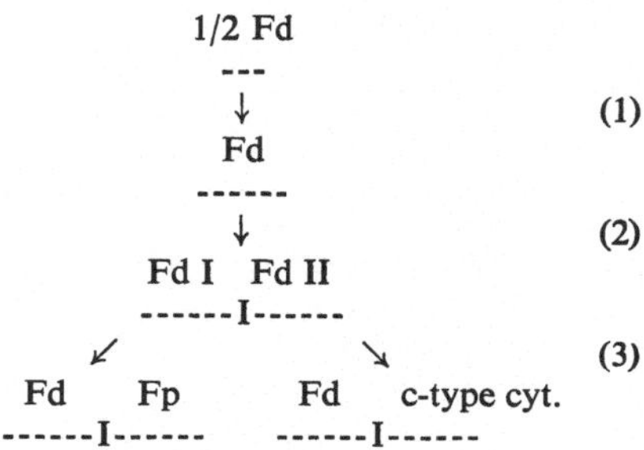

Fig. 2. Tentative minimum scheme for the possible evolution of flavoprotein and c-type cytochrome from ferredoxin. The dashed lines may be visualized as gene material which is duplicating in steps (1) and (2). Fd = ferredoxin, Fp = flavoprotein, and c-type cyt. = c-type cytochrome.

from nonheme iron protein). (1)–(2) in the figure indicate gene duplication steps, starting with the duplication of the 'half' which is assumed to have led to the early ferredoxin molecules and continuing with the duplication which supposedly gave, with point mutations, deletions, and insertions along the way, the transformation necessary for making possible the binding of heme and flavin, respectively.

In order to find out if additional experimental indications may be found which bear on the question of whether molecular evolution of electron carriers has proceeded along and across electron transport chains, one may turn to look for information obtained from X-ray studies on tertiary structures of proteins.

Very recently, thanks to the successful efforts in many laboratories to unravel primary, secondary, and tertiary structures of various dehydrogenases, it has been found by Rao and Rossman (1973) in connection with work on the tertiary structure of lactate dehydrogenase, that a structure, consisting of three consecutive strands of a parallel pleated sheet and two joining α-helices, exists to varying degrees of conformity in a number of proteins. One such structure was discovered in *Clostridium* MP flavodoxin and two in dogfish lactate dehydrogenase subunit. Brändén *et al.* (1973) found a very similar structural arrangement in liver alcohol dehydrogenase subunit and concluded that a substructure, built up from six parallel strands in a pleated sheet arrangement and joined by helices or loops may be a general one for the binding of nucleotides and, in particular, the coenzyme NAD. Buehner *et al.* (1973) have extended this approach in connection with a study of D-glyceraldehyde-3-phosphate dehydrogenase and have presented an evolutionary model for the relations between the flavodoxin and the various dehydrogenases. Figure 3 is a slightly extended version of their model, with the single insertion of glutamate dehydrogenase, which would seem to fit in as indicated, according to the study of Engel (1973).

These data as visualized in their evolutionary context, first of all, strongly support the concept of evolution across the potential scale, at the dehydrogenase level of electron transport chains. Furthermore, they indicate evolutionary relationship between the two groups of the nucleotide binding electron carriers, the flavoproteins and the nicotinamide adenine dinucleotide binding proteins. It still seems premature to

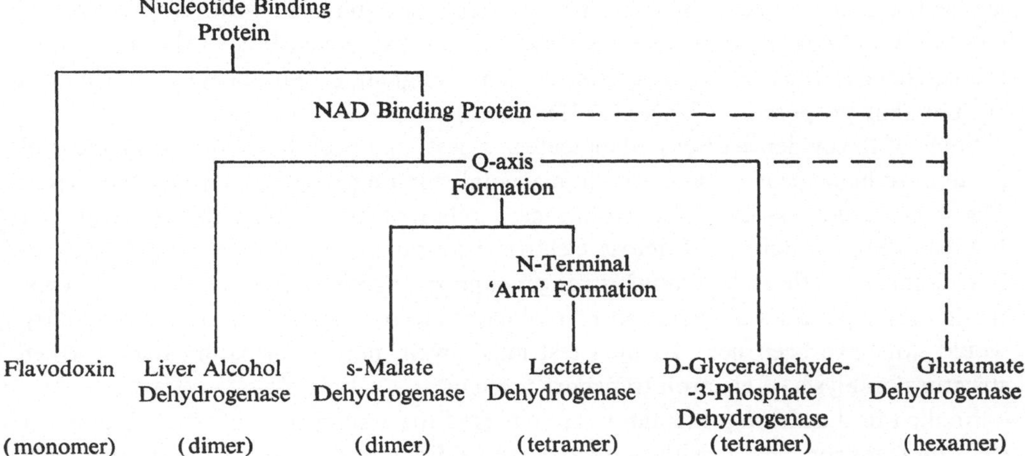

Fig. 3. Model for evolutionary relationships between flavodoxin and various dehydrogenases. After Buehner *et al.* (1973) and Engel (1973).

consider the question of whether these groups had a common ancestor or whether the latter type of protein, being the more complex one, evolved from the former. Finally, the data underline the power of X-ray crystallography in the search for evolutionary relationships which may otherwise (*i.e.* sequencewise) have been obscured with the passing of time since actual divergences in evolution, and motivate a closer look at earlier X-ray data in the light of the current development.

The question which immediately arises is whether X-ray data provide any indication *pro* or *contra* the above assumption that rubredoxins may constitute direct or indirect evolutionary links between ferredoxins and flavodoxins. In the already classical study by Herriott *et al.* (1970) on the structure of *Clostridium pasteurianum* rubredoxin it was found that the molecule contains two regions of antiparallel pleated sheet structure, one of which involves three chain sections and the other two short ones. This rubredoxin has about the same molecular weight as the corresponding ferredoxin, and also shows some gross similarity with respect to three-dimensional structure. In the ferredoxin there is only a very short section of antiparallel pleated sheet, involving the two terminal parts of the chain (Adman *et al.*, 1973). It should be pointed out that the flavodoxin and the dehydrogenase pleated sheet structures are parallel, not antiparallel, and that no particular evolutionary conclusions can be drawn at this stage from the fact that both rubredoxin and flavodoxin contain extensive pleated sheet regions.

4. Concluding Remarks

A number of indications have been brought forward to show the basis which at the present time exists for the hypothesis advanced, that biological electron transport has emerged from the potential region of the hydrogen electrode and evolved along and across the potential scale in small discrete steps, in agreement with the Principle of Continuity advanced some years ago by Orgel (1968). According to the new hypothesis, there is thus no need to consider any 'singular evolutionary discontinuity', such as that proposed by Singer (1971).

Some of the evidence discussed indicates, at best, that evolution from nonheme iron proteins to heme proteins and even to nucleotide binding proteins may have occurred. The very recent evidence for evolutionary relationship between flavoprotein and NAD-binding protein, and across electron transport chains at the dehydrogenase level is most exciting. With much stronger support than what exists today, the hypothesis presented for the evolution of biological electron transport in all probability would not have been new. Its final test must await more information from further structural analysis of electron transport proteins.

Implicit in the above presentation has emerged the assumption that early biological electron transport chains with as many as three different functional groups may have consisted of nonheme iron protein, heme protein, and flavoprotein. Such a complex would be able to accomodate the ribose phosphate structure, as in the apoprotein of flavodoxin. Coupling of electron transport and energy conservation in energy-rich phosphate compounds may well have orginated approximately simultaneously with flavoproteins, provided that adenosine phosphates are considered as necessary participants in such coupling, but possibly much earlier, if on the other hand the inorganic ortophosphate-pyrophosphate system was a preceding functional part of biological electron transport phosphorylation, as may indeed have been the case (Baltscheffsky, 1971). Considerations such as these may become increasingly valuable as the picture about the three-dimensional structures of proteins involved in phosphate metabolism becomes clearer. Evolutionary relationships may well be found between proteins involved in electron transport and in electron transport coupled energy conservation.

Thus, the evolutionary approach as applied to both electron transport and electron transport phosphorylation may perhaps be hoped to elucidate not only their origin and biochemical evolution but also their elusive molecular mechanisms.

References

Adman, E. T., Sieker, L. C., and Jensen, L. H.: 1973, *J. Biol. Chem.* **242**, 3987.

Baltscheffsky, H.: 1971, in R. Buvet and C. Ponnamperuma (eds.), *Chemical Evolution and the Origin of Life*, North-Holland Publ. Co., p. 466.

Baltscheffsky, H.: 1972, *VI International Congress of Photobiology, Abstracts*, no. 368.

Baltscheffsky, H.: *Quart. Revs. Biophys.* (to be published).

Barker, W. C., McLaughlin, P. J., and Dayhoff, M. O.: 1972, *Federation Proc.* **31**, 837 (abs.).

Brändén, C.-I., Eklund, H., Nordström, B., Boiwe, T., Söderlund, G., Zeppezauer, E., Ohlsson, I., and Åkesson, Å.: 1973, *Proc. Nat. Acad. Sci.* **70**, 2439.

Buehner, M., Ford, G. C., Moras, D., Olsen, K. W., and Rossman, M. G.: 1973, *Proc. Nat. Acad. Sci.* **70**, 3052.

Dayhoff, M. O.: 1971, in R. Buvet and C. Ponnamperuma (eds.), *Chemical Evolution and the Origin of Life*, North-Holland Publ. Co., p. 392.

Dayhoff, M. O.: 1972, in *Atlas of Protein Sequence and Structure* **5**, 101.

Dubourdieu, M., Le Gall, J., and Fox, J. L.: 1973, *Biochem. Biophys. Res. Commun.* **52**, 1418.
Egami, F.: 1973, these proceedings, p. 405.
Engel, P. C.: 1973, *Nature* **241**, 118.
Fitch, W. M. and Margoliash, E.: 1967, *Science* **155**, 279.
Fox, J. L. and Brown, J. R.: 1971, *Federation Proc.* **30**, 1971, 1242 (abs.).
Hall, D. O., Cammack, R., and Rao, K. K.: 1971, *Nature* **233**, 136.
Hall, D. O., Cammack, R., and Rao, K. K.: 1973, in A. W. Schwartz (ed.), *Theory and Experiment in Exobiology*, Wolters-Noordhoff Publ., p. 67.
Hall, D. O., Cammack, R., and Rao, K. K.: 1974, these proceedings, p. 363.
Herriott, J. R., Sieker, L. C., Jensen, L. H., and Lovenberg, W.: 1970, J. *Mol. Biol.* **50**, 391.
Jensen, L. H., Sieker, L. C., Watenpaugh, K. D., Adman, E. T., and Herriott, J. R.: 1973, *Biochem. Soc. Transactions* **1**, 27.
Kretsinger, R. H.: 1972, *Nature New Biology* **240**, 85.
McLachlan, A. D.: 1972, *J. Mol. Biol.* **64**, 417.
Ohnishi, T.: 1973, in press, and personal communication.
Orgel, L. E.: 1968, *J. Mol. Biol.* **38**, 381.
Rao, S. T. and Rossman, M. G.: 1973, *J. Mol. Biol.* **76**, 241.
Sieker, L. C., Adman, E., and Jensen, L. H.: 1972, *Nature* **235**, 40.
Singer, T. P.: 1971, in E. Schoffeniels (ed.), *Biochemical Evolution and the Origin of Life*, North-Holland Publ. Co., p. 203.
Yamanaka, T.: 1972, *Advan. in Biophys.* **3**, 227.

PATHWAYS OF CHEMICAL EVOLUTION OF PHOTOSYNTHESIS

A. A. KRASNOVSKY

A. N. Bakh Institute of Biochemistry, Academy of Science of the U.S.S.R., Moscow, U.S.S.R.

Abstract. The primary metabolism of protobionts was probably based on the electron transfer reactions regulated by catalysts or photosensitizing pigments. The action of photoreceptive pigments was inevitable in the case of electron transfer leading to light energy storage in the reaction products. The primitive tetrapyrrolic pigments formed abiogenically (porphin, chlorin) as well as their more complicated biogenic analogs (chlorophylls) are capable to photosensitize electron transfer in systems, having various degree of molecular complexity. The inorganic photosensitizers (titanim dioxide, zinc oxide, etc.). being excited in near UV are able to perform the same reactions as porphyrins – electron transfer from donor to acceptor molecule (including photoreduction of viologens) or water molecule photooxidation (oxygen liberation), coupled with reduction of ferric compounds and quinones. The inorganic photosensitizers are not used in biological evolution; actually the inorganic ions entered into tetrapyrrolic cycle, forming effective photocatalysts. Inclusion of pigments into primary membranes led to elaborated coupling between pigments and enzymatic systems. The involvement of the excited pigments into the biocatalytic electron transfer chain served as prerequisite of effective function of photosynthetic organisms.

The life on our planet depends on photosynthesis of plants providing organic matter and oxygen to all the organisms living on the Earth. Now the question arises: does the Origin of Life on the Earth depend on photosynthesis too?

The study of the pathways of biological evolution is based on the data of comparative biology of contemporary organisms and on paleontology, providing information on the development of ancient species. The most ancient photosynthetic organisms blue-green algae and photosynthetic bacteria are usually considered; more simple photoautotrophes are not found on the Earth.

A. I. Oparin presented convincing arguments that photosynthesis was developed after the primary heterotrophes were originated, their metabolism being based on the use of organic matter of abiogenic origin (Oparin, 1957).

To trace the pathways of prebiotic evolution experimental data derived from different sources of information are used (Gaffron, 1962; Calvin, 1969).

Chemical analysis of ancient rocks and meteorites would provide data on the nature of primary carbonaceous matter; due to the success of space exploration the samples from the Planets will be available for chemical analysis; the spectroscopy of stars and interstellar space provides us with valuable information on the nature and transformations of organic and inorganic substances in the Universe.

Impressive data were obtained by modeling in laboratory the reactions in primary reducing atmosphere under the influence of UV, visible, and corpuscular radiation of the Sun, radioactivity and electric discharges (Ponnamperuma, 1971).

The model experiments sustained the hypothesis on the accumulation of highly active organic and inorganic substances in primary ocean.

Diverse organic substances of abiogenic origin in primary soup could chemically interact one with another in aqueous solution, in adsorbed state or in structures of protobionts. These reactions could be accelerated in general by catalysis or photochemical activation. I wish to present here some model experiments which may be useful to construct a hypothetic picture of the photochemical evolution.

1. The Photochemical Activation of Coenzymes

In the case of abundance of chemically active substances in the primary soup there was probably no need in the photochemical mode of substrate activation. The primary energetic metabolism was controlled by catalytic systems probably similar to contemporary coenzymes.

At the first International Symposium on the Origin of Life held in Moscow in 1957 we assumed that the photochemical activation of some coenzymes might be an intermediary metabolic step in light energy utilization (Krasnovsky, 1959). The photochemical activity of flavine coenzymes was known at that time. Later we studied the photochemical activation of reduced pyridinenucleotides and analogs excited by near UV radiation (365 nm) (Krasnovsky and Brin, 1964). Photochemical activation enhanced NADH oxidation by electron acceptor molecules including viologens which have more negative redox potential as pyridinenucleotides. In the latter case it was possible to store some light energy in the reaction products.

$$\text{NADH}^* + \text{Viologen} \longrightarrow \text{NAD} + \text{reduced Viologen}$$

Recently we revealed photochemical activity of some biocatalytically active quinones in cytochrome *c* reduction.

In the course of evolution the photosensitizing and biocatalytic function become specialized. A good example is to compare the properties of iron and magnesium porphyrines, the latter being an effective photosensitizer inactive in the dark and extremely active when excited. On the contrary the biocatalysts – iron porphyrine complexes – being extremely active in the dark are practically nonactivated by light.

The significance of photosensitized substrate activation becomes more important in the course of exhaustion of active substances in the primary ocean.

It is generally considered that on the primary Earth there was no ozone layer in the upper parts of atmosphere absorbing now the shortwave UV radiation of the Sun. But the water layer absorbs shortwave UV too. So, in the case of reactions occurring in the ocean the primary photochemical reactions required sensitization to the more longwave part of the solar spectrum. In general, the primary photoreceptors-photosensitizers could be either of inorganic or organic nature.

2. Inorganic Photosensitizers

In the literature on chemical evolution the use of inorganic photoreceptors was not yet considered. But some inorganic substance which are components of the Earth's crust possesses photosensitizing activity. They could play a role of primary photosensitizers. It is surprizing that the oxides of titanium, zinc and tungsten possesses high photosensitizing activity in redox reactions comparable with the activity of porphyrines and chlorophylls. These compounds are capable to sensitize reactions with light energy storage in terminal stable products (Krasnovsky and Brin, 1970).

It was shown in our laboratory that titanium, zinc and tungsten oxides under action of UV radiation (365 nm) are capable in water media to photosensitize oxygen evolution coupled with exogeneous electron acceptors reduction (ferric ions, ferricyanide

TABLE I

Photosensitization of redoxreactions by TiO_2, ZnO and WO_3 (UV 365 nm)

Electron donor	Electron acceptor	Reaction products
H_2O	Fe^{+++}, $Fe(CN)_6^{+++}$	Oxygen, Fe^{++}, $Fe(CN)_6^{++}$
H_2O	p-benzoquinone	Oxygen, hydroquinone
H_2O	Viologens	Reduced viologens, dyes
Glycin	Redox-dyes	Oxidized electron donor
Hydrazin		
Thiourea, etc.		

ions, p-benzoquinone were used; see Table I. The quantum yield of the reactions studied run up to 1%. These experiments demonstrate working models of Hill's reaction or the so called Photosystem II of photosynthesis.

We have also created models of the Photosystem I using viologens as final electron acceptors in experiments accomplished *in vacuo*.

These observations point to the feasible mode of involvement of inorganic photosensitizers into the structure of protobionts. Nevertheless, since none of the existing organisms includes inorganic sensitizers in the patterns of their metabolism, they being probably 'blind alleys' of the evolutionary tree.

In the course of evolution the most effective catalysts and photosensitizers had been developed by bonding inorganic ions to organic ligands, the porphyrin ring being most widespread.

3. Abiogenic Synthesis of Porphyrins

Rothemund (1936) described the synthesis of porphin from pyrrole and formaldehyde. This reaction is thermodynamically spontaneous, proceeding slowly at room temperature in the dark in the presence of some catalysts. The presence of oxidants (oxygen, quinones) enhances the synthesis of porphin since the condensation of pyrrol and formaldehyde proceeds with hydrogen abstraction.

In the solution of pyrrole and formaldehyde not only porphin but also more reduced compounds are formed: chlorin and bacteriochlorin (Krasnovsky and Umrikhina, 1972). The synthesis and analysis of the most reduced substances requires anaerobiosis as these pigments are spontaneously oxidized in the presence of air oxygen. Chlorin may be formed from porphin not only by condensation of pyrrole and formaldehyde but via photoreduction of porphin too. The following scheme resumes the processes listed above:

$$\text{pyrrole} + \text{formaldehyde} - \begin{cases} \rightarrow \text{porphin} \\ \quad \downarrow \quad h\nu, \text{Red} \\ \rightarrow \text{chlorin} \\ \quad \downarrow ? \quad h\nu, \text{Red} \\ \rightarrow \text{bacteriochlorin} \end{cases}$$

Our studies show that interaction between pyrrole and formaldehyde is catalysed by some Earth crust components: silica, titanium dioxide and zinc oxide and some organic substances among them tryptophane which is most active.

When analyzing ancient rocks and samples from planets it is necessary to search simultaneously with porphin, the more reduced pigment – chlorins and bacteriochlorins.

All these compounds are photochemically active; they possess photosensitizing activity and are capable of undergoing reversible photooxidation and photoreduction.

However, in organisms free porphyrines are found mostly as intermediates of biosynthesis; among photosynthetic organisms magnesium complexes of porphyrines dominate and among catalysts–iron complexes.

The possibility of choosing the appropriate central metal atom in the porphyrine ring in the process of evolution is illustrated in Table II.

The compounds of silicon and titanium are practically insoluble in water this is what probably prevented the incorporation of these metals into porphyrine complexes. Na, K and Ca complexes of porphyrines are hydrolysed in water media. There are only complexes of magnesium and iron which are sufficiently stable to win the competition. These complexes have auxiliary coordination vacancies perpendicular to the plain of the porphyrine ring thus are capable of bounding reacting molecules. In the course of evolution, they became the property which made it possible to bound specific proteins.

TABLE II

Possible incorporation of the Earth's crust metals into a porphyrine ring

Element	% % in Earth crust	Activity of porphyrine complex	
		Photochemistry	Catalysis
Si	27.6	insoluble in water media	
Al	8.5		
Fe	5.0	–	+
Ca	3.5		
Na	2.6	hydrolysed in water	
K	2.5		
Mg	2.0	+	–
Ti	0.6	insoluble in water media	

The principle difference of magnesium porphyrine complexes to the ferric complexes is that the first are active only after excitation by light quanta and are converted into singlet or triplet excited states.

Studies in our laboratory showed that excited porphyrines and their magnesium and zinc complexes are capable of reversible photochemical reduction or oxidation, that is, to accept or donate an electron to a partner molecule (Krasnovsky, 1972), see Table III.

Reversible photoreduction: $P^* + \mathrm{Red} \rightleftarrows \cdot P^- + \cdot \mathrm{Red}^+$
Reversible photooxidation: $P^* + \mathrm{Ox} \rightleftarrows \cdot P^+ + \cdot \mathrm{Ox}^-$

TABLE III

Electron donors (Red) and acceptors (Ox) in photoredox reactions of porphyrines

Electron donors	Electron acceptors
Dienols	Oxygen
SH-compounds	Ferric compounds
Ferrous compounds	Quinones
NADH, NADPH	Nitrocompounds
Hydroquinones	Dyes
Hydrazines	NAD
	Flavins, etc.

In the triple system composed of electron donor (Red), electron acceptors (Ox) and pigment-excited molecules (P*) the photosensitization of the electron transport from Red molecules to Ox molecules proceeds at the expense of light energy absorbed by pigment molecules.

$$\mathrm{Red} \xrightarrow{e} \text{excited pigment} \xrightarrow{e} \mathrm{Ox}$$

In the case of more complicated protobionts having lipoproteinic membranes, the ability of porphyrines to incorporate into this membrane required introduction of the lipophilic tail into pigment molecules. In the case of chlorophyll this tail belongs to phytol and in the case of bacterioviridin to farnesol.

It should be noted that the lipophylic tail does not influence the photochemical properties of the pigment determinated mainly by the porphyrine ring of the molecule.

4. Evolution of Excitation Energy Migration

The ancestors of photosynthetic organisms probably possessed small quantities of pigments which directly participated in the photochemical electron transfer since the effective pigment biosynthetic system had not yet developed. As far as the catalysts-enzymes are concerned they are active in very low concentrations; on the contrary, the quantity of pigment-photosensitizer must be very high to absorb effectively the Sun radiation so as to win the competition for survival.

In contemporary organisms the energy of light quanta absorbed by aggregated pigments migrate to a minor part of the pigment to function in the so called 'active center'.

The ancient photosynthetic organisms – blue-green algae already possess a fully developed energy migration system from phycobilins and a bulk of chlorophyll to the chlorophyll *a* in active center.

The development of metabolism in the course of biological evolution gradually led to the appearance of effective biosynthesis of chlorophylls, phycobilins, and carotenoids.

As the capacity of pigment biosynthesis increased in the course of evolution, the function of excitation energy migration inevitably developed. This energy migrated from the bulk of the pigment to the minor part of the pigment that existed previously.

It is remarkable that in the case of inorganic photosensitizers (TiO_2, ZnO) there exists a mechanism of energy migration from the crystal lattice to the active center where the primary photoprocess takes place.

5. Types of Photochemistry and Catalysis Coupling

Thus it is possible to suppose that primary photocatalytic block of chemical and biological evolution comprised the ternary system: electron donor-pigment-electron acceptor. This system works efficiently not only in models but in biologically organized structures, too (Krasnovsky, 1971a, b).

The photochemical and catalytic electron transport chains were improved by diverse coupling of the primary photocatalytic blocks incorporated into the primary protobiont membranes. In the processes of biological evolution diverse combinations of photocatalytic blocks were realized.

It is possible to trace several types of primitive photocatalytic electron transfer: we begin with noncyclic electron transfer. The excited pigment molecule decreases the activation energy and the overall reaction is thermodynamically possible ($-\Delta F$) here the photosensitizer acts as an photocatalyst.

$$\text{Red} + \text{Ox} + (\text{P})^* \rightarrow \text{Red}_{\text{ox}} + \text{Ox}_{\text{red}} + (\text{P})$$

Another type of reaction which is more important for the use of an inactive hydrogen donor, that is the water molecule, is the overall process which is accompanied by the increase of free energy of the system ($+\Delta F$). This case is possible to reproduce in a model system – solutions of chlorophyll; here pyridinenucleotides and other electron acceptors are photoreduced at the expense of ascorbic acid as an electron donor. This case was studied in our laboratory:

$$\text{Ascorbate} + \text{NAD} + (\text{CHL}) \xrightarrow{\text{red light}} \text{Ascorbate}_{\text{ox}} + \text{NADH} + (\text{CHL})$$

A modelling of the Hill reaction photosensitized by inorganic semiconductors also proceeds with a gain of free energy (see above).

It is possible to presume a primitive cyclic electron transfer coupled with phosphorylation. In the cyclic processes Mg and Fe porphyrine may participate; the first cycle is reproduced in a model system: chlorophyll and porphyrins effectively photosensitize oxido-reductive transformations of cytochromes. But the coupling of such cyclic electron transfer with ATP-formation was not yet achieved in any model systems.

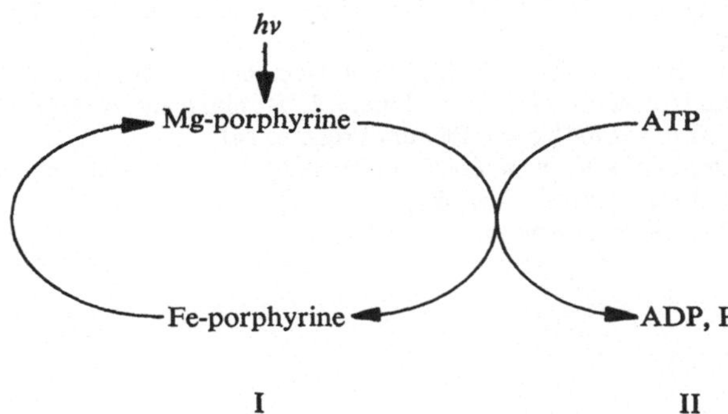

On the other hand the formation of high energy phosphate esters in model reactions was described by Ponnamperuma, Wang, etc. In the experiments of Lohrmann and Orgel the urea and inorganic phosphate mixture was an efficient phosphorylating agent.

It was found by Arnon and Kamen that ancient photosynthetic organisms (photosynthetic bacteria) possess both types of electron transfer: noncyclic and cyclic coupled with photophosphorylation. So we may conclude that the primitive 'bricks' of electron transfer may be used in biological evolution.

The use of water as an ultimate electron donor in more perfect photosynthetic organisms required the coupling of two photoreactions for electron transport from water to pyridinenucleotides via ferredoxins. In other words the coupling of two photocatalytic 'bricks' which is denoted, according to contemporary ideas, as photosystems I and II.

$$H_2O \xrightarrow{e} \underset{\text{II}}{\overset{h\nu \downarrow}{(\text{CHL})}} \xrightarrow{e} \text{Cytochromes} \xrightarrow{e} \underset{\text{I}}{\overset{h\nu \downarrow}{(\text{CHL})}} \xrightarrow{e} \text{Fd} \xrightarrow{e} \text{NADP}$$

The chemical and biological evolution from simple photocatalytic reactions to the perfectly organized systems of photosynthetic electron transfer required probably more than two 2 b.y. The model experiments represent plausible explanation of the various stages of chemical evolution; however only the future experimental data derived from geology and exobiology would reveal the real nature of the process.

References

Calvin, M.: 1969, *Chemical Evolution*, Clarendon Press, Oxford.

Gaffron, H.: 1962, in M. Kasha and B. Pullman (eds.), *Horizons in Biochemistry*, Acad. Press, p. 59.

Krasnovsky, A. A.: 1959, *The Origin of Life on the Earth*, Pergamon Press, p. 606.
Krasnovsky, A. A.: 1971a, in R. Buvet and C. Ponnamperuma (eds.), *Molecular Evolution, I*, North Holland Publ., p. 279.
Krasnovsky, A. A.: 1971b, in A. Kimball and J. Oró (eds.), *Prebiotic and Biochemical Evolution*, North Holland Publ., p. 209.
Krasnovsky, A. A.: 1972, *Biophysical J.* **12**, 749.
Krasnovsky, A. A. and Brin, G. P.: 1969, *Problems of Evolutionary and Technical Biochemistry* (in Russian), p. 221.
Krasnovsky, A. A. and Brin, G. P., 1970, *Molecular Photonics* (in Russian), p. 161.
Krasnovsky, A. A. and Umrikhina, A. V.: 1972, in D. L. Rohlfing and A. I. Oparin (eds.), *Molecular Evolution, Prebiological and Biological*, Plenum Press, p. 141.
Oparin, A. I.: 1957, *Origin of Life on the Earth* (in Russian), U.S.S.R. Academy of Sciences, Moscow.
Ponnamperuma, C.: 1971, *Quart. Rev. Biophys.* **4**, 77.
Rothemund, P.: 1936, *J. Amer. Chem. Soc.* **58**, 625.

INORGANIC TYPES OF FERMENTATION AND ANAEROBIC RESPIRATIONS IN THE EVOLUTION OF ENERGY-YIELDING METABOLISM

FUJIO EGAMI

Mitsubishi-Kasei Institute of Life Sciences, 11 Minamiooya, Machida-shi, Tokyo 194, Japan

Abstract. We proposed long ago the following sequence as one of the main pathways in the evolution of energy-yielding metabolism: fermentation → nitrate fermentation → nitrate respiration → oxygen respiration. In the present report our concept is presented in a more general form: (1) fermentation → → (2) fermentation with H_2 release → (3) inorganic types of fermentation → (4) anaerobic respirations → (5) oxygen respiration, based upon recent biological and physical information. The energy-yielding efficiency increased gradually together with the evolution. (2) is characterized by the participation of ferredoxin, (3) by the establishment of electron transfer chain, and (4) by the participation of cytochrome and oxidative phosphorylation.

The close relationship between the primary structure of ferredoxins of anaerobic bacteria and that of a cytochrome (cytochrome c_3) was demonstrated. It reveals that the transition from inorganic types of fermentation to anaerobic respirations was direct and accompanied by the transition from ferredoxins to cytochromes, and it further supports our concept that the cytochrome system, and consequently the oxidative phosphorylation, were induced at this evolutionary step.

Our concept based upon biological obseıvations is consistent with a physical theory recently proposed by M. Shimizu.

We studied long ago (1946–1961) the enzymatic mechanism and physiological meaning of biological nitrate reduction and in the course of the study we proposed the principal pathway in the evolution of energy-yielding metabolism as follows:

Fermentation → nitrate fermentation → nitrate respiration → oxygen respiration. It was summarized or reviewed by Egami (1957), Ishimoto and Egami (1959), Taniguchi (1961) and Takahashi *et al.* (1963). Similar views have been presented by several microbiologists (Nason, 1962; Hall, 1971).

In the present report I should like to revise the hypothesis, based upon the recent information.

Since Oparin and Haldane it is generally considered that the most primitive form of energy-yielding mechanism was anaerobic decomposition of organic compounds. We don't know what kind of anaerobic decomposition it was. But it may be presumed that the common ancester of all present organisms had evolved to have an enzyme system containing enzymes transferring hydrogen and phosphate and coenzymes such as NAD, ATP, etc. and that it could perform the fermentation similar to the E.-M. pathway, which is found as the basis of energy-yielding metabolism of all present organisms.

So I should like to regard it as a starting point of the evolution of energy-yielding metabolism.

The fermentation can be carried out by the soluble enzyme system. It may mean that it was the energy-yielding mechanism by the organisms without any cellular structure. In fermentation only substrate level phosphorylation can take place, for example, the phosphorylation of triose phosphate coupled with the reduction of NAD:

$$\text{Triose phosphate} + NAD^+ + Pi \rightarrow \text{1,3-Diphosphoglycerate} + NADH + H^+ \quad (1)$$

and the energy yield is evaluated as 2ATP/mole of glucose.

Generally speaking, fermentations gave rise to highly oxidized substances such as carbon dioxide and highly reduced substances such as ethanol, butanol, lactate, butyrate. These reduction products of pyruvate contain higher potential energy, but were dumped as a waste in such typical fermentations.

The second step in the evolution is characterized by the escape of pyruvate from reduction processes. It was the fermentation with H_2 release. Such a fermentation with H_2 release was regarded by Oda (1959) as a transition step from anaerobiosis to aerobiosis. Here we observe the advent of ferredoxin-hydrogenase system and phosphoroclastic enzyme system. Thus the reducing power was eliminated as hydrogen gas and the substrate level phosphorylation as illustrated in the reaction (1) could be enhanced; furthermore pyruvate could be utilized as an energy source giving rise to acetyl phosphate, and consequently ATP. The energy yield may be evaluated as 4ATP/mole of glucose:

$$\begin{array}{l} C_6H_{12}O_6 + 2ADP + 2Pi + 2NAD^+ \rightarrow \\ \qquad 2CH_3COCOOH + 2ATP + 2NADH + 2H^+ \quad (2) \\ \qquad \downarrow \qquad\qquad\qquad\qquad\qquad \downarrow \\ \qquad \text{2 acetyl phosphate} \qquad\qquad 2NAD^+ + H_2 \\ \qquad \text{or } 2ATP \end{array}$$

The third step in the evolution is 'inorganic types of fermentation', in which SO_4^{2-}, Fe^{3+} or oxidized nitrogen compounds participate as a final oxidant. The term 'inorganic types of fermentation' was coined by Lipmann (1965).

In nitrate fermentation, a typical example of inorganic types of fermentation, observed in strict anaerobes such as *Clostridium perfringens,* reducing power (NADH $+H^+$), H_2, and reduction products of pyruvate may be oxidized by nitrate. For example, in *Cl. perfringens* both ethanol and hydrogen gas are oxidized by nitrate (Katsura *et al.*, 1954; Koyama, 1956) probably through the following electron transfer chain (Chiba and Ishimoto, 1973):

$$\begin{array}{lcl} \text{ethanol} \rightarrow \text{alcohol dehydrogenase} \rightarrow & & \text{NADH: ferredoxin oxidoreductase} \\ & & \downarrow \\ \text{hydrogen} \rightarrow \text{hydrogenase} \rightarrow & & \text{ferredoxin} \qquad (3) \\ & & \downarrow \\ & & \text{nitrate reductase} \\ & & \downarrow \\ & & NO_3^- \end{array}$$

Here we recognize the establishment of an electron transfer chain somewhat similar to that in respiration. It should be noted that nitrate reductase in *Clostridium perfringens* is an inducible enzyme (Katsura *et al.*, 1954). It means that nitrate reduction is not a nonphysiological secondary activity of any constitutive enzyme, but an activity of a specific inducible nitrate reductase. It suggests that a strict anaerobe, before the accumulation of atmospheric oxygen, had a gene specific for nitrate reductase and further suggests that nitrate accumulated before the accumulation of atmospheric oxygen.

The fourth step in the evolution is anaerobic respiration, in which sulfate or active sulfate such as APS, nitrate or nitrite played the role of final electron acceptor. The anaerobic respirations are characterized by the participation of a cytochrome-system. A typical example of anaerobic respiration is nitrate respiration such as observed in *E. coli* and denitrifiers such as *Pseudomonas aeruginosa.*

The term 'nitrate respiration' was coined by us more than 20 years ago to denote anaerobic reduction of nitrate, the significance of which is energy-yielding rather than nitrogen assimilation. An extensive series of study has now established that the electron-transfer mechanism functioning in nitrate respiration in *E. coli* is closely similar to that operating in oxygen respiration by the same organism. Both involve flavoproteins, a lipophilic quinone(s), and a cytochrome B (cytochrome b_1). As the terminal component, however, the former contains a molybdenum protein, while the latter a cytochrome(s) A. Another resemblance between the two is the fact that both are bound to a membrane structure of the cell. Furthermore, in contrast to nitrate fermentation, in nitrate respiration, owing to the presence of a phosphorylation system coupled to the electron transfer chain, oxidative phosphorylation takes place efficiently and thus the yield of ATP is greatly increased.

We don't know the exact value of energy-yielding efficiency, but it may be tentatively evaluated as 8∼10 ATP per mole of glucose. A typical example of energy-yielding metabolism in nitrate respiration may be expressed as follows:

	ATP yield
$C_6H_{12}O_6 + 2NAD^+ \rightarrow 2CH_3COCOOH + 2NADH + 2H^+$	2
$2NADH + 2H^+ + 2KNO_3 \rightarrow 2NAD^+ + 2KNO_2 + 2H_2O$	2∼4
$2CH_3COCOOH + 2Pi \rightarrow 2$ acetyl phosphate $+ 2HCOOH$	2
$2HCOOH + 2KNO_3 \rightarrow 2CO_2 + 2KNO_2 + 2H_2O$	2
$C_6H_{12}O_6 + 4KNO_3 \rightarrow 2CH_3COOH + 2CO_2 + 4KNO_2 + 2H_2O$	8∼10 (total)

I have discussed mainly nitrate fermentation and nitrate respiration as an example of inorganic types of fermentation and anaerobic respirations respectively. The transition from inorganic types of fermentation to anaerobic respirations may be expressed by a simplified general formula:

Inorganic Types of Fermentation

Substrate →	Dehydrogenase → --- →	Ferredoxin → X reductase → X	
(Hydrogen Donor)	(Hydrogenase in the case of H_2)		(Ferredoxin: X oxidoreductase)

Anaerobic Respirations

Substrate →	Dehydrogenase → --- →	Cytochrome → X reductase → X.
(Hydrogen Donor)	(Hydrogenase in the case of H_2)	or Cytochrome system

Then it is rather natural to search for the structural relationship between ferredoxins and cytochromes in anaerobic organisms. It was tried by Yamanaka (1972), and I would like to further extend the structural comparison between them. The existence of certain similarities between them is evident as shown in Table I. Both cytochrome c_3 and ferredoxins have a pair of homologous sequences which are very similar to each other. The remarkable differences between them are (1) histidine residues found in cytochrome c_3 but not in ferredoxins, and (2) residual cysteine residues other than the corresponding Cys–Cys sequence in ferredoxins. These differences might be related to the difference between cytochrome c_3 as a heme protein and ferredoxin as nonheme-iron protein. The substitution of amino acid fragments was accompanied or followed by the exchange of prosthetic groups. If the sequences are compared (except for those sequences in rectangles in Table I) the close relationship between the amino acid sequences of ferredoxins and that of cytochrome c_3 is evident. Different amino acids found in corresponding positions of both sequences are mostly amino acids of one nucleotide change in codons. Thus the close relationship between the primary structure of ferredoxins of anaerobic bacteria and that of primitive cytochrome reveals that the transition from inorganic types of fermentation to anaerobic respirations was direct and accompanied by the transition from ferredoxins to cytochromes, and it further supports our concept that the cytochrome system and consequently the oxidative phosphorylation were induced at this evolutionary step.

Now the transition from anaerobic respirations to oxygen respiration must be considered. But the transition is rather a variation in details. Once such anaerobic respirations had been established in the process of evolution, there must have been no difficulties in modifying these mechanisms to develop a more efficient machinery suitable for oxygen respiration. This modification seems to have been induced by the gradual increase of atmospheric oxygen, an event which was then occurring probably by the appearance of photosynthetic green plants. At the same time, this modification must have been accompanied by a considerable increase in the efficiency of energy production. Table II is a summary of the comparative biochemistry of fermentation, nitrate fermentation, nitrate respiration, and oxygen respiration.

Recently, Yamanaka and Okunuki (1968) and Yamanaka (1972) have discussed the evolution of respiratory systems based on their extensive studies on comparative biochemistry of cytochrome systems. They have essentially supported our view, summarized by Takahashi *et al.* (1963), and further suggested that an evolved nitrate re-

TABLE I

Comparison of the amino acid sequences of ferredoxins and cytochrome c_3

Ferredoxins						
Clostridium pasteurianum	1	5		10	15	20
	AlaTyrLysIle	Ala	AspSer Cys	ValSerCysGlyAlaCys	AlaSerGluCys	ProValAsnAla—
	30		35	40	45	50
Tanaka *et al.* (1966)	—IlePheValIle Asp	Ala	AspThr Cys	IleAspCysGlyAsnCys	AlaAsnValCys	ProValGlyAla—
	1 5			10	15	20
Clostridium butyricum	AlaPheValIleAsn		AspSer Cys	ValSerCysGlyAlaCys	AlaGlyGluCys	ProValSerAla—
	30		35	40	45	50
Benson *et al.* (1966)	—GlnPheValIleAsp	Ala	AspThr Cys	IleAspCysGlyAsnCys	AlaAsnValCys	ProValGlyAla—
	20	25		30	35	
Cytochrome c_3	—ValValPhe Asn HisSer	Thr His	LysSer	ValLysCysGlyAspCys	HisHis	ProValAsnGly—
Desufovibrio vulgaris						
			75	80		85
Ambler (1968)	—AspLys Asn	Thr	LysPhe	LysSerCysValGlyCys	HisVal	GluValAlaGly—

TABLE II

Direction of evolution	Fermentation →	Nitrate fermentation →	Nitrate respiration →	Oxygen respiration
Localization of enzyme systems	Soluble part	Soluble part	Soluble part (Glycolysis) and membrane bound (Electron transfer to nitrate)	Soluble part (Glycolysis) and membrane bound (Electron transfer to oxygen)
Nature of phosphorylation	Substrate level	Substrate level	Substrate level and oxidative	Substrate level and oxidative
Participation of cytochrome system	−	−	+	+
Maximum ATP yield/ mole Glc	2	4	probably 8 ∼ 10	38 in higher organisms; less in prokaryotes

spiration or 'nitrate-nitrite respiration' observed in denitrifying bacteria might be the direct precursor of oxygen respiration. In fact, *Pseudomonas aeruginosa,* a denitrifier, contains nitrite reductase besides nitrate reductase, and the former has been shown to act as a cytochrome oxidase. A likely hypothesis derived from this finding is that this enzyme acted, by reducing nitrite, as the terminal component of the anaerobic respiration before the advent of atmospheric oxygen; as molecular oxygen accumulated it could function, without any modification, as the oxidase in oxygen respiration. However, since the nitrite reductase contains heme *d* as one of its prosthetic groups, it must have been extremely labile to hydrogen peroxide which had to be produced in the presence of molecular oxygen. This circumstance as well as the other environmental conditions might have been the cause leading to the development of cytochrome oxidases having heme *a* which is stable to hydrogen peroxide.

From the considerations discussed above, it does not seem unreasonable to conclude that the evolution of energy-yielding mechanism from fermentation to oxygen respiration has undergone the following sequence:

Fermentation → fermentation with H_2 release → inorganic types of fermentation → anaerobic respirations (sulfate, nitrate, and nitrate-nitrite respiration etc.) → oxygen respiration.

This scheme which is exclusively based upon biological considerations requires the existence of oxidized forms of sulfur and nitrogen before the accumulation of atmospheric oxygen.

For the formation of these compounds on the primitive earth before the accumulation of molecular oxygen, two possibilities have been taken into consideration. One is biogenic or photosynthetic, such as the sulfate formation from sulfide or sulfur by colored sulfur bacteria and the formation of oxidized nitrogen compounds from reduced compounds as suggested independently by Olson (1970) and Taniguchi

(1971). The other is abiogenic. Taking into consideration the fact that a strict anaerobe such as *Cl. perfringens* has a specific and inducible nitrate-reducing enzyme system, I would rather like to support the abiogenic accumulation of sulfate, nitrate, and nitrite on the primitive earth before the accumulation of atmospheric oxygen. Of course the view does not exclude the possibility of biogenic formation of these compounds.

It is generally accepted that an abundant accumulation of oxygen in the atmosphere was caused by green plants. But I believe that it does not exclude the possibility of the existence of sulfate, nitrate, and nitrite before the accumulation of free oxygen. A small amount of oxygen (Van Valen, 1971) or any other oxidants produced nonbiologically by water photolysis or any other mechanisms might produce nitrate, nitrite, or sulfate from less oxidized precursors. An essentially similar view on the occurrence of nitrate on the early Earth was recently presented by Hall (1973).

How do physical sciences consider the problem? So far I know, there are two quite contradictory hypotheses. One considers that the surface of the primitive Earth before the origin of Life was in quite reduced state (Miller and Urey, 1959), which remained for a long time and excludes the existence of oxidized nitrogen compounds, and the other considers the existence of considerable amounts of oxidized sulfur and nitrogen compounds. Recently Broda (1970, 1971) based upon the former consideration rejected our concept.

I would of course like to support the latter consideration consistent with our concept. Recently Shimizu (1973), extending the hypothesis by Rubey (1968), proposed a hypothesis that the primitive formation of certain oxidized compounds of sulfur and nitrogen such as SO_2, SO_3, NO and NO_2 did not necessarily require molecular oxygen for their formation and these compounds either directly, or after oxidation, were dissolved and gave rise to a relatively large amount of soup containing SO_4^{2-}, NO_2^- and NO_3^- before the accumulation of molecular oxygen.

In this connection I would like to refer to a finding by Sugawara *et al.* (1944, 1949). It is an aqueous inclusion in nepheline-basalt discovered by S. Tsuboi in Shimane Prefecture in Japan and chemically investigated by Sugawara *et al.* (1944, 1949). The rock has lots of cavities with a kind of aqueous inclusion. According to them the inclusion owes its origin to magma and it represents an example of the so-called 'residual water from magmatic differentiation'. The water phase and the dissolved gas in the inclusion were analyzed: the gas consisted exclusively of nitrogen and carbon dioxide, and no oxygen, hydrogen, and methane were detected. The composition of the solutes in the aqueous phase was qualitatively similar to that of sea water (Na, K, Ca, Mg, Fe, Al, NH_4, chloride, nitrate, nitrite, sulfate, silicate), but quantitatively the composition was different from sample to sample, (SO_4^{2-} 173, NO_2^- 21, NO_3^- 44 mg l^{-1}; SO_4^{2-} 326, NO_2^- 0.54, NO_3^- 97 mg l^{-1}) and quite different from sea water. Anyhow the finding also supports the possibility of the accumulation of nitrate, nitrite, and sulfate in the absence of molecular oxygen or at least in the absence of residual molecular oxygen.

I have presented my actual concept on a pathway of metabolic evolution of energy-yielding mechanism. But it should be pointed out that the evolutionary pathway here discussed is nothing but one of the main pathways to various types of energy-yielding mechanism. Indeed I have not considered the position of different photosynthesizers and chemoauthotrophs in evolution, because the important problem is far beyond the scope of the present report. But I insist that the evolution from fermentation to anaerobic respiration leading to the formation of cytochrome systems and the development of oxidative phosphorylation, was realized independently of photosynthesis and that the evolution from anaerobic respiration to oxygen respiration and the gradual development of higher oxygen-respirers were forced by the increasingly oxidizing atmosphere caused by blue-green algae and higher photosynthetic organisms. As for the interrelation between the advent of photosynthesis and that of respiration, the consideration that anaerobic respirations preceded photosynthesis is supported also by Yamanaka and Okunuki's comparative studies on cytochromes (1968, 1972). Furthermore the view that photosynthetic organisms came relatively late, preceded by the chemosynthesis, has been suggested since long ago and even recently by several authors including Lipmann (1965) and Rabinowitch and Govindjee (1969). Our concept of the presence of oxidized forms of sulfur and nitrogen and the presence of cytochrome-containing organisms before the accumulation of atmospheric oxygen does not exclude the probable advent of certain primitive chemoautotrophic organisms before the advent of photosynthetic organisms.

I have presented my actual concept on the metabolic evolution of energy-yielding mechanism. I hope that it will be further investigated from the standpoints of both biological sciences and physical sciences.

Acknowledgements

A part of the present report was recently published as a short communication (Egami, 1973) in reply to the criticisms by Prof. E. Broda. The present report describes mainly the development of my concept made after that, especially the development made possible by the discussions with Drs M. Ishimoto, R. Sato, M. Shimizu, and T. Yamanaka, to whom my thanks are due.

References

Ambler, R. P.: 1968, *Biochem. J.* **109**, 47 P.
Benson, A. M., Mover, H. F., and Yasunobu, K. T.: 1966, *Proc. Nat. Acad. Sci. U.S.* **55**, 1532.
Broda, E.: 1970, *Prog. Biophys. Mol. Biol.* **21**, 143.
Broda, E.: 1971, in E. Schoffeniels (ed.), *Molecular Evolution*, Vol. 2, North-Holland Publ. Co., p. 224.
Chiba, S. and Ishimoto, M.: 1973, *J. Biochem.* **73**, 1315.
Egami, F.: 1957, *Svensk Kem. Tydskrift* **69**, 652.
Egami, F.: 1973, *Z. Allg. Mikrobiol.* **13**, 177.
Hall, J. B.: 1971, *J. Theor. Biol.* **30**, 429.
Hall, J. B.: 1973, *Space Life Sciences* **4**, 204.
Ishimoto, M. and Egami, F.: 1959, in *The Origin of Life on the Earth*, Pergamon Press, New York, (*Acad. Sci. U.S.S.R.*), p. 555.
Katsura, T., Ito, H., Nojima, T., Nomoto, N., and Egami, F.: 1954, *J. Biochem.* **41**, 745.
Koyama, J.: 1956, *Seikagaku* **28**, 74 (in Japanese).

Lipmann, F.: 1965, in S. W. Fox (ed.), *The Origin of Prebiological Systems and of Their Molecular Matrices*, Academic Press, New York, p. 259.
Miller, S. L. and Urey, H. C.: 1959, *Science* **130**, 245.
Nason, A.: 1962, *Bact. Rev.* **26**, 16.
Oda, Y.: 1959, in *The Origin of Life on the Earth*, Pergamon Press, New York (*Acad. Sci. U.S.S.R.*), p. 593.
Olson, J. M.: 1970, *Science* **168**, 438.
Rabinowitch, E. I. and Govindjee: 1969, *Photosynthesis*, John Wiley and Sons Inc., p. 20.
Rubey, W. W.: 1968, *Bull. Geol. Soc. Amer.* **160**, 729.
Shimizu, M.: 1973, *Viva Origino* **2**, 1 (in Japanese).
Sugawara, K., Oana, S., and Koyama, T.: 1944, *Proc. Imp. Acad. Tokyo* **20**, 721.
Sugawara, K., Oana, S., and Koyama, T.: 1949, *Proc. Japan Acad.* **25**, 103.
Takahashi, H., Taniguchi, S., and Egami, F.: 1963, in M. Florkin and H. S. Mason (eds.), *Comparative Biochemistry*, Vol. V, Academic Press, New York, p. 91.
Tanaka, M., Nakashima, T., Benson, A. M., Mower, H. F., and Yasunobu, K. T.: 1966, *Biochem.* **5**, 1666.
Taniguchi, S.: 1961, *Z. Allg. Mikrobiol.* **1**, 341.
Taniguchi, S.: 1971, in S. Suzuki, M. Ishimoto, and M. Kageyama (eds.), *Different Slices of Biochemistry* (in Japanese), Kodansha Ltd., Tokyo, p. 122.
Yamanaka, T.: 1972, *Adv. in Biophys.* **3**, 227 (Japan).
Yamanaka, T. and Okunuki, K.: 1968, in K. Okunuki, M. D. Kamen and I. Sekuzu (eds.), *Structure and Function of Cytochromes*, p. 390, Univ. of Tokyo Press, University Park Press.
Van Valen, L.: 1971, *Science* **171**, 439.

PART VI

EXOBIOLOGY

TEST RESULTS ON THE VIKING GAS CHROMATOGRAPH-MASS SPECTROMETER EXPERIMENT

K. BIEMANN

Dept. of Chemistry, Massachusetts Institute of Technology, Cambridge, Mass. 02139, U.S.A.

Abstract. The gas chromatograph-mass spectrometer instrument to be utilized in the Viking 1975 Molecular Analysis experiment has undergone preliminary testing in its flight configured version. A synthetic mixture of 24 components as well as a sample of the Murchison meteorite has been used for this purpose. The resulting data did not only allow the identification of most of the organic compounds known to be present, but also revealed the identity of a few unexpected ones. Thus, the sensitivity and reliability of the instrument and data system are satisfactorily demonstrated. Shortcomings revealed by these tests are in the process of being remedied.

Since the last conference on the Origin of Life where the basic principle and design parameters of the instrument to be used in the Viking 1975 molecular analysis experiment had been discussed (Biemann, 1971), its development has reached the flight hardware state. Briefly, the experiment is designed (Anderson *et al.*, 1972) to deter-

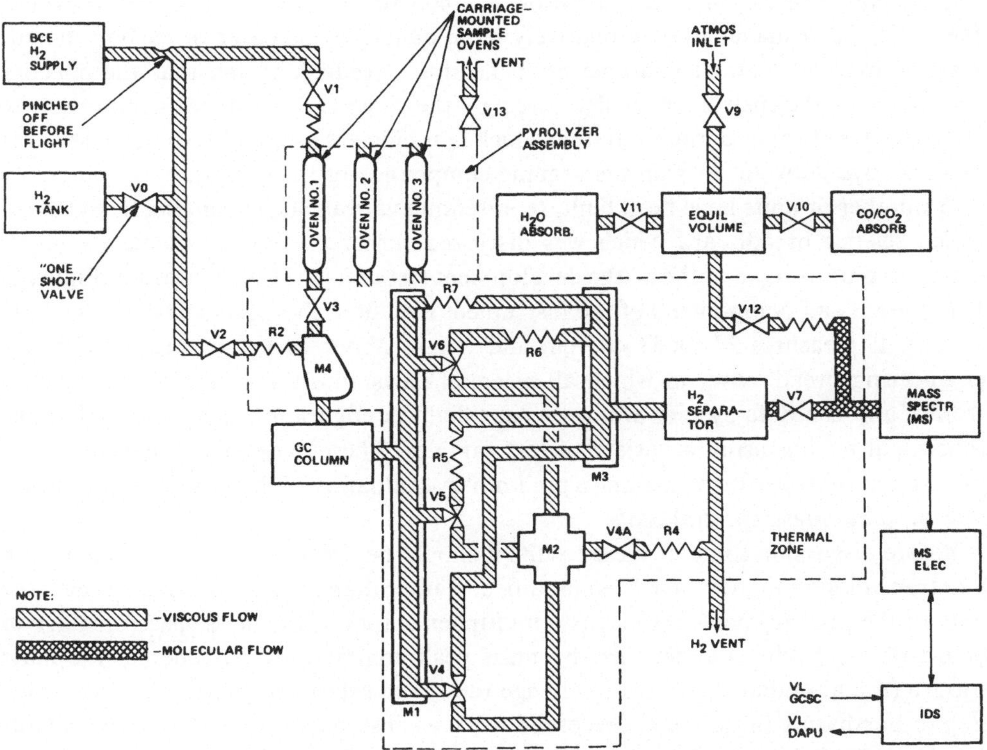

Fig. 1. Gas Flow Block Diagram. Dotted line delineates the heated zone (4 in Figure 2). M1, M2, V4-6 and R4-7 represent the effluent divider (see text).

mine the organic constituents of surface samples by expelling them at 200 °C, 350 °C and 500 °C, respectively into a gas chromatographic column and recording the mass spectra of the components. The mass spectrometer covers the mass range from m/e 12 to m/e 200 with adequate resolution. Water and other inorganic volatiles are detected at the same time. For the analysis of the atmosphere the mass spectrometer without the gas chromatograph is utilized. Practical considerations involving a realistic assessment of the feasibility within the time and funding constraints led to a simplification of the original plans which involved mainly the reduction of the number of samples to be analyzed (from eight to three) and the elimination of the capability to analyze one single sample directly rather than through the gas chromatographic system.

A block diagram of the final design is shown in Figure 1. In addition to the deletions referred to above, the most important sub-component that had been developed since our previous report is the so-called effluent divider. It serves to protect the mass spectrometer and its capacity-limited pump from sudden overpressures that would arise when a large component elutes from the chromatograph and, if permitted to enter the mass spectrometer in its entirety, would temporarily exceed the pumping capacity of the instrument. The effluent divider represents, in effect, a multi-stage gas stream splitter which automatically vents certain fractions of the effluent to the atmosphere. The divider ratio is controlled by the ion pump current to the values 1:0, 1:20, 1:400 and 1:8000. The very high divide ratios of 1:400 and 1:8000 are designed to protect the mass spectrometer in case relatively large quantities of water or carbon dioxide may be evolved from the sample upon heating. Needless to say that the effective sensitivity of the instrument is decreased by the same ratio. It is thus important to choose a gas chromatographic system which is as efficient as possible in the separation of those two substances from the organic compounds of the sample.

A number of units have been built, representing various stages ranging from the so-called science breadboard which was discussed earlier to almost completely flight-configured units. One of those, the development test unit (DTU), is shown in Figure 2. It represents all components of the instrument including the electronics and the data system. It measures 27.5 × 33 × 25 cm and weighs about 20 kg. Its predecessor, the engineering breadboard, in which all essential parts with the exception of the electronics and the data system are in or near flight configuration, has now undergone functional testing using a variety of test samples. These experiments permitted the assessment of its performance, and a preliminary evaluation of its capability to produce interpretable mass spectral data.

Before discussing some of these results it should be noted that the data acquisition and transmission modes have been modified. Rather than greatly condensing the large bulk of the primary data (the electron multiplier output of the mass spectrometer) on board of the Viking Lander, first by mass peak computation to generate the mass spectra (ion abundance vs mass-to-change ratio) and selectively storing and transmitting to Earth only those spectra recorded when a substance elutes from the chromatograph, all 3800 data points accumulated during each of up to 500 mass spectral scans (10.3 s each) of a gas chromatogram are recorded on a digital tape on board of the

Fig. 2. Development Test Unit of the Viking GC-MS instrument (side view). (1) Sample oven housing. (2) Hydrogen tank. (3) GC-column. (4) Valving, effluent divider, separator (in housing held at 200°C). (5) Ion source housing. (6) Electric sector. (7) Magnet. (8) Ion pump. (9) Electron multiplier.

lander and transmitted to Earth via the Orbiter link. Similarly, all data of an atmospheric experiment are being transmitted. While this may appear to result in the transmission of a large number of unimportant data bits it has the overriding advantage that no information can be lost in the automatic condensation of the original data which can be refined later to any extent desired. Thus, if for one reason or another, the mass calibration would have changed, it could be recognized and corrected at any later time. Transmission of all mass spectra continuously recorded during the gas chromatogram assures that a complete record of every single compound emerging from the gas chromatograph, regardless of its resolution characteristics, will be obtained and one will be able to bring to bear all the computer based evaluation techniques which have been developed over the recent years in our laboratories for very complex gas chromatograph-mass spectrometer data (Hites and Biemann, 1970; Hertz *et al.*, 1971). A unique feature of the Viking mass spectrometer is its wide dynamic range which covers seven orders of magnitude. This range is retained in the data which are recorded and transmitted on a logarithmic scale and converted to linear data on Earth where the data handling computers can easily cope with such a wide dynamic range. This is particularly important in view of the requirement that the instrument has to be capable of detecting minute amounts of organic compounds in the presence of relatively large amounts of others, such as water or carbon dioxide.

TABLE I

24-Component Test Mixture

No.	Formula	Compound
1.	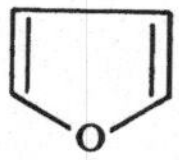	Furan
2.	$CH_2{=}CH{-}CN$	Acrylonitrile
3.	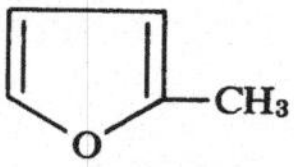 $-CH_3$	2-Methylfuran
4.	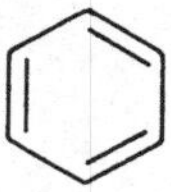	Benzene
5.	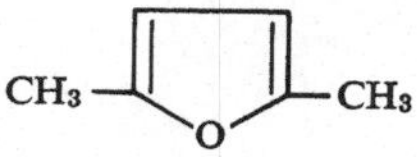 CH_3- $-CH_3$	2,5-Dimethylfuran
6.	$CH_3CH_2CH_2CH_2CH_2CH_2CH_2CH_3$	Octane
7.	$CH_3{-}CH(OH){-}CH_3$	Isopentanol
8.	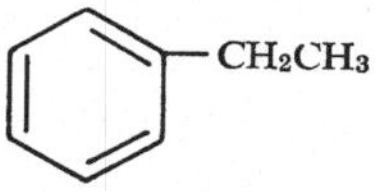 $-CH_2CH_3$	Ethylbenzene
9.	$CH_3(CH_2)_7CH_3$	Nonane
10.	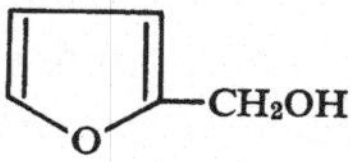 $-CH_2OH$	Furfurylalcohol
11.	$CH_3(CH_2)_8CH_3$	Decane
12.	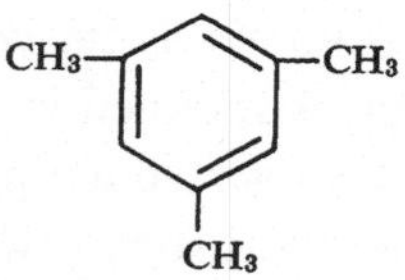 CH_3, CH_3, CH_3	Mesitylene
13.	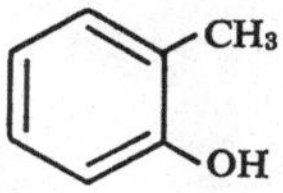 CH_3, OH	ortho-Cresol
14.	$CH_3(CH_2)_9CH_3$	Undecane
15.	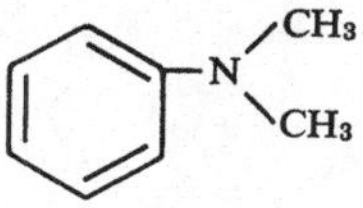 $-N(CH_3)_2$	Dimethylaniline
16.	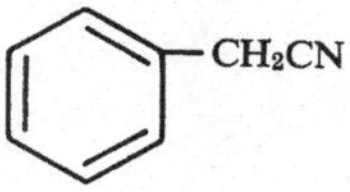 $-CH_2CN$	Benzylcyanide
17.	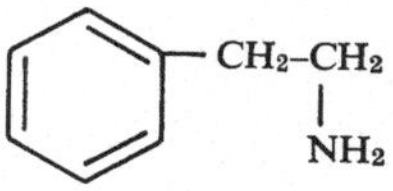 $-CH_2{-}CH_2{-}NH_2$	Phenylethylamine
18.	$CH_3(CH_2)_{11}CH_3$	Tridecane
19.	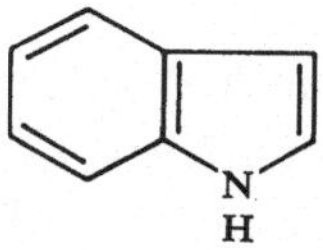 N H	Indole
20.	 CH_3, OH	para-Cresol
21.	$CH_3(CH_2)_{13}CH_3$	Pentadecane
22.	$CH_3(CH_2)_{13}CH{=}CH_2$	Hexadecene
23.	$CH_3(CH_2)_{14}CH_3$	Hexadecane
24.	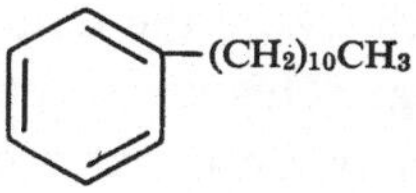 $-(CH_2)_{10}CH_3$	Phenylundecane (Solvent)

For the atmospheric experiment both pure gases and mixtures thereof have been admitted through the atmospheric inlet system and their mass spectra measured. These experiments have shown that sensitivity and resolution of the system are adequate and that the carbon monoxide-carbon dioxide adsorption system incorporated in the atmospheric inlet is sufficiently efficient to remove 99% of carbon monoxide and 99.9% of carbon dioxide from an atmospheric sample and thus should permit one to detect nitrogen down to a level of 100 ppm. For the organic experiment both a synthetic test mixture that can be injected into the gas chromatographic system as well as a sample of the Murchison meteorite has been used. The latter permits, in addition to the injection experiments, the evaluation of the performance of the sample oven.

The components of the test mixture (Table I) were chosen to cover the range of organic compounds which the gas chromatograph-mass spectrometer system should be capable of handling and thus to test the adequacy of its performance rather than to represent plausible components of the Martian surface. Phenylundecane was chosen as the solvent because it would elute at the end of the gas chromatogram rather than at the beginning. The gas chromatogram (representing about 5×10^{-8} g of components 1–23 of Table I) is obtained upon plotting the sum of all data points per mass spectrum

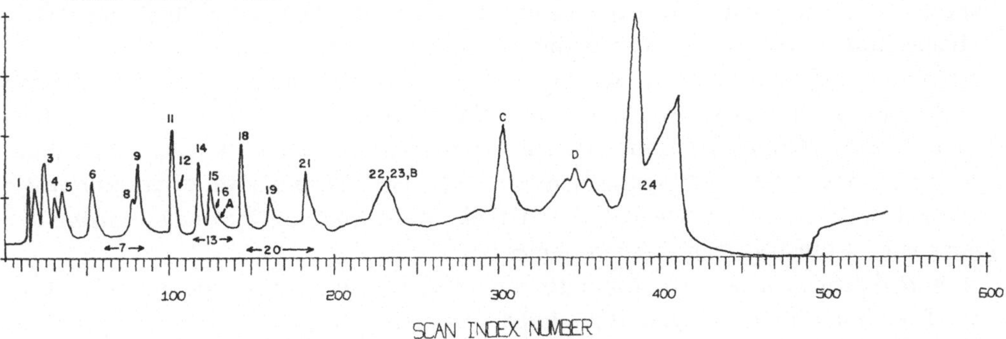

Fig. 3. Gas chromatogram of the 24 component test mixture. Carrier gas flow (hydrogen) 2 ml min^{-1}. Column temperature: 10 min at 50°C, 18 min programmed to 180°C, isothermal at 180°C for 54 min (scan index number × 10.3 = time elapsed in s).

vs the scan index number of the mass spectrum as shown in Figure 3. Upon conversion of the original data to mass vs intensity values for each mass spectrum as well as further conversion to mass chromatograms, that is plots of individual masses throughout the gas chromatogram (Hites and Biemann, 1970), the various components could be identified. The arabic numbers in Figure 3 relate to those in Table I. Comparison of Figure 3 with Table I reveals, however, that at the 50 ng level three of the components (acrylonitrile, furfurylacohol and phenylethylamine) did not appear in the gas chromatogram. Only a relatively small amount of benzyl cyanide reached the mass spectrometer and other polar compounds like isopentanol and the two cresols emerged as broad peaks. Particularly the latter two could easily be detected by inspec-

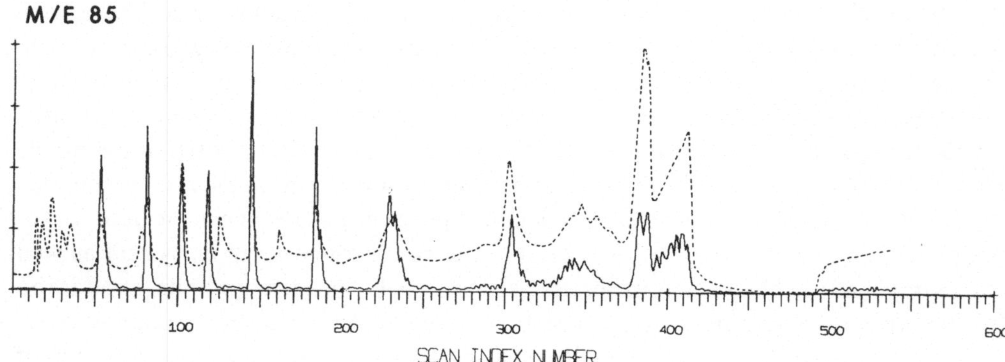

Fig. 4. Mass chromatogram of m/e 85 superimposed on GC-trace of 24-component mixture.

tion of the mass chromatograms of their molecular ions and abundant fragment ions but the same technique did not reveal any of the more polar components (2, 10 and 17 of Table I). Thus, this test revealed that substances of that type are not transmitted through the gas chromatographic column used in these experiments (Chromosorb coated with DEXSIL 300 as the liquid phase and a small amount of a polyester, High-F8, to increase its polarity). These shortcomings revealed by the test experiments are expected to be eliminated by a change of the material to TENAX coated with PMP.

Inspection of the data recorded during the emergence of those portions of the chromatogram labeled A, B, C and D in Figure 3 revealed the presence of four components not previously known to be part of the mixture. Their identification, which is outlined below demonstrates the capability of the Viking-instrument to produce data which can indeed be unambiguously interpreted. The interpretation of continuously recorded mass spectra, particularly after their transformation into mass chromatograms is best illustrated in these examples.

Figure 4 represents a mass chromatogram of m/e 85 which corresponds to $C_6H_{13}^+$, one of the homologous alkyl ions in aliphatic saturated hydrocarbons and other substances having a long saturated alkyl chain. The first few maxima of the solid line (the dotted trace always corresponds to the gas chromatogram that is shown in Figure 3)

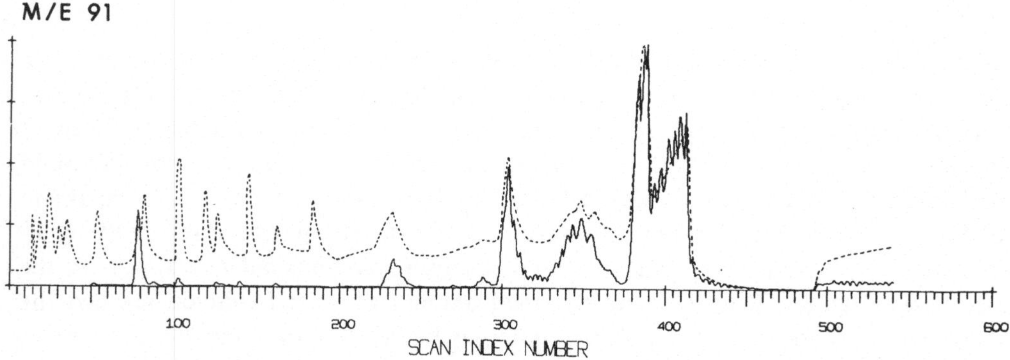

Fig. 5. Mass chromatogram of m/e 91 superimposed on GC-trace of 24-component mixture.

are due to the seven aliphatic hydrocarbons of the test mixture and an inspection of related mass chromatograms of other alkyl ions (m/e 43, 57, 71 and 99, etc.) confirm this identification. The last peak in the gas chromatogram which ranges from scan number 375 to 415 (an appears as a multiplet but only because this large component caused the effluent divider to switch into the 1:20 and 1:400 mode at about scan 385 and 390 respectively) is also paralleled by the m/e 85 trace due to the C_{11} chain in this molecule. The gas chromatographic peaks near scan 302 and around 350 also appear to contain such an aliphatic chain. While none of the components of the test mixture were expected to elute in that area of the gas chromatogram inspection of the mass chromatogram of mass 91 (Figure 5) immediately suggested their identity. The $C_7H_7^+$ ion is, of course, a very abundant one in alkyl benzenes and thus the species where mass 91 maximizes must have such a partial structure. Obviously, phenylundecane does and so does ethylbenzene which is component 8 emerging shortly before nonane. The other maxima of the mass chromatogram of mass 91 which are found around scan 235, 302 and 350 reveal alkyl benzenes not expected in the test mixture. Inspection of mass spectrum number 235 (Figure 6) clearly shows the presence of an aliphatic saturated hydrocarbon as evident from the series of alkyl ions (m/e $43+14\,n$, where $n=0, 1, 2\ldots$) in decreasing intensity as well as of an phenylalkane as deduced from the abundant ions of m/e 91 and 92. The unexpected peak at m/e 204 corresponds to an alkyl benzene with a saturated C_9 substituent.

On the basis of these data one can conclude that phenylnonane elutes almost unresolved from hexadecane and hexadecene, but that the data clearly reveal its presence. While this at first was surprising because phenylnonane had not been added to the test mixture its source is quite obvious. Considering the high sensitivity of the instrument and the fact that phenylundecane has been used as a solvent, i.e., in relatively large quantities, a trace of phenylnonane present in the solvent would of course be

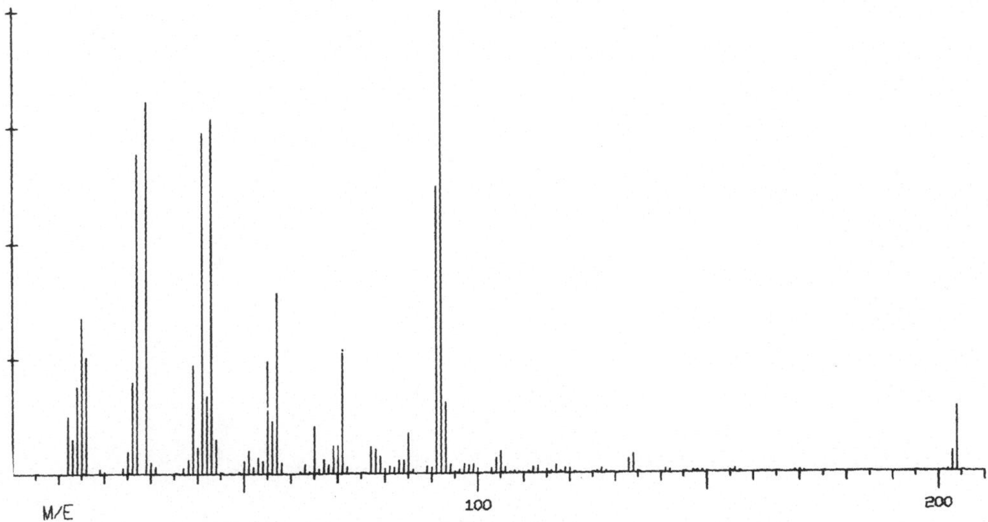

Fig. 6. Mass spectrum corresponding to scan 235 in Figure 3.

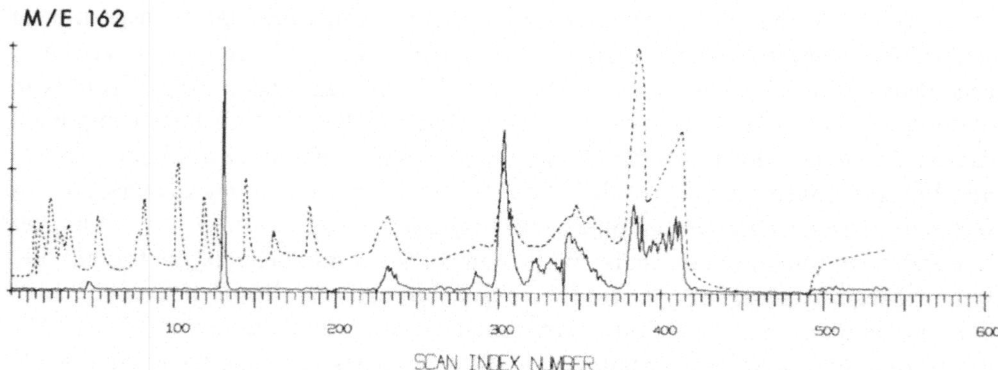

Fig. 7. Mass chromatogram of m/e 162 superimposed on GC-trace of 24-component mixture.

detected. In a similar way the fractions emerging around scan index number 302 and 340 were identified as phenyldecane (component C) and a mixture of isomeric phenylundecanes (component D). Since the latter must have branched alkyl substituents they emerge before the unbranched phenylundecane (solvent, component 24) and are not well resolved thus giving rise to a broader peak.

The utility of mass chromatograms, all of which are generated automatically and available for inspection, is further demonstrated by the detection of another component (A) of the test mixture. Figure 7 represents a mass chromatogram of mass 162 and indicates that a component giving rise to an ion of that mass emerges sharply around scan 130 although this ion is not expected of any of the components of the test mixture (except as the ^{13}C-isotope peak of $C_6H_5{-}C_7H_{13}^+$ of the phenylalkanes). Inspection of the mass spectra recorded in that region of the gas chromatogram (which is where dimethylaniline is eluted) showed indeed the sudden appearance of a peak

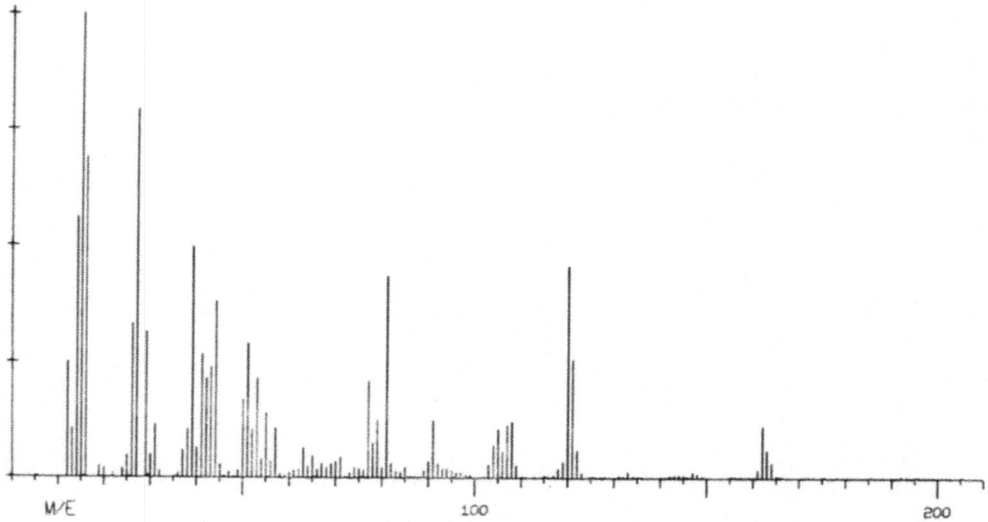

Fig. 8. Mass spectrum corresponding to scan 130 in Figure 3.

at mass 162 (Figure 8) and the only other peak that is not associated with the mass spectrum of dimethyaniline is that at mass 81. Thus, there must be present a trace of a compound which gives rise to abundant ions of mass 162 and mass 81 and both their mass and intensity indicates that it should be a symmetrical aromatic compound with an easily cleaved central bond. Mass 81 is an abundant ion in alkyl furanes and difurylethane is a compound that would be expected to give rise to this mass spectrum. Again it must have been an impurity present in one of the alkyl furanes (most likely methyl furane) of the test mixture. The structures of components A, B, C and D are summarized in Table II.

TABLE II

Additional Components detected in Test Mixture

A. Difurylethane

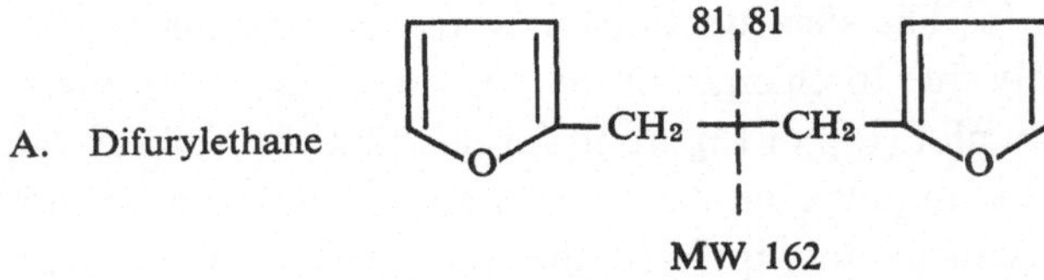

B. Phenylnonane $C_6H_5-(CH_2)_8CH_3$

C. Phenyldecane $C_6H_5-(CH_2)_9CH_3$

D. Branched Phenylundecanes $C_6H_5-C_{11}H_{23}$

These examples indicated that the data generated by the Viking gas chromatograph-mass spectrometer system are such that traces of rather complex organic molecules can be identified without prior knowledge of their presence although the purpose of the test originally was merely to show whether or not the components listed in Table

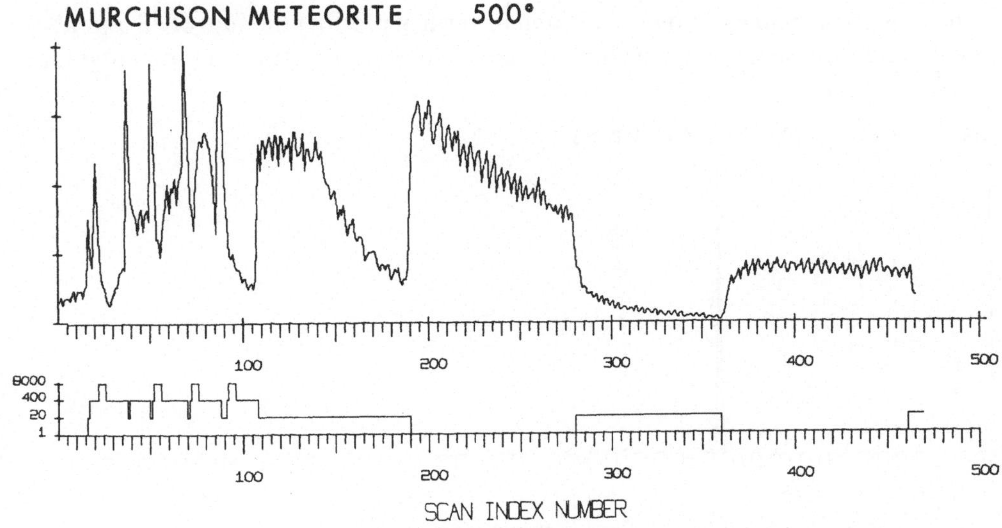

Fig. 9. 'Gas chromatogram' obtained from the effluent of a sample of Murchison meteorite heated to 500 °C for 30 s. (top). Effluent divider status (bottom). For explanation see text.

I are transmitted by the system and that the mass spectrometer produces useful mass spectra. In order to evaluate the performance of the instrument on the next level of resemblance to the actual Martian soil sample, a specimen of the Murchison meteorite (approximately 100 mg) was analyzed by placing it into the sample oven of the instrument and heating it to 500 °C over a period of 30 s in a stream of hydrogen, the carrier gas for the gas chromatograph.

The 'gas chromatogram' obtained on plotting total ion intensity (summation of individual data points) is shown in Figure 9. Although it appears like a well structured gas chromatogram over the first 100 scans the function of the effluent divider represented on the lower trace in Figure 9 should be taken in to account. It is clear that the sharp peaks around scan index numbers 39, 51, 71 and 90 are due to the divider briefly switching into the 1:20 ratio. The step functions at scan index numbers 108, 190, 280, 360 and 461 are similarly due to changes in divider ratio. Factoring these changes into the total ion intensity plot (top of Figure 9) reveals that the bulk of the signal reaches a maximum around scan index number 80 and decays first rapidly and then slowly during the entire experiment (the ripple on the trace is an artifact introduced by the test equipment). Inspection of the mass spectra reveals the major component to be water released from the sample and eluted from the gas chromatograph in this manner. The data processing techniques developed for such situations permit, however, the detection of individual gas chromatographic fractions eluting 'under' such tails. As an example, the mass chromatogram of m/e 128 is shown in Figure 10 (solid line) superimposed on the total ion plot (broken line) corresponding to Figure 9. Clearly, a substance generating an abundant ion of mass 128 emerges as a sharp peak around scan index numbers 140–144, and the mass spectra recorded at this point are clearly that of naphthalene.

The gas chromatographic trace can be enhanced by removing the contribution of water. The broken line in Figure 11 is obtained when plotting the sum of all ions above m/e 45 and now some fine structure appears. The solid line (mass chromatogram of

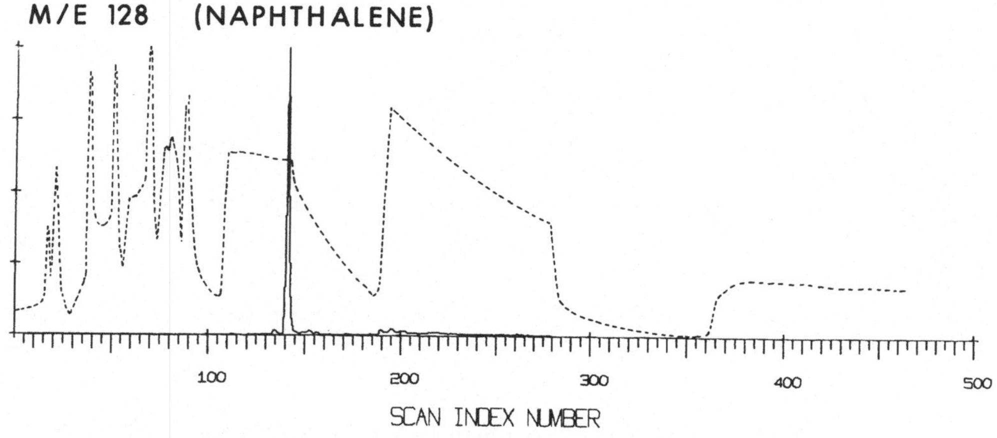

Fig. 10. Mass chromatogram of m/e 128 superimposed on Figure 9.

TABLE III

Viking GCMS	Mol. Mt.	Literature (by benzene extraction and GC; Pering and Ponnamperuma, 1971)
Acetone	58	
Benzene	78	
Thiophene	84	
Toluene	92	
Methylthiophene	98	
C_2-Benzene (2 isom.)	106	
C_2-Thiophene	112	
C_3-Benzene	120	
Indene	118	
	√[a] 128	Naphthalene
	? 132	?
	√ 142	2-Methyl Naphthalene
	√ 142	1-Methyl Naphthalene
	152	? Not Acenaphthylene or Biphenylene
	180	? Not Stilbene
	√ 154	2,6 Dimethyl Naphthalene
	√ 156	Diphenyl
	√ 156	Dimethyl Naphthalene
	√ 156	1,3 Dimethyl Naphthalene
	168	Diphenyl Methane
	156	1,4 and/or 2,3 Dimethyl Naphthalene
	170	C-3 Naphthalene
	√ 154	Acenaphthene
	168	? Not a Methyl Biphenyl
	√ 166	Fluorene
	184	C-4 Naphthalene
	184	C-4 Naphthalene
	√ 178	Phenanthrene
	178	Anthracene
	192	Methyl Phenanthrene
	192	Methyl Phenanthrene
	192	1-Methyl Phenanthrene
	202	Fluoranthene
	202	Pyrene

[a] √ indicates compound listed in left column also detected by Viking GCMS.

m/e 141 due to loss of H from methylnaphthalene and loss of alkyl from higher homologes thereof) represent the elution of the two isomeric methylnaphthalenes around scan index numbers 160 and 164. Mass 156 indicates the emergence of five dimethylnaphthalenes (or, generally speaking, C_2-naphthalenes) at scan index numbers 173, 185, 190, 195 and 201 and inspection of the corresponding mass spectra confirms this conclusion. Using this approach, the components indicated by a check mark in Table III were identified.

Inspection of the mass chromatograms in the region before the emergence of naphthalene revealed the presence of a series of compounds (see Table III) not pre-

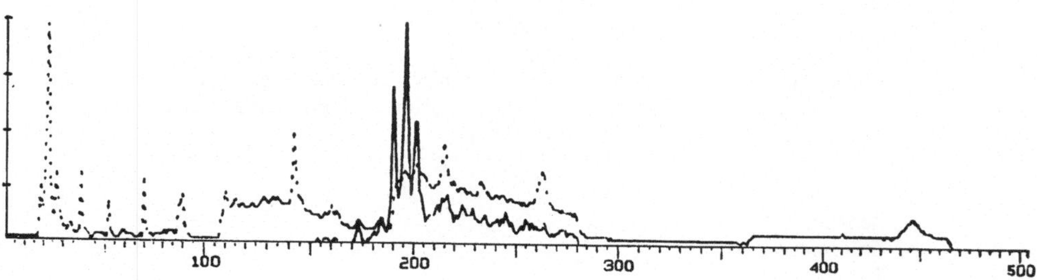

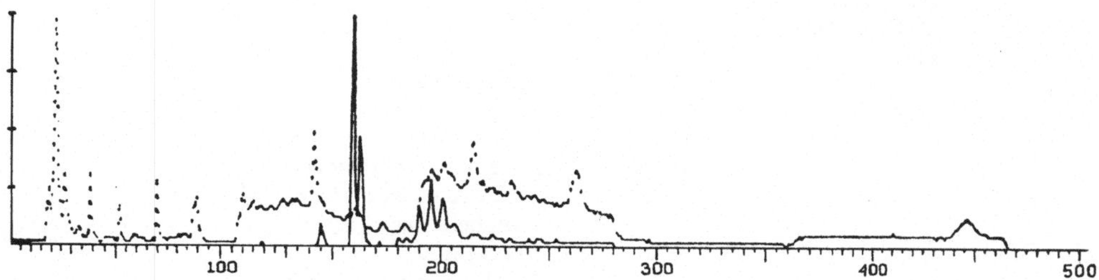

Fig. 11. Mass chromatogram of m/e 141 (bottom) and 156 (top) superimposed on plot of summed ion intensity above m/e 45 (dotted lines) in the Murchison experiment.

viously identified by the wet chemical analysis of the Murchison meteorite (Pering and Ponnamperuma, 1971). Certain ions clearly maximized during the earlier scans and the mass chromatogram of m/e 84 (Figure 12), although obsviously enhanced by the effluent divider, is a typical example. Inspection of the mass spectra emerging in this region reveal the presence of thiophene. Similarly, a methylthiophene is detected at scan index number 61 (Figure 13). By deletion of m/e 16, 17, 18, 28 and 44, before plotting the spectrum, the wide dynamic range of the instrument and of the logarithmic recording is taken advantage of.

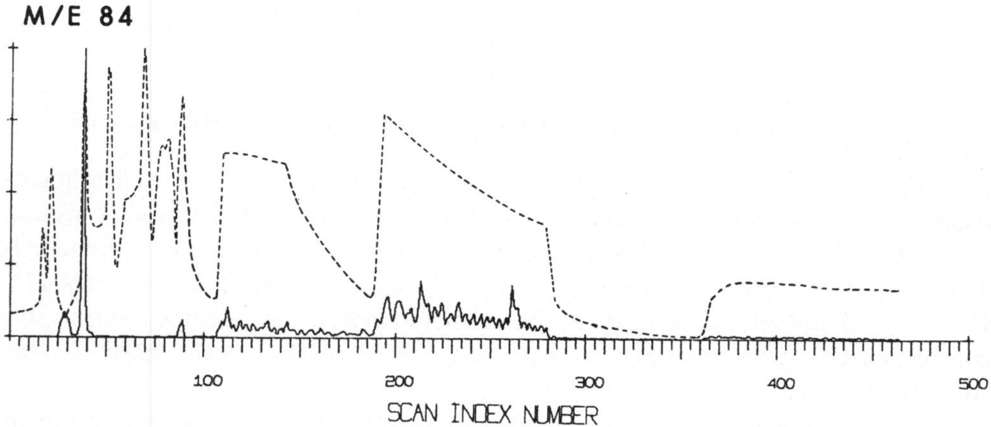

Fig. 12. Mass chromatogram of m/e 84 superimposed on the Figure 9.

As outlined in Table III, benzene and thiophene, along with their lower homologes are released upon heating of the meteorite samples. The good coincidence of the identification of the naphthalenes and higher aromatics by heating as well as extraction suggests that under the conditions of the Viking experiment, no thermal synthesis occurred. The fact that the components more volatile than naphthalene were not

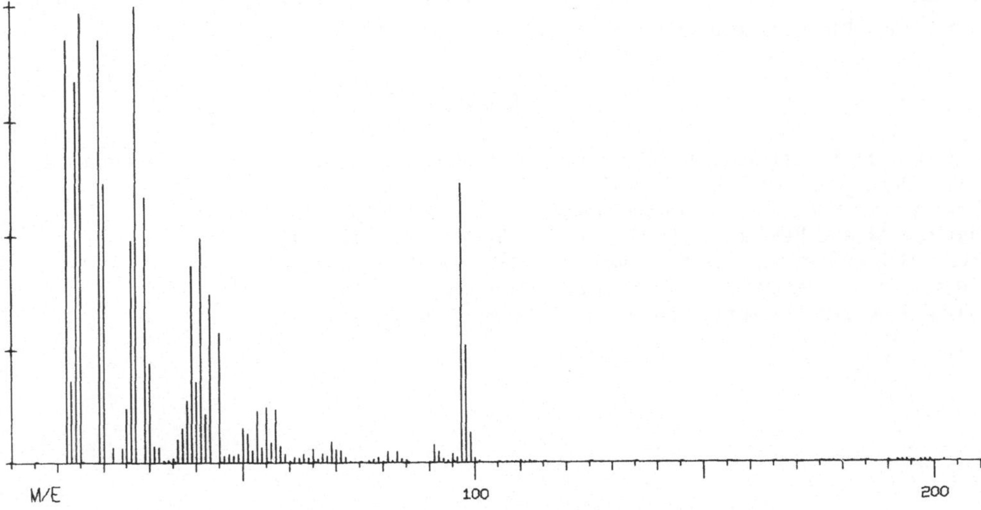

Fig. 13. Mass spectrum corresponding to scan 61 in Figure 9.

detected in the extraction experiment is, of course, due to their loss during solvent removal. Their presence in the Murchison meteorite is not surprising; thiophenes have been found previously in the Murray Meteorite (Hayes and Biemann, 1968).

Both types of test experiments, injection of mixtures and heating of solids, demonstrates that the flight-configured Viking-GC-MS system generates data which make it possible to identify reasonably complex compounds present at the ppm level, or somewhat below, in a 100 mg soil sample. These preliminary tests also revealed a number of shortcomings of the present design, such as the limitations of the gas chromatographic column (DEXSIL 300) with respect to very polar substances, particularly primary and secondary amines and a severely tailing water peak. As pointed out earlier, this is already being remedied. Refinement of the data processing techniques, for example, by incorporation of the effluent divider status (which obviously causes large, abrupt changes in ion intensity) will further facilitate the interpretation of the spectra of minor components emerging simultaneously with major ones. Finally, an extended series of experiments is planned for the very near future, which are designed to evaluate the performance of the instrument in much greater detail (and with the exercise of all the operational modes) than was possible in the tests described here which had to be carried out within the framework of engineering performance tests.

Acknowledgements

The author is indebted to the other members of the Viking Molecular Analysis Team (D. M. Anderson, A. O. C. Nier, L. E. Orgel, J. Oro, T. Owen, G. P. Shulman, P. Toulmin III, and H. C. Urey) Associates of the Team (D. A. Flory, R. A. Hites, A. Lafleur, and P. G. Simmonds) and the personel of the NASA Contractors (Litton/Ind., Beckman Inst. and Perkin-Elmer Corp.), particularly D. Rushneck and A. Fine, for their contributions, and to the Viking Project for support.

References

Anderson, D. M., Biemann, K., Orgel, L. E., Oro, J., Owen, T., Shulman, G. P., Toulmin, III, P., and Urey, H. C.: 1972, *Icarus* **16**, 111.
Biemann, K.: 1971, *Chemical Evolution and the Origin of Life*, p. 541.
Hayes, J. M. and Biemann, K.: 1968, *Geochim. Geophys. Acta* **32**, 239.
Hertz, H. S., Hites, R. A., and Biemann, K.: 1971, *Analyt. Chem.* **43**, 681.
Hites, R. A. and Biemann, K.: 1970, *Analyt. Chem.* **42**, 855.
Pering, K. L. and Ponnamperuma, C.: 1971, *Science* **173**, 237.

AUTOMATED LIFE-DETECTION EXPERIMENTS FOR THE VIKING MISSION TO MARS

HAROLD P. KLEIN

Ames Research Center, NASA, Moffett Field, Calif. 94035, U.S.A.

Abstract. As part of the Viking mission to Mars in 1975, an automated set of instruments is being built to test for the presence of metabolizing organisms on that planet. Three separate modules are combined in this instrument so that samples of the Martian surface can be subjected to a broad array of experimental conditions so as to measure biological activity. The first, the Pyrolytic Release Module, will expose surface samples to a mixture of $C^{14}O$ and $C^{14}O_2$ in the presence of Martian atmosphere and a light source that simulates the Martian visible spectrum. The assay system is designed to determine the extent of assimilation of CO or CO_2 into organic compounds. A small amount of water can be injected into the gas phase during incubation upon command. The Gas Exchange Module will incubate surface samples in a humidified CO_2 atmosphere. At specified times, portions of the incubation atmosphere will be analyzed by gas chromatography to detect the release or uptake of CO_2 and several additional gases. A rich and diversified source of organic nutrients and trace compounds will be available as further additions to the incubating samples. The Label Release Module will incubate surface samples with a dilute aqueous solution of simple radioactive organic substrates in Martian atmosphere, and the gas phase will be monitored continuously for the release of labeled CO_2. Each module, in addition to its gas and nutrient sources, incubation chambers, and detector systems, contains heaters capable of sterilizing surface samples to serve as controls. Since the instrument is designed to operate under Martian conditions and to detect Martian, not terrestrial, organisms, and because the final flight instruments can perform only four assays for each module, formidable problems exist in testing the hardware. The implications of this situation are discussed.

1. Introduction

The planet Mars has been the object of speculation concerning the existence of extraterrestrial life for a long time. During the past hundred years, scientific interest has waxed and waned as new information accrued about the planet. After the flight of Mariner 4 in 1964, and Mariners 6 and 7 in 1969, estimates of the probability of life on Mars were pessimistic (Horowitz, 1971) based in part on the data returned by these spacecraft. However, Mariner 9, which has been orbiting Mars since November 1971, revealed that planet to be much more complex than has been assumed and also considerably more interesting as a target in the search for extraterrestrial life (Mariner Mars, 1971).

In the late summer of 1975, two more U.S. spacecraft are planned for launch to Mars. These spacecraft, carrying identical payloads, constitute the Viking mission. They will also orbit the planet for prolonged periods but, of more interest in connection with understanding how far chemical evolution has proceeded on Mars, they will deliver two landers to the surface, approximately 6 weeks apart. Tentative primary landing sites have already been selected – the first one being at about 20° N and 34° W within a large basinlike area, the second, at 44° N and 10° W within a latitude band where the highest quantities of water are regularly observed in the Martian atmosphere and where liquid ground water may exist for brief periods (Farmer, 1973).

2. General Considerations about Viking Biology Experiments

Many of the scientific experiments on the Viking lander should contribute, either directly or indirectly, to answering questions about the current stage of chemical evolution of Mars (Soffen and Young, 1972). Of these, only the so-called 'direct biology' investigation is described here, the instrumentation for which is based on the concepts and experimental work of a group of biologists* who have been formulating this aspect of the mission for several years. This group early established a number of general principles that have guided the development of the designs for the Viking Biology Instrument (VBI). First, it was agreed that, with the current lack of information about the Martian surface, together with almost total ignorance about local environmental 'niches', and because of the extreme importance of the question being addressed, a *mix* of experiments was imperative, based on different assumptions about the characteristics of Martian organisms. Of the many techniques for 'life detection' under development at that time (Bruch, 1966), those based on metabolic measurements were considered the most desirable for an initial mission. Second, the initial search was to be for microbial soil organisms since, based on terrestrial analogy, these should be most ubiquitous, most adaptive, and most hardy as a group. Third, samples were to be obtained from the upper few centimeters of the soil** where both chemosynthetic and photosynthetic forms might be available. Sampling, furthermore, was to be conducted several times during the lifetime of the landers to take maximal advantage of any seasonal variations in accessibility to Martian organisms. Samples were to be incubated at temperatures no higher, and preferably lower, than the highest known temperatures anywhere on Mars (approximately 35 °C). The incubations were to extend over periods of several days. Finally, it was stipulated that the capability to sterilize soil samples – to serve as controls in the event of positive results – must be incorporated into the experiment.

The deliberations of this group interacting with the mission planners resulted in the final choice of three separate biological experiments (discussed below). Each experiment is based on different assumptions about Martian biota. In addition, they incorporate a number of internal options, thus further extending the flexibility of the combination.

3. VBI Experiments

Samples for the VBI will be acquired several times during each 90-day Viking lander mission through the use of a surface sampler that will also serve other experiments,

* The Viking Biology Team consists of N. Horowitz (Cal. Tech.), H. P. Klein (Ames), J. Lederberg (Stanford Univ.), G. Levin (Biospherics Corp.), V. I. Oyama (Ames), A. Rich (M.I.T.), and W. Vishniac (Univ. of Rochester).

** The term 'soil' throughout this discussion denotes the Martian surface material. The precise nature of the surface material is, of course, unknown at this time.

including an inorganic analysis experiment (Toulmin *et al.*, 1972) designed to yield data on the elemental composition of the soil, and an organic analysis experiment (Anderson *et al.*, 1972). Surface samples will enter the VBI through a sieving device so that particles larger than 2 mm will be excluded. The soil will pass into a soil

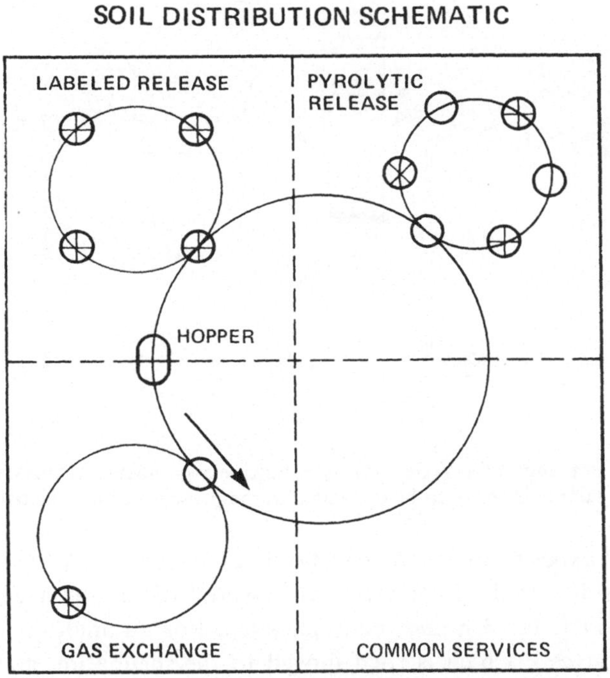

Fig. 1. Schematic drawing of the three Viking biology modules.

distribution assembly within the VBI, which will then automatically deliver measured volumes of soil to each of the three experiments. The latter are to be housed in individual modules, which together with a common services module and an electronics subassembly, comprise the VBI. Figure 1 is a schematic drawing of these modules (see also Figure 7).

4. Pyrolytic Release (PR) Experiment

The PR experiment (Horowitz *et al.*, 1972) is designed to measure either photosynthetic or chemosynthetic fixation of CO_2 or CO. The main rationale for this is that the Martian atmosphere consists primarily of CO_2, with CO as a trace component, and that the Martian biota would include organisms capable of assimilating one or both of these gases. Furthermore, it seems reasonable that a sustained biota on Mars would include photosynthetic organisms. The PR experiment incubates soil in a Martian atmosphere to which $^{14}CO_2$ and ^{14}CO are added and then, by pyrolysis and the use of an organic vapor trap, determines whether ^{14}C has been fixed into organics. The experiment can be conducted in the dark or in light.

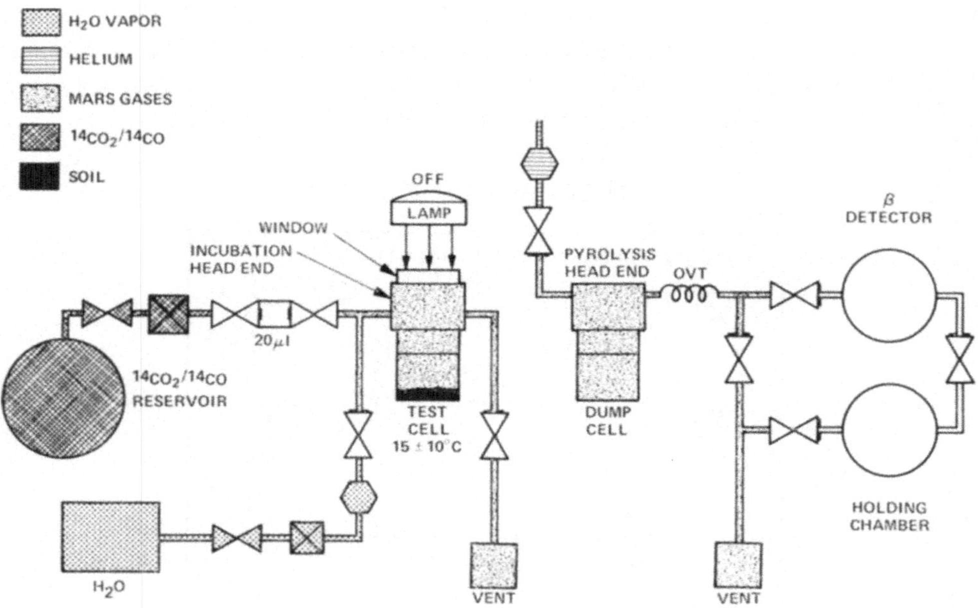

Fig. 2. Schematic drawing of the Viking Pyrolytic Release experiment. At this stage of the sequence, the soil sample is ready to be incubated in the presence of radioactive gasses.

Within the PR experiment are three test cells, at least one of which may be used for more than one soil sample. In the test cell reserved for use as a control, the soil is first heated at 160 °C for 3 h and then incubated. For an analysis, 0.25 cc of soil is loaded into a test cell, which is then moved to the incubation station and sealed. Figure 2 is a diagram of this experiment at this stage. After establishing the incubation temperature of $15° \pm 10\,°C$, one of the PR options may now be exercised. Water vapor can be introduced by ground command or omitted. Then 20 μl of a mixture of $^{14}CO_2$ (95%)–^{14}CO (5%) are provided from a gas reservoir, and a xenon arc lamp (12 V, 6-W maximum power) would normally be turned on. However, the option exists to command this lamp not to be turned on during the ensuing incubation, which lasts for 5 days. After incubation, the test cell is heated to 120 °C to remove the residual incubation gases that are vented to the outside. Background counts are made at about this time in the sequence, after which the test cell is moved from the incubation station to the pyrolysis station. As shown in Figure 3, pyrolysis is accomplished by heating the test cell to 625 °C while purging the test cell with helium. The purged gases pass through the organic vapor trap (OVT), packed with Chromosorb P coated with CuO, into a ^{14}C detector. Since the OVT is designed to retain organic compounds and fragments, the ^{14}C detector at this stage will sense a 'first peak' consisting mainly of unreacted (bound?) $^{14}CO_2/^{14}CO$ (and also some $^{14}CO_2$ from decarboxylation reactions that occur during pyrolysis). This first peak is regarded as nonbiological in origin. After this operation, the test cell is moved away from the pyrolysis station, the detector is heated and purged with helium, and background

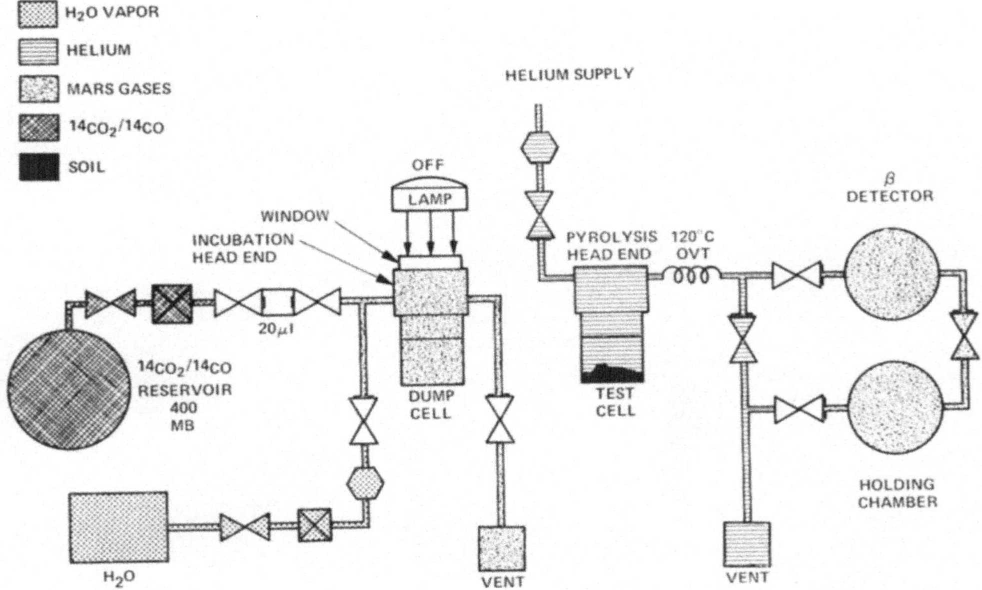

Fig. 3. Pyrolytic Release experiment. At this stage of the analysis, the incubation chamber has been moved to the pyrolysis station.

counts are taken once more to verify that the background radiation is down to pre-pyrolysis levels. The organics are then released from the OVT by heating it to 700 °C, at which time they are simultaneously oxidized to CO_2. As these are flushed into the ^{14}C detector, a second radioactive peak at this point would indicate biological activity.

A series of cleanup operations follows these steps, bringing the PR module to a stage of readiness for the next sample.

5. Gas Exchange (GE) Experiment

This experiment (Oyama, 1972) measures the production or uptake of CO_2, N_2, CH_4, H_2, and O_2 during the course of incubation of a Martian soil sample by means of gas chromatography. The GE experiment can be conducted in one of two modes: in the presence of water vapor, without added nutrients, or in the presence of a complex source of nutrients. The first option is based on the assumption that substrates may not be limiting in the Martian soil (Hubbard *et al.*, 1971) and that biological activity may be stimulated when water becomes available. The second option assumes the presence of significant numbers of anaerobic heterotrophs in the Martian soil.

For the GE experiment, only a single test cell is available, but this can be used sequentially for a number of samples if necessary. The test cell can also be heated to 160 °C for 3 h to serve as a control.

After receiving 1 cc of soil from the distribution assembly, the test cell is moved to the incubation station and sealed. After a helium purge, a mixture of helium, krypton,

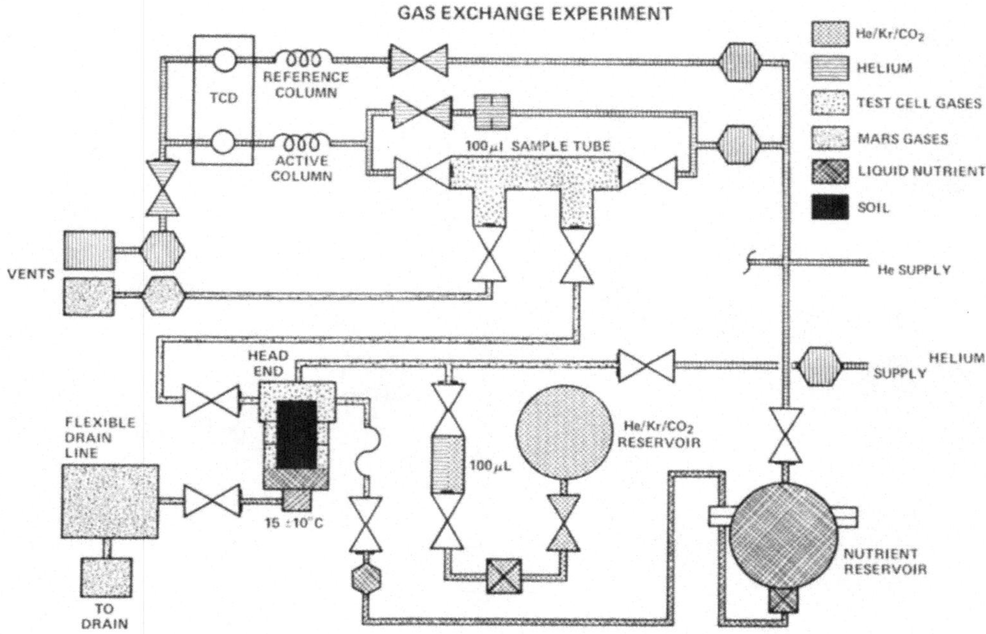

Fig. 4. Schematic drawing of the Viking Gas Exchange experiment incubated in the 'humid' mode.

and CO_2 is introduced and this becomes the initial incubation atmosphere*. At this point, the option exists to introduce either 0.5 or 2.5 cc of the nutrient solution provided for this experiment. Using the lesser quantity, as in Figure 4, the soil does not come into contact with the solution and incubation proceeds in a 'humid' mode. An additional 2.0 cc allows contact between the soil and the nutrients. The latter consists of an aqueous mixture of d- and l-amino acids, vitamins, other organic compounds, and inorganic salts. As currently planned, incubation at 15° $\pm$10°C (in the dark) will initially be in the humid mode for 7 days, after which additional nutrient solution is added by command. For gas analyses, 100 μl of the atmosphere above the soil are removed through the use of a gas sampling tube. This occurs at the beginning of each incubation and after 1, 2, 3, 5, 9, and 15 days. The sampled gas is placed in a stream of helium flowing through the chromatograph column (25 ft long, packed with 100–120 mesh Poropak Q) into a thermal conductivity detector. The system used in the GE experiment separates the gases of interest with a resolution of at least 95% between adjacent peaks and detects changes of the order of 1 nanomole over the range of 1 to 1000 nanomoles, except for hydrogen which is detectable in the range of from 10 to 10000 nanomoles.

After a 15-day incubation cycle, the option exists to add a fresh soil sample to the test cell and begin a new incubation cycle or to drain the medium from the test cell, replacing it with fresh nutrients, and also replacing the original atmosphere with

* Krypton serves here as an internal standard, and helium is used to bring the pressure in the test chamber to approximately 200 mb.

fresh incubation atmosphere. The latter procedure will be used if gas changes are noted in the initial incubation, on the assumption that if these changes are due to biological activity, they should be repeatable and also be enhanced, while if of non-biological origin, they should not reappear.

6. Label Release (LR) Experiment

The LR experiment (Levin, 1972) is designed to test metabolic activity in a soil sample moistened with a dilute aqueous solution of ^{14}C-labeled simple organic compounds. The rationale for this experiment is that Martian organisms should be in equilibrium with the atmospheric CO_2 on Mars and that at least some of them should be able to

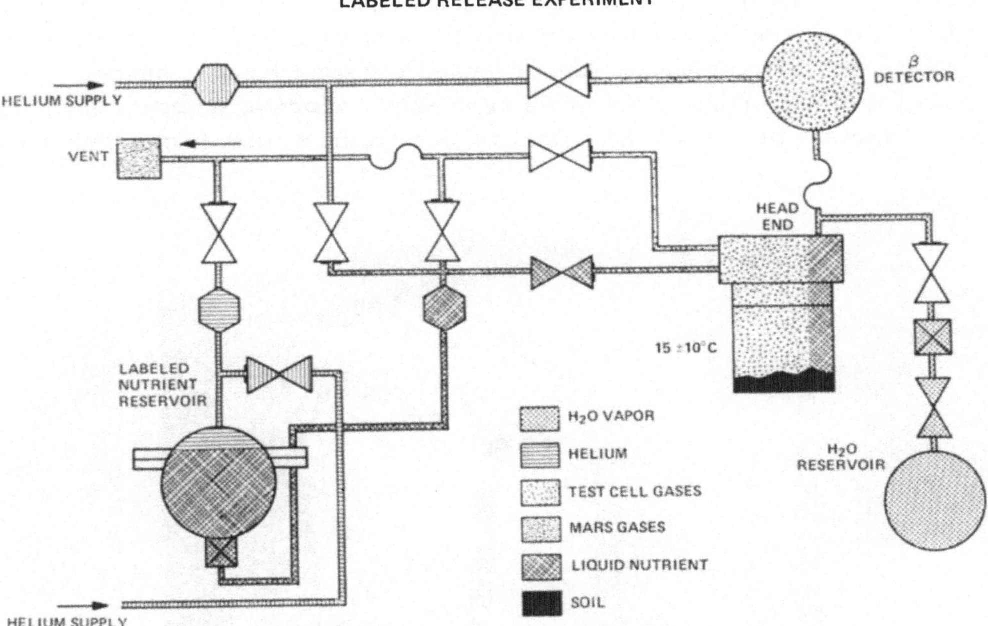

Fig. 5. Schematic drawing of the Viking Label Release experiment.

catabolize organic compounds to CO_2. The experiment then depends on the biological release of radioactive gases from a mixture of simple compounds supplied during incubation.

For this experiment (Figure 5), one of the four test cells provided receives 0.5 cc of soil sample and is moved to the incubation station and sealed, the Martian atmosphere being established in the test cell in this process. Before the labeled nutrients (a mixture of ^{14}C-formate, ^{14}C-glycine, ^{14}C- d- and l-lactate, ^{14}C- d- and l-alanine, and ^{14}C-glycolic acid; all carbons labeled) are added, a background count is taken, and the nutrient solution is degassed by passing helium through the nutrient reservoir. Approximately 0.15 cc of nutrient is then added, and incubation proceeds in the dark at 15° ±10°C for 11 days. The atmosphere above the soil sample is monitored by a ^{14}C detector continuously throughout the incubation, after which the test cell and detectoı are purged with helium. Additional cleanup operations are performed to bring the remaining radioactivity of the detector down to background levels in preparation for the next analysis. As with the other experiments, the test cells can be heated at 160°C for 3 h when a control analysis is to be performed.

7. Testing the VBI

Figures 6 and 7 show the VBI brought to a flightlike configuration. Unlike the instrument shown, however, the actual flight instruments, which will be very similar in appearance, will not be subjected to testing with biological (soil) samples since the final VBI's must be pristine with respect to soil samples. The main reason for this is that each of the final instruments can only test a maximum of four samples, after which some of the consumable supplies or test cell space becomes limiting. Even without these constraints, there would be inherent hazards in subjecting the final flight hardware to terminal heat sterilization procedures after testing with soils

Fig. 6. Viking Biology Instrument (Development unit). Larger box, approximately 1500 cu³, contains the mechanical subsystem including the four modules described in the text. Smaller box contains the electronic subsystem, which includes sequencers and data processing equipment.

Fig. 7. VBI Development Unit with side panels removed. Individual modules can be seen; the inverted U-shaped tube (left center) is the OVT for the PR experiment; the gas chromatograph columns are coiled at top, right.

because unsuspected soil particles might later plug nutrient, vent, or other lines in the VBI or interfere with proper sealing of the test chambers at their incubation stations. Thus, it follows that confidence in the final instruments must rely on systematic prior testing of precursor instruments, like the one shown, together with rigorous control over the manufacture and assembly of each component used in the VBI.

In this regard, several difficulties immediately become apparent in devising appropriate test procedures. The most formidable is that the VBI *is designed to operate on Mars* and not on Earth. Proper operation of the instrument itself – quite apart from the incubation conditions – requires appropriate thermal and atmospheric environments as well as ambient pressures equivalent to those expected on Mars (i.e., about 5 mb). Another major testing problem stems from the fact that positive biological signals cannot be obtained for all three experiments, even with a properly functioning instrument, from valid positive terrestrial soil samples, *because the experiments are designed for Martian*, not terrestrial, *organisms*. Thus, performing the PR experiment on terrestrial soils without water, or even in the presence of the small amount of water vapor that can be commanded in this experiment, would be futile as a test procedure since a body of laboratory evidence exists showing that *terrestrial* organisms require the frank addition of water before significant CO_2 fixation is observed. Similarly, the GE experiment, when conducted in the humid mode, yields virtually no biologically derived gas changes with terrestrial soils – the latter apparently being substrate-limited, rather than water-limited. Only the LR experiment gives reasonable data with terrestrial soil when these are tested under conditions approximating those to be encountered on Mars.

From the above, it is evident that no single soil type, nor any known terrestrial organism, can serve as a standard test object for all three VBI experiments. Consequently, the following procedures have been developed as testing 'standards' to gain confidence in the metamorphosis of the VBI from laboratory 'breadboards', working with terrestrial samples in a terrestrial environment, through several precursor instruments, to the final flight articles. First, a set of test standard gases has been selected for each experiment. For the LR and PR experiments, a known quantity of $^{14}CO_2$ of known radioactivity is used as the standard against which the operation of OVT's and ^{14}C detectors can be tested. When injected into the test chambers, the $^{14}CO_2$, of course, serves also to check for leaks. Development of the GE experiment is guided by the use of several known standard mixtures of the gases of interest. These serve to calibrate gas chromatograph column performance and detector performance.

In addition to the test standard gases, soil standards are also used. For the PR experiment, a standardized sample of 'Chatsworth' soil, which had previously fixed $^{14}CO_2$ under optimal (i.e., wet) conditions, is used to test the complete system beginning with pyrolysis. This 'preincubated' soil is remarkably uniform in its properties and is stable over long periods of time when stored dry and at low temperature. Table I illustrates the reproducibility of data obtained in the PR experiment using this preparation. For the LR and GE experiments, another soil (Aiken soil),

TABLE I

PR experiment: performance data and the reproducibility of 'standard' pre-incubated chatsworth soil

Run	Pyrolysis peak (1st peak P_1)	Organic peak (2nd peak P_2)
1	5.8×10^4	1.77×10^4
2	4.8×10^4	1.43×10^4
3	5.9×10^4	1.71×10^4
4	6.1×10^4	1.70×10^4
5	5.6×10^4	1.69×10^4
6	5.5×10^4	1.69×10^4
Mean values	$5.3 \pm 0.45 \times 10^4$ (8%)	$1.66 \pm 0.1 \times 10^4$ (7%)

containing about 10^6 bacteria per gram, is used to challenge the complete systems. It, too, can be maintained for long periods without any significant change in the biological 'signals' obtained.

By the use of these test standards, it has in fact been possible to relate the results obtained with conventional laboratory equipment to those obtained with the hardware being built for eventual use on Mars. Several versions of the VBI already have been manufactured and tested and more will be assembled and tested during the evolution of the final flight hardware. After all the planned testing has been successfully accomplished, we will have reasonable assurance that the landed instruments will perform satisfactorily on Mars.

References

Anderson, D. M., Biemann, K., Orgel, L. E., Oro, J., Owen, T., Shulman, G. P., Toulmin, P., III, and Urey, H. C.: 1972, *Icarus* **16**, 111.

Bruch, C. W.: 1966, in C. S. Pittendrigh, W. Vishniac and J. P. T. Peterman (eds.), *Biology and the Exploration of Mars*, Publ. no. 1296, National Academy of Sciences, National Research Council, Washington, D.C., p. 487.

Farmer, C. B.: 1973, *Icarus*, in press.

Horowitz, N. H.: 1971, *Bull. Atomic Sci.* **27**, 13.

Horowitz, N. H., Hubbard, J. S., and Hobby, G. L.: 1972, *Icarus* **16**, 147.

Hubbard, J. S., Hardy, J. P., and Horowitz, N. H.: 1971, *Proc. Nat. Acad. Sci. U.S.* **68**, no. 3, 574.

Levin, G. V.: 1972, *Icarus* **16**, 153.

Mariner Mars: 1971, Project Science Report, National Space Science Data Center, NASA Goddard Spaceflight Center, Greenbelt, MD.

Oyama, V. I.: 1972, *Icarus* **16**, 167.

Soffen, G. A. and Young, A. T.: 1972, *Icarus* **16**, 1.

Toulmin, P., Baird, A. K., Clark, B. C., Keil, K., and Rose, H. J.: 1973, *Icarus* **20**, 153–178.

ORGANIC CONTAMINATION PROBLEMS IN THE VIKING MOLECULAR ANALYSIS EXPERIMENT

DONALD A. FLORY and JOHN ORÓ

Dept. of Biophysical Sciences, University of Houston, Houston, Tex. 77004, U.S.A.

and

PAUL V. FENNESSEY

Martin Marietta Corporation, Denver, Colo. 80201, U.S.A.

Abstract. The combination of analytical instrumentation selected for the molecular analysis experiment can carry out a survey of the organic compounds present on Mars regardless of their origin. The high sensitivity of this analysis, the limited number of samples which can be analyzed, the close proximity to the landed spacecraft on the surface of Mars which is accessible to the sampling device, the implications of the positive detection of indigenous organic matter in the Martian soil, and our previous experience with meteorites and lunar samples point to the need for a carefully designed program to maintain the integrity of the analyzed Martian surface samples.

A principal problem in interpreting the results of an organic analysis of an extraterrestrial sample is that of distinguishing contaminating material from indigenous material when unknown types and amounts of contaminants make their way into the sample being analyzed. An approach for control of sample integrity in the Viking molecular analysis experiment has been devised which we believe will eliminate such problems. Basically this involves (1) placing an upper limit on the amount of terrestrial contamination that can be tolerated and still allow scientifically meaningful analyses, (2) identifying the potential sources of contamination and analyzing their relative significance, (3) establishing methods to control these sources, and (4) obtaining complete information on the chemical composition of potential contaminants. Our previous experience in the Apollo mission has been of great value in developing the Viking program, perhaps the most important carryover being the recognition of the importance of establishing a comprehensive contamination control program in the early stages of mission planning and hardware design.

The upper limit of total allowable organic contamination has been established as 1 μg g^{-1}. The principal source types, or modes, which contribute to the contamination load have been identified, each requiring a different approach to control. Spacecraft outgassing is controlled by materials selection to minimize outgassing and hermetic sealing whenever possible. Particulate fallout is controlled by selection of materials, particulate seals, cleaning of the spacecraft exterior, and clean room handling. The cleanliness of the direct sample path is controlled by severe materials limitations, ultracleaning, and pressurized sealing of the assembled hardware.

Analysis of the relative probabilities of the sources contributing to the allowable contamination and consideration of the practical aspects of achieving a desired level of control for a particular source has resulted in an allocation 'tree' whereby fractions of the total allowable contamination are distributed to the various individual sources. These efforts have pointed out the need for more information concerning some of these sources and have actually dictated certain design changes in the spacecraft. Additional information was obtained experimentally on descent engine exhaust characteristics which led to the use of an organically cleaner fuel. In summary, the early recognition in the Viking mission of the importance of organic contamination control has allowed the evolution of a complete contamination control program encompassing spacecraft design, mission operations, flight operations, and the design of the science instrumentation for the molecular analysis experiment.

1. Introduction

The molecular analysis experiment has been designed to contribute to the central theme of the Viking Mission, the search for life on Mars. This experiment has been planned to obtain valuable information about the composition of the Martian surface and atmosphere around a landed spacecraft. The combination of analytical instrumentation selected for the experiment can carry out a survey of the organic compounds present on Mars regardless of their nature. The instrumentation has been described elsewhere in detail (Anderson *et al.*, 1972) and in general involves volatilization and/or pyrolysis of the organic material, fractionation of the volatile and/or pyrolosate compounds by gas chromatography, and identification of the chromatographic effluent by mass spectrometry. The projected sensitivity of the gas chromatograph-mass spectrometer system (GC-MS) is such that identification of 1–10 ng (10^{-9} g–10^{-8} g) amounts of many organic compounds will be possible. The limitations on size, weight and complexity inherent to spacecraft instrumentation have resulted in a volatilization-pyrolysis oven assembly capable of accepting three separate 60–100 mg suıface samples for organic analysis or a total of six samples in the two Viking Landers.

The surface samples to be analyzed for organic content will be collected by a sampling scoop mounted on the end of a furlable boom. This sampling scoop and boom has been designated as the Surface Sample Acquisition Assembly. The surface

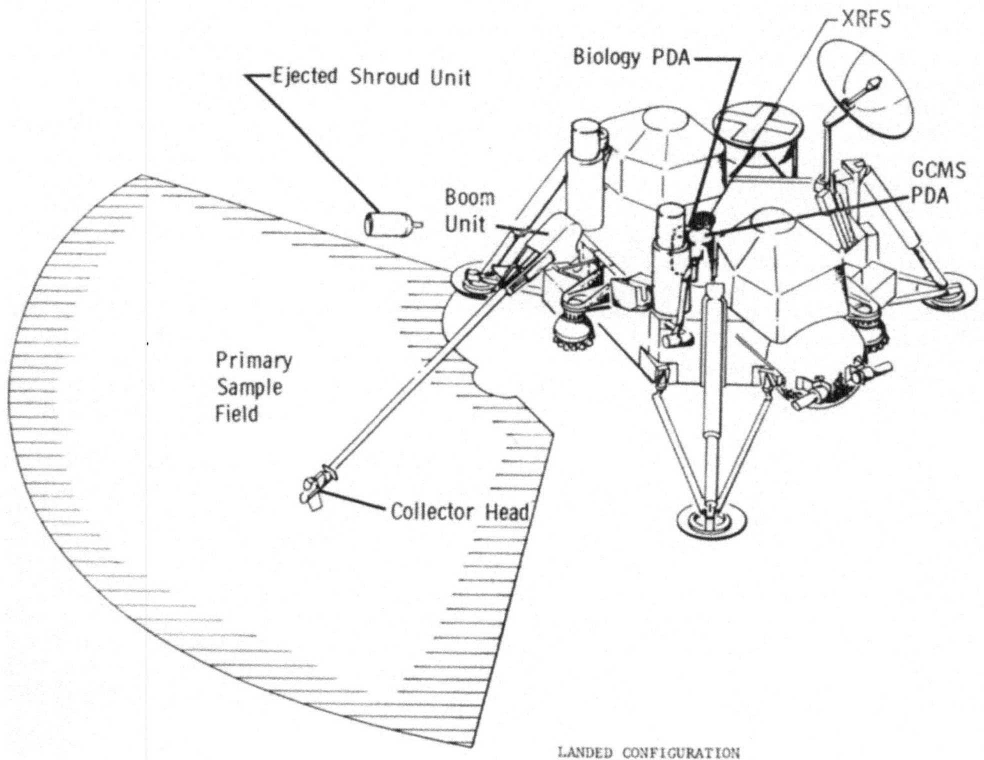

Fig. 1. The Viking Lander sampling area is located on one side of the spacecraft and is accessible by a 3⅓ m long boom located on the top edge of the lander.

sample acquisition assembly delivers the sample and distributes it to the ovens for subsequent volatilization and/or pyrolysis. Figure 1 shows the surface area accessible to the collector which amounts to approximately 10 m^2 ranging from 1–3 m distance from the spacecraft over an arc of about 120°. The high sensitivity of the analysis, the limited number of samples which can be analyzed, the close proximity of the sampling area to the landed spacecraft, and the implications of the positive identification of indigenous organic matter in the Martian soil all point to the need for a carefully designed program to maintain the integrity of the samples. The very serious problems which have been encountered in work with meteorites (Nooner and Oró, 1967; Hayes, 1967; Gelpi and Oró, 1970) and lunar samples (Flory and Simoneit, 1972) serve to make us further aware of the importance of positive organic contamination control as a major factor of the molecular analysis experiment.

A principal problem in the interpretation of the results of an organic analysis of an extraterrestrial sample is obtaining the data to distinguish contaminating material from indigenous material. In addition to having a basic experience in the interpretation of terrestrial mass spectra, this discrimination involves (1) placing an upper limit on the amount of terrestrial contamination that can be tolerated and still obtain scientifically meaningful analyses, (2) identifying the potential sources of contamination and analyzing their relative significance, (3) establishing methods to control these sources, and (4) obtaining as much information as possible on the chemical composition of potential contaminants. This paper will discuss some elements and corresponding rationale of the control of organic material on the Viking lander, and summarize the present status regarding information obtained on potential contaminants.

2. Maximum Allowable Terrestrial Contamination

The molecular analysis experiment should be able to analize samples of the Martian surface with total organic material contents in the 0.1–2500 ppm (by weight) range. The range 0.1–2500 ppm encompases the expected total organic content for an organic poor soil such as a lunar type and an organic rich soil such as a carbonaceous chondrite or terrestrial soil. A sensitivity requirement has been set for minor compounds present at concentrations of 10^{-8} g g^{-1} sample in the organic poor soil.

The maximum allowable terrestrial contamination level which will not interfere with the GC-MS detection of organic compounds in concentrations of 10^{-8} g g^{-1} depends on the nature of the contamination. A single organic compound present as a contaminant at a concentration of 10^{-6} g g^{-1} is acceptable since its presence should not confuse the interpretation of the data. This, of course, assumes one has sufficient background information to recognize the compounds as a contaminant. Ten or one hundred organic compounds present at concentrations of 10^{-7} and 10^{-8}g g^{-1} respectively, and totalling 10^{-6}g g^{-1} is an unacceptable situation since the compounds would be spread over a large portion of the gas chromatogram and the intensities of their spectra nearly equal to those expected from an organic poor sample. A similar argument shows that total contamination levels higher than 10^{-6}g g^{-1} are not acceptable because of their potential interference with organically richer samples. Contamination levels lower than 10^{-6}g g^{-1} could not be justified unless instrument sensitivity was much better than 10^{-9}g g^{-1}. The sensitivity of the prototype GC-MS has been shown to be about 10^{-9}g g^{-1} for the 0.1 g sample size expected in the molecular analysis experiment. Based on these considerations, the total allowable terrestrial contamination has been set at 10^{-6}g g^{-1} (1 ppm) for each sample.

3. Potential Contamination Sources and Their Relative Significance

There are four principal source types or modes from which organic material may find its way into the molecular analysis experiment. These include (1) those surfaces with which the sample comes into direct contact during transfer from the Martian surface to the analysis ovens; (2) fallout of particulate matter or other material from the spacecraft onto the Mars surface; (3) condensation of volatile material due to outgassing of organic materials used in the spacecraft; and (4) other sources such as the retrorockets (site alteration), parachute, and aeroshell (the ablative cover used for protection during atmospheric entry). The relative significance of these four types of sources has been determined by constructing an organic contamination model which identifies all individual sources in each of the four categories that can potentially contribute to the total allowable terrestrial contamination. Analyses have been carried out utilizing known parameters of the Martian environment and spacecraft characteristics, whenever possible, to demonstrate how these sources singly or in combination may contribute to the overall sample contamination. When no discrete parametric data has been available probabalistic operations have been performed on individual or combinations of sources. These analyses are continually updated as new

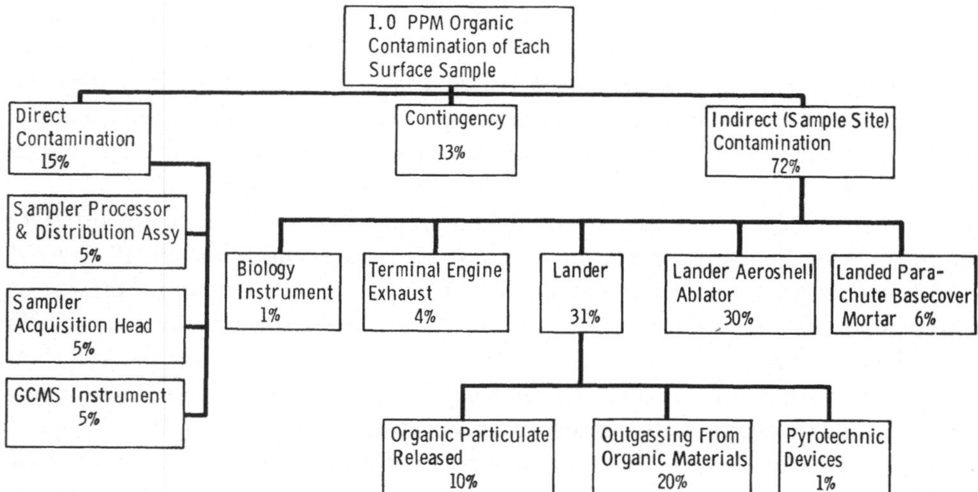

Fig. 2. The Viking Organic Contamination Tree shows the major potential sources of contamination and the percent of the 1×10^{-6} g g^{-1} total allocated to each. This type of presentation is used to guide the analytical effort.

and better data become available. These analyses have resulted in an allocation tree which is shown in Figure 2. Each source has been allocated a fraction of the total allowable contamination and the supporting analyses demonstrate that the combined sources do not contribute organic contaminants in excess of the overall requirement when each source meets its individual allocation.

4. Approach to Control and Present Status

A general approach to control of organic contamination has been imposed on the entire Viking Lander System in order to meet the allocations described previously. The basic concepts of this approach are as follows:

(a) Design and Material Selection – Constraints on the design of each component and subsystem and the specific application of each material relative to the sources identified in the model shall be developed. These constraints shall consider such factors as the location within the lander and the environment to which the material will be subjected from manufacture through operation. Constraints on material selection, expressed in terms of material type, properties and amounts for each potential contaminating source, shall be identified.

(b) Fabrication and Processes – Constraints on fabrication techniques and processes, such as lubrication, cleaning methods or surface preparation shall be identified as to type and where applied.

(c) Testing – Component, subsystem and system level tests required to verify compliance of each source with its allocation shall be identified. Where these requirements constrain an engineering test at any level, such constraints shall be identified and justified.

(d) Operation – Constraints on the operation of any component or subsystem required to provide compliance shall be described. This shall include potential constraints, if so indicated, such as the sequence of venting of biology experiment wastes in relation to collection of samples for the molecular analysis experiment. Additionally, each of the four principal types of contamination mentioned previously requires variations to this general approach to control. These variations are briefly described in the following paragraphs and are in large measure derived from previous experience in the Apollo program (Simoneit *et al.*, 1973 and Reynolds *et al.*, 1973).

4.1. Direct Sample Path

During the early design phases, trade-off studies were conducted which led to a configuration of the direct soil path (those surfaces which come into direct contact with the analytical samples) which affords maximum protection from possible contaminating matter. The present design includes a sealed and pressurized system which is maintained in this condition from final cleaning until it is opened after landing on the surface of Mars, a period of approximately 2 yr. Figures 3 and 4 show the prototype for the direct soil path in its sealed condition immediately after final cleaning and pressurization with CO_2. One additional design factor not immediately apparent

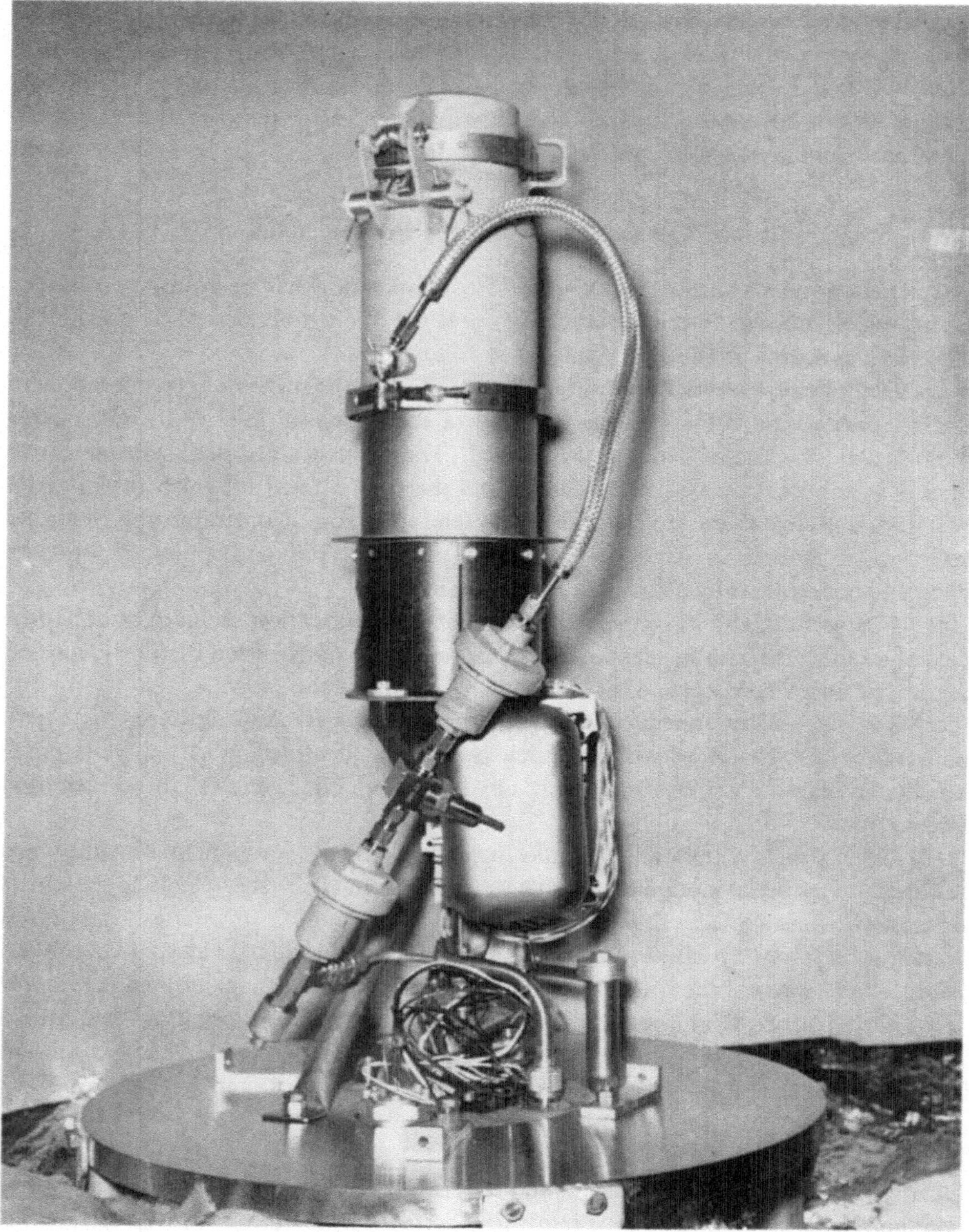

Fig. 3. This photograph shows the Processor and Distribution Assembly mounted to the Loader and Pyrolyzer Assembly after cleaning and pressurization with carbon dioxide. A pressurization unit (tube seen on side of the assembly) will remain attached until final encapsulation of the spacecraft on the launch pad.

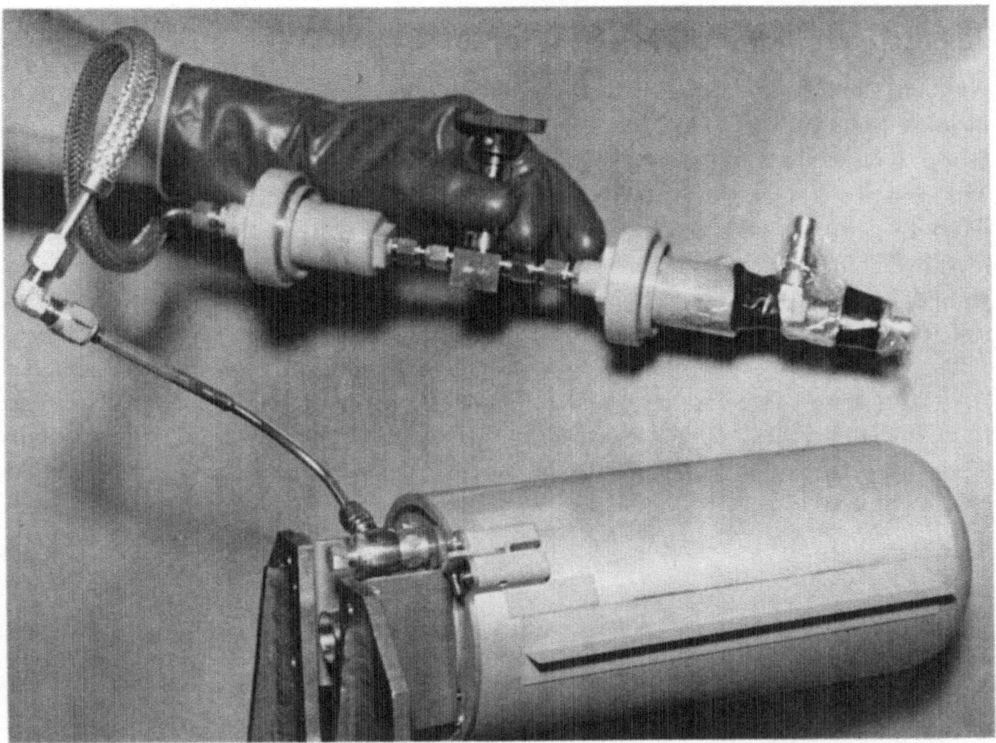

Fig. 4. A photograph of the surface samples collector head which will be attached to the end of the sampling boom in its final encapsulated state after cleaning and pressurization with carbon dioxide.

upon inspection of the sample path hardware are the small quantities of organic material used in all of this hardware. There is only one compound in direct contact with the surface sample and this is teflon used in an O-ring in the scoop. Further, there are a very few other compounds within the pressurized area and these are all very stable compounds i.e. silicon, mylar, teflon, vespel.

The cleaning of the direct soil path is also considered of prime importance in the control of possible contamination and great emphasis has been placed on the procedures used for this purpose. The cleaning methods used for Viking have been tested on prototype hardware and verified by passing 'blank' soil over the cleaned surface and pyrolyzing it at 500 °C to prove that the soil path is clean. The method used involves a series of precleaning steps of the detailed parts to remove traces of organics used during fabrication (cutting oil, marking inks, etc.), a final cleaning and rinse using extremely pure freon ($<1\times10^{-10}$g g^{-1} of total impurities). This is followed by a hot helium purge at reduced pressure to remove all traces of residual solvents (125 °C for 96 hr at 10 mm of Hg). A final set of data will be obtained by sampling the gas in the pressurized unit periodically during the storage phase between cleaning and launch.

4.2. Fallout of particulate matter

Any particulate matter such as dirt, lint, paint, thermal control coating, peeled decal or loose parts such as a chip or wire cutting, shaved insulation, or shred of parachute fabric may be shaken free upon landing, or be blown by wind or wind blown sand and come to rest on the surface of Mars in an area over which the soil sampler is likely to gather a sample. The approach to control requirements here is twofold. It involves, first, control of all such materials to limit their organic chemical content, and second, extreme care in cleanliness control in manufacture, assembly, inspection, and test to ensure minimizing the amount of such potentially loose particulate matter. These requirements are satisfied by cleaning the outer surfaces of the assembled lander spacecraft (see Figure 5) and subsequently maintaining the lander in a clean room meeting class 100000 specifications (Federal Specification 209). Particulate levels are periodically monitored on the outer surfaces and recleaning will be carried out as required.

Fig. 5. The complete Viking lander showing the complex mixture of materials necessary for the construction of such a spacecraft.

4.3. CONDENSATION OF VOLATILE MATERIAL

This mode of contamination derives from the environmental conditions on Mars where the ground surface may be very cold compared to the landed vehicle. Any outgassing of organic chemicals from any part of the Lander may recondense on the cold ground surface to be picked up by the soil sampler scoop as it gathers and delivers a sample of Mars soil to the instrument.

This condensation may involve depolymerized monomers of plastics used in the Lander such as wire insulation or potting compounds, residues of solvents or plasticizer entrapped with in these plastics, residues of solvents used in cleaning, volatiles evaporated from lubricants or greases, any fuel dripping from the propulsion system valves, etc.; in short, any volatile chemical from the Lander which may accumulate as a condensate on the soil over an area where the sampler is likely to probe. The approach to control requirements here involves a quantitative analysis of the evaporation and condensation characteristics of all of the likely materials in a conservatively assumed set of Martian environmental conditions, and the establishment of upper limits of outgassing rates and total amounts of such materials which would assure that the allocation limit for this type of source is not violated.

A very careful material screening process was adopted in the early phases of the Viking program. This process required an organic signature test on all organic material intended for any application in the program. This screening process has helped immensely in the final design; for many materials which could have outgassed in large quantities or decomposed in other ways were not allowed to even be con-

TABLE I

Materials inside direct sample path

Material	Amount	Use	Comments
Kapton H	3 g	Level detector (microswitch) and soil grinder (motor)	Behind dynamic seals
Kapton	50 g	All wire used inside unit	
Teflon	36 g	Dynamic seals, washers grommets spacer, etc.	Makes direct contact with soils
Silicon	15 g	O-ring on collector head cover	
Kel-F	10 g	Locking for screw threads	Only used in a few screws
Vespel	7 g	Insulation plate on end of grinder motor	
Diallylphthalate	3 g	Connector pin isolation	
Kapton polyimide	45 g	Terminal insulation and wire bundle control	Heat shrinkable tubing

TABLE II

Material on lander in large quantities

Material	Amount	Use	Comments
Teflon sheet with Kapton cover film	15 kg	Insulation of electrical lines	Additional covering for sand and dust protection
Epoxy laminate	10 kg	Conformal coating	Almost all is inside of black boxes
Solithane 113	4 kg	Conformal coating	Almost all is inside of black boxes
Thermal paint	6.5 kg	Thermal control paint	Covers exterior of lander
Resin (phenolic) and cork	26 kg	Entry ablator	Lands remote from lander
Dacron and Nomex cloth with fiber *Bx* lines	47 kg	Parachute	Lands remote from both lander and ablator
Phenolic fiberglass	6.3 kg	Basecover	Remainder attached to parachute

sidered for use in the program. Material rejection was not possible in all material applications, but in the few cases where materials with high out gassing rates were used the application was brought into the open and identified as a special concern to the contamination control group.

This orderly process of identification of materials and the cataloging of their application has been a primary tool in understanding and dealing with the overall contamination control program. The data shown in Tables I and II show this type of presentation. The material is identified by name (chemical or trade) and data is presented showing where it is used, how much is used and other pertinent information. This type of matrix is not only important for keeping a log of material and its application, but will also be useful during mission operation when one is interpreting the data from Mars and wants to make sure that a certain mass spectrum could not be an impurity.

One additional safeguard for the interpretation of the data returned from Mars has been the storage of all compounds which could possibly reach the surface sample. Although this is considered a remote possibility there is always a chance for a surprise on the surface of Mars, and in 1976 or 1977 it may be difficult to find an authenic sample of a material on the spacecraft. For this reason materials samples are being collected now, in 1973, and will be stored in sealed containers until the scientific data has been returned and interpreted.

4.4. Other sources

Additional sources of particulate and organic contamination are present in the Viking Lander Capsule. They are the aeroshell ablator, basecover, parachute and mortar, terminal engine exhaust, and pyrotechnic and explosive devices. These potential sources can add contamination to the Lander or the landing site and, therefore, to the surface sample. However, due to the nature of their operation with respect to performance, the amount of organic material they contain is not directly controlled. Contamination control for these sources is either to isolate or eliminate the source or to provide an analysis which shows the extent of the contamination under prescribed sets of conditions consistent with their allocation limit.

The ablator used in the Viking mission is composed of organic compounds which could chip off during the assembly of the Viking Lander Capsule, outgas during sterilization, decompose and redeposit during entry, or contaminate the surface around the Lander on Mars. The first three of these problem areas occur while the Lander and aeroshell are attached, the latter after the aeroshell impacts on the Martian surface (The aeroshell is the outer covering and ablator which protects the lander during the first stage of atmospheric entry.).

The control of ablator particulate material during the assembly of the Lander Capsule is twofold; (a) the fabrication of the ablator shall be carried out in an area separate from the Lander; (b) the ablator shall be covered such that chips or flakes will be contained during assembly and test. The outgassing from this large quantity of organic material will be controlled during the sterilization phase by sweeping a clean, inert and inorganic gas over the ablator surface and away from the internal areas of the Lander Capsule. This does not eliminate the contamination but does substantially reduce the likelihood of the organic compounds reaching those surfaces that will come into direct contact with the analysis samples or remain as organic residue on Lander surfaces.

During the entry phase of the mission, aeroshell ablator products with a wide range in volatility will flow into the wake around the Lander Capsule. Control of this contamination is derived from a Viking Lander Capsule design that eliminates the flow of ablative material onto the Lander. This involves both a carefully designed union between the basecover and the aeroshell, an appropriate venting system and an operational sequence which considers contamination.

The aeroshell will impact on the Martian surface and act as a source of particulate contamination. The location of the aeroshell on Mars with respect to the Lander and the transport of particulates to the sampler site depends on winds and is therefore a matter of probabilities. The probability of contamination of the surface sampling area, to a level above its allocation, by the landed ablator organics has been shown by analysis to be less than three in one thousand.

The basecover, parachute and its noncontained mortar are potential contamination sources, which are less in magnitude but identical in nature to the aeroshell ablator, during both the entry phase of the mission and after they land on the Martian surface.

The probability of contamination of the surface sampling area to a level above its allocation by these sources after they land has also been shown by analysis to be less than three in one thousand.

A further aspect of the control of organic contamination introduced into the landing site surface involves assurance that the terminal engine fuel and its exhaust byproducts do not produce more than a specified amount of terrestrial organic contaminant in the surface sample. This source of organic contamination has been examined both by analysis and special testing and special requirements or special mission strategy evolved to deal with it. For example, the monopropellent fuel, hydrazine was checked for organic content and found to contain 0.1–0.3% concentrations of aniline. Analysis of test soils exposed to the exhaust plume of a full scale lander engine during test firings in a simulated Martian atmosphere indicated deposition of several ppm of aniline and HCN in what would be the sampling area on Mars. Following these tests a process was developed to purify the hydrazine. This purification procedure reduced the aniline content by 2–3 orders of magnitude. The purified fuel was subsequently tested in a similar engine firing test. Test soils exposed in this manner contained no detectable aniline or HCN (less than 0.1 ppm) when pyrolyzed at 500 °C.

Pyrotechnic and explosive devices are an additional source of contamination. The approach to control here has been to minimize the use of such devices, and when such devices must be used to minimize the proportion of organic material consistent with pyrotechnic propellants that are stable. Where there is a possibility of the combustion products (solid or condensible organic) reaching the sampling area, the device shall be contained to eliminate that source. Containment of these devices is being verified by actual tests.

5. Conclusions

In an endeavor as complex as the definition and manufacture of a Lander to explore an extraterrestrial surface for organics (as well as conduct other equally complex experiments), the control of organic contamination had to begin at the earliest possible stages of the mission. This early start may be the most significant factor in the apparent success of the control of organic contamination, and has resulted in the incorporation of a very aggressive and meaningful program for the control of organic material on the Viking lander. This program has resulted in an optimization of spacecraft design, mission operations, flight operations, and science instrumentation design which greatly facilitates reliable contamination control. Although the approaches to control are extensive in some cases, they provide a large amount of confidence that a very complex and expensive molecular analysis experiment will provide the scientific community with basic information concerning the organic composition of the surface of an extraterrestrial planet that is essentially free of the ambiguities arising from uncontrolled contamination.

Acknowledgements

This work was supported by NASA contract NASI-9685 to the University of Houston and the spacecraft fabrication contract NAS1-9000 to the Martin Marietta Corporation. The authors wish to thank the many employees of the NASA centers and the spacecraft contractors who have worked to establish and maintain the contamination control program described.

References

Anderson, D. M., Biemann, K., Orgel, L. E., Oró, J., Owen, T., Schulman, G. P., Toulmin, P., and Urey, H. C.: 1972, *Icarus* **16**, 111.

Flory, D. A. and Simoneit, B. R.: 1972, *Space Life Sci.* **3**, 457.

Gelpi, E. and Oró, J.: 1970, *Geochim. Cosmochim. Acta* **34**, 981.

Hayes, J. M.: 1967, *Geochim. Cosmochim. Acta* **31**, 1395.

Nooner, D. W. and Oró, J.: 1967, *Geochim. Cosmochim. Acta* **31**, 1359.

Reynolds, M. A., Turner, N. L., Hurgeton, J. C., Barbee, M. F., Flory, D. A., and Simoneit, B. R.: 1973, 'Environmental Control of Lunar Samples in the Lunar Receiving Laboratory', ASTM Special Publication (in press).

Simoneit, B. R., Flory, D. A., and Reynolds, M. A.: 1973, 'Organic Contamination Monitoring and Control in the Lunar Receiving Laboratory', ASTM Special Publication (in press).

MODEL SYSTEMS FOR LIFE PROCESSES ON MARS

M. A. MITZ

National Aeronautics and Space Administration, Washington, D.C., U.S.A.

Abstract. In the evolution of life forms non-photosynthetic mechanisms have developed. The question remains whether a total life system could evolve which is not dependent upon photosynthesis. In trying to visualize life on other planets, the photosynthetic process has problems. On Mars, the high intensity of light at the surface is a concern and alternative mechanisms need to be defined and analyzed. In the UV search for alternate mechanisms, several different areas may be identified. These involve activated inorganic compounds in the atmosphere, such as the products of photodissociation of carbon dioxide and the organic material which may be created by natural phenomena. In addition, a life system based on the pressure of the atmospheric constituents, such as carbon dioxide, is a possibility. These considerations may be important for the understanding of evolutionary processes of life on another planet. Model systems which depend on these alternative mechanisms are defined and related to our presently planned and future planetary missions.

Present knowledge of the environment and the chemistry of Mars (Mars Scientific Model, 1972) would lead one to suspect that life, if it exists on the planet, may have taken another path than the life processes on Earth. To identify these alternatives is a challenge but as any one who is trying to define a life detection experiment knows, it is necessary to study all possible alternatives in addressing this problem. The rationale for a search for terrestrial-like organisms was reviewed by Ponnamperuma and Klein (1970). This paper identified several possible alternatives to photosynthesis for a basis of life on Mars and discusses the impact on planning for future missions.

The Martian atmosphere is composed mainly of carbon dioxide with small amounts of carbon monoxide and oxygen (Owen and Sagan, 1972). Low concentrations of water vapor are detectable in the atmosphere on a seasonable basis from time to time. In contrast to Earth, nitrogen in any form is not sufficiently high in the atmosphere to be measurable with the instrumentation used to date; although there remains the possibility that it is present at lower detection levels.

Photodissociation of carbon dioxide by solar UV light reaching the planet is believed responsible for the carbon monoxide in the atmosphere. Based on calculations of UV intensity on the planet, a large portion of the carbon dioxide is subject to the dissociation annually but the amount of carbon monoxide present at any given time is relatively small (less than 0.1%) at the surface and the corresponding free oxygen is about the same order of magnitude. Complex reverse reactions to return that carbon monoxide to carbon dioxide are being considered because it is not obvious that the direct reversal can take place readily (McElroy and Donahue, 1972). In fact, biological oxidation of carbon monoxide is one of the mechanisms which has been proposed to account for the observed low concentration of CO and as a possible energy source for life (Wolfgang, 1970). An energetic intermediate of this type over comes the problem of organisms adapting with ability to receive light for photosynthesis in the presence of known destructive fluxes of UV light on the surface of Mars. There is also the problem of extended periods (months) of global dust storms which reduces the light level reaching the surface. Once liberated from direct contact with sunlight, Earth organisms do develop which live at great depths. On Mars this opens the possibility for life to exist under more favorable conditions of temperature, and water concentration than is possible on the surface.

This paper considers possible alternative mechanisms for life which depend on carbon monoxide, carbon dioxide, and geothermal energy. These factors are considered individually as well as in combinations. Using these proposed mechanisms the question is discussed as to whether the experiment on the Viking mission to explore the surface of Mars in 1976 can detect these types of organisms if indeed they do exist. In addition, the model systems identified in this paper are examined for consideration in the design of future experiments.

1. Possible Mechanisms

1.1. Carbon monoxide

As a source of energy, carbon monoxide can yield considerable energy on conversion of carbon dioxide. Wolfgang pointed out the possibility of organisms making this conversion and thereby gaining the energy chemically rather than returning it to the atmosphere as heat (Equation (1)). Lederberg (1969) earlier speculated that carbon dioxide could react directly with water to synthesize carbohydrates (Equation (2)).

$$CO + \tfrac{1}{2}O_2 \rightarrow CO_2 + 67\ \text{Kcal} \tag{1}$$

$$2\,CO + H_2O \rightarrow CH_2O + CO_2 \tag{2}$$

In addition, there are other reactions of carbon monoxide and carbon dioxide which transfer energy and provide the cell with useful products. For example, nitrates and nitrites react with carbon monoxide under the proper conditions to produce ammonia (Equations (3) and (4)) which in turn reacts exothermically with carbon dioxide to produce ammonium carbamate (Equation (5)).

$$HNO_3 + 4CO + H_2O \rightarrow NH_3 + 4CO_2 \tag{3}$$

$$NO_2^- + 3CO + 2H_2O \rightarrow NH_3 + 3CO_2 + OH^- \tag{4}$$

$$2NH_3 + CO_2 \rightarrow NH_2COONH_4 + 44\,\text{Kcal} \tag{5}$$

Under suitable conductions ammonium carbamate is converted to urea (Equation (6)) then to ammonium cyanate (Equation (7)). The latter can also be converted to cyanic acid.

$$NH_2COONH_4 \rightleftarrows NH_2CONH_2 + H_2O \tag{6}$$

$$NH_2CONH_2 \rightarrow NH_4CNO \tag{7}$$

Most of these reactions will take place slowly at room temperature and more rapidly at elevated temperatures. For example, ammonium carbamate is converted to urea in 53% yield at temperature of 150 °C in 30 min. Once the ammonium cyanate is present, it may be involved in several interesting reactions. Of particular interest is the reaction with phosphate (Equation (8)) to produce an active ester.

$$KCNO + KH_2PO_4 \rightarrow NH_2COOPO_4H_2 + 2KOH \tag{8}$$

According to Jones and Lipmann (1960) this reaction takes place at room temperature at pH 5 to 6 with 40% conversion. They indicate that equilibrium is reached 'fairly rapidly'. The carbamyl phosphate produced in Equation (8) is a phosphorylating agent which reacts with adenosine diphosphate (ADP) to form adenosine triphosphate (ATP).

$$NH_2COO\text{–}P + ADP \rightarrow NH_3 + CO_2 + ATP \tag{9}$$

The ammonia and carbon dioxide generated can be recycled. However, the carbon dioxide would be diluted by the pool of carbon dioxide in the process. Except for reaction 2 and 6, the original carbon monoxide may transfer its energy to several new compounds which do not *per se* involve incorporation of the carbon or oxygen into the material produced.

1.2. Carbon dioxide

Another potential source of enei gy is the atmosphere of carbon dioxide. Assuming that any moisture present will be saturated with CO_2, then the first reaction to consider is the solubilization and subsequent reaction of CO_2 with water (or ice) (Equation (10)). This assumes that there is liquid water somewhere on the planet.

$$CO_{2\,gas} + H_2O \rightleftarrows CO_{2\,liq.} + H_2O \rightleftarrows H_2CO_3 \rightleftarrows HCO_3^- + H^+ \tag{10}$$

The heat of solution of CO_2 in water alone is $+4.7$ Kcal mole^{-1}. The dissolved carbon dioxide then reacts slowly with water to form bicarbonate which may react with other ions in contact with the solution. If insoluble salts are present, they may be solubilized. For example, in the presence of carbon dioxide saturated water insoluble calcium carbonate is converted to slight more soluble calcium bicarbamate (Equation (11)).

$$\underline{CaCO_3} + HCO_3^- + H^+ \rightleftarrows + Ca(HCO_3)_2 \tag{11}$$

In this way the water saturated with carbon dioxide slowly leaches out some of the soluble minerals needed by the organism.

If the organism possesses even a simple primitive ionic membrane, the extraction process might be accelerated and extended. In this case the carbon dioxide may be used for generating stronger acid to extract the needed nitrogen and phosphorous containing minerals without depending on large volumes of watei for leaching at great distances from the organisms. In the past, the extraction has been postulated based on excreted organic acids or biologically derived chelating agents which are generally also organic in nature. It is theoretically possible (Mitz, 1971) to produce significantly high hydrogen ion concentrates using the pressure of carbon dioxide and the appropriately charged membrane. The overall reaction can be written as salt and carbon dioxide going to sodium bicarbonate and a strong acid, Equations (12) and (13).

$$NaCl + CO_2 + H_2O \rightleftarrows NaHCO_3 + HCl \quad (12)$$

$$MNO_3 + CO_2 + H_2O \rightleftarrows MHCO_3 + HNO_3 \quad (13)$$

The mechanism is much like that found in the human stomach which produces one normal hydrochloric acid from salt. With a negatively charged membrane, the negative ions outside the cell are exchanged for the bicarbonate ions generated from carbon dioxide reacting with water from within the cells. The driving force for this reaction is the pressure of carbon dioxide working with the enzyme, carbonic anhydrase inside the cell. The enzyme accelerates the normally slow reaction of CO_2 and H_2O to form bicarbonate ions. The increased ion concentration also increases the osmotic pressure which attracts water into the cell. The net effect is to actively 'pump' needed nitrate, nitrite, etc., and water into the cell. The dilute nitric acid is then available to react with carbon monoxide in the cell as indicated above to produce ammonia (Equation (3)) shifting the pH toward neutrality and providing a reactive form of nitrogen (Equation (5)). By removing the nitrate as it enters the cell, the maximum gradient between inside and outside the cell is maintained. Phosphate needed for activated phosphate may also be mobilized by this mechanism. This type of reaction, if it exists at all, can be visualized as acting on the mineral in close contact with the organism. Water can be externally attracted or recycled from within the cell. Small amounts of moisture supplied by the organism on its exterior with intermittent periods of reabsorption favor the concentration of the acid produced by the cell. In either case the amounts of water required in the biosphere is minimal in this model.

1.3. Geothermal energy

If the interior of Mars has any geothermal activity it may be warmer deep within the crust than at the top. The surface may be a dehydrated porous layer which overlays a permafrost layer. Trapped below the permafrost may be an area saturated in water vapor and, if the pressure is high enough, even liquid water. Because of the porosity and the lack of liquid water on the surface, the atmospheric gases may penetrate deep into the permafrost region. The organisms not dependent on photosynthesis should be able to take advantage of this situation. If the temperature is hot enough, urea and other organic compounds could be generated by many of the same reactions indicated above. By a process of 'gardening of the soil' the organics may move up towards the surface. In which case, conditions may be suitable for organisms to develop which live on these compounds as a primary source of energy. It is also possible to think about a life process which combines geothermal abiogenic synthesis as a source of a few compounds like urea with carbon monoxide as a major energy source, augmented by the carbon dioxide-acid extraction of needed minerals.

2. Discussion

It would be fortuitous if one or a combination of the mechanisms indicated were the basis for life on Mars because the possible reactions described above were selected to fit a working model. On the other hand, there may be many other reactions using many of the same starting materials and/or principles which are equally reasonable and lead to the same result under the same or similar conditions. The specific reaction is not as important at this point as the possibility that there are at least two energy sources for biota outside of photosynthesis.

What does this model tell us about the presently planned Viking and future missions? There are a number of things. First, we have already seen that if carbon monoxide or carbon dioxide is an energy source, one does not necessarily expect the carbon of that molecule to be incorporated into the organic backbone of the compounds synthesized. The energy may be derived without direct incorporation. However, if one monitors both the atmospheric carbon dioxide and carbon monoxide as one does in two of the three reactions in the Viking biology experiment package (Horowitz, 1972, Oyama, 1972) one should detect incorporation or changes if the atmosphere is a direct source of carbon. However, if the carbon is first converted into organic compounds abiogenically deep within the crust and then utilized by organisms, these same two experiments might be negative but the third biology Viking experiment (Levin, 1972) which depends on releasing a label gas from supplied organics should show that organisms have developed which depend on such compounds. In addition, any organic material present will be detected and may be identified by the Viking gas chromotograph-mass spectrometer (Biemann, 1971). Furthermore, the model tells us something about the water concentration, temperature, and pressure limitations. Specifically, it allows for a wide range of water, pressure, and temperature conditions. Certainly it is not limited to the condition of the surface or in the atmosphere of the planet.

3. Conclusion

A model has been developed for an alternative non-photosynthetic basis for life which leads to several conclusions about the search for life on Mars. (1) The Viking investigations are a good first step towards a search for Martian biota. (2) If we do not find life on the surface, we need to look for it deep within the crust, near or below the permafrost layer with the same or similar instruments. (3) A search for organic materials at various depths in the crust should also prove extremely rewarding. (4) The models developed in this paper indicate that subsurface profile measurements of temperature, pressure, and water vapor, as well as the presence of nitrates, nitrites, and phosphates are important measurements that should be considered in the design of future missions.

References

Biemann, K.: 1971, in R. Buvet and C. Ponnamperuma (eds.), *Chemical Evolution and the Origin of Life*, North Holland Publ. Co., pp. 541–547.

Horowitz, N. H., Hubbard, J. S., and Hobby, G. L.: 1972, *Icarus* **16**, 147.

Jones, M. E. and Lipmann, F.: 1960, *Proc. Nat. Acad. Sci.* **46**, 1194.

Lederberg, J.: 1969, *Appl. Opt.*, **8**, 1269.

Levin, G. V.: 1972, *Icarus* **16**, 153.

Mars Scientific Model: 1972, *JPL Document*, No. 606-1.
McElroy, M. and T. Donahue, 1972, *Science* **177**, 986.
Mitz, M. A.: 1971, in R. Buvet and C. Ponnamperuma (eds.), *Chemical Evolution and the Origin of Life*, North Holland Publ. Co., p. 355–362.
Owen, T., and Sagan, C.: 1972, *Icarus* **16**, 557.
Oyama, V.: 1972, *Icarus* **16**, 167.
Ponnamperuma, C. and Klein, H. P.: 1970, *Quart. Rev. Biol.* **45**, 235.
Wolfgang, R.: 1969, *Nature* **225**, 876.

AN AUTOMATICALLY-RETURNED MARTIAN SAMPLE BY 1985?

G. EGLINTON
Organic Geochemistry Unit, University of Bristol, Bristol, England

and

S. TONKIN
British Aircraft Corporation, Bristol, England

Abstract. The success of the lunar sample analysis programs underscores the desirability of a returned Martian sample. A Mission which would bring back about 1 kg of soil is outlined. The vehicle would have a mass of about 15 tonnes on departure from Earth and would make extensive use of Viking and Mariner technology. Russian experience in the field of automatic soil sampling and automatic rendezvous would be invaluable and the Shuttle would make possible a tidier launch. Sterilisation or quarantine will be necessary to preclude back-contamination of Earth by hypothetical Martian micro-organisms. A prime quarantine facility designed to detect biogenic organic compounds and life processes could be set up at a Lunar base or in a Sky-lab.

A single soil sample could be informative as to the general surface composition of Mars. Life detection would be a major task, followed closely by the chemistry of carbon and other life-related elements. However, knowledge of the detailed physics, chemistry and mineralogy of the Martian sample would be of inestimable value to planetary studies.

1. Introduction

The purpose of this article is to examine the desirability and feasibility of an unmanned mission to Mars to return a soil sample to Earth for laboratory analysis. We have not attempted to make a detailed case in this brief review, which is of necessity more superficial than the treatment normally given to specific areas by committees and mission planning teams. Much should emerge between now and the early 1980's when such a mission might be flown (Dwornik, 1973). By 1977 Russian probes and the American Viking Spacecraft may have analysed the Martian surface and reported on the processes occuring thereon. The choice will then be between further automatic analysis missions of the Viking type as against missions designed to return samples, though appropriate combinations may also be considered, including orbital missions. However, the long lead times require that discussion begin now. The relative merits of these two approaches are summarized in Table I.

As a sequence for planetary exploration the Mariner, Viking and Sample Return Missions make good sense. A Sample Return Mission should not be vastly more expensive than a Post-Viking Remote Analysis Mission and it is our contention that the scientific return would be much greater. That detailed analyses in terrestrial laboratories are much more effective than *in situ* is well seen from the lunar programme wherein the vast output of data from the Apollo and Luna samples can be contrasted with the very limited return from the Surveyor and Lunik soft landers that preceeded the Apollo and Luna missions. Furthermore, flying sophisticated instruments designed

TABLE I

The choice between returned sample and post Viking remote analysis missions

For returned sample	Against returned sample
The mission would be a dramatic step forward in planetary exploration, arousing world-wide interest.	Possibility of back-contamination by Martian organisms renders sterilisation and/or quarantine essential.
Most sectors of the Space Programme are usefully harnessed – manned and unmanned space exploration programmes and laboratory analytical facilities. The analytical involvement would be world-wide, as for the Apollo and Luna programmes, ensuring immediate and continuing benefit to bio- and geosciences.	The complexity and long duration of the mission necessitates elaborate managerial structures on an international basis.
Hardware can be based extensively on that employed in other projects – Viking, Luna, Shuttle etc. The Lander is itself not unduly complex.	Major technical difficulties to be overcome – e.g. automatic docking in Mars orbit and collection in Earth orbit.
Laboratory analyses give full scope for experiments which meet the real nature of the samples. Experiments can be sequential and continually modified so as to give *definitive* answers. A clear 'life' or 'no life' statement should be possible.	The interaction between the sample and the Martian surface environment would be an important feature not available for study on this mission. Would Martian life survive the interplanetary journey?
The laboratory techniques of the 1980's would likely permit a detailed grain-by-grain study of the returned sample. Together with Mariner and Viking 75 and data from any other missions it should be possible to begin to decipher the complex history of the Martian surface.	If almost any portion of the mission goes wrong then there would be no sample return. The scientific return would then be very small or zero depending on any instruments carried on the lander.
A portion of virgin sample set on one side would afford long-term research possibilities, including future checking of results and theories.	Important changes may occur in the sample during storage (orbit, return flight and quarantine etc.)

to make limited tests has additional disadvantages; thus, data returned from Mars by an amino acid analyser would be positively misleading if amino acids were absent but other compounds giving a positive response were present. The evidence so far available from Mariner 9 has been interpreted by Masursky and others (1973) as indicating that "the Martian crust has undergone a complex evolution" and that the "interaction of crustal and surface processes is still in progress." Based on experience with lunar samples, it is clear that much of this complex history and contemporary scene will only be deduced after extremely detailed examination of individual particles from the Martian regolith. However, Viking type analyses are based on bulk samples.

Information on the nature of the Martian surface and the physics and chemistry of the surface processes will assist in understanding other planets, including our own. There is a general resurgence in the study of the evolution of the solar system and in the development of comparative planetology. The Viking missions have been oriented mainly around the possibility of life detection and this will remain a major interest in subsequent missions since it is doubtful that Viking will give a definitive answer. Hence, we can take the study of the surface of Mars in two ways. First, if life is absent it would be of great interest to see what had happened in the interaction of the atmosphere and the surface material on this planet which is otherwise somewhat similar to our own. Second, if life is present, the comparison between Martian an terrestrial biochemistries is likely to revolutionise the life sciences and our view of the Universe.

2. The Martian Surface

A single sample of Martian soil is likely to contain wind-borne particles ultimately derived from the whole surface of the planet and hence to be somewhat of an average sample. In this aspect, there is a close similarity between Martian and lunar dusts as a result of aeolian and impact transport, respectively.

The recent Mariner findings indicate that Mars is geologically active. Thus, there are marked differences in the order of tens of kilometers in elevation between parts of its surface and a variety of non-aquatic terrain is to be found in the Mariner photographs – mountain, plateau, valley, plain, volcanic caldera, lava flow, impact crater and fault. There are also features which may indicate fluid flow at some time, such as sinuous valleys, braided channels, aıroyos and canyons. At the poles curious laminated features suggest sequential layers of dust and ice (probably CO_2 and/or water ice) and elsewhere there are extensive regions of great erosion interpreted as mass wasting down to a uniform level. A few features may have been mud slides but aeolian erosion is paramount. Some impact craters and volcanic structures clearly show the effects of long-term wind erosion. The seasonal light and dark changes which have been seen on Mars from Earth must have their origin in planet-wide dust storms, one of which was observed in great detail by Mariner 9. As Masursky (1973) has said "aeolian, glacial, fluviatile and mass-wasting processes are paramount in the surface modification of Mars." Wind erosion is scouring away the equatorial regions and depositing the dust in the polar and intervening regions. The fine dust of 1–10 μ size is carried to heights of over 50 km. Hence, the material mantling high latitudes of the Northern and Southern hemisphere must be of mixed local and distant origin. From the limited information we have about the history of Mars, the climate should vary periodically (e.g. the 50000 year cycle for the differential temperature of North and South hemispheres, due to the conical precession of the sin axis with respect to the line of apsides of the rather eccentric orbit). The very fine dust may have been in and out of polar and desert deposits many times *via* transport through the upper atmosphere and should contain material from most regions of the planet. However, mineral and particle size sorting will have occurred. The terrestrial equivalent of these wind-deposited sediments would be loess deposits, which have silica (SiO_2) contents of between 59 and 74% and a grain size of between 30 and 60 μ. The silica content of the much finer dust in the Martian atmosphere has been estimated at $60 \pm 10\%$ by infra-red measurements made by the Mariner 9 orbiter. Terrestrial clays are closer in size (less than 5 μ) but they have a much lower silica content: in any case, they originate in aquatic alteration of rocks, a process which is unlikely to have made much contribution to the Martian dust. Incidentally, the planet-wide dust storms, such as those of Summer 1971 to February 1972, involve wind speeds of 50m a second at an atmospheric pressure of about 7mb, which is about 1% of the Earth's atmospheric pressure at the surface. Physical and chemical equilibria, involving dust and atmosphere, must control this situation. (Goody, 1973).

The fine dust of the mantled and Polar regions should contain particles derived from all main rock types; e.g. igneous fragments from the basaltic or andesitic lava flows, metamorphic fragments as a result of the tectonism visible in Mariner 9 pictures, impact generated breccias and glass and particles of sediment of aeolian and epiglacial origin. The epiglacial deposits at the poles may exist as permafrost involving CO_2 ice or possibly water ice. Meteorite fragments should be abundant: Mars is close to the asteroid belt; also, calculations show that deceleration by the atmosphere should result in a large proportion of free falls. Meteorites above 1 kg on entry at typical velocities should deposit mass at the surface, resulting in fragments of iron and stone meteorites and carbonaceous chondrites amongst the dust (Gault, 1973). The corollary that micro-meteorites do not reach the surface and that larger stones (up to certain masses) reach it only at low velocities means that micrometeorite gardening of the lunar type should not occur. The small scale morphology of the martian surface will be controlled by aeolian forces rather than meteorite impact. The presence of the atmosphere eliminates the solar wind as a factor in the surface chemistry and physics.

Some fluid erosion seems to have occurred on the planet in the not too-distant past. Water is the best candidate but it is more likely to have arisen by sudden melting of permafrost regions than by cloud burst in view of the very low atmospheric pressures. Water vapour occurs in the Martian atmosphere, especially over the non-polar regions. Ten to twenty precipitable microns partial pressure have been detected by infra-red measurements. (Hanel *et al.*, 1972; Hammond, 1973). There are extensive regions such as Hellas, where the total atmospheric pressure at the surface exceeds the triple point pressure for water vapour. However, the present atmosphere is almost entirely carbon dioxide, and hence liquid water could only exist if extremely local high concentrations of water vapour were generated. But, water-ice clouds have been detected and these findings have led to the selection of new Viking sites at low altitudes and Northern latitudes, chosen to give greater possibility of liquid water being available to micro-organisms. Incidentally, the planet must be losing water continuously as the result of photoionisation and the loss of hydrogen and oxygen atoms from the upper atmosphere to space. Volcanic activity may replace these losses periodically. Thus, Nix Olympica is believed to have been formed about 10^8 yr ago, giving rise to major emissions of water, CO_2 etc. Interglacial periods during the recession of he poles would cause the amount of water vapour to rise in the Martian atmosphere. However, silicate minerals can hold water and hence the dispersal of Martian dust into the atmosphere as dust storms followed by the falling-out process could abstract moisture from the atmosphere. The infra-red measurements from Mariner 9 indicate that this does take place since the water vapour pressure fell after the 1971–1972

TABLE II
Sampling considerations

Site and sampling	Comment
Site	
Between latitude 30° N and edge of N polar cap	Aim is to choose site with optimum chance of trapped water and a temperature regime favourable to life.
Very low-lying, flat, dust-mantled region	Low altitudes have pressures >6 mb. Dust mantle should not be too deep or unconsolidated.
Firm loess undergoing deposition on old eroded site	Fine dust should be representative of planet-wide geological surface units. Best possibility for micro-organisms and adsorbed compounds. Rock undesirable as difficult to core and only locally representative. Coarse sand in dune regions also undesirable.
Near successful Viking (or Russian) lander site	Prior assurance of suitability – smoothness, firmness etc. Provides 'ground truth' for 1976 Viking analyses. Viking craft could act as beacon or target (if not destroyed by winds etc).
Sampling	
Not during, or soon after, major dust storms	Spacecraft hazard. Even a brief undisturbed depositional record would be valuable.
Sample at night or dawn	Volatiles trapped at low temperatures; photography possible at dawn.
Core or successively deeper scoops	Subsurface protection from sun and extreme temperature variations important to bioscientists and geoscientists.
~ 1 kg unsieved dust	Mission costs not very sensitive to sample size

storm and then rose slowly. If clay minerals have been formed then water could be held in the clay mineral lattices along with organic compounds and other volatile or unstable materials. Mars is a differentiated planet and localised concentrations of minerals as bulk materials are possible. Mineralisations should also be revealed as concentration effects in the wind-blown dust. As a side effect, salts suitable for eutectic mixtures could be employed by microorganisms.

In summary, the available information suggests that the Martian soil at sites currently under discussion for Viking will contain considerable proportions of fine dust derived by aeolian erosion and transport from distant regions. Bulk *in situ* analyses of this averaged sample are unlikely to be very informative whereas detailed laboratory study of a returned sample could include a grain-by-grain analysis. Some of the sampling considerations are given in Table II. Wind erosion of particles in the carbon-dioxide-rich atmosphere will require special consideration.

3. Engineering Considerations for Automatic Return of a Mars Sample

The mission is outlined schematically in Figure 1. For this preliminary study we have plagiarised extensively, especially from Viking (Martin and Sibbers, 1972). This is, in fact, the way in which we think the mission should be done. There is a great deal of Viking technology which can be used directly on this project. For example, the Viking orbiter is largely a nerve centre containing solar power, navigation, communications, propulsion etc. It was used on Mariner and with some growth will navigate this project to Mars and bring it back. Again, the Viking propulsion system is used as far as possible, not only in principle but also as actual components. Russian co-operation would be valuable, especially in the fields of automatic sampling (Gatland, 1971), sample preservation, and automatic rendezvous. Use of these existing technologies will keep the cost within bounds.

The vehicle to be launched from Earth is too heavy for a Titan Centaur. A completely expendable launcher could be based on a Saturn, but with the timescale which would probably obtain, a shuttle launch seems appropriate. The vehicle to go into transfer orbit to Mars would have to be parked in Earth orbit and a booster taken up as a second shuttle payload and attached in orbit. The size of this orbit-orbit booster would be of the order of Centaur.

If the vehicle is to be carried by the shuttle its propulsion system must be suitably designed for abortive landing by the shuttle. This should be possible if a pressurised fuel system similar to Viking's is to be used. As to the Centaur booster, its tanks would be dumped into the shuttle tanks.

A single vehicle going to Mars, landing, lifting off, returning to Earth, and landing

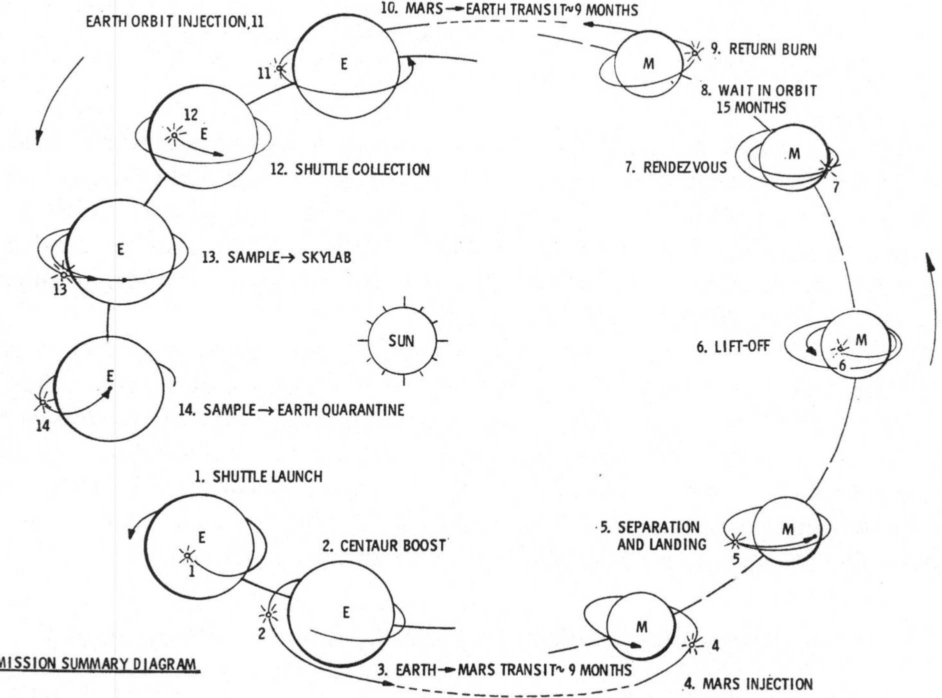

Fig. 1. Mission summary diagram.

on Earth would be of a formidable size when it left the Earth. The main assumption made to keep the launch size down is that a flight plan similar to Apollo can be used, i.e. the return vehicle waits in orbit whilst a land and lift-off vehicle descends and then comes back to rendezvous with the orbiter. This will not be easy to engineer. Whilst manned rendezvous has been done remote from Earth and unmanned rendezvous has been done close to Earth, unmanned remote rendezvous is quite a different matter. Control will have to be largely autonomous because signal transit time is of the order of half an orbit period of the rendezvous vehicle. On the approach to Mars, the lander could conceivably enter Mars' atmosphere directly, thereby saving about 15% of the mass leaving earth, mainly in propellants. However, the lander would be blind to the weather conditions and it is for this reason that a Viking-style approach has been adopted in this study. As a compromise solution the injection burn could be divided. The first burn could put the orbiter plus lander into a highly eccentric orbit. After a few quick looks, the Lander could enter after a modest burn, and the orbiter go into its synchronous orbit.

The energy requirement for journeys to and from Mars falls to a minimum about every 2 yr. These minima vary considerably with the epoch because of the eccentricity of Mars' orbit. The minimum-energy time for return will not occur till some fifteen months after arrival. There is then the question of how this time should be spent. Descent to Mars need not be delayed long after arrival. A landing site will have long since been chosen using Mariner and Viking experience. Take-off should take place as soon as possible after landing and sampling. This is to lessen the risk of damage or deterioration by unknown hostilities in the Martian environment. Having lifted off and achieved low orbit the lift-off vehicle and orbiter have plenty of time for rendezvous. This fact will reduce the energy requirement for correction of gross rendezvous errors and allow these corrections to be controlled from Earth, as signal transit time, though long (*ca* $\frac{1}{2}$ hr for 2-way) is not too long for this purpose. After rendezvous, the return launch window must be patiently awaited in orbit. The wait may be of the order of $1\frac{1}{4}$ earth years.

Whilst Mariner and Viking will pave the way for the project, it is doubtful that they can play a direct part. The most useful role for any spacecraft still operational would be to provide stand-by communications and some contact between the orbiter and lift-off vehicle when occulted by Mars.

Some examination of the sample requirements has been made and no serious difficulties seem to exist. Landing near aphelion between spring and summer seems to make dust storms less likely. As the wait for the return window is likely to be long, a wait in orbit for a suitable landing time should not present difficulty. Direct entry will allow no such choice. The site slope should probably be limited to 15°. A low altitude site must be chosen to give good atmospheric braking. To limit contamination there must be a delay of several days after landing and no venting before sampling. The sampling arm should extend at least 2 to 3 m. Some telemetry and imaging will be needed. Preferably there should be several sample containers with very positive seals. The temperature of the sample container should be kept low by placing them on the shady side of the spacecraft, which should be insulated well. The final earth orbiter will only be spin stabilized and it may be difficult to obtain suitable axis direction.

The approach to spacecraft sterilisation to avoid contamination of Mars can be very similar to Viking. The lander can be sterilised and kept in a jettisonable bioshield until entry. The hydrazine used for final braking should not alter the water content of the soil before sampling. No such consideration need apply for take-off. The spent hardware left lying about on Mars will be pre-sterilised like the Viking lander. Various things will fall in from orbit. The Mars injection propulsion stage can take much longer to decay than Mariner and Viking orbiters because any residual propellant can be used to raise the periapsis of the final orbit. The decay time will be much shorter for the discarded lift-off vehicle and various other items, but these will all be presterilised and it is unlikely that any contamination could survive entry heating.

The obvious alternatives for the returning vehicle are Earth-orbit with subsequent collection, or direct entry. The latter would be a much cheaper operation, but introduces an immediate risk that quarantine could be accidentally broken and the consequences just might be quite appalling. We have therefore selected Earth-orbit collection. If the Earth orbiting vehicle is equipped with a very simple CW beacon, its orbit can be determined, and it can be collected by a shuttle. Some mass can be saved if the final Earth orbiter is braked aerodynamically before a small circularisation burn. However, it is felt that this increases the risk of accidentally breaking quarantine whilst the mass penalty of an alternative system is not terribly serious. It is proposed instead that insertion into Earth orbit from the approach hyperbola should be done by solid fuel motor. Unfortunately, the only small motors presently available are long and thin and therefore useless as spinners. A special motor must be developed. Bad guidance could result in accidental re-entry but burn-up should then practically eliminate the chance of contaminating the atmosphere. Because of the fixed mass of the 'Nerve Centre' borrowed from Viking the launch mass is relatively insensitive to sample mass.

A study has been made of a suitable vehicle for this project. Masses have been based on Viking's component masses, making extrapolations for propellants, structures, aerodynamic decelerators etc. Figure (2) summarises the components and Figure (3) is a rough sketch of the proposed vehicle. A distinct similarity to the Viking vehicle can be seen. In the waist in the 'Nerve Centre' which is very similar to the Viking Orbiter. It uses the same power, communications and nagivation equipment, etc. It omits some experimentation used by Viking Orbiter. Attached in a similar position to the Viking Lander is the Descent Vehicle and Lift-Off Rocket contained in a bioshield with an entry heatshield. Sandwiched between these two is the Final Earth Orbiter with its solid motor, and the radio beacon with its solar power supply and antenna. In a similar position to that on Viking is the main propulsion system, using monomethyl hydrazine (MMH) and nitrogen tetroxide (N_2O_4) in tanks pressurised by nitrogen. The same 1340 N motor as used on Viking would do. However, in this case the motor would be used for two main burns (rendezvous

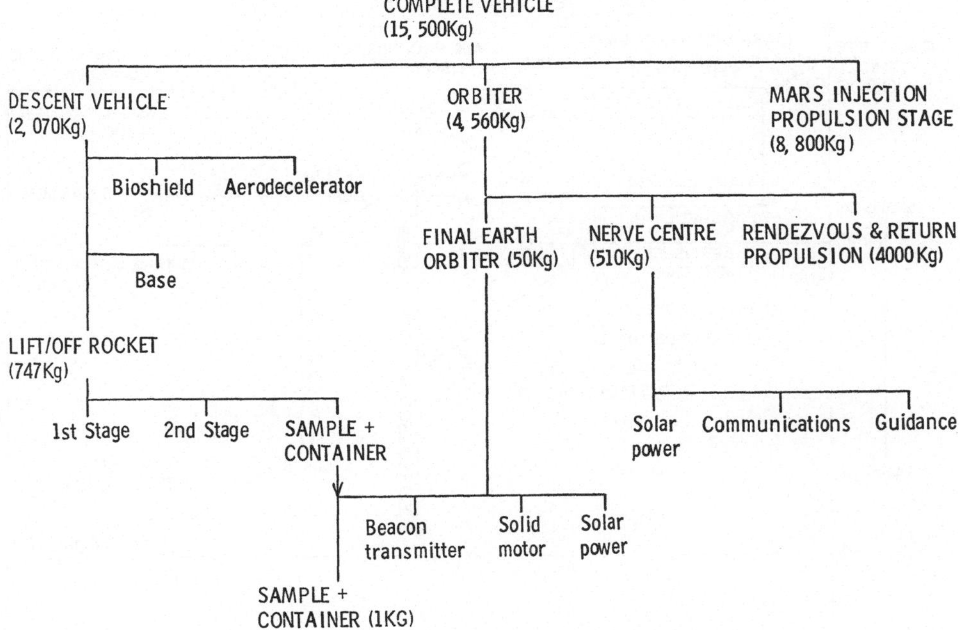

Fig. 2. Vehicle component breakdown.

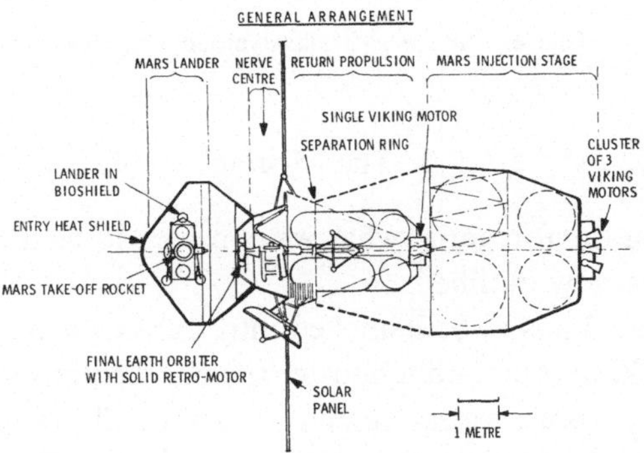

Fig. 3. General arrangement of Mars vehicle.

and earth return) with a total impulse greater than that of Viking's main injection burn. Consequently the propellant tanks are shown a little larger than those of Viking. Peculiar to this mission is another much enlarged propulsion stage for Mars injection. This uses Viking technology (MMH and N_2O_4) and a cluster of three 1340 N motors.

The mass of the whole vehicle as shown in Figure 3 is about 15 tonnes (cf. Viking 3.5 tonnes). In Figure 4 is shown the mass of the vehicle as it progresses through the mission. The scale on the horizontal axis provides a measure of the various velocity changes.

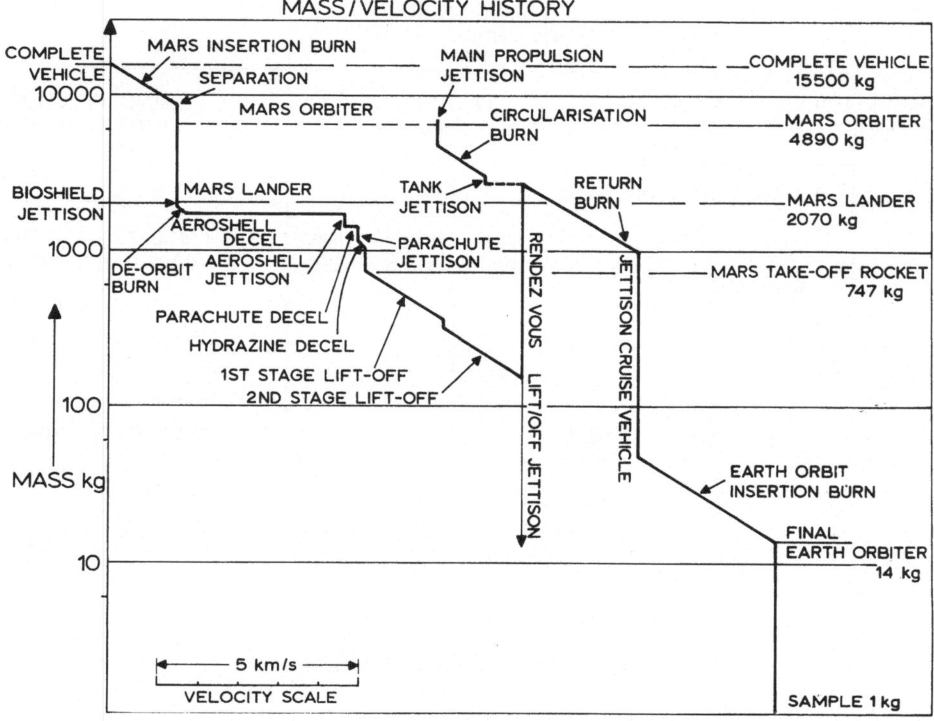

Fig. 4. Spacecraft mass/velocity history.

4. The Mission

We have selected from the foregoing what we consider to be a reasonable option for the mission and this is now outlined.

The vehicle of Figure 3 has to be launched into transfer orbit to Mars. The vehicle is placed in a high parking orbit and a booster such as Centaur attached. This booster gives the vehicle a hyperbolic excess velocity substantially tangential to the Earth's orbit: this velocity must be between 3 and 4.5 km s^{-1} depending on the epoch. After injection into the transfer orbit, the Centaur booster is jettisonned. The reaction control system (RCS) borrowed from Viking is then used to orient the vehicle with its axis pointing to the Sun. In this attitude the solar panels can charge batteries, and solar and Canopus sensors provide accurate attitude when gyroscopes are switched off. The transit time to Mars is 7–11 months. For mid-course correction gyros are started, and an inertial system used to control attitude and velocity during the burn.

With ten days remaining before Mars arrival any latitude can be reached. After this time only a few degrees variation in landing site latitude is possible. As Mars is approached the attitude is acquired for Mars injection burn which commences just before the peri-apsis of the approach hyperbola. The velocity change is of the order of 1.45 km s^{-1} including gravity loss, and the burning time for the three motors just over an hour. After this burn the orbit is observed and perhaps trimmed by up to 50 m s^{-1} by the Mars injection stage. When it is not being used for orbit trimming the whole vehicle reverts to its Sun-pointing attitude.

Later on, when the Mars injection stage is completely finished with, it is jettisoned in an attitude for an accelerating burn at apoapsis, with the addition of a slow spin. When separation is sufficient, a burn of residual propellants is made at apoapsis, to lengthen the decay time.

The synchronous orbit is the same as that used on Viking. The orbiter/lander waits in this orbit for the signal to land. When the signal comes, the lander separates mechanically, at the appropriate part of the orbit, discards its bioshield and does its de-orbit burn of about 150 m s^{-1} using a number of attitude control jets (N_2H_4) round the periphery of the heat shield. Orbiter and Lander masses are about 5 and 2 tonnes, compared with 1 and 0.85 tonnes, respectively for Viking. The atmosphere is entered at a height of about 250 km and the aeroshell decelerates the Lander until about 6.5 km altitude is reached.

Viking uses a lifting entry with a lift/drag ratio of 0.18 and this would seem to be suitable for this project also. A problem which is more severe than on Viking is to get the mass centre of the lander close enough to the heat shield for stability. The method suggested is to use a very squat lift-off rocket, and to lay this across a diameter of the heat shield. At about 6.5 km altitude, a parachute will be deployed by mortar and the aeroshell separated. Before deployment the Lander and Lift-off Rocket is on its side, but after deployment it hangs up by its nose. After the aeroshell has gone, the landing feet splay out to form a tripod.

At an altitude of a little over a kilometer the parachute is jettisonned and the hydrazine terminal propulsion engines take over. They reduce the vertical speed from a little over 100 m s^{-1} to about 2.5 m s^{-1} and the horizontal speed to zero from a wind speed as high as 65 m s^{-1}. Like Viking, the last few meters would be in free fall, to reduce contamination of the soil by the hydrazine engines. All this is controlled by a radio altimeter and a landing radar. A homing beacon on a well placed Viking lander could be useful. As soon as the lander is down it determines its position by means of Sun and star sensors and sensing of the local vertical against a clock. This information is sent to the orbiter and relayed to Earth. After several days, the sample arm is extended and a sample taken and lifted to the sample containers in the nose of the lift-off rocket. When the containers are full they are closed up and sealed. The lander is based on Viking and Luna 16. Lift off commands from Earth are relayed by the orbiter, and without unnecessary delay, the ascent stage of the vehicle takes off. There is a central second stage and on each side, a booster is fitted to form the first stage. To achieve a good mass ratio Viking propulsion techniques may be unsuitable and a pumped system may be necessary. This vehicle lifts the payload into a fairly low, roughly circular orbit, which is not very precise. Only the second stage will have attitude control, which will remain operational after burn-out. It will be equipped with a beacon so that it can be tracked by the orbiter, and the orbiter will command its attitude during rendezvous.

The orbiter spends a little time observing the orbit of the new satellite. It then descends to rendezvous by means of a braking burn, followed by several other burns. 1200 m s^{-1} is allowed for all this. After the burns the empty propellant tanks are discarded.

When rendezvous takes place the payload is transferred to the Final Earth Orbiter portion of the vehicle and the second stage of the lift-off rocket discarded. Any equipments in the Orbiter not to be used again are also discarded. These will have been carefully sterilized before the mission because eventually they will fall into the atmosphere, albeit to be burnt up.

After a wait of about 1¼ earth years, the return burn of about 2750 m s^{-1} is made. This takes about 55 min with the single 1340 N thruster. Midcourse corrections are commanded from Earth and executed in the same way as they were on the outward journey. The course should be corrected if possible so that the Sun is on the tangent at the apex of the hyperbola as the Earth is approached. Transit time is again 7–11 months. The return vehicle is oriented for the final braking burn. If the trajectory is in the right place relative to the Sun this manoeuvre should involve little angular motion. The RCS is then used to impart a spin to the vehicle. When the rate of spin is sufficient the solid motor is fired and the Final Earth Orbiter separates. Its burn gives a velocity reduction of about 3.5 km s^{-1}, which is sufficient to achieve a circular orbit at about 1000 km altitude. The return vehicle goes into heliocentric orbit.

On board the satellite is a solar powered C.W. transmitter which will enable the shuttle to find it. The sample container is still on the shady side, to maintain the low temperature.

When a shuttle rendezvous with the Final Earth Orbiter (mass about 14 kg) the outside of the sample container will be contaminated with Martian dust. The entire Orbiter would be enclosed in a collapsible container before securing it to the outside of the shuttle and transporting it to a Skylab.

5. Sample Quarantine and the Search for Life

Some information about the carbon chemistry and life detection experiments should be forthcoming from the Viking mission in 1976. However, the Viking biological and chemical tests may not be really appropriate for any Martian microorganisms. Negative results from Viking will not guarantee the absence of pathogens in a returned Martian sample but if hypothetical Martian organisms have a biochemistry so alien that they do not respond to Viking experiments, then they are also unlikely to be able to interact with Earth organisms and cause disease. There is, however, some risk in an unsterilised Martian sample, if returned to Earth. Any quarantine barrier system has a finite risk of being accidentally broken, as was seen with the lunar quarantine at Houston. Table III gives some options open to mission planners.

The return of lunar samples to Earth was the first calculated risk of this type. The lunar quarantine was for Apollo samples, equipment and astronauts originally stipulated by the Space Science Board of the National Academy of Science (U.S.A.).

TABLE III

Prevention of back-contamination of Earth by a returned Mars sample

Option and method	Comment
1. *Sterilise sample and container before return to Earth*	
(i) Heat (400 °C?)	Thermal degradation would result in loss of volatiles and of biological information.
(ii) Chemical (e.g. ethylene oxide for duration of return cruise?)	Loss of carbon chemistry information might be limited by use of isotopically labelled ethylene oxide. Damage to sample otherwise slight. Ethylene oxide may not be an effective sterilisant.
(iii) Irradiation (e.g. with ^{60}Co)	Some modification to carbon chemistry. Loss of information *re* tracks and other physical features due to radiation damage.
2. *Examination in a sky lab. or moon base*	
(i) Automated tests in unmanned laboratory	Quite elaborate automatic equipment could be set up by astronauts who would then leave the spacecraft prior to return of the sample. Careful examination for bioorganics, microorganisms etc. would then be made by remote control. Such tests would be much more effective than any that could be conducted on Mars.
(ii) Laboratory tests conducted behind biological barriers by astronauts	Crew now formally at risk. If long-term incubation for pathogens is insisted on, then the crew might have to stay up for a long time.
3. *Examination on Earth in a quarantine facility*	
Long and detailed biological testing behind quarantine barriers. The barriers need to be much more efficient than those used in the lunar programme. However, information obtained by Viking and by prior preliminary examinations (Items i and ii) might obviate some tests.	Personnel and biosphere at risk. Note that there were numerous failures of the barriers of the Lunar Receiving Laboratory during the quarantine for Apollo 11, 12 and 14. Major efforts would be needed to improve sensitivity and reliability of tests for pathogenicity and to devise new ones.

(Fox *et al.*, 1972) Arbitrary decisions had to be taken as to what should comprise the quarantine tests and how long they should operate. For the Mars sample international agreement could be sought on a quarantine sequence such as (Table III) 2 (i) → 2 (ii) → → 3, with appropriate contingency planning: for example, if in stage 2(i) complex bio-organic polymers were shown to be undetectable at extremely low levels then one might specify restricted tests in 2(ii) and 3. No doubt astronauts would risk prolonged quarantine in orbit just as they are currently willing to risk failure of essential hardware.

The prime aims of the quarantine facilities (orbital and terrestrial) must be first, containment of the sample and second, detection of any pathogenic, toxic or competitive response. Experience with terrestrial biota has shown that organisms and environments evolve together and that most organisms specialise in a particular environment and do not readily adapt to others. The common biochemicals of the Earth may not in any way match those of a hypothetical Martian biota even though it be based on carbon. Hence, there may be little likelihood of interaction but the tests would have to be made. Naturally, no facility can be set up to expose examples of the whole biota of the Earth but the fact that there is a common terrestrial biochemistry is a considerable help. Again, long term incubations might be insisted upon, for detection of certain effects such as carcinogenesis may require many years of exposure by test organisms (Lederburg and Alexander, 1971). Thus, incorporation or release of $^{14}CO_2$ would not be enough to satisfy determined proponents of the quarantine tests. Parenthetically, the same difficulty arises for the Viking life-detection tests as these depend on rapid responses to incubation tests.

Life detection experiments would be in parallel with the main quarantine tests. The bioscience experiments currently planned for Viking itself are discussed in other papers in this volume. The tests used in the quarantine facility would be much more extensive and sophisticated. Furthermore, many of the difficult decisions regarding choice of growth media, inhibitors, temperature regimes etc., which presently have to be made in advance for Viking could be made in real time in a Skylab laboratory, using information afforded there and then by the sample. Because of the extensive new apparatus and up-dated Viking equipment which would be available much could be learned of any contemporary or fossil microbiota by microscopic observation and by chemical tests. Thus, microorganisms which had died in storage should still be detectable by staining techniques and by chemical responses afforded by their biochemicals and metabolites. Such tests using automated equipment have indeed been proposed for the post-Viking lander.

One problem with the returned sample would be any effect of the long period of sealed storage during Mars orbit and trans-Earth cruise. A relatively constant low temperature – say less than $-10\,^{\circ}C$ – would be desirable to avoid degradation of biochemicals and death of organisms. Some tests would be conducted in cabinets holding a simulated Martian atmosphere.

An orbital quarantine facility would have to be very limited in size and scope if it is not to be extremely expensive. However, the miniaturization already achieved in the design of the Viking Biology experiments offers hope that an effective automated laboratory could be set up. The development work for the experiments and hardware would need several years of intensive effort and the extensive involvement of experts in cell culture, viral and biochemical techniques. Due allowance would be made for difficulties associated with weightlessness.

6. Analysis of the Returned Sample

An organisational framework for the reception and analysis of the returned Martian sample could be based on experience of the Apollo sample analysis programme (Chamberlain, 1972; Ponnamperuma, 1972). However, the limited sample return of the Mars programme justifies even greater commitments to it by the analysts during the first two years. Management should be of the type associated with major technological projects, in which some pooling of resources, results and tasks would be required of the participating scientists. As Wasserburg (1972) has put it: "The scientific community must be willing to participate in both the planning, engineering and decision-making functions".

A scheme could be organised so that the Preliminary Examination Quarantine Facility would be primarily concerned with the quarantine tasks and with obtaining the basic minimum of information necessary to decide on a sensible first round of sample distribution. The analyses would be conducted in the laboratories of individual Principal Investigators (P.I.'s) and in a small number of specially equipped Consortium Laboratories. Each Consortium Laboratory would house several (e.g. 2–4) small research groups and the Consortium P.I.'s would have made a prior agreement to work together full-time at that site for a defined period (e.g. 2 yr). The research would be in a particular area (e.g. dating by several techniques). Coordination and control of contamination would be the responsibility of the Consortium leader.

Working conferences would be for specific areas of research and participants would be required to produce a joint document detailing the scientific outcomes of the Conference. The more infrequent major conferences would be of the 'Lunar Rock Festival' type held annually at Houston.

The arrangements outlined above would seem appropriate for a few hundred gram sample of fine dust and small mineral fragments. Experience with the valuable core samples returned by the Apollo and Russian Luna missions, has shown that best results are obtained by careful and coordinated distribution of the samples, accompanied by dead-lines for publication of preliminary results. The recent publication of some 40 papers describing the analytical data from 623 mg of Luna 20 material provides, as the editors (Anders and Albee, 1973) put it: "a demonstration that highly significant information and conclusions can be obtained from even a very small sample returned by an unmanned spacecraft." These papers may be consulted for details of the distribution scheme and for the information that was obtained from the small samples. Numerous advances involving existing and completely novel techniques are sure to be made by the 1980's and a detailed prediction of the possible scientific outcome of a returned Mars sample would be very lengthy. However, by way of illustration, a scheme based on contemporary technology (Table IV), is outlined in Figure 5

Amongst the assumption made are:

(i) The Martian soil samples would be mainly fine dust ($<10\,\mu$) containing a sizeable fraction of sand-sized particles ($100\,\mu$–1 mm) and a small number of larger rock fragments (>1 mm).

(ii) Considerable emphasis would be given to describing, analysing and understanding the nature of *individual* particles. A statistically significant number should be studied, such that their origin and history could be fitted into an overall picture.

(iii) Comparatively large quantities of bulk samples are utilized in the quarantine and life detection tests and also set aside as posterity sample.

(iv) The several subsequent rounds of allocation would be based on results from the earlier rounds.

TABLE IV

Techniques for the morphological, chemical and isotopic examination of sand- and dust-sized particles. (Single or small numbers.)

Technique	Damage to particles	Data	Information
Optical microscopy			
Transmission (thin sections) Reflection (visible, near i.r. etc.)	None or partial	Shapes, texture, erosion and damage features. Location of minerals, glasses, inclusions, gas bubbles, shock and track damage. Optical behaviour.	Gross morphological, mineralogical and petrological descriptions of particles and phases; e.g. lithic fragments, glasses, devitrified glass, degraded rock (breccias etc.). Microfossils. History of particles: – origin, exposure and weathering. Modal analyses and correlations with orbital data.
Electron microscopy			
Transmission (low and high voltage) Scanning (Stereoscan) (sections, etching, coatings, replicas and use of diffraction and other accessories, e.g. electron microprobe)	Partial to complete	As for optical but especially suitable for very small (μm size) particles. Tracks: – flare and cosmic ray damage.	As for optical. Radiation damage (metamictization) and radiation history (tracks). History of shielding of Mars surface by atmosphere
Luminescence			
(thermo-, electron-excited etc.)	Partial	Glow curves/temperature	Recent exposure history
Electron microprobe	Small	Location of major elements at μm scale.	Mineralogy of grains and major element assay of surface & bulk.
Ion microprobe (Ion sputtering/mass spectrometry)	Small	Location and isotopic composition of many elements (especially low atomic number)	Mineralogy of grains: chronology for some elements e.g. $^{207}Pb/^{206}Pb$ on 10^{-14} g (a few μ across)
Mass spectrometry			
Heat (gas extraction)	Partial to complete melting	Trapped/adsorbed gases, volatiles and pyrolysis products (CO, CO_2 etc.). Rare gas analysis, e.g. Ar isotopes	Early atmospheres. Geochronology – Crystallisation, metamorphic and exposure ages. Differentiation and volatilization/condensation fractionation. Neutron, cosmic ray and other fluences.
Extracts (stable isotope dilution) analysis/'spike')	Dissolution	Elemental and isotopic abundances (e.g. Pb, Rb/Sr, U, Th, REE, Sm/Gd) Carbon compounds	Trace elements – meteoritic influx and geochemical differentiation.
Spark source	Dissolution	Elemental abundances	
Auger electron spectroscopy (AES)	Slight	Surface analysis, limited to atomic layers	Surface chemistry -nm scale light elements especially.

Table IV (continued).

Technique	Damage to particles	Data	Information
(*ESCA*) Electron spectroscopy chemical analysis	Slight	As for AES	Chemical binding
Magnetic Susceptibility	None	Response to magnetic field/ temperature	Paramagnetism and intermediate ferromagnetics. Valency states.
X-ray diffraction (XRD)	None	Lattice structures for crystals (10 μm across)	Crystal structure, mineral recognition, zones and defects.
X-ray fluorescence (XRF)	None	Elemental abundances	Bulk chemistry
Optical spectroscopy Atomic Absorption Spectroscopy (AAS) Colorimetry	Dissolution	Abundances of major and minor elements.	Bulk chemistry (major and minor elements).
Radiochemistry e.g. Instrumental neutron activation analysis (INAA) Radiochemical neutron activation analysis (RNAA)	Dissolution	γ-ray spectrometry of neutron irradiated sample	Bulk chemistry (major, minor and trace elements, including REE). Bulk chemistry (trace elements – meteoritic origin)
Counting		Natural radioactivity	Natural abundances of radionuclides – cosmic ray, exposure differentiation history.

Note: All methods alter or contaminate a sample in some way. Sequential analysis can be conducted with a limited number of techniques arranged in the appropriate order.
The listings of techniques is illustrative and not exhaustive. Thus, some additional methods include SCANNIR, APS, ESR, NMR, LEED, elipsometry and electron scatter spectrometry, and Mössbauer.

(v) Some redundancy of analyses to ensure cross-checking.

(vi) Consortia and P.I.'s are selected for ability to make *and* interpret analyses.

(vii) Particular attention to be given to determining such parameters as would assist in providing 'ground truth' for planet-wide observations made by Mariner, Viking and subsequent orbiters (here improved methods are needed for dealing with micron-size particles – separation, identification etc.).

(viii) Optical selection under the microscope and hand-picking with a micro manipulator would provide 'particle separates', aliquots of which would be used to conduct a number of destructive analyses. In other cases methods could be used sequentially on single particles (c.f. Luna 20 Consortia) to obtain information on their origin, exposure and burial histories.

(ix) The end result of such schemes would be to conduct lunar rock consortium type studies with 'rocks' in the μg–mg mass range.

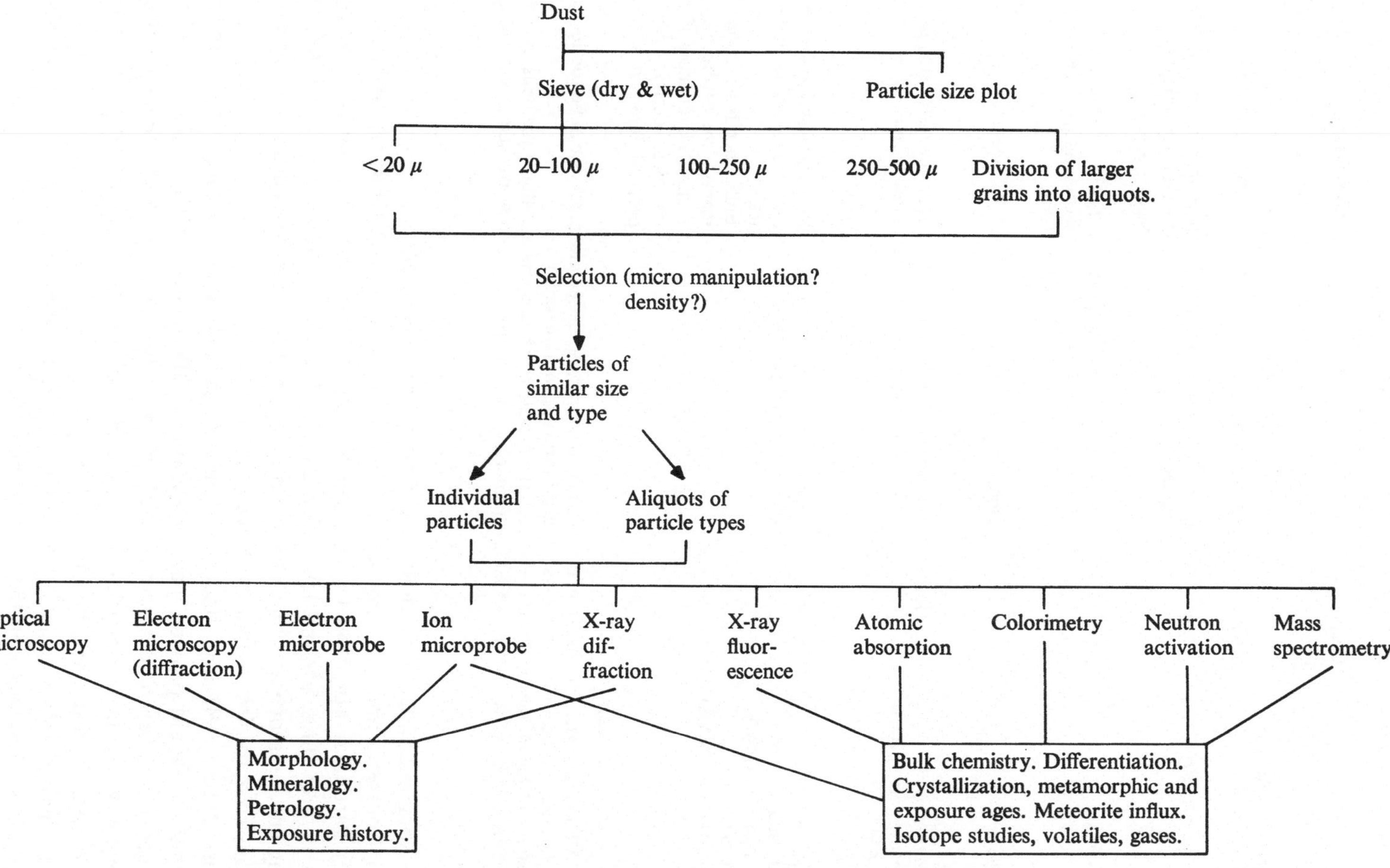

Fig. 5. Ultramicronalysis of micron-sized dust particles.
(Note a 100 μ diameter particle would have a mass of ~ 3 μg.)

One example of the multiple technique analysis of small fragments is provided by the Luna 16 studies of the group of investigators led by Wasserburg (Craig *et al.*, 1972). A 62 mg 'basaltic boulder' was split into smaller fragments which were then examined optically and by scanning electron microscopy before and after sectioning so as to define the gross morphology, mineralogy and petrology. Electron microprobe analyses provided data on the location of 13 elements and the common minerals, while study of fission tracks with the electron microscope revealed the uranium distribution. Chemical, isotopic dilution, thermal and neutron irradiation treatments coupled with mass spectrometry then afforded the major interest in providing geochronological information – model ages (Rb/Sr), gas retention ages ($^{40}Ar/^{39}Ar$) and cosmic ray exposure ages ($^{38}Ar–^{37}Ar$).

7. Conclusions

The 1973 NASA Mission Model for planetary exploration lists a post-Viking Mission for 1979 launch and a Surface Sample Return Mission for 1983 launch. From our limited study it would seem to us that high priority should be given to the Surface Sample Return Mission.

A kilogram of fine Martian soil would be ample foı study. Multitechnique (physical, chemical and mineralogical) analysis could be conducted on thousands of individual small dust particles. Particle descriptions, when combined with orbital observation of patterns of aeolian erosion and deposition, might allow extension of the results for one site over considerable portions of the Martian surface. It should be possible to infer the composition and history of major exposed units of the Martian crust.

Definitive life-detection experiments would have the highest priority of all studies for herein lies the greatest possibility for scientific return. However, here too is the nub of the concern associated with a returned sample – the possibility of back-contamination of the Earth by a Martian biota. This paper is presented in the hope that it will stimulate informed discussion of this problem and of the analytical challenge which a returned Mars sample would present.

Acknowledgements

We thank the following for helpful discussions over the past three years: Dr A. L. Burlingame, Dr M. Calvin, Mr J. L. Crowder, Dr E. I. Hamilton, Dr N. H. Horowitz, Dr K. Keil, Dr H. Nissen, Dr C. T. Pillinger, Dr R. S. Young; and Dr Paul Davies and Dr J. Chamberlain, Director of the Lunar Science Institute, for facilities. We are most appreciative of the information and advice provided by Dr J. M. Hayes of NASA's Post Viking Mars Sample Working Group and Dr L. Hochstein of NASA's Ames Research Center concerning their study of the 'Quarantine of Returned Mars Samples'. We thank the British Aircraft Corporation for time and facilities provided; however, the views expressed are the personal ones of the authors and do not represent an official view of the company.

References

Anders, E. and Albee, A. L. (eds.): 1973, *Geochim. Cosmochim. Acta* **37**, 719.

Chamberlain, J. W. (ed.): 1972, *Post-Apollo Lunar Science*, Lunar Science Institute, Houston.

Craig, H., Geiss, J., Gertner, W., Moorbath, S., Simmons, G., and Wasserburg, G. J. (eds.): 1972, *Earth Planetary Sci. Letters* **13**, 466 pages.

Dwornik, S.: 1973, NASA Mission Model, NASA Planetary Project Office.

Fox, D. G., Hall, L. B., and Bacon, E. J.: 1972, in W. Vishniac (ed.), *Life Sciences and Space Research* **X**, Akademie Verlag, Berlin, G.D.R.

Gatland, K. N.: 1971, *Spaceflight*, Jan., p. 2.

Gault, D.: 1973, 'Results of Spacecraft Missions to Mars', Meeting at the Royal Society, London, 15–16 May.

Goody, R.: 1973, *New Scient.* **58**, 602.

Hammond, A. L.: 1973, *Science* **179**, 463.

Hanel, R., Conrath, B., Hovis, W., Kunde, V., Lowman, P., Maguire, W., Pearl, J., Pirraglia, J., Prabhakara, C., Schlachman, B., Levin, G., Straat, P., and Burke, T.: 1972, NASA X-620-72-280.

Lederburg, J. and Alexander, M.: 1971, in A. W. Schwartz (ed.), *Theory and Experimenting in Exobiology* **2**, Norhoff, p. 123.

Martin, J. S. and Sibbers, C. W.: 1972, Viking 75 Project M75-145-0 (NASA) 23rd Congress of International Astronautical Federation, Vienna. See also NASA: 1972, Viking Mars Engineering Model (Langley Document M75-125-2) and NASA: 1969, Viking Baseline Orbiter Conceptual Design Description (Project Document 611-2).

Masursky, H.: 1973, 'Results of Spacecraft Missions to Mars', Meetings at the Royal Society, London, 15–16 May.

Ponnamperuma, C. (ed.): 1972, *Space Life Sci.* **3**, 311.

Wasserburg, G. J.: 1972, *Astronaut. Aeronaut.* 16 April.

LIFE ON JUPITER?

W. F. LIBBY

Dept. of Chemistry, Institute of Geophysics, and Planetary Physics, University of California, Los Angeles, Calif., U.S.A.

Abstract. The possibilities of life on Jupiter are discussed from the point view of life as we know it. That is, we assume that any life on Jupiter would not involve new principles foreign to us. Proteins would be a constituent as would fats and the other building blocks of living organisms on Earth. This leads us to a set of limiting parameters, such as pressure. Studies in the laboratory have shown that proteins and other essential molecules are denatured by pressures of 4000 atm and higher. Thus, we must not expect life in the great depths of the Jovian atmosphere. It could exist only at depths of several hundred kilometers in the atmosphere. Since no solid surface could possibly exist at such altitudes, any organisms present must be small enough to be buoyed up by the turbulent atmospheric currents or must fly or both. Such possibilities, however, seem to be real. The necessary nutrients to preserve life and foster growth could be furnished by the Miller-Urey type reactions of ionizing radiation on the reducing atmosphere undoubtedly present. There can, of course, be no possibility of oxygen on Jupiter, and so the life forms, if they exist, must be anaerobic. Such possibilities are real and have often been cited in connection with the origin of life on Earth.

1. Introduction

The question of whether or not life could exist on Jupiter is fascinating. The answers would seem to require four separate discussions. First, are there regions in which the intensity of sunlight is adequate to promote the excitation characteristic of living molecules?; second, are there regions where the temperatures are tolerable?; third, are there regions where the pressures are not too high?; and fourth, are there regions where bodies of the density characteristic of living system we know could be supported?

2. Sunlight

It appears likely that the cloud tops of Jupiter occur at about 2.8 atm pressure (Michaux, 1967). The temperature at the cloud tops apparently is about 150° absolute but could rise rapidly with descent so that at some depth where the pressure is still tolerable, the temperature could be in the tolerable range of 0 to 100 °C, the pressures being a few hundred atmospheres. The Jovian atmosphere is very active, so considerable stirring undoubtedly occurs. This could suspend the particles and so expose them to necessary sunlight. The dense clouds of Venus still transmit some sunlight to the surface (Venera, 1972) through some ninety atmospheres so it seems very likely that some sunlight will permeate the tolerable temperature layers on Jupiter.

3. Temperature

If one assumes that the atmosphere of Jupiter in the higher levels where light might be possible is essentially pure hydrogen with the appropriate amounts of ammonia, methane and water, it is straightforward to show that there must be a level of tolerable temperatures. The thickness of the cloudy warm hydrogeneous atmosphere of tolerable temperatures, 0 to 100 °C, will probably be several tens of kilometers so we envisage the life zone, if any exists, as being in the clouds at pressures ranging from 10's to 100's of atmospheres and densities ranging up to 0.19 cc^{-1}. Under such conditions any life forms either have to fly or be supported by wind and updrafts. The gravity force on Jupiter is some 2.65 times that on Earth, so the settling velocity of matter of unit density would be considerable. The very existence of clouds, however, shows that there is an inversion layer at the cloud top with convective mixing below. Therefore, there are currents that transport vertically, and it is obvious from visual observation that there are strong horizontal currents, particularly in the east west direction. Therefore, it seems clear that the possibility exists for organisms to comfortably survive if they are small enough to be supported by these currents, or if they can fly, or both.

Siegel and Giumarro (1965) and Molton and Ponnamperuma (1972) have shown that certain microorganisms can survive and flourish in ammonia rich atmospheres containing gases normally thought to be toxic such as hydrogen cyanide. Thus a Jovian microbiosphere would seem to be a definite possibility (Oparin, 1957; Ponnamperuma and Molton, 1972).

There is a strong argument that life as we know it originated in chemically reducing conditions (Oparin, 1957) and there is abundant evidence that reducing atmospheres are far more likely to form polymeric and complex molecules in the presence of ionizing radiation than are oxidizing atmospheres (Sheridan *et al.*, 1972). The classical experiments of Miller (1953) and others since have shown that amino acids can be produced in the type of atmosphere we envisage that Jupiter possesses. More recently Molton and Ponnamperuma (1972) have shown that colored bodies, similar to the red spot on Jupiter can be produced by a glow discharged in a Jovian atmosphere (Miller, 1953). Thus, it seems quite clear that there are definite possibilities of nutritious food molecules being produced which could be used to nurture living organism (Michaux, 1967; Ponnamperuma and Molton, 1972).

4. Pressure

Where the temperatures in the Jovian atmosphere are quite indeterminate the pressure is known with more certainty. Peebles and De Marcus (Michaux, 1967) conclude that the top segment of 750 km contains little except hydrogen and that the pressure at this level is some 20000 atm corresponding to a density of nearly 0.2 g cc^{-1}. At higher levels, of course, the pressure is less. At 143 km depth, the calculated pressure is 230 bars and the density 0.032 g cc^{-1}. Similarly, at twice this depth 286 km the pressure is calculated to be 2700 bars and the density 0.103 g cc^{-1}. Since we have quite good evidence of the cloud tops being at some 3 bars and temperature of about 156 °C it would seem likely that temperatures somewhere in the top 300 km could reach comfortable levels of 0 to 100 °C. We know from laboratory experimentation that proteins are denatured by pressures of some 4000 atm. DNA, desoxyribonucleic acid, is uncoiled at about 4300 atm (Solomon *et al.*, 1965). We know that organisms live in the deepest part of the ocean which corresponds to about 1000 atm. Therefore, it would appear to be reasonable to expect that life forms could survive under the pressures on Jupiter, perhaps as deep as 750 km.

5. General Overview

The possibility of life on the planets has been debated often. Sagan (1964) suggested some time ago that turbulence in the atmosphere could support life in the clouds. Our own group has considered at length the possibility of life on Venus and suggest that warm polar seas may indeed harbor life on this planet. We studied (Seckbach and Libby, 1970) the tolerance for high CO_2 of hot spring bacteria and found that indeed they could tolerate considerable pressures of CO_2 and consequently acidic media. The work of Siegel and Giumarro (1965) and Molton and Ponnamperuma (1972) referred to earlier, further bolsters the notion that the great adaptability of living organisms make it reasonable that life could exist on Jupiter. There is of course no possibility of moving into an oxidizing phase as it did on earth, but so long as energy from the sun is absorbed the energy rich molecules apparently necessary for all living systems could be generated, giving food which could be used to nurture living systems. It is completely clear from the intense coloration of the Jovian atmosphere that enormous chemical operations are underway which is true of course with the other giant planets as well. The colors probably are due, almost without doubt, to large organic molecules generated by the kind of plasma chemistry possible in reducing environments. It is clear also, of course, that there is a possibility that the planet is sterile. This could be true if the temperatures are excessive. If the atmosphere were strictly adiabatic there would be only a relatively narrow altitude range in which life would be possible at all. But the strong stirring that is evident from the movements of the clouds and the rapid changes in markings and invisible features indicate that violent mixing, possibly even turbulent, is occurring.

Jupiter is the strongest radio source among the planets. It has been suggested that this is due to the organic chemistry of the reducing environment (Marshall and Libby, 1967).

What would be the best way of establishing the existence of life? It would seem that more laboratory experiments are in order. The kind of test that Siegel and Giumarro (1965) and Molten and Ponnamperuma (1972) have done, if extended, could give a considerable degree of confidence. The possibility of sampling the Jupiter atmosphere seems remote. Further work on the temperatures would be important. It is not clear, however, that this is possible although it might be conceivable. Occultation work, especially from orbiting satellite, could give valuable information as to the scale height and therefore to the temperatures. It seems, however, that from the information now at hand, it would be very difficult to get below the cloud tops without having a direct parachute lander, a very difficult task indeed.

The type of experiment in the laboratory which would be most revealing would be to test the viability of terrestrial organisms in their ability to adapt to possible Jovian atmospheres. This would be particularly revealing in the case of microbiological systems. We know well from the information available at present that soil bacteria and marine bacteria are quite capable of living in reducing environments.

A program to develop life in reducing environments illuminated by sunlight possibly is the shortest and surest way to an answer to the prime question as to the possibility of life on Jupiter and the other giant planets. The question of life on Mars is likely to be answered by landers in the next few years. Mars, however, is much more like the Earth in that it possesses an oxidizing atmosphere, although a very thin one. Venus, also, probably is highly oxidizing although the Venera data (Otroshchenko and Surkov, 1974) on oxygen leave the question somewhat open. The vast dominance of CO_2 in its atmosphere speaks strongly for an oxidizing environment. Mercury undoubtedly is barren because it is so hot.

Thus only the outer planets could harbor life in a totally reducing environment so experimental studies of these possibilities should be encouraged.

Acknowledgements

Research Sponsored by the National Aeronautics and Space Administration under NASA Grant Number 05-007-003. Contribution No. 3120 from the Department of Chemistry, UCLA.

References

Marshall, L. and Libby, W. F.: 1967, *Nature* **214**, 126.
Michaux, C. M.: 1967, in *Handbook of the Physical Properties of the Planet Jupiter*, NASA SP-3031, Office of Technology Utilization.
Miller, S. L.: 1953, *Science* **117**, 528.
Molton, P. and Ponnamperuma, C.: 1972, *Nature* **238**, p. 217.
Oparin, A. I.: 1957, *The Origin of Life*, McMillan *et al.*, Academic Press, 3rd edition.
Otroshchenko, V. and Surkov, Y.: 1974, this volume, p. 487.
Ponnamperuma, C. and Molton, P.: 1972, *Space Life Sci.* **2**, 14.
Sagan, C.: 1964, in M. Florkin and A. Dollfus (eds.), *Life Sciences in Space Research*, North Holland Publ. Co., Amsterdam, p. 35.
Seckbach, J. and Libby, W. F.: 1970, *Space Life Sci.* **2**, 121.
Sheridan, M. E., Greer, E., and Libby, W. F.: 1972, *J. Am. Chem. Soc.* **94**, 2614.
Siegel, S. M. and Giumarro, C.: 1965, *Icarus* **4**, 37.
Solomon, L., Zeegen, P., and Eiserling, F. A.: 1965, *Biochim. Biophys. Acta* **112**, 102.
Venera 8 Data: 1972, as reported in Pravda, September 10.

THE POSSIBILITY OF ORGANIC MOLECULE FORMATION IN THE VENUS ATMOSPHERE

V. A. OTROSHCHENKO and Yu. A. SURKOV

V. I. Vernadsky Institute of Geochemistry and Analytical chemistry, USSR Academy of Sciences, Moscow, U.S.S.R.

Abstract. Based on the detection of ammonia in the Venus atmosphere, and the suggested presence of hydrogen chloride, a structure for the Venus atmosphere was suggested as having 3 cloud layers, consisting of ammonium chloride (30 to 50 km above the ground), a mixture of ammonium bicarbonate and ammonium carbamate (NH_2COOHN_4) from 50–60 km, and water ice crystals above this. There is a strong possibility of electrical discharge in the atmosphere as a result of thermal convective turbulence, which in the case of the slightly reducing atmosphere outlined above could lead to organic compound formation.

The hypothesis was tested experimentally by passing a 60 KV spark from platinum electrodes through a gas mixture of composition: N_2(0.2%), NH_3 (2%), Water (5%), O_2 (0.6%), CO_2 (remainder), for 8 hr. The products were analysed by mass spectrometry and amino acid analysis by ion exchange. Methane and formaldehyde were identified by MS, and glycine and alanine by the amino acid analyzer.

The presence of organic compounds in the Venus atmosphere is therefore a strong possibility.

The presence of complicated compounds, analogous to terrestrial ones, the base of which is carbon, is on planets of the solar system of principal importance from the point of view of the prebiological evolution of the matter.

Results of research in this field must be taken into account when methods of detecting extraterrestrial forms of life are elaborated (Hubbard *et al.*, 1971; Otroshchenko, 1973; Glasston, 1968).

In many model experiments (Hubbard *et al.*, 1971; Ponnamperuma *et al.*, 1969; Chada *et al.*, 1971; Sagan *et al.*, 1971) the possibility of organic molecule formation in the atmospheres of Mars and Jupiter has shown.

According to results of astronomic observations the presence of water and complicated polyatomic organic molecules even outside the solar system has been established (Bune, 1971; Rank *et al.*, 1971).

Based on experimental data, namely results of direct measurements of the chemical composition of the Venus atmosphere out by the automatic stations 'Venera 4–8' (Vinogradov *et al.*, 1968, 1970a, b; Surkov *et al.*, 1973b), Table I, an attempt was made to study the possibility of organic molecule formation in the atmosphere of this planet.

From the chemical composition of the atmosphere, as well as from a number of indirect reasons concerning the evolution of the whole planet and the interaction of the lithosphere with the atmosphere (Otroshchenko and Muchin, 1971), it may be considered that the Venus atmosphere must have slightly reducing properties. Under such conditions processes of organic compound synthesis may go on (Hubbard and Hardy,

1971; Otroshchenko and Muchin, 1971; Garrison *et al.*, 1951; Abelson, 1957a, b).

One of the most interesting results obtained with the aid of the space station 'Venera-8' is the presence of ammonia in the atmosphere of the planet (Surkov *et al.*, 1973b). In the atmosphere it may easily associate with water vapours and carbon dioxide forming various ammonia compounds. There are also data (Lewis, 1968) on the presence of HCl and HF vapours in the atmosphere. Therefore some of the ammonia may form compounds of the NH_4Cl type.

TABLE I

Chemical composition of the Venus atmosphere (Data obtained by the automatic stations 'Venera 4–8')

Component	Content in weight %
CO_2	97
N_2	$\leqslant 2$
H_2O	2.5–3[a]
O_2	$\leqslant 0.1$
NH_3	0.01–0.1[b]

[a] at a pressure of 2–0.6 atm.
[b] at a pressure of 2–10 atm.

The recently proposed model of the Venus cloud layer (Surkov *et al.*, 1973a) mainly calculated on the basis of direct experimental data, assumes the existence of three cloud zones passing one into the other.

From the 30 km height over the average level of the planet surface and to 50 km the existence of clouds consisting of ammonium chloride is assumed. Further, up to 60 km height the main component of the clouds must be ammonium compounds NH_4HCO_3 and NH_2COONH_4 and above 60 km – fine ice crystals.

According to existing ideas (Ginzburg and Feigelson, 1968; Avduevsky *et al.*, 1971; Kertanovich, 1972) on heights corresponding to the cloud layer (50–60 km) intensive convective displacements of atmospheric currents with speeds up to 50 m s^{-1} must occur. This process can probably lead to the formation of electric discharges in the depth of the cloud layer.

Based on these reasons an installation was created which allowed simulation of the Venus atmosphere in the region of the cloud layer, both in chemical composition and in the temperature and pressure gradient. Through a mixture of gases the analogue of the Venus atmosphere, a spark discharge of about 60000 V voltage was passed. The distance between the electrodes made of platinium was about 15 mm.

The mass-spectrometric analysis of gas mixtures corresponding to the Venus atmosphere at the level of the cloud layer confirms the possibility of the existence at a 50–60 km height of clouds which consist of ammonium compounds.

In the presence of great amounts of carbon dioxide during ammonium bicarbonate formation:

$$NH_3 + H_2O + CO_2 \rightleftarrows NH_4HCO_3$$

and ammonium carbamate formation:

$$2NH_3 + CO_2 \rightleftarrows NH_2COONH_4$$

as is shown by calculations (Surkov *et al.*, 1973b) ammonium compounds form compact clouds. Atmosphere parameters at their lower boundary are $P \sim 1.1$ atm, $T \sim 315$ K. At this level precipitations of ammonium salts with subsequent dissociation into ammonia, water and carbonic dioxide must be expected. Indeed, our measurements show that the mass-spectrum corresponding to the gaseous mixture (93% CO_2 and 7% NH_3) at $P \sim 1$ atm and $T \sim 315$ K qualitatively differs from the mass-spectrum of the same mixture at $P \sim 0.56$ atm and $T \sim 286$ K by the presence of a peak corresponding to NH_3 (17 mass). This is most probably caused by the dissociation of ammonium compounds which form a white precipitate on the walls of the containers vessels at a temperature of 286 K.

A mass-spectral analysis of the gas mixture was carried out consisting of 0.2% nitrogen, 2% ammonia, 5% water, 0.6% oxygen and 92,74% carbonic dioxide, through which for 8 hours an electric discharge was passed. During this analysis we succeeded in revealing the formation of cianic hydrogen, methane and formaldehyde which is in agreement with results of preceding work (Hubbard and Hardy, 1971; Garrison *et al.*, 1951).

With the aid of an amino acid analyser an investigation of the precipitate that formed on the walls during the action of the electric discharge was carried out. Glycine and alanine have been identified.

Yet this result is only preliminary.

Thus, it apparently may be considered that at heights of 50–60 km the existence in the cloud layer of ammonium compounds in the form of a solid phase is possible. When electric discharges are passing through the cloud layer (though energy may be supplied in any form (Oró, 1963), the formation of complicated organic molecules of the type of aminoacids presumably according to strecker's mechanism is possible.

The detection of some organisms in the atmosphere scarcely can be expected (Young, 1971) though such assumptions had been expressed earlier (Sagan, 1966; Morowitz and Sagan, 1967). But the formation and further evolution of organic molecules and in particular of amino acids in the Venus atmosphere is of great interest.

Acknowledgement

We thank Dr O. M. Kalinkina, Dr A. N. Pudov and Dr E. P. Senchenkov for their assistance.

References

Abelson, P. H.: 1957a, *Science* **124**, 935.
Abelson, P. H.: 1957b, *Ann. N.Y. Acad. Sci.* **69**, 274.

Avduevsky, V. S., Zavelevich, F. S., Marov, M. Ja., Naikina, A. I., and Polezhaev, V. I.: 1971, *Kosmich. Issled.* **9**, 2.
Bune, D.: 1971, *Nature* **234**, 5328.
Chadha, M. and Ponnamperuma, C.: 1971, in R. Buvet and C. Ponnamperuma (eds.), *Molecular Evolution, I. Chemical Evolution and the Origin of Life*, North-Holland Publ. Co., Amsterdam, pp. 143–154.
Garrison, W. M., Morrison, D. C., Hamilton, J. G., Benson, A. A., and Calvin, M.: 1951, *Science* **114**, 2964.
Ginzburg, A. S. and Feigelson, E. I.: 1968, *Dokl. Akad. Nauk S.S.S.R.* **182**, 1.
Glasston, S. (ed.): 1968, *The Book of Mars*, NASA SP-197, Washington D. C.
Hubbard, J. S., Hardy, J. B., and Horowitz, N. H.: 1971, *Proc. Nat. Acad. Sci.* **68**, 3.
Kertanovich, V. V.: 1972, *Kosmich. Issled.* **10**, 2.
Lewis, J. S.: 1968, *Icarus* **8**.
Morowitz, H. and Sagan, C.: 1967, *Nature* **215**, 5107.
Oró, J.: 1963, *Nature* **197**, 4871.
Otroshchenko, V. A.: 1973, in A. I. Oparin (ed.), *Problemy proishozdenia i Sushchnosti Zhizni*, 'Nauka', Moscow.
Otroshchenko, V. A. and Muchin, L. M.: 1971, in R. Buvet and C. Ponnamperuma (eds.), *Molecular Evolution, I. Chemical Evolution and the Origin of Life*, North-Holland Publ. Co., Amsterdam, pp. 516–522.
Ponnamperuma, and Woeller, F.: 1969, *Icarus* **10**, pp. 386–392
Rank, D. M., Townes, C. H., and Welch, W. I.: 1971, *Science* **174**, 4014.
Sagan, C.: 1966, *Astron. Aeronaut.* **4**, 7.
Sagan, C. and Khare, B. N.: 1971, *Astrophys. J.* **168**, 3.
Surkov, Yu. A., Andreichikov, B. M., and Kalinkina, O. M.: 1973a, *COSPAR XVIth Meeting*, Konstanz, West Germany.
Surkov, Yu. A., Andreichikov, B. M., and Kalinkina, O. M.: 1973b, *Dokl. Akad. Nauk S.S.S.R.* **213**, 2.
Vinogradov, A. P., Surkov, Yu. A., and Andreichikov, B. M.: 1970a, *Dokl. Akad. Nauk S.S.S.R.* **190**, 3.
Vinogradov, A. P., Surkov, Yu. A., Andreichikov, B. M., Kalinkina, O. M., and Grechishcheva, I. M.: 1970b, *Kosmich. issled.* **8**, 4.
Vinogradov, A. P., Surkov, Yu. A., Florensky, K. P., and Andreichikov, B. M.: 1968, *Dokl. Akad. Nauk S.S.S.R.* **179**, 1.
Young, R.: 1971, in R. Buvet and C. Ponnamperuma (eds.), *Molecular Evolution, I. Chemical Evolution and the Origin of Life*, North-Holland Publ. Co., Amsterdam, pp. 510–515.

PLANETARY SYSTEMS AND EXTRATERRESTRIAL LIFE

SHIV S. KUMAR

Dept. of Astronomy, University of Virginia, Charlottesville, Va. 22903, U.S.A.

Abstract. The paper reviews the present status of the problem of the existence of other planetary systems in the Galaxy. Observational data and theoretical results are presented to show that the occurrence of planetary systems is, most probably, not a universal phenomenon. Study of the stability of planetary orbits in the vicinity of double stars indicates that, in general, planetary systems can not survive around them over long periods. Therefore, we should rule out the possibility of the existence of planetary systems similar to our own in the neighborhood of double stars. In the solar neighborhood, at least 60% of the stars are known to be members of double systems. The nature of the 'dark' companions is discussed and it is concluded that they are stellar objects and not planets. Recent work on the absence of a perturbation in the motion of Barnard's star is discussed. Comments are made on the existence of extraterrestrial life in the solar system and around other stars in the Galaxy.

1. Introduction

This paper deals with the question: Is the occurrence of planetary systems a universal phenomenon? During the last 25 yr, the general feeling among scientists, particularly astronomers, has been that practically all stars in the Galaxy (the stellar system of which our Sun is a member) possess planetary systems similar to our own. This belief was based on the assumptions that the formation of a planetary system is intimately connected to the formation of a star and that, in general, planets are formed whenever a star is formed somewhere in space. Although not all theories of the origin of our solar system subscribe to this view, the general theme in many such theories is that our planetary system was formed by processes which must take place or have already taken place in the vicinity of most stars in the Galaxy.

This belief has not been substantiated thus far by any direct observational evidence. We have not been able to observe a planetary system in the process of formation around any star. Also, there is no clear observational evidence for the presence of a second planetary system (besides our own) in the Galaxy. Further, the origin of our own solar system remains a mystery despite the recent tremendous advances in our knowledge of the structure and evolution of the Sun, planets, and other members of the solar system. In the following discussion, I shall try to show that the occurrence of planetary systems is most probably not a universal phenomenon. I shall also make some general remarks concerning the existence of extraterrestrial life in the solar system and around other stars in the Galaxy.

2. The Observational Data

The observationally known number of planetary systems in the Galaxy continues to be one! Other planetary systems most probably exist in the Galaxy, but we have no direct evidence for the existence of any other system similar to our own. Our system contains nine planets which are moving in nearly coplanar, nearly circular orbits around a central star (the Sun). The mass of the most massive planet (Jupiter) is only 0.001 $M_{\odot}$, where $M_{\odot}$ stands for the mass of the Sun and its numerical value is 2×10^{33} g. Further, the eccentricity of Jupiter's orbit is only 0.048, and the inclination of Jupiter's orbital plane to the ecliptic is also very small (1°.3).

The age of the solar system (from measurements of the ages of the Earth, Moon and the meteorites) is approximately 4.5×10^9 yr. Our solar system has survived over such a long period, and it appears that it will remain dynamically stable for several billion years in the future. The high degree of stability of the system is most probably due to the small mass, small orbital eccentricity and small orbital inclination for Jupiter (Kumar, 1971). If any one or more of these three quantities (the mass, the orbital eccentricity and the inclination of the orbital plane) for Jupiter had numerical values much higher than the present ones, then the stability of the orbits of all the other planets would most probably have been destroyed by now. It appears that the Sun and the planets make up a system which is very different from the typical double star. Double stars generally have mass ratios M_A/M_B, where M_A is the mass of the primary and M_B the mass of the secondary, in the range 1–10, while in the case of the solar system the ratio of the mass of the Sun to that of any planet ($M_{\odot}/M_P$) exceeds the number 1000. Further, double stars generally have eccentric orbits. From a study of the stability of planetary orbits in the three-body problem (Shelus and Kumar, 1970; Kumar, 1971; Shelus, 1972), it has been found that a planetary system such as our own most probably cannot survive over long periods in the vicinity of a double star.

Worley's (1969) study of the observational data concerning the duplicity of stars in the solar neighborhood clearly indicates that at least 60% of the stars are members of double systems. Since the detection of faint companions is an extremely difficult problem, it is clear that a large fraction of the so-called single stars are accompanied by them. It should be noticed that, only 34 yr earlier, the percentage of the stars occurring in double systems was estimated to be approximately 30% (Aitken, 1935). The manner in which this percentage has been going up leads me to the tentative conclusion that practically all stars are members of double and multiple systems. Perhaps the percentage of single stars (a single star is defined here as a star unaccompanied by any other star) is 1% or lower. Even if 1% of all stars are single, the number of stars accompanied by planetary systems will be less than 10^9, for there is no known physical law which says that every single star has to have a planetary system around it. A single star may exist all by itself without being accompanied by one or more planets. This point is discussed further in the next section.

3. The Processes of Star and Planet Formation

The processes of star formation are poorly understood at the present time. Star formation is generally considered to take place in interstellar clouds, and, consequently, stars get formed in groups. It is very difficult to form a star by itself, especially if its mass is as low as one solar mass. After a group or cluster of stars is formed from a cloud, the formation of binaries may take place within the cluster. Some single and double stars will be ejected due to the dynamical evolution, and the residual system will generally be a multiple star system with two or more members.

The processes of planet formation, on the other hand, are thought to take place in a nebula somehow formed in the vicinity of a single star. Once a dust-gas cloud is formed around a single star (such as our Sun), then the collisional accretion will lead to the formation of small rocks which eventually grow into planetary sizes. Thus stars are formed as extended, gaseous objects by instabilities in the interstellar clouds, while the planets start as small, dense rocks by collisional accretion in a nebula somehow formed around a single star. The formation mechanism for such a cloud around a single star continues to be a puzzle. It has generally been assumed that, when a single star is formed, some left-over material will be the source of a dust-gas cloud. However, only the very massive stars ($M > 50\ M_{\odot}$) may form as single objects; the rest of the stars most probably get formed in groups or clusters. It is hard to form a stable planetary system around a star which is a member of a group or cluster. A planetary system will therefore get formed around a star of mass 1.0 $M_{\odot}$ after such a star gets ejected from a stellar group. If this argument is correct, then the formation of a planetary system may take place as a result of evolutionary effects (accretion of interstellar matter by a star, internal instabilities, etc.). Therefore, a planetary system may originate if certain special conditions are met during the course of the evolution (in space and time) of a single star. For example, the accretion of interstellar matter will take place only if the star encounters a very dense cloud moving at a low relative velocity. It appears that the formation of a planetary system around a star is a special and rare process. The percentage of single stars accompanied by planetary systems similar to our own may be as low as 1%.

4. The Nature of the Dark Companions

There exists definite evidence that some nearby stars are accompanied by 'dark' companions of low mass ($M \simeq 0.01\ M_{\odot}$). Some workers have suggested that these objects are similar to the planets in the solar system. However, on closer examination (Kumar, 1966, 1972), it turns out that they are very different from the planets. Since the minimum mass that a star can have is around 0.01 $M_{\odot}$ (Kumar, 1972), these objects are stellar. It should be noted again that the mass of the most massive planet in the solar system is only 0.001 $M_{\odot}$. From the theory of stars of very low mass (Kumar, 1963), we know that a star with mass in the neighborhood of 0.01 $M_{\odot}$,will go through a contraction phase soon after its formation and it will be a luminous object during this phase. As the contraction proceeds, the object becomes degenerate and cannot generate energy by hydrogen burning. The dark companions are nothing but stars of very low mass that are gradually cooling toward the stage of complete degeneracy (black dwarfs). These objects are still far from the zero-luminosity stage, but we cannot see them in the telescope because of the glare from the more-luminous primaries in these systems. The mass ratios and eccentricities in the systems containing dark companions are also typical of double stars, and there is very little doubt that these systems are very different from our solar system.

Some time ago, it was reported by van de Kamp (1963) that he had observed a perturbation in the motion of Barnard's star and he interpreted this perturbation as being due to the presence of a dark companion of mass around 0.002 $M_{\odot}$ (twice the mass of Jupiter). The existence of an object of such a low mass is obviously of great interest, but the evidence for perturbation reported by van de Kamp is very weak. Recently Gatewood and Eichhorn (1973) have independently analyzed the astrometric data for Barnard's star. They have carefully studied the 161 plates of Barnard's star, obtained with the Thaw refractor of the Allegheny Observatory during 1916–71, and 80 plates obtained with the Van Vleck Observatory refractor between 1923 and 1971, and their conclusion is that no perturbation is present in the motion of this star! Therefore, it is very difficult to say if anything (a star of low mass or a planet) is moving around Barnard's star. It is clear that more observational work needs to be done to conclusively establish the presence or absence of a perturbation in the motion of Barnard's star.

5. Planetary Systems and Extraterrestrial Life

I have already presented enough arguments to show that the occurrence of planetary systems is not a universal phenomenon in the Galaxy. The number of planetary systems in the Galaxy is certainly not as large as 10^{11} or 10^{10}. Therefore, the estimates of habitable planets in the Galaxy given by various workers (see, for example, Brown, 1964) are highly exaggerated. It is extremely difficult to estimate the exact number of habitable planets in the Galaxy. If one starts with a total of 10^7 planetary systems in the Galaxy, then, at a given time, the number of inhabited planets may not exceed a few hundred. It should be noted that not every habitable planet is necessarily inhabited. It is, therefore, quite likely that the nearest planet with any kind of evolved life on it is located at a distance of several hundred light years from the solar system. If this is correct, then the task of interstellar communication between us and any other technologically advanced civilization, if it exists, is almost impossible, unless signals can somehow be made to travel through interstellar space with speeds exceeding the speed of light. I am aware of the current interest in the field of interstellar communication (Kaplan, 1971; Troitskii *et al.*, 1971; Oliver, 1973), and I am not completely against such ventures. However, it should be clearly understood that the chances of success in this kind of experiment appear to be very slim. This should be kept in mind before starting a program of search which involves huge expenditures of time and money.

On the other hand, the probability of success in detecting life (in some stage of biological evolution) on other planets in the solar system appears to be quite high. All the planets have existed for approximately five billion years, and it is quite likely that the processes leading to the origin of life on the Earth have also taken place on one or more of the other eight planets. Whether or not biological evolution ever proceeded on them to a level comparable to our present level is very hard to say at the present time. But this is something that can be settled by the unmanned and manned explorations. It is true that the environment on and around the giant planets is not very favorable for the evolution of life, but even there the possibility of chemical and biological evolution cannot and should not be ruled out. The search for living or dead biological systems on planets such as Mars is within our grasp and should be given a very high priority. Obviously, the detection of life, even in a very primitive form, on any other planet in the solar system will be of great significance for the study of the origin of life on the Earth. The search for life around distant stars should be attempted only after we have explored the solar system and improved our understanding of its origin and the origin of life within it.

References

Aitken, R. G.: 1935, *The Binary Stars*, McGraw Hill, New York.

Brown, H.: 1964, *Science* **145**, 1177.

Gatewood, G. and Eichhorn, H.: 1973, *Astron. J.* **78**, 769.

Kamp, P. van de; 1963, *Astron. J.* **68**, 515.

Kaplan, S. A.: 1971, *Extraterrestrial Civilizations*, NASA TT F-631.

Kumar, S. S.: 1963, *Astrophys. J.* **137**, 1121.

Kumar, S. S.: 1966, in M. Hack (ed.), *Colloquium on Late-Type Stars*, Trieste, Italy.

Kumar, S. S.: 1971, *Nature* **233**, 473.

Kumar, S. S.: 1972, *Astrophys. Space Sci.* **17**, 219.

Oliver, B. M.: 1973, *Project Cyclops*, CR 114445.

Shelus, P. and Kumar, S. S.: 1970, *Astron. J.* **75**, 315.

Shelus, P.: 1972, *Celes. Mech.* **5**, 483.

Troitskii, V. S., Staroduhtsev, A. M., Gershtein, L. I., and Rakhlin, V. L.: 1971, *Soviet Astron.* **15**, 508.

Worley, C.: 1969, in S. S. Kumar (ed.), *Low Luminosity Stars*, Gordon and Breach, New York.

THE ORIGIN OF LIFE IN A COSMIC CONTEXT

CARL SAGAN

Laboratory for Planetary Studies, Cornell University, Ithaca, N.Y. 14850, U.S.A.

Abstract. The significance of examinations of the planets and their satellites, asteroids, comets and the interplanetary medium for the origin of life is assessed. It appears that the deprovincialization of biology must await the search for extraterrestrial life.

In our present profound ignorance of exobiology, life is a solipsism. So far as we know, all organisms on the Earth use nucleic acids for self-replication. All organisms use proteins for enzymatic catalysis. Small molecules such as porphyrins and nucleoside phosphates are used over and over again for functionally diverse tasks. Of the billions of possibilities, life seems to have used around 50 simple molecular building blocks for all its chemical functions. There are even remarkable commonalities in such structures as the $9+2$ cross-sectional structure which is present from the flagella of protozoa to the spermatazoa of human beings. Some of these chemical or functional units – for example the particular porphyrins used in contemporary biology – may be mere historical accidents. The substitution of vanadium or niobium for iron in the circulatory fluids of ascidians; the rejection of sunlight in the frequency range where it is most plentiful by chlorophyll; and the development of accessory photosynthetic pigments to absorb in the yellow/green while chlorophyll remains the ultimate photon acceptor all seem to speak in favor of this idea. On the other hand, nucleic acids may be a unique molecule for self-replication; at least we know of no alternatives to date. But these arguments are weak. There is no aspect of contemporary biology in which we can distinguish the evolutionary accident from the biological *sine qua non*. We cannot distinguish the contingent from the necessary. The time scale of the evolutionary process is so long that it is unlikely that such questions will ever be answered by terrestrial experimentation. For this reason biology suffers from a deadening provincialism, an inevitable parochialism – like the physics of falling bodies before Newton showed that the same laws applied to the motion of apples in England, the moon in circumterrestrial space and the planets about the Sun. The deparochialization of biology can only come in the same way: by comparisons with examples elsewhere, by the pursuit of extraterrestrial life.

Not only does the understanding of the evolution of living systems after the origin of the first self-replicating system require investigations in an astronomical context, but understanding the origin of the molecular precursors of life does so as well. The universe is mostly hydrogen, oxygen, nitrogen, carbon and noble gases. Biology is mostly hydrogen, oxygen, carbon, nitrogen, phosphorous and sulfur. Materials which are of major abundance in the crust of the Earth such as iron and silicon are relatively underabundant and unimportant for cosmic abundances and for biology.

TABLE I

Comparison of the relative abundances of the elements in the cosmos, in biology, and in the crust of the Earth

Atom	Universe	Life (terrestrial vegetation)	Earth (crust)
Hydrogen	87	16	3
Helium	12	0	0
Carbon	0.03	21	0.1
Nitrogen	0.008	3	0.0001
Oxygen	0.06	59	49
Neon	0.02	0	0
Sodium	0.0001	0.01	0.7
Magnesium	0.0003	0.04	8
Aluminum	0.0002	0.001	2
Silicon	0.003	0.1	14
Sulfur	0.002	0.02	0.7
Phosphorus	0.00003	0.03	0.07
Potassium	0.000007	0.1	0.1
Argon	0.0004	0	0
Calcium	0.0001	0.1	2
Iron	0.0007	0.005	18

After Sagan (1970).

There is at least a qualitative sense in which the atomic constitution of life on Earth is closer to the cosmic abundances of the elements than it is to the elemental composition of the planet on which it resides (see Table I). It is the Earth which has an anomalous chemistry, the reason being that the light elements were able to escape from the Earth during its formation and hydrogen has been able to escape since. The Jovian planets – Jupiter, Saturn, Uranus and Neptune – have much larger accelerations due to gravity and much colder exosphere temperatures. They have retained their original cosmic abundances. Their atmospheres are comprised – apart from noble gases – of the fully saturated hydrides of the cosmically most abundant atoms: H_2, OH_2, CH_4, and NH_3. This is also the presumptive composition of the primitive terrestrial atmosphere at the time of the origin of life. Other constituents of the early atmosphere probably included N_2, CO_2, and H_2S. An argument for ammonia in the Earth's atmosphere up to 2×10^9 yr ago which is independent of direct appeals to cosmic abundances has recently been put forth; NH_3 appears to be required in mixing ratios $\sim 10^{-5}$ to keep the Earth's temperature above the freezing point of water through the atmospheric greenhouse effect to offset the lower luminosity of the Sun (Sagan and Mullen, 1972). As Abelson (1957) and many others have shown, organic compounds are produced if and only if the net conditions are reducing. The production of the simplest molecular precursors of life depends upon the overall cosmic abundances of the elements.

Other astronomical, paleontological and geological evidence argues pursuasively that the origin of life happened quickly. There seems little doubt that the Earth was populated by a wide variety of microorganisms in the period between 2 and 3×10^9 yr ago; and at least moderately convincing evidence for organisms like bacteria and blue-green algae as long ago as 3.4 or 3.5×10^9 yr ago (Schopf, 1970). In the context of the origin of life, such organisms are highly evolved. A long period of time is required for the evolution of such organisms from the atmosphere and waters of the primitive Earth. Also there is no reason to think that the oldest organisms have been found; there is a clear observational selection effect which makes it increasingly difficult to discover older and older organisms. At the same time, the oldest rocks on the Earth are about 4×10^9 yr old, as are the oldest rocks on the Moon returned by the Apollo lunar missions. In the lunar chronology revealed by radioactive dating (Wasserburg, 1973) there is clear evidence for a set of catastrophic events which reset radioactive clocks on the Moon no later than 3.9 or 4.0×10^9 yr ago. The most likely such event is the final sweeping up by the Moon of large objects and miscellaneous debris which had been formed in the early solar system. In fact, the regularity of the planetary and satellite orbits which is observed in the solar system today is very likely the result of such a collisional natural selection in which only objects not on intersecting orbits survive (see, for example, Dole, 1970; Lecar and Franklin, 1973). Some of these objects may have been stored for a few times 10^8 yr in lunar orbit (Reid, 1973). It seems unlikely that the Earth would have been protected from these catastrophes which were inflicted upon the Moon, and it is as least plausible that the Earth – because of endogenous volcanism as well as exogenous impact melting – was unfit for the origin of life until about 4.0×10^9 yr ago. That leaves only a few hundred million years for the origin of life, perhaps much less. But if the origin of life happened fast, the origin of life is probably easy.

This conclusion is to some extent borne out by the enormous success of laboratory experiments on the recreation of primitive environments and the production of prebiological organic molecules as is discussed elsewhere in this Symposium. Typical experiments require only a few days for the production of significant quantities of organic molecules. A wide range of amino acids, nucleotide bases, five and six carbon sugars and their polymers are produced under rather general cosmic reducing conditions. The yields are startlingly high. For example, from the long wavelength ultraviolet quantum yields and the early luminosity of the Sun, some 200 kg cm^{-2} of amino acids would have been produced in the first 10^9 yr of Earth history (Sagan and Khare, 1971a). This is more carbon than there is in the sedimentary column and suggests that the production of organic molecules – at least of certain types – may have been precursor-limited and destruction-rate-limited rather than production-rate-limited. The carbon content of the sedimentary column, now in the form of carbonates, is the equivalent of 50 to 100 atmospheres of carbon dioxide or some tens of atmospheres of methane. But the synthetic rates seem to be so high that most of this carbon, once outgassed, must have been in the form of organic compounds on the early Earth. Since there is nothing in such calculations unique to the Earth – liquid water clouds for example are expected throughout the outer solar system – such highly efficient organic synthetic processes must be common throughout the universe.

TABLE II

Comparison of interstellar and cometary compounds

Microwave identifications of interstellar molecules	Optical identifications of cometary radicals
OH, H_2O	O, OH, OH^+
CO	CO, CO^+, CO_2^+
NH_3, HN_3	NH_2, NH, N_2^+
CN, HCN, HNC	CN
HCHO, CH_3CN, HC_2CN, HC_2CH_3, many other organics	CH, CH^+, C_2, C_3

TABLE III

Simple organic compounds of H, C, N, and O in the interstellar medium and in experiments on prebiological organic chemistry

	Interstellar microwave or optical identification	Laboratory identification
CO	√	√
C_2H_2	No accessible lines	√
C_2H_4	No accessible lines	√
C_2H_6	No accessible lines	√
HCHO	√	√
CH_3CHO		√
HCN, HNC	√	√
CH_3CN	√	√
NH_3	√	
HNCO, HCNO	√	
CH_3OH	√	√
C_2H_5OH		√
HC_2CN	√	√
HC_2CH_3	√	
HC_2NH_2		
HC_2CHO		
NH_2CHO	√	√
CH_3NH_2		
HCOOH	√	√
CH_3COOH		√
⋮		
Higher nitriles and polynitriles		√
Sugars		√
Amino acids		√
Nucleotide bases		√
Porphyrins	?	√
Polycyclic aromatics		√
'Intractable polymers'		√

Where the table shows no entry, the molecule in question has not been reported.

Radio astronomical line searches have uncovered a large number of simple organic compounds in interstellar space. These compounds closely resemble those formed in experiments in prebiological chemistry and those deduced in comets, as Tables II and III indicate. Particularly because of the intense interstellar UV radiation field, the lifetime of such molecules is measured in centuries. Because of the low densities in the interstellar medium, the molecules in question cannot be produced under typical interstellar conditions. The most likely sources of these molecules are the interstellar grains, in which case the grains must have a significant organic component. Organics may be made on the grains or the grains may themselves be ejected during early stages of solar system formation around other stars – either of which indicates the ubiquity of organic compounds in the galaxy. Because of the high temperatures attendant to planetary formation, interstellar organic molecules are unlikely to contribute to the origin of planetary life; because of the low interstellar densities, interstellar indigenous biology appears to be an exceedingly remote contingency (Sagan, 1972).

The origin of simple organic compounds and their polymers, including molecules resembling polynucleotides and polypeptides, is obviously relevant for the origin of life. But it is not quite the same thing as the origin of life. If polypeptides were weakly self-replicating or polynucleotides were weakly catalytic, the production of such molecules might be the same as the origin of life. In the absence of evidence for such cross-functional properties, it is clear that the critical unresolved problem in the origin of life is the origin of the genetic code. The genetic code that we note today is what the molecular biologists unselfconsciously describe as 'universal'. By this they mean that all organisms on the Earth share a common dictionary for converting polynucleotide information into polypeptide information. The use of the word 'universal' in this inappropriate context is itself an indication of the provincialism of biology. Do we have the twenty biological amino acids involved in proteins today because they perform functions which no other amino acids will perform; or because of some frozen evolutionary accident? An early genetic code much simpler than the present one would certainly have strong selective advantage even if it were substantially less accurate and more ambiguous in its coding than the present code.

A related problem is the fact that the present code is a triplet code. A primitive code could very well have been singlet or doublet. But singlet or doublet codes cannot accommodate twenty amino acids and various punctuation marks. At the same time there seems no way to convert from an initial single or doublet code to a triplet code without losing all of the painfully acquired genetic information of the preceding evolutionary process. Why therefore is the present code a triplet code? Some new insights into these questions have recently been provided by L. Orgel (these proceedings) but they remain largely unresolved. It is only the discovery and characterization of extraterrestrial life – even if very 'simple' forms are found – which can deprovincialize biology. This is the reason for the great importance of the biologically oriented experiments on the Viking Mars landers, even though such experiments are hardly the most general and sensitive conceivable experiments for the purpose. And if after an extended biological reconnaissance of Mars the planet proves lifeless, we must face a different problem critical for studies of the origin of the life: why, on two nearby planets with rather similar conditions, did life arise on one and not the other?

Just as the primitive transcription of the genetic code may have been much cruder than the contemporary code, primitive catalysis could have been much less efficient than contemporary enzymatic catalysis and still provide significant selective advantage over competing organisms which had even less efficient catalyses. For example, the functional group or active site of many contemporary enzymes is only five amino acids long, although these amino acids need not be (in fact none of them are) in consecutive order in the primary structure of the protein. But with 20 amino acids there are only about 3×10^6 possible pentapeptides. It is therefore conceivable that very short strands of primitive polynucleotides, crudely directing the synthesis of a small number of simple polypeptides nearby, could have provided a great variety of biochemical functions. The critical requirement is that the accuracy of replication be high enough that advantageous phenotypes previously selected for are not rapidly lost to subsequent generations by low-accuracy replication (see M. Eigen, these proceedings).

While amino acids are made in very high yields and nucleosides in at least modest yields in such experiments, there are a wide range of other molecules also produced. In seeking to understand the origin of life on Earth, we naturally are biased towards the molecules with which we are familiar. But very little attention has been paid to characterizing and contemplating the alkanes, the polycyclic aromatic hydrocarbons, and the wide variety of other molecules which consititute some tens of percent by mass of the total yield in such experiments. Might these molecules provide alternative starting points for self-replication and catalysis in alternative biological systems? Might it be for example that the nucleic acid/protein way of life is best fit for temperatures roughly in the terrestrial range, where hydrogen bonds have just the right energies to provide just the right compromise between inflexible stability and extreme fragility? On very low temperature worlds might van der Waals forces play the role of hydrogen bonds on Earth? On high temperature worlds, might the lower energy ionic or covalent bonds play the role of hydrogen bonds?

The carbon content of the Earth's crust, mentioned above, corresponds roughly to some 10^{-4} to 10^{-5} of the mass of the Earth. Approximately one percent of meteorites falling on the Earth are carbonaceous chondrites. Several percent by mass of carbonaceous chondrites are organic matter, largely aromatics and straight-chain alkanes but with a significant content of amino acids, possibly nucleoside bases and other molecules of terrestrial biological interest (see for example Anders *et al.*, 1973). Thus the carbon content of meteorites which fall on the Earth is of the same order as (or perhaps an order of magnitude larger than) the carbon content of the Earth itself. But recent investigations have shown that the ordinary chondrites, among the most abundant meteorites to be found on the Earth, do not correspond in their photometric or polarimetric properties with the most abundant main belt asteroids (Egan *et al.*, 1973; Chapman and Salisbury, 1973). Instead the chondrites appear to resemble such Apollo objects as Icarus. The Apollo objects lie in orbits which cross the orbit of Mars. One tantalizing possibility, stressed by Öpik, is that the Apollo objects are not strayed asteroids but, rather, dead cometary nuclei. It may therefore be that the ordinary and carbonaceous chondrites are of cometary rather than of asteroidal origin, and that their content of organic matter is not typical of the asteroid belt, but rather of the interstellar environment in which comets mainly reside and may have been produced. Because of their Mars-crossing orbits, they would impact the Earth with a much greater frequency than their mere numbers would imply. However I stress that this is a speculation; and that other interpretations of the data are possible.

While an *in situ* search for life on Mars is the most direct method at our command for deparochializing biology, it may not shed much light on the prebiological organic syntheses which preceded the origin of life. Primordial organic molecules are unlikely to have survived for several $\times 10^9$ yr on the Martian surface or immediate subsurface. The place for such investigations is the outer solar system. The Jovian planets and at least one of their satellites – Titan, the largest moon of Saturn – have reducing atmospheres and energy sources reminiscent of those conjuctured to have played a major role in the early production of organic molecules on the primitive Earth. While the upper clouds of Jupiter are at temperatures of the order of 120° K, the atmosphere is convective and the adiabatic lapse rate implies that, less than 100 km below, typical terrestrial surface temperatures are reached. While methane, ammonia and hydrogen are known to exist on Jupiter, water has not been detected. But all the water would be condensed out at 120K and there is no way for oxygen atoms to have escaped from Jupiter. It seems very likely that near the 280K level in the Jovian atmosphere there is an extensive cloud of liquid water (or, because of the presence of ammonia, ammonium hydroxide solution). A range of energy sources for organic synthesis are available in the Jovian atmosphere, including solar ultraviolet radiation, the precipitation of weakly relativistic particles from the Jovian van Allen belts; and electrical discharges in the clouds. There thus seems little escape from the conclusion that organic molecules fall from the skies of Jupiter like manna from Heaven. Similar lines of argument apply to Saturn, Uranus, Neptune and Titan.

The belts and great red spot of Jupiter, the equatorial zone of Saturn, and Titan have a similar and very red coloration (see, for example, McCord *et al.*, 1971). In the case of the Jovian reddish colored regions, we must ask why the coloration is not uniformly distributed over the disc. The most likely explanation is preferential production. This production might be from the outside or from the inside. The only plausible reason for preferential production of red chromophores from the outside would be precipitation from the van Allen belts; but the mirror points of the Jovian trapped charged particles correspond neither to the Great Red Spot nor to the North Equatorial Belt. It seems much more likely therefore that the red chromophores are produced at depth and carried up or exposed to view in these locations. Indeed, the Red Spot appears to be a great storm system propagating high into the Jovian atmosphere and carrying deep material up by convection within it (Sagan, 1971a). It is therefore of some interest that when the Jovian atmosphere is simulated and energy sources applied, a complex mixture of organic molecules with an overall reddish coloration is produced (Sagan *et al.*, 1967; Woeller and Ponnamperuma, 1969; Sagan, 1971b; Sagan and Khare, 1971b; Khare and Sagan, 1973). Upon acid hydrolysis this polymer yields amino acids in great abundance (Sagan and Khare, 1971b). It also contains high carbon number, straight chain alkanes with amino and probably hydroxyl and carbonyl groups; there are similarities between this polymer mixture and organic compounds recovered from carbonaceous chondrites and precambrian sediments (Khare and Sagan, 1973). The visible and near UV transmission spectrum of the polymer shows its optical depth to be redder than λ^{-2}. Somewhat redder polymers would be able to explain the visible and near UV reflection spectrum of the Jovian planets and Titan as a function only of the polymer/gas ratio. The near UV absorption coefficient is of the order of 10^3 cm^{-1}.

Because of their atmospheric structures and high gravitational accelerations, the Jovian planets are not easy objectives for *in situ* organic chemistry by space vehicle. Titan is a much more accessible objective. Moreover the carbon-to-hydrogen ratio on Titan is likely to be larger than for any of the Jovian planets. We know directly from polarimetric observations (Veverka, 1973; Zellner, 1973), and indirectly from requirements for a Titanian greenhouse effect (Sagan and Mullen, 1972; Sagan, 1973a; Pollack, 1973), that in the visible and UV we are observing an almost completely cloud-covered satellite, probably of similar composition to the reddish polymers described above. The surface pressures are probably at least a few tenths of a bar and the surface temperature perhaps 150K. Temperatures as high as 200K are not excluded. However the pace of plans for examination of Titan is not breathless. Apparently the earliest time for flyby spectroscopic examination of Titan from a distance of a few tens of thousands of kilometers is 1981, and a Titan landing mission probably not until many years after that. But by the late 1980's or early 1990's direct investigations of the organic chemistry of the outer solar system – the clouds of Jupiter, the surface of Titan, the heads and comas of comets – may be expected, and the results of 5×10^9 yr of prebiological organic chemistry uncovered for the first time.

There is one further cosmic context of the origin of life. In discussions of the likelihood of extraterrestrial intelligence a critical parameter is the likelihood that life arises on planets of other stars, given several $\times 10^9$ yr of planetary evolution (see, for example, Shklovskii and Sagan, 1966; Sagan, 1973b). In the view of some thoughtful investigators (for example, Francis Crick, in Sagan, 1973b) the development of technical civilizations may occur readily on planets given the origin of life there; but, in their view, the probability of the origin of life is an unknown quantity. It seems clear that a serious search for extraterrestrial radio communication is a time-consuming and moderately expensive enterprise, but one with the greatest significance for mankind whether it succeeds or fails. It is however much more likely that such a search will be mustered if the chance of success seems high. This probability can begin to be assessed by examination of the organic chemistry and biology, if such there be, on Mars, Titan, the comets, and the Jovian planets.

Acknowledgement

This research was supported by NASA Grant NGR 33-010-101.

References

Abelson, P. H.: 1957, *Ann. N.Y. Acad. Sci.* **69**, 276.
Anders, E., Hayatsu, R., and Studier, M. H.: 1973, *Science*, in press.
Chapman, C. R. and Salisbury, J. W.: 1973, *Icarus* **20**, in press.
Dole, S. H.: 1970, *Icarus* **13**, 494.
Egan, W. G., Veverka, J., Noland, M., and Hilgeman, T.: 1973, *Icarus* **19**, 358.
Khare, B. N. and Sagan, C.: 1973, *Icarus* **20**, 311.
Lecar, M. and Franklin, F. A.: 1973, *Icarus* **20**, in press.
McCord, T. B., Johnson, T. V., and Elias, J. H.: 1971, *Astrophys. J.* **165**, 413.
Pollack, J. B.: 1973, *Icarus* **19**, 43.
Reid, M. J.: 1973, *Icarus* **20**, 240.
Sagan, C.: 1970, 'Life', *Encyclopaedia Britannica.*
Sagan, C.: 1971a, *Comm. Astrophys. Space Phys.* **3**, 65.
Sagan, C.: 1971b, *Space Sci. Rev.* **11**, 73.
Sagan, C.: 1972, *Nature* **238**, 77.
Sagan, C.: 1973a, *Icarus* **18**, 649.
Sagan, C. (ed.): 1973b, in *Communication with Extraterrestrial Intelligence*, MIT Press, Cambridge, Massachusetts.
Sagan, C., Dayhoff, M. O., Lippincott, E. R., and Eck, R.: 1967, *Nature* **213**, 273.
Sagan, C. and Khare, B.: 1971a, *Science* **173**, 417.
Sagan, C. and Khare, B.: 1971b, *Astrophys. J.* **168**, 563.
Sagan, C. and Mullen, G.: 1972, *Bull. Am. Astron. Soc.* **4**, 368.
Schopf, J. W.: 1970, *Biol. Rev.* **45**, 319.
Shklovskii, I. and Sagan, C.: 1966, *Intelligent Life in the Universe*, Holden-Day, San Francisco, California.
Veverka, J.: 1973, *Icarus* **18**, 657.
Wasserburg, G. J.: 1973, Paper presented to the Division for Planetary Sciences at the American Astronomical Society Meeting, Tucson, Arizona, March, 1973.
Woeller, F. and Ponnamperuma, C.: 1969, *Icarus* **10**, 386.
Zellner, B.: 1973, *Icarus* **18**, 661.

LIST OF PARTICIPANTS

Akaboshi, D. M., Research Reactor Institute, Kyoto University, Kumatori, Sennan, Osaka, Japan

Anders, E., Enrico Fermi Institute, 5630 Ellis Ave., Chicago, Ill. 60637, U.S.A.

Aranalde Forto, J. Ma, Victor Pradera, 32 Manresa, Barcelona, Spain

Arluciaga Esnal, M., Asken-Portu 9, A 5° derecha, Zarauz, Guipuzcoa, Spain

Asensio, C., Instituto de Enzimología, Velázquez, 144, Madrid-6, Spain

Aviles Puigvert, F. J., Instituto de Biología Fundamental, Universidad Autónoma de Barcelona, Barcelona, Spain

Bakardjieva, N. T., bul. Totleben 19, Sofia, Bulgaria

Ballester, A., Instituto Investigaciones Pesqueras, Paseo Nacional, s/n°, Barcelona, Spain

Bahadur, K., Chemistry Dept., Allahabad University, 68 Dilkusha, New Katra, Allahabad, India

Baltscheffsky, H., Botaniska Institutionen, Lilla Frescati, Stockholm 50, Sweden

Bar-Nun, A., Dept. of Physical Chemistry, The Hebrew University, Jerusalem, Israel

Bartrons, R., Rio Rosas, 10 entlo. 1ª Barcelona, Spain

Beaus Codes, R., Laboratorios del Norte de España S. A., Carretera de Francia, Masnou, Barcelona, Spain

Becker, R. S., Dept. of Chemistry, University of Houston, Houston, Tex. 77004, U.S.A.

Bennedetti, E., Istituto Chimico-Università, Via Mezzocannone 4, 80134 Napoli, Italy

Biemann, K., Dept. of Chemistry, Room 56-010 Massachusetts, Institute of Technology, Cambridge, Mass. 02139, U.S.A.

Biermann, L., Max-Planck Institute for Physics & Astrophysics, Fohringer Ring 6, 8000 Munich 40, W. Germany

Bogdanski, C. A., Hotel Balmoral, 9 rue Reitzer, Casablanca, Morocco

Bonner, W. A., Chemistry Dept., Stanford University, Stanford, Calif. 94305, U.S.A.

Brack, A., Centre de Biophysique Moléculaire CNRS, Ave. de la Recherche Scientifique, 45045 Orléans-Cedex, France

Brito, R., Boters, 10 2°, Barcelona-2, Spain

Broda, E., Institut für Physikalische Chemie, Währinger Strasse 42, Universität, Vienna IX, Austria

Buhl, D., National Radio Astronomy Observatory, Edgemont Road, Charlottesville, Va. 22901, U.S.A.

Buvet, R., Laboratoire d'Energétique Biochimique, Université Paris-Val de Marne, Ave. Gén. de Gaulle, 94000 Créteil, France

Caballin Fernandez, M. R., Buxarons 14, 2°2ª Barcelona, Spain

Cabezas, J. A., Departamento de Bioquímica, Facultad de Ciencias, Universidad de Salamanca, Salamanca, Spain

Cabre, O., Instituto de Biología Fundamental, Universidad Autónoma de Barcelona, Barcelona, Spain

Cairns-Smith, A. G., Dept. of Chemistry, University of Glasgow, Glasgow 912 8QQ, Scotland

Calvet, F., Departamento de Bioquímica, Facultad de Ciencias, Universidad de Barcelona, Barcelona, Spain

Calvin, M., Lab. of Chemical Biodynamics, University of California, Berkeley, Calif. 94720, U.S.A.

Camp, J., Instituto de Inv. Pesqueras, Paseo Nacional, s/n°, Barcelona, Spain

Camprubi, J., Castillejos, 406, atico 2ª, Barcelona-13, Spain

Carbo, R., Div. de Matemáticas, Facultad de Medicina, Universidad Autonoma de Barcelona (Hospital de San Pablo), Barcelona-13, Spain

Carreras Planells, R., Enrique Granados 155, Barcelona, Spain

Casas Alvero, C., Depto. de Biología, Laboratorio Microbiología, Universidad Autónoma de Barcelona, Barcelona, Spain

Chadha, M. S., Bio-Organic Division, Bhabba Atomic Research Centre, Bombay 4000, 85, India

Chang, S., Amers Research Center-NASA, Chemical Evolution Branch 239-9, Moffett Field, Calif., 94035, U.S.A.

Christophersen, O. A., Institute of Geology, Oslo University, Blindern, Oslo 3, Norway

Cid Capella, A. Mª., Benedicto Mateo 26, 2°2ª, Barcelona-17, Spain

Ciutat Falco, D., Marques de Villa Antonia, 5 1°, Lérida, Spain

Climent Romeo, F., Instituto de Biología Fundamental, Universidad Autónoma de Barcelona, Barcelona, Spain

Comamala de Florensa, C., Trafalgar, 6, Barcelona, Spain

Comerma, J., Milton, 1 Barcelona-12, Spain

Corcoy, F., Valencia, 25 1°2ª Barcelona, Spain

Creaser, E. H., Genetics Dept. RSBS, The Australian National University, P.O. Box 475, Canberra City, A.C.T., 2601, Australia

Crusa Font Pairo, M., Colón 13, Sabadell, Barcelona, Spain

Cruz Cumplido, M., Dept. of Geology, University of Illinois, Urbana, Ill. 61801, U.S.A.

Cuchillo, C. M., Instituto de Biología Fundamental, Universidad Autonoma de Barcelona, San Antonio Mª Claret, 171, Barcelona-13, Spain

Danielli, J. F., Center for Theoretical Biology, State University of New York at Buffalo, 4248 Ridge Lee Rd., Amherst, New York, N.Y. 14226, U.S.A.

Darge, W., Friedensstrasse, D 5175 Juelich-Koslar, W. Germany

Daufi, L., Departamento de Patología, Hospital de San Pablo, San Antº, Maria Claret, Barcelona, Spain

Dayhoff, M. O., National Biomedical Research Foundation, Georgetown University Medical Center, 3900 Reservoir Road, N.Y. Washington, D.C. 20007, U.S.A.
Deborin, G. A., A. N. Bach Institute of Biochemistry, Academy of Sciences of the U.S.S.R., Leninsky prosp. 33, Moscow, U.S.S.R.
Decker, P., Stammestr. 74 A, 3 Hannover, W. Germany
Del Valle, J. A., Instituto de Enzimología (CSIC), Dto. Bioquímica Facultad de Medicina, Universidad Autónoma, Madrid-3, Spain
Dierckesen, G., Max-Planck Institute for Physics & Astrophysics, Foringer Ring 6, 8000 Munich 40, W. Germany
Diez-Caballero Arnau, T., Conde Salvatierra 35, 5ª, Valencia, Spain
Diver, W. L., Dept. of Geology, Royal School of Mines, Imperial College of Sciences and Technology, Prince Consort Road, London S.W.7. 2BP England
Dose, K., University of Mainz, Institute for Biochemistry, 65 Mainz, W. Germany
Drozdova, N. N., A. N. Bach Institute of Biochemistry, Leninsky pr. 33, Moscow, U.S.S.R.
Dwyer, M. J., Graduate Group on Molecular Biology, 116 Wistar, University of Pennsylvania, Philadelphia, Pa. 19104, U.S.A.
Egami, F., Mitsubishi-Kasei Institute of Life Sciences, 11 Minamiooya, Machida-shi, Tokyo, Japan
Eglinton, G., School of Chemistry, University of Bristol, Cantacok's Close, Woodland Rd. Bristol, England
Esteve Martinez, M. I., Avda. de Madrid, 138 sobreático 2ª, Esc. B, Barcelona, Spain
Estrada Miyares, M., Departamento de Ecología, Facultad de Ciencias, Universidad de Barcelona, Barcelona, Spain
Étaix, E., Laboratoire d'Energetique Biochimie, Université de Paris-Val de Marne, Ave. de Gén. de Gaulle, 94000 Creteil, France
Ettore, B., Istituto Chimico-Università, Via Mezzocannone 4, 80134 Napoli, Italy
Feijoo Melle, J., Saltos del Sil, La Rua, Orense, Spain
Fernandez-Alonso, J. I., Facultad de Ciencias, C-XIV Universidad Autonoma de Madrid, Madrid, Spain
Ferris, J. P., Dept. of Chemistry, Rensselaer Polytechnic Institute, Troy, N.Y. 12181, U.S.A.
Flegmann, A. W., School of Biological Sciences, The University, Claverton Down, Bath, BA2 7AY, England
Fox, S. W., Institute for Molecular & Cellular Evolution, University of Miami, Coral Gables, Fla. 33134, U.S.A.
Fripiat, J. J., 42 Decroylaan, 3030 Heverlee, Belgium
Gandia, V., Decano Facultad de Ciencias, Universidad de Barcelona, Bellaterra, Barcelona, Spain
Garay, A. S., Institute of Plant Physiology, Biological Research Center, Hungarian Academy of Sciences, Szeged, P.O. Box 521, Hungary
Garcia, M., Rbla. Montanya, 82 1°2°, Barcelona, Spain
Gelpi, E., Instituto Biología Fundamental, Universidad Autonoma, San Antº Mª,

Claret 171, Barcelona, Spain

Gibert, J., Instituto Biología Fundamental, Universidad Autonoma, San Ant° Mª, Claret 171, Barcelona-13, Spain

Gil-Av, E., Weizmann Institute, Rehovot, Israel

Giner Sorolla, A., Sloan-Kettering Institute, 145 Boston Post Road, Rye, N.Y. 10580, U.S.A.

Golpe-Posse, J., Elisabets 19, Barcelona-1, Spain

Gracia Alonso, C., Calle del Centro, 21, entlo 2ª, Barcelona-13, Spain

Gregoire, Ch., Dieweg 292, 1180 Bruxelles, Belgium

Guerrero Moreno, R., Rosellón 104, Barcelona-15, Spain

Hall, D. O., King's College, 68, Half Moon Lane, London, S.E. 24, England

Halmann, M. M., Isotope Department, Weizmann Institute of Sciences, Rehovot, Israel

Harada, K., Institute for Molecular and Cellular Evolution, University of Miami, 521 Anastasia Avenue, Coral Gables, Fla. 33134, U.S.A.

Hasegawa, M., Dept. of Biophysics & Biochemistry, Faculty of Sciences, University of Tokyo, Hongo, Tokyo 113, Japan

Hawtrey, A. O., Dept. of Biochemistry, University of Rhodesia, Bag M.P. 167, Salisbury, Rhodesia

Hayes, J. M., Depts. of Chemistry & Geology, Indiana University, Bloomington, Ind. 47401, U.S.A.

Herrera, E., Cátedra de Fisiología General Facultad de Ciencias, Universidad de Barcelona, Avda. José Antonio 585, Barcelona, Spain

Hochstim, A. R., College of Engineering, Research Institute for Engineering Sciences, Wayne State University, Detroit, Mich. 48221, U.S.A.

Holland, H., Dept. of Geological Sciences, University of Harvard, Cambridge, Mass. 02138, U.S.A.

Hubbard, J. S., California Institute of Technology, Biology Division, Pasadena, Calif. 91109, U.S.A.

Irrure-Perez, J., Mayor de Sarriá, 184 2°1ª, Barcelona-17, Spain

Isern Viñas, A., Estanislao Rico Ariza 5, Barcelona, Spain

Ishigami, M., Laboratory of Biology, Jichi Medical School, Tochigi, Japan

James, H. L., U.S. Geological Survey, Menlo Park, Calif. 94025, U.S.A.

Jaume, M., Inst. de Investigaciones Pesqueras, Pseo. Nacional, s/n, Barcelona, Spain

Jofre, J., Departamento de Microbiología, Facultad de Ciencias, Universidad de Barcelona, Plaza Universidad, Barcelona, Spain

Jukes, T. H., University of California, Space Sciences Laboratory, Berkeley, Calif. 94720, U.S.A.

Karapetyan, N. V., A. N. Bach Institute of Biochemistry, Academy of Science of the U.S.S.R., Leninsky pr. 33, Moscow 71, U.S.S.R.

Keosian, J., Marine Biological Laboratory, Woods Hole, Mass. 02543, U.S.A.

Klein, H. P., 1022 N. California Ave., Palo Alto, Calif. 94303, U.S.A.

Krasnovski, A. A., A. N. Bach Institute of Chemistry, Academy of Sciences of the

U.S.S.R., Leninsky pr. 33, Moscow 71, U.S.S.R.

Kritsky, M. S., A. N. Bach Institute of Biochemistry, Academy of Sciences of the U.S.S.R., Leninsky pr. 33, Moscow 71, U.S.S.R.

Kumar, S. S., Dept. of Astronomy, Box 3818, University Station, Charlottesville, Va. 22903, U.S.A.

Kutyurin, V. M., Vernadskii Geochemistry Institute, Academy of Sciences of the U.S.S.R., Moscow 117334, U.S.S.R.

Kvenvolden, K. A., NASA-Ames Research Center, N-239-9, Moffett Field, Calif. 94035, U.S.A.

Lacey, J. C., Laboratory of Molecular Biology, University of Alabama in Birmingham, Birmingham, Ala. 35294, U.S.A.

Lasaga, A. C., Dept. of Chemical Physics, Harvard University, Cambridge, Mass. 02139, U.S.A.

Latorre, C., Parroco Ubach 7, Barcelona 6, Spain

Le Port, L., Laboratoire d'Energétique Biochimique, Université Paris-Val de Marne, Ave. Gén. de Gaulle 94000, Créteil, France

Ley De, J., Lab. Microbiology, Lendenganchstraat 35, 94000 Gent, Belgium

Libby, W. F., University of California, Dept. of Chemistry, 405 Hilgard Ave., Los Angeles, Calif. 90024, U.S.A.

Llobet Zubiaga, J., Borrell 189, 6°1ª, Barcelona, 15, Spain

Loew, G. B., Dept. of Genetics, Stanford University, Medical Center, Stanford, Calif. 94305, U.S.A.

Lohrmann, R., The Salk Institute Biological Studies, P.O. Box 1809, San Diego, Calif. 92112, U.S.A.

Lopukhin, A. S., Academy of Sciences of the U.S.S.R., Institute of Geology, Dzerdzinskogo Ave., 30.720040, Kirghiz SSR, Frunze, U.S.S.R.

Loren, J. G., Dep. de Microbiología, Facultad de Ciencias, Universidad de Barcelona, Plaza Universidad, Barcelona, Spain

Llobet, J., Borrell, 189, 6°1ª, Barcelona 15, Spain

Macelroy, R. D., N-239-10, Ames Research Center, Moffett Field, Calif. 94035, U.S.A.

Margalef, R., Instituto Investigaciones Pesqueras, Paseo Nacional, s/n, Barcelona 3, Spain

Margulis, L., Biological Science Center, Boston University, 2 Cummingham St., Boston, Mass. 02215, U.S.A.

Martin Barrientos, J., Departamento de Bioquímica, Facultad de Ciencias, Universidad de Salamanca, Salamanca, Spain

Martin Vicente, E., Vallespir 117, 2°4ª, Barcelona, Spain

Martinell Callico, J., Enrique Granados 32, 1°1ª, Barcelona, Spain

Mas, R. L., 311 Braddock Avenue, Daytona Beach, Fla. 32018, U.S.A.

Matheja, H. J., Institut für Phys. Chemie, Kernforschungsanlage, 517 Juelich, W. Germany

Matthews, C. N., Dept. of Chemistry, University of Illinois at Chicago, Circle, Chicago, Ill. 60091, U.S.A.

Maurette, M., Laboratoire René Bernas, 91406 Orsay, France

Mayor, F., Consejo Superior de Investigaciones Científicas, Serrano, 117, Madrid 4, Spain

McCabe, M., Institute of Medical Chemistry, University of Uppsala, Uppsala, Sweden

Medrano gil, H., Calle del Centro 21, entlo. 2ª, Barcelona 13, Spain

Mensua, J. L., Dto. de Genética, Facultad de Ciencias, Universidad Autónoma, Barcelona, Spain

Mikhailova, H., A. N. Bach Institute of Biochemistry, Academy of Sciences of the U.S.S.R., Leninsky pr. 33, Moscow, U.S.S.R.

Miller, S. L., Dept. of Chemistry, University of California, San Diego, La Jolla, Calif. 92037, U.S.A.

Miquel Quintanilla, J., Laboratorios Miguel Calle Musitu s/n, Barcelona, Spain

Miravitlles, L., Ayuntamiento de Barcelona, Barcelona, Spain

Mitz, M. A., NASA Headquarters, Washington, D.C. 20546, U.S.A.

Morimoto, S., Dept. of Chemical Engineering, Faculty of Engineering, Tokushima, University, 2 Minamijosanjima, Tokushima, 770, Japan

Morita, R. Y., Dept. of Microbiology, Bioscience Bldg., Oregon State, Corvallis, Oreg. 97331, U.S.A.

Muir, M., Dept. of Geology, Royal School of Mines, Imperial College, Prince Consort Rd., London SW7 2BP, England

Nagy, B., The University of Arizona, 533 Space Sciences, Dept. of Geoscience, Tucson, Ariz. 85721, U.S.A.

Navarro, D., Dept. of Biophysical Sciences, University of Houston, Houston, Tex. 77004, U.S.A.

Niño Larru, F., Cátedra de Biología, Fctad. de Veterinaria, Universidad de Córdoba, Córdoba, Spain

Noda, H., Dept. of Biophysics & Biochemistry, Faculty of Science, University of Tokyo, Bunkyo, Tokyo, Japan

Nogues, R., Rda. San Pablo 80, 3º2ª, Barcelona 15, Spain

Novak, V., Dept. of Physiology, Institute of Entomology, CSAV, Na Folimance 5, Praha 2, C.S.S.R., Czechoslovakia

Nuñez de Castro, I., Departamento de Bioquímica, Facultad de Farmacia, Granada, Spain

Ochoa, S., Dept. of Biochemistry, New York University Medical Center, 550 First Avenue, New York, N.Y. 10016, U.S.A.

Oparin, A. I., A. N. Bach Institute of Biochemistry, Academy of Sciences of the U.S.S.R., 33 Leninsky pr. 117071, Moscow, U.S.S.R.

Oren, R., Israel Institute for Biological Research, Tel-Aviv University, Medical School., P.O. Box 19, Ness-Ziona, Israel

Orgel, L. E., The Salk Institute for Biological Studies, P.O. Box 1809, San Diego, Calif. 92112, U.S.A.

Oró, J., Dept. of Biophysical Sciences, University of Houston, Houston, Tex. 77004,

U.S.A.

Otroshchenko, V. A., Vernadski Institute of Geochemistry & Analytical Chemistry of the Academy of Sciences, Moscow, U.S.S.R.

Owen, T., 6 Iry Lane, Setanket, N.Y. 11790, U.S.A.

Paecht-Horowitz, M., The Weizmann Institute of Science, Polymer Department, Rehovot, Israel

Palacios, J., Avda. San Antonio M[a] Claret, 124–126, Barcelona 13, Spain

Palau Albet, J., Instituto Biología Fundamental, Universidad Autónoma, San Antonio M[a], Claret, 171, Barcelona 13, Spain

Paolillo, L., Instituto Chimico-Universitá, Via Mezzocannone 4, 80134 Napoli, Italy

Pares, R., Cátedra Microbiología, Facultad de Ciencias, Universidad de Barcelona, Avda. Generalisimo s/n. Barcelona, Spain

Parr, W., 7400 Tubingen, Lehrstuhl für Org. Chemie, Aufder Morgenstelle, Germany

Pavlovskaya, T. E., A.N. Bach Institute of Biochemistry, Academy of Sciences of the U.S.S.R., Leninskii pr. 33, 117071, Moscow, U.S.S.R.

Pedro de, Departamento de Bioquímica, Facultad de Ciencias, Universidad de Salamanca, Salamanca, Spain

Peralta, E., Nápoles 352, sobreático 2[a], Barcelona 13, Spain

Pereda de la Reguera, A., Paseo de Canalejas, 50 Santander, Spain

Perez Garcia, A., San Andrés 194, 1°3[a], Barcelona 16, Spain

Petitpierre Vall, E., Mallorca 316, Barcelona 9, Spain

Philippovich, I., A.N. Bach Inst., Lenin prospekt, 33 Moscow, U.S.S.R.

Pla, J. M., Colegio Oficial Farmaceuticos Gerona, Ultonia, 13, Gerona, Spain

Ponnamperuma, C., Laboratory of Chemical Evolution, Dept. of Chemistry, University of Maryland, College Park, Md. 20742, U.S.A.

Pons, A., Regás 3, II, Barcelona 6, Spain

Ponsa, M., Campo Vidal 16, Barcelona 6, Spain

Puig Muset, P., Laboratorios Pevya, Santiago Ramón y Cajal 6, Molins de Rey, Barcelona, Spain

Prevosti, A., Centro de Genética Animal y Humana, Facultad de Ciencias, Plaza de la Universidad, Universidad de Barcelona, Barcelona, Spain

Prieto, J., Departamento de Microbiología, Facultad de Ciencias, Universidad de Barcelona, Spain

Puigdollers, J. M., Avda. de Sarriá, 36, Barcelona, Spain

Querol, E., Lauria 36, 3°1[a], Barcelona 9, Spain

Rasool, I., National Aeronautics & Space Administration, Washington, D.C. 20546, U.S.A.

Raulin, F., Lab. d'Energétique Biochimique, Université Paris-Val de Marne, Ave. du Gén. de Gaulle 94000, Creteil, France

Renzi, M., Córcega 220, 3°3[a], Barcelona 11, Spain

Revuelta, J., Osio 46, Barcelona, Spain

Rho, J., Jet Propulsion Laboratory, California Institute of Technology, 4800 Oak

Grove Drive, Pasadena, Calif. 91103, U.S.A.

Ribo, G., Cardenal Reig, B-7, 6°3ª, Barcelona 14, Spain

Rich, A., Dept. Biology, Massachusetts Institute of Technology, Cambridge, Mass. 02139, U.S.A.

Riera, J., División de Matemáticas, Facultad de Medicina, Universidad Autónoma de Barcelona, Hospital San Pablo, Barcelona 13, Spain

Rios, E., Paseo de las Delicias, 129, Madrid 5, Spain

Rivera, L., Conde Borrell 308, 1°2ª, Barcelona 15, Spain

Robinson, R., Shell Chemicals U.K., Limited, Downstream Building, Shell Centre, London, S.E. 1, England

Robert, M., Colegio Mayor Virgen de Nuria, Ganduxer 122, Barcelona 6, Spain

Rodriguez, E., Departamento de Farmacología, Patronato 'Juan de la Cierva' CSIC, Calle Jorge Girona Salgado s/n., Barcelona 17, Spain

Rohlfing, D. L., Dept. of Biology, University of South Carolina, Columbia, S.C. 29209, U.S.A.

Roldan, S., Juan Bravo 42, Madrid 6, Spain

Romaña, A., Presidente del Patronato Alfonso el Sabio, Consejo Superior de Investigaciones Científicas, Serrano 117, Madrid, Spain

Rubio, A., Generalísimo Franco, 41, 4°1ª, Masnou, Barcelona, Spain

Rull, M., Lourdes, 27 'La Miranda', Esplugas, Barcelona, Spain

Sadron, Ch., Directeur, Centre de Biophysique Moléculaire, Ave. de la Recherche Scientifique, 45045, Orléans, Cédex, France

Sagan, C., 302, Space Science Building, Cornell University, Ithaca, N.Y. 14850, U.S.A.

Sagarra, E., Provenza 329, 2°2ª, Barcelona 9, Spain

Sanguinetti, F., Via Tommaso Gulli 28, 48100, Ravenna, Italy

Santos, A., Departamento de Bioqu'imica, Facultad de Farmacia, Ciudad Universitaria, Madrid 3, Spain

Segal, J., 1017, Berlin (Rda), Strausberger Platz 1, W. Germany

Schidlowski, M., Max-Planck Institut für Chemie (Otto-Hahn Institut Abt. Luftchemie), 65 Mainz, W. Germany

Schofield, P., Instituto de Enzimología del CSIC, Facultad de Medicina, Universidad Autonoma, Madrid 34, Spain

Schopf, J. W., Dept. of Geology, Paleobiological Laboratory, University of California, 405 Hilgrad Avenue, Los Angeles, Calif. 90024, U.S.A.

Schuster, P., Institut für Theoretische Chemie der Universität Wien, Währingerstrasse 38-A-1090, Wien, Austria

Schwartz, A. W., Dept. of Exobiology, University of Nijmegen, Toernooiveld, Nijmegen, The Netherlands

Serratosa, F., Patronato Juan de la Cierva, Instituto Química Orgánica, P° Jorge Girona Salgado s/n, Barcelona 17, Spain

Sherwood, E. S., Dept. of Biophysical Sciences, Univ. of Houston, Houston, Tex. 77004, U.S.A.

Shimizu, M., Institute of Space and Aeronautical Sciences, University of Tokyo,

Komaba, Meguroku, Tokyo 153, Japan

Shrago, E., Thorstrand Rd, Madison, Wis. 537705, U.S.A.

Soffen, G. A., 801 s. Armstead Ave., Hampton, Va., U.S.A.

Soler Colomer, J., Oliana, 16, 1°1ª, Barcelona, Spain

Sols, Alberto, Facultad Medicina de la Universidad Autónoma, Instituto Enzimología del CSIC, Herederos de Navas, s/n, Madrid 34, Spain

Subirana, J. A., Departamento Química Macromolecular, ETSII, Diagonal 99, Barcelona, Spain

Suess, H. E., Dept. of Chemistry, University of California, San Diego, la Jolla, Calif. 92037 U.S.A.

Temussi, P. A., Instituto Chimico, Via Mezzocannone 4, Napoli 80134, Italy

Thiemann, W., Gutenbergstr 22, D-517 Juelich, W. Germany

Tirado, A., Rosellón 502, entlo 1°, Barcelona, Spain

Torrella, F., Departamento de Biología, Laboratorio Microbiología, Universidad Autónoma Barcelona, Barcelona, Spain

Toupance, G., Laboratoire d'Energétique Biochimique, Université Paris-Val de Marne, Avenue du Général De Gaulle 94000, Creteil, France

Trueta, J., Sociedad Catalana de Biología, Rbla. Cataluña 74, 1°, Barcelona, Spain

Urey, H. C., Dept. of Chemistry, University of California, La Jolla, Calif. 92038, U.S.A.

Usher, D., Baker Laboratory, Cornell University, Ithaca, N.Y. 14850, U.S.A.

Vallespinos, F., Departamento de Ecología, Facultad de Ciencias, Universidad de Barcelona, Barcelona, Spain

Vdovykin, G. P., V.I. Vernasdky Institute of Geochemistry and Analytical Chemistry, U.S.S.R. Academy of Sciences, Moscow, U.S.S.R.

Velarde, M., Departamento de Física C-III-603, Universidad Autónoma, Cantoblanco, Madrid, Spain

Victory, J., Instituto Químico de Sarriá s/n, Barcelona-17, Spain

Vidal, R., Paseo de Gracia, 2, Barcelona, Spain

Villanueva, J. R., Departamento de Microbiología, Facultad de Ciencias, Universidad de Salamanca, Salamanca, Spain

Villar Palasi, V., Universidad Autónoma de Barcelona, Bellaterra, Barcelona, Spain

Vishniac, W. V., Dept. of Biology, University of Rochester, Rochester, N.Y. 14627, U.S.A.

Vitorovic, D., Dept. of Chemistry, University of Beograd, P.O. Box 550, 11000 Beograd, Yugoslavia

Wald, G., Harvard University, Cambridge, Mass. 02138, U.S.A.

Walker, G. W., Genetics Department, University of Alberta, Edmonton, Alberta, Canada

Weber, H., University of Zurich, Institute for Molecular Biology, Honggerberg 8049, Zurich, Switzerland

Weiss, A., Institut für Anorg. Chemie der Universität München 8, München 2, Meiserstr 1, W. Germany

Wolman, Y., Dept. of Organic Chemistry, The Hebrew University, Jerusalem, Israel

Young, R. S., 7927 Falstaff Rd., McLean, Va. 22101, U.S.A.

Yuasa, S., Dept. of Biophysical Sciences, University of Houston, Houston, Tex. 77004, U.S.A.

INDEX OF SUBJECTS

Abiogenic
  origin, 397
  synthesis, 303, 304, 460
Abiotic synthesis, 239
Abundances
  in outer solar systems, 42
  of atoms in molecular clouds, 38
  of elements in the Cosmos, 498
  of elements in the Earth, 498
  of molecules in molecular clouds, 38
Acrolein, 146, 147
Acylphosphate, 162
Adaptor hypothesis, 347
Adenine, 197
Adenosine, 165
Adenosine-phosphate, 160
Adenyl-adenosine, 161
Adenylyl-uridine, 209
Alanine, 73, 306, 314
Alkaline earths, 174
Alkanes, 56, 60, 61, 64
Alloisoleucine, 142, 145, 146
Allothreonine, 142, 144
Aluminium sylicates, 174, 176
Aminoacetaldehyde, 145
Aminoacyl adenylates, 228, 235, 240
Amino acid (*see also* poly-amino acids)
  adenylates, 177, 178, 180
  analysis (Mars samples), 476–481
  arrangement of – residues, 243
  code, 337
  codon, 189, 338, 342, 361
  composition of polymers, 242 (*see also poly-amino acids*)
  homocodonic, 235
  mixture, 198, 228, 243
  in Murchison, 62, 75–77, 139, 149, 150
  in Murray, 73, 149
  in Orgueil, 72, 73
  prebiotic synthesis of, 139, 140, 141, 142, 147
  synthesis of, 141, 145, 303, 305, 307
α-aminoisobutyric acid, 145, 149
α-amino-n-butyric acid, 145
AMP, 232
Anaerobic
  decomposition, 405
  organisms, 94
  process, 10
  respiration, 407, 408, 410
Angiospermy, 119
Anticodon, 317, 337, 346, 354, 355, 358, 361
  loop, 213
  sequences, 355
Antigenic determinants, 341
ApU crystal, 210
Arginine, 194, 338–341
Aspartic acid, 306, 342
Assimilative power, 10
Asymmetric
  amino acids in meteorites, 76
  factors, 270
  molecules, 263
Asymmetry and life, 264
Asymmetry effect, 271, 272, 273, 280, 282
Atmosphere
  Earth, 172, 305
  Jupiter, 41
  Martian, 457, 466
  Neptune, 52
  of primitive Earth, 139
  Saturn, 41, 52
  Titan, 41
  Uranus, 52
ATP, 10, 230, 231, 235, 405, 459
Autocatalysis, 286
Autotrophy, 119
Azotobacter, 369

Bacteria
  colored sulfur, 410
  photosynthetic, 363, 368, 397
  sulphate reducing, 368
Bacterial ferredoxins, 370
Bacteriochlorin, 399
Bacterioviridin, 401
Barney Creek formation, 105
Batten subgroup, 105–107
Biochemical redox process, 253
Biological electron
  carriers, 388
  transfer, 363
Biomacromolecules, 236
Biopoesis, 287
Biota (Martian), 432, 433, 461
Bitter Spring
  formation, 108, 109, 111, 117, 125, 128, 130, 131
  microflora, 105

Biuret values, 242, 243
Blue-green algae, 84, 85, 98, 105, 119–125, 127, 130, 363, 368, 377, 397, 412, 499
Bonds
 in CO_2, 8
 in SiO_2, 8
Bulawayan
 group, 122
 series, 88
 system, 79, 80, 83, 85
 time, 89

Carbamyl phosphate, 157, 459
Carbon
 complex tarry compounds, 59
 in a solar gas, 58
 isotope fractionations, 60
Carbon dioxide
 biological oxidation of, 457
 source of energy, 458
Carbon monoxide
 photodissociation, 457
 source of energy, 459
Carbonaceous chondrites
 Allende, 76, 77
 Mars, 466
 Murchison, 62, 71, 74–76, 149
 Murray, 71, 73, 74, 149
 Orgueil, 61, 72, 73
 type I, 58, 72
 type I, II, III, 71
Carbonaceous meteorites (*see* Carbonaceous chondrites)
Carboniferous conifer seeds, 127
β carotene, 15, 18
Carotenoids, 13, 14, 16, 18, 19
Cassopeia A, 30
CCA ends, 216, 218
CCA stem, 215, 216, 218
CERES, 256, 258, 260
Chemiosmotic theory, 253
Chemoanthotrophs, 412
Chemosynthesis, 10
Chlorin, 399, 400
Chlorophyll, 399, 402
 spectra, 10, 14
 types, 10, 11
Chromatin-like granules, 127
Chromatium, 370, 373–375
Chromophores
 in outer solar systems, 48
 Jovian, 45
Clay minerals, 174
Climate (in Mars), 465
Coacervates, 201–205, 225, 227, 285, 360
Coding triplets, 193
Codon
 anticodon pairing, 334, 343
 catalogue, 352
 initiator – in E. Coli, 335
 mutated, 343
 quartets, 331, 351
Codons, 342, 343
Coenzymes
 contemporary, 398
 flavine, 398
Cometary
 activity, 299
 atmosphere, 297–300
 compounds, 500
 nuclei, 298
 spectra, 300
Comets
 Kohoutek, 54, 298, 300
 nucleus of, 42, 300
 photochemistry in, 303
Complex III, iron-proteins, 379
Contamination
 in Mars samples mission, 469, 475
 in Mars spacecraft, 470
 terrestrial, 73
Cosmic distribution of elements, 12
Cosmic rays, 38
Cotton effect of ORD, 276
Cyanamide, 159, 160, 165, 170
Cyanohydrin synthesis, 140
Cyanophytes, 129
Cyanophytic
 evolution, 124
 phylogeny, 123
Cyclothymidine-phosphate, 162
Cysteine, 314, 382
Cytochrome, 316, 325, 338, 387, 389–392, 407–410, 412

DAMN, 153, 201
Darwinian
 interpretation, 223, 225
 selection, 227, 228, 338
DEAE cellulose, 230
DHU loop, 213
Deoxynucleoside cyclic phosphate, 167
α-γ-Diaminobutyric acid, 144, 145, 146, 147
α-β-Diaminopropionic acid, 145
Diaminosuccinic acid, 155, 156
Diaminomaleonitrile (see DAMN)
Dianisidine, 203, 204
Dihydro U loop, 215
Dihydro U stem, 215–217
Dihydrofolic acid, 341
Diploid-dominant, 132
DL-camphor, 269
DL-ephedrine, 269

DL-glutamic acid, 269
DL-lysine, 269
DNA, 159, 165, 170, 235, 338, 344
  polymerase, 164
DNP, 232
Double bonds (in CO_2), 8
Dust
  clouds (interstellar), 32, 34, 493
  grains (interstellar), 30
  in Mars, 465, 466
  meteoritic, 58

Earth
  Early, 79
  oldest rocks on the, 499
E. Coli, 317, 318, 321, 322
  anticodon in, 334
  base sequences of, 332, 333
  codon in, 336
  initiator codon in, 335
Electron
  carriers, 387, 388, 390, 392
  transport, 387, 388, 394
Elementary particles, 8, 25
Elements
  abundances, 498
  cosmic distribution of, 12
  enzymatic trace, 349
Enantiomers, 263, 266, 270
  bond strengths, 267
  racemization, 273
  thermodynamic date of, 267
Energy (geothermal), 458, 460
Enzymatic trace elements, 349
Eukaryote, 126, 127, 128, 132, 311, 320, 321
Eukaryotic
  cell, 124, 218
  organization, 119, 127
  proteins, 339, 340
Eumetazoans, 132, 133
Evolution of genetic code, 331
Evolutionary selections, 349
Extraterrestrial
  life, 494
  radio communication, 504

FAD, 389
Fatty acids (in Orgueil), 61
Fe protoporphyrin, 11
Fermentation, 405–408, 410
Ferredoxin, 170, 311, 312, 322–324, 363–368, 388–392
Fig Tree group, 79, 80, 121, 122
Flavine coenzymes, 398
Flavines, 18
Flavoproteins, 389, 391, 392, 394
FMN, 389

Formaldehyde, 399, 400
Fossil taxa, 123

Gaia hypothesis, 93, 102
Galactic plane, 33, 491
Galaxy, 7, 29, 32, 38, 303
Gas chromatography-mass spectrometry (*see* GC-MS)
Gas exchange experiment, 435
GC-MS, 59, 73–76, 144, 155, 417, 419
Gene
  elongation, 320
  proteins, 190
  duplication, 387
Genetic calendar, 351
Genetic code, 331, 332, 344, 347, 351, 355, 357, 501
Glutamic acid. 142, 306, 342
Glycine, 306
Glycosideic bonds, 209
Glyoxal, 305, 306
Granular cells, 130, 132
Guanidine, 159
Guanylyl-cytidine, 209, 210
Gunflint chert, 105, 108, 117

Hairpin structure, 169
Halobacterium halobium, 17
Haploid-diploid coequal, 132
Haploid-dominant, 132
H/C ratio of outer solar systems, 42
HCN, 153–156
H/D ratio, 34
Heme proteins, 388, 394, 408
Heterocatalysis, 286
Heterocoders, 351, 352
Heterocoding quartets, 351, 354
Hexose-pentose-phosphate cycle, 10
Homeostasis, 93, 94, 99–101
Homoiothermy, 119
Hooggenoeg formation, 81–83, 85
Hyaline sheath, 109
HYC pyritic shale, 105–107, 111
Hydantoin, 156
Hydrogen cyanide (synthesis), 140 (*see also* HCN)
α-hydroxy-γ-aminobutyric acid, 144, 145, 146, 147
β-hydroxyacyl thioesters, 254

Imidazole, 165, 170, 305, 306
Indole, 307
Immunoglobulins, 341, 342
Interglacial periods of Mars, 467
Interstellar
  clouds, 30, 40, 65, 297
  compounds, 500

molecules, 37, 65, 297, 303
Iron
binding proteins, 388
protein (High potential), 374
sulfur proteins, 363
sulphide compounds, 368–370
Isoserine, 145
Isotope abundances in molecular clouds, 38
Isovaline, 145

Jovian
atmosphere, 483, 503
colored regions, 503
microsphere, 484
planets, 498, 504
radiation belts, 48
Van Allen Belts, 503
Jupiter
atmosphere, 41
clouds, 483
composition, 43
energy sources, 45
galilean satellites of, 46
H/C ratio, 42
life on, 483
pressure, 52
radio spectrum, 50
red spot on, 45, 503
temperature, 52
yellowish tint, 43

Kemion's system, 175
β-Ketothioester, 255
Kohoutek (comet), 54
Kromberg formation, 79

Label release
experiment, 437
module, 431
Labile sulfur, 388
Lander of Mars vehicle, 473
Landing of Mars vehicle, 469
Left- and right-handed molecules, 264
Leucine, 306
Life
on Jupiter, 483
on Mars, 458, 476
Lipoproteinic membranes, 401
Lysine, 194, 231, 232, 235, 338–341

Main sequence, 1, 20–22
m-Aminophenyl, 146
Mallapunyah formation, 106, 107
Marine evaporite sequences, 87
Mars, 457, 458, 460
carbonaceous chondrites of, 466
climate, 465
contamination in ... missions, 469
dust, 465, 466
dust storms, 466
engineering considerations on ... missions, 467
lander of ... vehicle, 473
meteorites, 466
minerals, 467
missions, 463
missions summary diagram, 468
orbit of ... vehicle, 470, 472
orbiter of ... vehicle, 473
samples returned from, 463, 464, 467, 476, 477–481
spacecraft for ... mission, 470–472
surface of, 465
vehicle to, 469–474
volcanic activity in, 466
water vapor in, 466
Martian
atmosphere, 457, 466
biota, 461
soil, 426, 467
surface, 464
McArthur group, 105, 106, 107, 117
Meiotic disjunction, 353
Mendelian logic, 353
Methane-hydrogen atmosphere, 90, 102
Metaphytes, 132
Metaphytic evolution, 132
Metazoan, 133
Metazoan evolution, 126, 127, 132
Meteorites (*see also* carbonaceous chondrites)
Bruderheim, 72
in Mars, 466
Murchison, 62, 71, 74–76, 149, 303, 417, 426, 429
Murray, 71, 73, 74, 149, 429
organic compounds in, 57
Meteoritic
dust, 58
hydrocarbons, 60
magnetite, 58
Methionine (prebiotic synthesis of), 146, 147
Micelles, 358–360
Microfossils, 105, 107, 117, 120, 121, 288, 377
Microspheres, 227, 236, 288, 360
Miller-Urey
reaction, 60, 483
synthesis, 59, 62, 64, 66
Mission
Mars, 463, 468, 472
Viking, 458, 464
Molecular clouds, 29, 30, 32, 33, 37, 40
Monocoders, 351
Monocoding quartets, 351
Monomethylurea, 154

Montmorillonite, 59, 175, 178, 240
Moon (old rocks on the), 499

NAD, 389, 392, 394, 405
NADP, 389
Natural selection, 9, 14, 25
Neptune
atmosphere, 52
H/C ratio, 42
temperature, 51
N-ethyl-β-alanine, 144
Neobiogenesis, 285, 287, 289
Neutron stars, 8
N-Kjeldahy content, 241, 242
Nonheme iron proteins, 388, 389, 394
Norleucine, 145
Norvaline, 145
N-Tri-fluoracetyl-D-2-butyl ester, 146
N-Trifluoroacetyl-sec-butyl-esters, 144
Nucleoproteinoid, 233
Nucleosides, 159
Nucleoside-monophosphate, 194
Nucleotide binding proteins, 388, 389
Nucleotide-amino acid, 194
Nucleotides, 159, 189

Ocean, primordial oil slick, 358
Oligonucleotides, 159, 161, 212
Oligouridylates, 161
Onverwacht (chert), 80, 82, 121, 122
Oparin's hypothesis, 173, 284
Optical activity on Earth, 263
Orbit of Mars vehicle, 470, 472
Orbiter of Mars vehicle, 473
Organic molecules, 29
Organic pigments, 12, 16
Orion, 34, 35
Ornithine nitrile, 149
Orthorhombic cell, 214
Outer solar system
chromofores in, 48
H/C ratio, 42
molecular abundance, 42
temperatures, 50
Oxidation-reduction potential, 388

Particles (elementary), 8, 25
Peptide synthesis, 173, 236
Peptidoglycan molecules, 345
Peroxidase, 203
Phanerozoic era, 87, 88, 94, 121, 123, 133
Phenylalanine, 139
Phosphoamidate, 162
Phosphodiester bonds, 164, 236
Phosphoimidazolide, 162
Phosphorylation, 159, 160, 256, 403, 407
Photoautotroph, 120, 121, 397
Photocatylitic
electron transfer, 402
reactions, 403
Photochemical
activity, 398
evolution, 398
Photophosphorylation, 10, 12
Photoreceptors-photosensitizers, 398
Photoreduction, 400
Photosynthesis, 10, 12, 13, 14, 19, 84, 85, 101, 124, 322, 355, 389, 397, 412
Photosynthetic
membranes, 377, 388
bacteria, 363, 365, 370, 397
Phototaxes, 18
Phototropism, 14, 18, 20
Phycobilins, 402
Phycobilins-phycoerythrin, 13
Phycocyanin, 13
Pi-electrons, 13
Pigment excited molecules, 401
Pigments, 19
light-harvesting, 13
organic, 12, 16
rhodopsin like, 17
Pipecolic acid, 147, 149
Pivaldehyde, 145
Planetary systems, 491
Planets
formation of, 492, 493
number of, 17
Pluto
temperature, 53
IR spectrum, 53
Polyamino acids
catalytic activities, 248–250
configuration, 247
ionic behavior, 246
IR spectra, 247, 248
linkages, 247
nonrandomness of, 244
Polybasic amines, 165
Polynucleotide, 230, 286
double stranded, 353
single stranded, 353
strands, 503
Polyphenoloxidase, 203
Polyphosphate, 162
Polythymidylic acid, 164
Porphyrins, 13, 57, 63, 64, 398, 400
Prebiotic
atmosphere, 208
evolution, 359
fore proteins, 240
synthesis, 207
Precambrian
cyanophytes, 124

eon, 119, 123
fossils, 120, 132
microorganisms, 119, 120
paleobiology, 119, 120, 125
rocks, 80
sea water, 89
sediments, 88, 122
stromatolites, 84
Swaziland sequence, 79
Pre-nucleic acid, 239
Pre-proteins, 239
Pressure in Jupiter, 52
Primordial
cell, 235
sequence, 227
soup, 189, 226, 256
Principle of continuity, 361
Probionts, 224, 227–236, 239, 244
Prokaryote, 126, 127, 166, 311
Prokaryotic microorganisms, 123, 127, 336, 338
Proline, 149
Proteinoid, 245, 265
Proteins
argenine content, 338–341
heme, 388, 394
iron binding, 388
iron sulphur, 364, 368, 369, 381, 382
lysine content, 338–341
nonheme iron, 388, 394
nucleotide binding, 388
structure, 387, 392
Protocell, 227, 228, 236
Protoprotein, 227, 228, 230, 236
Protoribosomes, 231
Protostar, 38
Proto-stellar clouds, 300
Purines, 57, 61, 156, 159
Pyridine nucleotides, 402
Pyrimidines, 57, 61, 156, 157, 159, 169
Pyrolitic release
experiment, 432
module, 431
Pyrrole, 399, 400

Quartets of codon, 331, 351
Quinones, 388, 398, 399

Racemic
amino acids, 263, 270
half-life, 273
mixture, 73, 76, 265, 268
molecules, 264
monomers, 270, 272
point, 273
poly-peptides, 263
rate of, 274
solution, 264
substrates, 263
Radiospectrum, 29
Redox
activity, 256
electrochemical properties, 256
potential, 95, 363, 382
processes, 256
protein, 365
reactions, 254, 399
Resonant frequency of a molecule, 31
Retinal, 15, 16, 17, 20
Retinol, 15, 16
R_f values of nucleotides, 359
Rhodopsin, 16
Rhodopsin-like pigment, 17
Riboflavine, 18
Ring protons, 195, 196, 197
RNA, 159, 168–170, 208, 210–218, 232, 236, 311, 312, 317, 318, 321, 332–338, 341, 343–349
polymerase, 229
ribosomal, 311, 312, 317, 321, 322
Roper group, 105, 106
Rubredoxin, 363, 379, 380, 381

Samples of Mars
analysis, 476–481
contamination, 469–475
of returned missions, 463, 464, 469, 475
storage, 476
Sarcosine, 149
Saturn
atmosphere, 41, 52
composition, 43
equatorial belt on, 45
H/C ratio, 42
Self-ordering reactions, 227
Self-reproduction, 228
Sequence of D. Gigas, 372
Serine, 73, 306, 314
Silicified algal unicells, 127
Single bonds (in SiO_2), 8
Sinks of nonnoble gaseous components, 96
Soil (Martian), 467
Solar
nebula, 64, 72
system, 492
wind, 298, 299
Sources of nonnoble gaseous components, 96
Spacecraft of Mars, 470, 472
Spontaneous generations, 285, 289
Spot cells, 130, 132
Stars
double, 492
formation of, 492
number of, 7

Sterilization of Mars spacecraft, 470
Stromatolitic structures, 122
Strong low velocity lines, 37
Succinic dehydrogenase, 384
D-Sugars, 263, 264
Surface (of Mars), 465
Survival of the fittest, 225
Swartkoppe formations, 79, 80, 81
Swaziland sequence, 79, 80, 83, 85
Synthesis
  FTT, 77
  hydrogen cyanide, 140
  MFT, 66
  Miller-Urey, 59, 62, 64, 66
  polypeptides, 207
  prebiotic, 141, 207

Temperature
  in Jupiter, 52
  in Neptune, 51
  in outer solar systems, 50
  in Pluto, 53
  in Titan, 53
  in Uranus, 51
Theespruit formations, 79, 80–83
Thiophenylesters, 198, 199
Thiouridine, 334
Threonine, 142, 144
Thymidine, 167
Thymidine-phosphate, 160, 167, 170
Titan
  atmosphere, 41
  H/C ratio, 42, 504
  spectroscopic observations of, 47
  temperature, 53
TPN-10
Transacylations, 253
Transpeptidation, 175
Transphosphorylation, 253
Trimetaphosphate synthesis, 160, 161
Troland's enzyme theory, 286
Tryptamine, 307
Tryptophan, 139, 314, 400
Tunicates, 349
Tyrosine, 139
TΨC loop, 213, 215, 216, 218
TΨC stems, 215, 216, 218

Umbolooga group, 105
Uranus
  atmosphere, 52
  H/C ratio, 42
  temperature, 51
Urea-ammonium chloride phosphate, 161
Uridine oxyacetic acid, 334
Uridine-phosphate, 160, 161
Uridylate, 358

Valine, 73, 306
Vehicle (Mars), 469–474
Venus
  atmosphere, 487, 488
  clouds of, 483
Viking, 417, 431
  Biology Instrument (UBI), 419, 425
  Gas Exchange (GE) Experiment, 435
  GC-MS Instrument, 419, 425, 461
  Label Release (LR) Experiment, 437
  Lander, 418, 432
  Pyrolytic Release (PR) Experiment, 433
  Pyrolytic Release Module, 431
Visual pigment, 15
Vitamin A
  Aldehyde, 14, 15
  retinal, 15, 16
  retinol, 15
Volcanic activity in Mars, 466

Water
  Sea ... in Precambrian, 89
  vapor (in Mars), 466
Watson and Crick
  central dogma, 239
  bases, 354
Weak binding theory (origin of code), 397
Wind (solar), 298, 299

Xanthine oxidoase, 381
X-ogen, 33
X-ray crystallography, 393

Yeast tRNA, 341

Zeolites, 176
Zwiteron, 342